Biomechanics of

Biomechanics of the Spine

Basic Concepts, Spinal Disorders and Treatments

Edited by

Fabio Galbusera
Head of the Laboratory of Biological Structures Mechanics,
IRCCS Galeazzi Orthopedic Institute, Milan, Italy

Hans-Joachim Wilke
Co-Director and Head of Spine Research, Institute of Orthopaedic
Research and Biomechanics, Centre for Trauma Research Ulm,
Ulm University, Ulm, Germany

ELSEVIER

ACADEMIC PRESS
An imprint of Elsevier

Academic Press is an imprint of Elsevier
125 London Wall, London EC2Y 5AS, United Kingdom
525 B Street, Suite 1800, San Diego, CA 92101-4495, United States
50 Hampshire Street, 5th Floor, Cambridge, MA 02139, United States
The Boulevard, Langford Lane, Kidlington, Oxford OX5 1GB, United Kingdom

Library of Congress Cataloging-in-Publication Data
A catalog record for this book is available from the Library of Congress

British Library Cataloguing-in-Publication Data
A catalogue record for this book is available from the British Library

ISBN 978-0-12-812851-0

For information on all Academic Press publications
visit our website at https://www.elsevier.com/books-and-journals

Publisher: Mara Conner
Acquisition Editor: Fiona Geraghty
Editorial Project Manager: Lindsay Lawrence
Production Project Manager: Kamesh Ramajogi
Cover Designer: Christian Bilbow

Typeset by SPi Global, India

Contents

SECTION 2 BIOMECHANICS OF SPINAL TISSUES

SECTION 4 SPINAL DISORDERS AND SPINE SURGERY

DEGENERATIVE DISORDERS

SPINAL DEFORMITIES

TRAUMA AND TUMORS

Contributors

Tito Bassani
IRCCS Galeazzi Orthopedic Institute, Milan, Italy

Marco Brayda-Bruno
IRCCS Galeazzi Orthopedic Institute, Milan, Italy

Gloria Casaroli
IRCCS Galeazzi Orthopedic Institute, Milan, Italy

Francesco Costa
Humanitas Clinical and Research Center, Rozzano, Italy

Fabio Galbusera
IRCCS Galeazzi Orthopedic Institute, Milan, Italy

Nicolas Graf
SpineServ GmbH & Co. KG, Ulm, Germany

Rami Haj-Ali
Tel-Aviv University, Tel-Aviv, Israel

Maryem-Fama Ismael Aguirre
IRCCS Galeazzi Orthopedic Institute, Milan, Italy

René Jonas
Institute of Orthopaedic Research and Biomechanics, Ulm University, Ulm, Germany

Annette Kienle
SpineServ GmbH & Co. KG, Ulm, Germany

Luigi La Barbera
Politecnico di Milano; IRCCS Galeazzi Orthopedic Institute, Milan, Italy

Christian Liebsch
Institute of Orthopaedic Research and Biomechanics, Ulm University, Ulm, Germany

Richard A. Lindtner
Medical University of Innsbruck, Innsbruck, Austria

Andrea Luca
IRCCS Galeazzi Orthopedic Institute, Milan, Italy

Andrea Malandrino
Institute for Bioengineering of Catalonia, Barcelona, Spain

Frank Niemeyer
Institute of Orthopaedic Research and Biomechanics, Ulm University, Ulm, Germany

Claudia Ottardi
Politecnico di Milano, Milan, Italy

Benedikt Schlager
Institute of Orthopaedic Research and Biomechanics, Ulm University, Ulm, Germany

Werner Schmoelz
Medical University of Innsbruck, Innsbruck, Austria

Luca M. Sconfienza
IRCCS Galeazzi Orthopedic Institute; University of Milano, Milan, Italy

Mirit Sharabi
Tel-Aviv University, Tel-Aviv, Israel

Elena Stucovitz
IRCCS Galeazzi Orthopedic Institute, Milan, Italy

Themis Toumanidou
Universitat Pompeu Fabra, Barcelona, Spain

Tomaso Villa
Politecnico di Milano; IRCCS Galeazzi Orthopedic Institute, Milan, Italy

Jacopo Vitale
IRCCS Galeazzi Orthopedic Institute, Milan, Italy

David Volkheimer
Institute of Orthopaedic Research and Biomechanics, Ulm University, Ulm, Germany

Kelly Wade
Institute of Orthopaedic Research and Biomechanics, Ulm University, Ulm, Germany

Hans-Joachim Wilke
Institute of Orthopaedic Research and Biomechanics, Ulm University, Ulm, Germany

About the Editors

Dr. Fabio Galbusera is head of the Laboratory of Biological Structures Mechanics, IRCCS Galeazzi Orthopedic Institute, in Milan, Italy. Dr. Galbusera has worked for the past 12 years in this field. His experience includes both experimental and numerical investigations, as well as clinical studies in the field of spinal surgery. He is author and coauthor of more than 70 peer-reviewed publications about the spine. The team he works with includes several researchers and doctors at the institute and colleagues from other institutions (notably Ulm University, with which he is actively cooperating).

Prof. Dr. Hans-Joachim Wilke is Co-Director and Head of Spine Research or Institute of Orthopaedic Research and Biomechanics, Centre for Trauma Research Ulm or Ulm University, Ulm, Germany, where he is responsible for the spine research group. He is a former president of both EUROSPINE, the Spine Society of Europe and the German Spine Society and is since 1998 coeditor of the *European Spine Journal,* which is dedicated to research papers related to all aspects of the spine. Prof. Wilke is also founder and chairman of the GRAMMER Award for spine research. He has been involved in several large-scale research projects funded by the European Commission.

Foreword by Jeffrey Lotz

Humans are designed for movement, for which the spine is the central musculoskeletal element. To fulfill this essential role, the spine has evolved complexity at multiple hierarchical scales—posture, anatomy, material properties, and cellular activities. Its proper, pain-free function requires synergy among the full spectrum of musculoskeletal tissues, including bone, cartilage, ligament, tendon, and muscle. Given this complexity, it's not surprising that spine-related pain is the leading cause of disability worldwide. What's more, there is a known crossover effect of exacerbation of other chronic conditions when spine health and mobility decline.

For over 50 years, biomechanics has been the fundamental tool used to generate insights into spine disease/injury causation and therapy development. Early leaders in the field, such as Nachemson, Panjabi, Markolf, Hirsch, and Morris, helped establish the functional/descriptive anatomy, concepts of mechanical stability and intrinsic versus extrinsic stiffness, injury tolerance, and their respective clinical implications, such as with external bracing and implants.

Today, the breadth and depth of concepts included under the heading of spine biomechanics has grown significantly to link macromechanics to subcellular biologic signaling pathways. Advances in diagnostics, therapies, and the tracking of clinical outcomes are providing new insights into disease mechanisms, and helping to bridge basic discoveries to patient care. These advances include computational techniques and models, functional imaging, genomics, proteomics, large dataset analyses, and wearable sensors that have the potential to quantify normative function and provide opportunities for advancing personalized medicine.

Despite growing knowledge, spine-related pathology continues to increase in its societal and economic burden. What are we still missing? The complexity of the spinal system necessitates multifactorial and multidisciplinary perspectives to integrate the growing volume of diverse data into new insights that underlie successful clinical solutions. We are increasingly cognizant of the extent to which biomechanics and biology are intrinsically linked, and the complex cross talk between tissues and organ systems that is critical for spine homeostasis and pain-free movement. For these reasons, substantive progress toward improving spine health will require tying diverse, fundamental concepts to contemporary clinical problems.

This is precisely why this book is so timely and valuable. Drs. Galbusera and Wilke are both international leaders in the field of spine biomechanics and have assembled an authoritative and comprehensive text of current research and perspectives of spine anatomy, composition, and function. These fundamentals are thoughtfully linked to clinical concepts such as diagnosis, spinal devices, fractures, tumors, and neurodegenerative conditions.

This book should be required reading for anyone interested in working at the leading edge of spine research and patient care. Researchers will find this book a trusted source for laboratory techniques, modeling approaches, and testing protocols. It is a valuable compendium of anatomic and material property data assembled by a diverse and respected team of experts. Clinicians will appreciate how theoretical concepts are bridged to current treatments, which demonstrates how state-of-the-art clinical techniques are rooted in solid biomechanical principles. It is particularly valuable for training programs that strive to provide a solid, comprehensive, and current foundation.

Thank you, Drs. Galbusera and Wilke, for delivering *Biomechanics of the Spine* to us. It is well written, beautifully illustrated, and a joy to read.

Jeffrey Lotz
University of California, San Francisco, California, United States

Foreword by Margareta Nordin

Biomechanics is the study of the mechanics of the living body. Knowledge of biomechanics is a base necessary to any individual interested in the complex structure of the spine. The title of the book, *Basic Biomechanics of the Spine: Basic Concepts, Spinal Disorders, and Treatment*, indicates a vast audience, and accurately so. The book is written primarily for clinicians and researchers, but also graduate students interested in spine biomechanics. The excellent content should attract physicians, surgeons, engineers, physiotherapists, chiropractors, osteopaths, clinicians at large, researchers, and others interested in improving their knowledge and advancing their understanding of the biomechanics of the human spine.

This book should be in all educational libraries, academic departments, and biomechanical laboratories interested in the topic as an educational tool and reference. Further, this book should be the basis for graduate educational programs in engineering, biomechanics, physiotherapy, chiropractic, osteopathy, orthopedics, neurosurgery, and spine surgery residency and fellowships, as well as in engineering programs interested in spine disorders and treatment. Spine biomechanics is a science which integrates multiple disciplines.

The book is well written, with instructive and excellent illustrations and thoughtfully considered reference lists for each section. The two editors, Fabio Galbusera and Hans Joachim Wilke, are world leaders in spine biomechanics, and they have assembled an impressive group of international and multidisciplinary co-authors, clinicians, and researchers, synthesizing current state-of-the-art knowledge and research.

Biomechanics of the Spine contains four major sections: the human spine, biomechanics of spinal tissues, methods, and lastly, spinal disorders and spine surgery. The first three sections are focused on fundamental concepts and new research applicable to the spine and spine ailments. The fourth section covers concepts related to the treatment of spinal disorders (including surgery) and the importance of biomechanics knowledge as applied to interventions. Each section has been carefully developed and contains four to seven chapters.

Section 1 deals with fundamental concepts of the spine, its evolution, and basic biomechanics of the cervical, thoracic, and lumbar spine; these fundamentals are necessary to understand further sections. Section 2, Biomechanics of Spinal Tissue, includes five comprehensive chapters on vertebral bone, intervertebral discs, collagen fibers, vertebral endplates, and spinal muscles, all of which are structures important in spinal disorders. Section 3 is a comprehensive overview of methods available today, ranging from imaging, in vivo and in vitro testing, standard testing, and modeling. Technological advancements are thoroughly described with illustrations, examples, and cases. Finally, Section 4 is geared toward spinal disorders and spinal surgery, and the reader will find case examples, reviews of literature regarding highly valuable treatments, and a discussion of biomechanical testing and its implications for the treatment of spinal disorders.

This international collaboration of clinicians and researchers is a success, and the reader will enjoy and learn from *Biomechanics of the Spine*. This book will serve as a valuable resource for any individual who is both interested in spinal disorders and eager to understand and pursue better clinical care, and who wants to improve scientific research. Enjoy your reading!

Margareta Nordin

New York University, New York, United States

SECTION

1

THE HUMAN SPINE

CHAPTER

THE SPINE: ITS EVOLUTION, FUNCTION, AND SHAPE

1

Fabio Galbusera
IRCCS Galeazzi Orthopedic Institute, Milan, Italy

THE EVOLUTION OF THE SPINE

During the Silurian Period (444 to 419 million years ago), the only animals with a bony skeleton were fish, which were structurally similar to the modern ray-finned fish that now comprise more 30,000 extant species. The skeleton of ray-finned fish includes a vertebral column consisting of a series of rigid bony vertebrae that protect the spinal cord (Fig. 1) (Nelson, 2006). For locomotion, ray-finned fish utilize fins consisting of bony rays covered by a layer of skin, which are not directly connected to the vertebral column (except for the tail) but are supported only by muscles. Indeed, ray-finned fish do not have a pelvic girdle, and the pectoral fins, from which the forelimbs evolved, are directly connected to the skull (Rockwell et al., 1938).

Anatomical structures similar to limbs appeared in lobe-finned fish, whose fins are fleshy and partially covered with scales (Benton, 2015). Fins in most lobe-finned fish did not, however, have a rigid connection to the spine; these features progressively emerged during the Devonian Period when the first amphibians with a tetrapod body plan appeared, such as the *Acanthostega* and the *Ichthyostega* (Fig. 1) (Pierce et al., 2013; Pierce and Clack, 2012). As these animals underwent a gradual transition from an aquatic to a terrestrial life, the design principle of a spine detached from the limbs fell short of being able to support their body weight outside an aquatic environment, and the pelvic girdle emerged as a mobile but sturdy connection between the spine and the lower limbs. Recent studies suggest that limbs appeared before the transition to terrestrial life began (Pierce and Clack, 2012; Pierce et al., 2013). The usefulness of the tetrapod plan for aquatic life remains, however, unclear (Benton, 2015). Besides, it is interesting to note that in later animals that reacquired an aquatic life, such as marine reptiles, the direct connection between pelvis and spine disappeared.

At the same time, the forelimbs detached from the skull and moved posteriorly forming the shoulder girdle and the neck (i.e., the region of the spine between the skull and the shoulders), allowing for head motion as well as for disconnecting the skull and the brain from the anatomical structures involved in locomotion, thus reducing the mechanical loads on the brain (Pierce et al., 2013). With these latter evolving steps, it can be said that the principles of the tetrapod body design with the vertebral column as its core, which was then maintained in all reptiles, birds, and mammals, were established.

Biomechanics of the Spine. https://doi.org/10.1016/B978-0-12-812851-0.00001-X

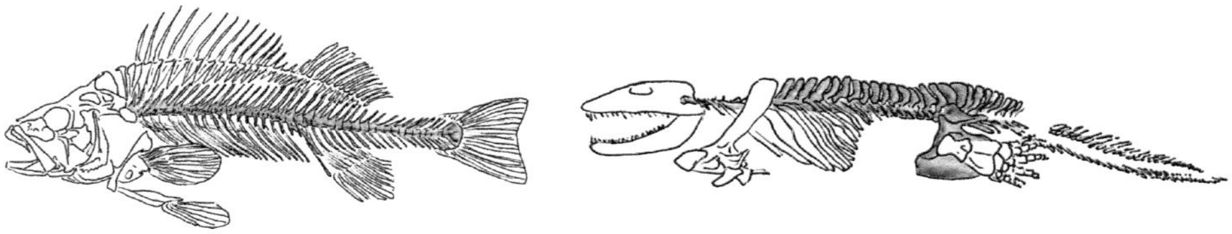

FIG. 1

Skeletons of a ray-finned fish (*left*, European perch) and of *Ichthyostega (right).* In the tetrapod design, the shoulder and the pelvic girdles connect the limbs to the spine and facilitate terrestrial life.

Adapted from Ahlberg, P.E., Clack, J.A., Blom, H., 2005. The axial skeleton of the Devonian tetrapod Ichthyostega. Nature 437, 137–140.

COMPARATIVE SPINAL ANATOMY IN VERTEBRATES

The global structure of bony fish vertebrae and that of the subsequently evolved terrestrial animals share many features. Fish vertebrae have a round vertebral body called centrum, endplates with a marked biconcavity, and a dorsally located neural arch that encloses the spinal cord (Rockwell et al., 1938). In most cases, a ventral vertebral arch called the hemal arch, or chevron, is also present (Fig. 2). The two arches commonly protrude in a spine. Together with the transverse processes, the neural and hemal spines serve as muscle insertions and articulate with the ribs (Bone and Moore, 2007). Zygapophysial joints limiting bending and flexion motions are also present in most fish. Instead of intervertebral discs, a segmentally constricted cartilage-like notochord running through the entire length of the spine provides flexibility to the body. The fish spine is generally subdivided into a pre-caudal region and a caudal region; no major regional differentiation is observable in the pre-caudal spine.

In comparison with fish and early amphibians, vertebrae of extant amphibians do not show any specific advancements and even tend to exhibit a simpler structure. However, amphibians show for the first time in the evolutionary scale the presence of a sacral bone, which consists of a series of fused vertebrae providing a solid interface to the pelvis and in turn to the lower limbs, and has therefore a

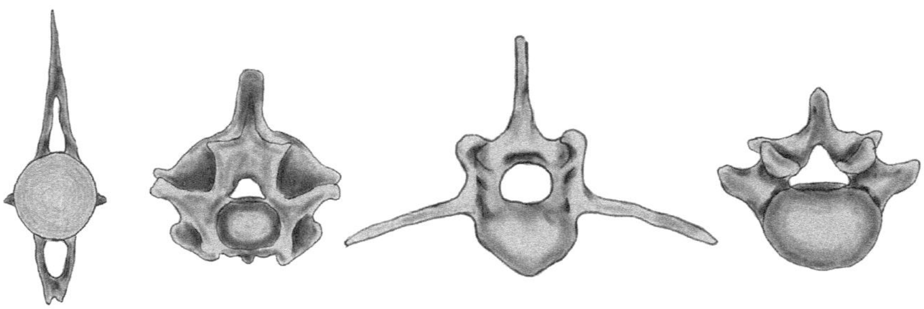

FIG. 2

Comparative vertebral anatomies. From left to right: a fish pre-caudal vertebra (pike), in which a hemal arch is also present; a reptile vertebra (python); a lumbar vertebra of a mammal quadruped (deer); a human lumbar vertebra.

fundamental mechanical role in terrestrial weight bearing and locomotion (Schoch, 2014; Romer and Parsons, 1977). Paleontological studies showed that the sacrum evolved in transitional animals such as the *Ichthyostega* that were still conducting an aquatic life (Pierce et al., 2013). Thus the development of the sacrum may not be related to terrestrial locomotion; however, this trait developed when the transition to terrestrial life initiated.

Vertebrae of reptiles show a very close resemblance to those of mammals and birds, the most evident difference being the concave socket of the anterior endplate fitting to the convex surface of the posterior endplate of the next vertebra (Fig. 2) (Romer and Parsons, 1977). Intercentra (i.e., small bony elements located between adjacent vertebrae that are also found in fish and amphibians but have been lost in mammals except in their tails) are often present in reptiles, as well as hemal arches. In birds, the only flexible region of the spine is the neck, whereas thoracic vertebrae are partially fused to provide for a robust attachment for the wings. Lumbar vertebrae and sacrum are fused into the synsacrum as well as the caudal vertebrae.

The anatomy of the spine of mammals exhibits a relatively low variability among the class; the number of vertebrae is almost constant, with all mammals having seven cervical vertebrae apart from sloths and manatees, and with all spine regions having a marked flexibility (Fig. 2) (Romer and Parsons, 1977; Kettler et al., 2007). Even in flying mammals such as bats and the aquatic cetaceans, the spine does not show any major changes with respect to the general mammalian spine design. Intervertebral discs progressively developed is such a way that the high stresses arising during terrestrial weight bearing and locomotion were distributed on the vertebral endplates (Boos and Aebi, 2008). The inner part of the mammalian intervertebral discs, the nucleus pulposus, directly developed from the notochord, whereas the annular rings were shown to originate from the perichordal sheath (Pattappa et al., 2012).

THE UPRIGHT POSITION

Although a full upright posture is one of the major features distinguishing humans from the other apes, bipedalism did not first evolve in humans, and is also commonly exhibited by several primates. Furthermore, several extant animals—such as penguins, ostriches, kangaroos—and extinct ones—such as many dinosaurs—did achieve an orthograde posture, although their posture was based on different biomechanical principles than those of humans (Le Huec et al., 2011; Berge, 1998). For example, dinosaurs used their tail to counterbalance the weight of their trunk and had an almost horizontal spinal orientation, like that of most extant birds.

Recent studies have shown that human bipedalism started to evolve in early homininae such as the *Sahelanthropus tchadensis*, which is believed to have lived shortly after the chimpanzee–human divergence (7 to 10 million years ago) (Harcourt-Smith and Aiello, 2004). More extensive bipedal adaptations such as a longer femur and the position of the muscle insertions have been noted in the *Orrorin tugenensis* (6 million years old); these features were well adapted to an arboreal life (Richmond and Jungers, 2008; Almécija et al., 2013). The ability to walk upright may have helped these species in their various African habitats, such as forests, grassland, and swampland.

Later species such as the *Australopithecus* (2.5 million years ago) also exhibited stronger bipedal adaptations to the spine and pelvis, which may have allowed for a gait quite like that of modern humans (Masao et al., 2016). The increased width of the pelvis provided a larger lever arm for the muscles responsible for maintaining the equilibrium of the trunk during locomotion (Gruss et al., 2017). At the same time, the spine assumed an S-shape, including the thoracic kyphosis and the lumbar lordosis that has been maintained in modern humans (Fig. 3).

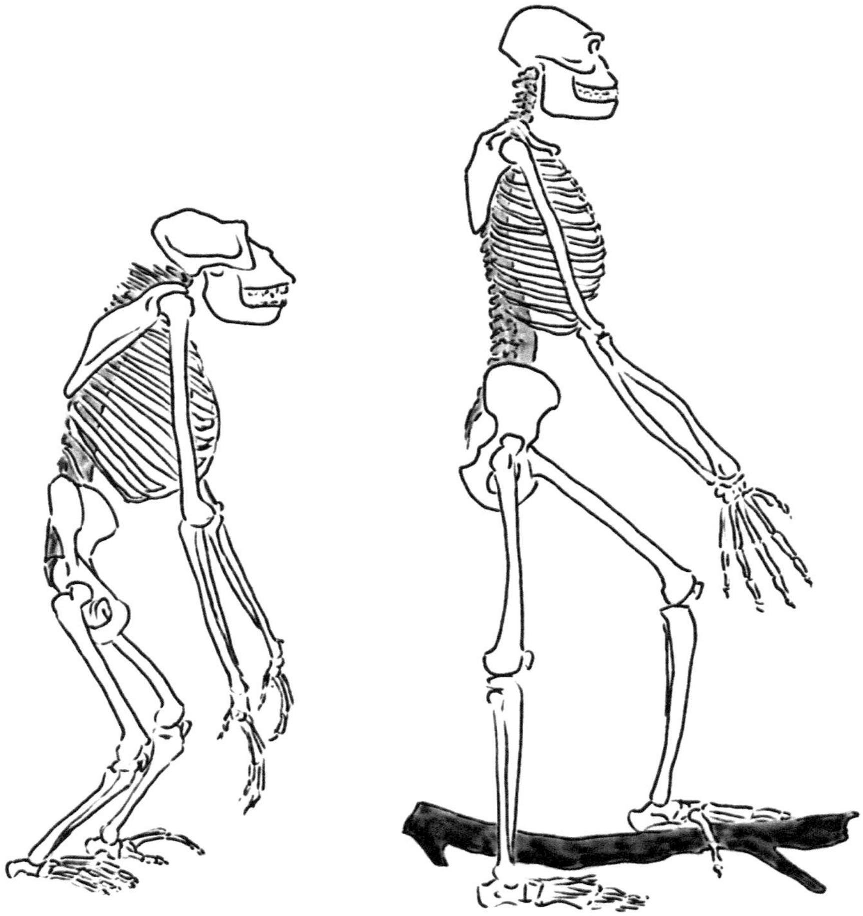

FIG. 3

Skeletons of a great ape (*left*, gorilla) and of a hominin (*right*, *Ardipithecus ramidus*) in the upright posture. Adaptations to the upright posture are evident (S-shape curvature of the spine, shorter and broader pelvis, extended rather than flexed knees while standing).

Adapted from Gibbons, A., 2009. A new kind of ancestor: Ardipithecus unveiled. Science 326(5949), 36–40.

FUNCTIONS OF THE SPINE

From its first appearance in early vertebrates, such as the *Myllokunmingia,* as local densifications of the notochord, or rudimentary vertebrae, the spine's main function was to protect the spinal cord (Shu, 2003). The spine's stiffness and strength, however, was also the basis for the development of the axial skeleton as the mechanical support for later animals, especially those that moved to a terrestrial environment where gravity loads were not alleviated by a buoyant force (Schindler and Heidtke, 2013). Thus in tetrapods, the functions of the spine can be summarized as follows: (1) to protect the spinal cord, by avoiding excessive straining; (2) to support the weight of the body, by transmitting the weight to the ground through the limbs; (3) to allow the motion of the trunk, by way of the spine's flexibility; and (4) to provide robust origins and insertions to the muscles of trunk and limbs.

THE HUMAN SPINE

The spine of extant humans has a straight appearance in the coronal plane, whereas it exhibits an S-shape in the sagittal plane (Fig. 4). Based on the sagittal curvatures, three spinal regions are defined: (1) the cervical spine connecting the head and the trunk, which has in most individuals a lordotic curvature (i.e., the concavity in the dorsal direction) in the standing posture; (2) the thoracic spine, which provides support for the rib cage and has a kyphotic curvature (i.e., a ventral concavity); and (3) the lumbar spine connecting the trunk to the pelvis and having a lordotic curvature. Additionally, the sacral bone comprising several fused vertebrae connects the lumbar spine and the pelvis.

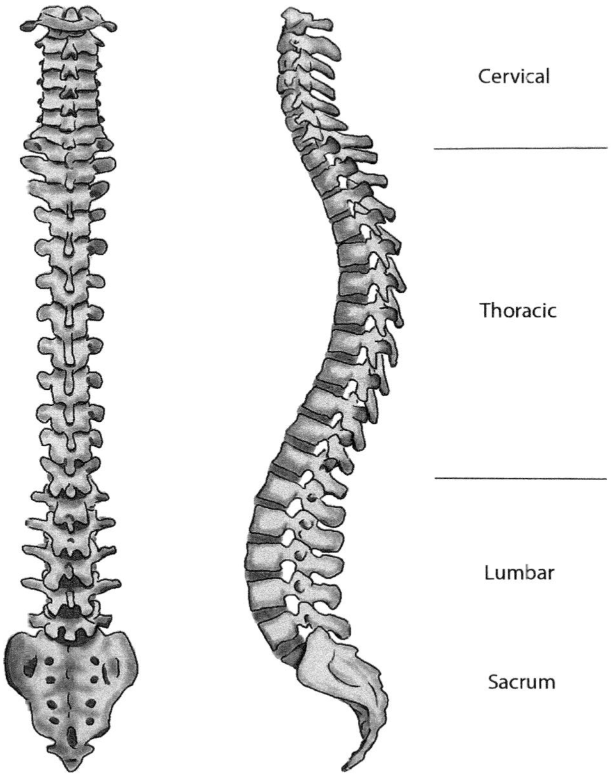

FIG. 4

The human spine in posteroanterior *(left)* and lateral *(right)* views.

The combination of such curvatures allows maintenance of the standing posture with minimal energy consumption (Le Huec et al., 2011). Indeed, the vertical projection of the center of gravity of the trunk passes through the sacrum; the curved sagittal profile of the spine contributes to this equilibrium by preventing buckling under gravity load (Lafage et al., 2008).

The main function of the cervical spine is to provide mobility to the head and to keep the line of sight horizontal during locomotion and daily activities. In the standing position, the cervical curvature compensates for the anterior inclination of the cervicothoracic junction to determine a correct alignment of the head and therefore has a general lordotic curvature. A straight or kyphotic alignment, however, is not uncommon (Gore et al., 1986); a study showed a prevalence of 13% to 34% of kyphotic cervical spines in asymptomatic subjects (Yukawa et al., 2012). The thoracic spine constitutes the main mechanical support of the rib cage and has a higher stiffness in comparison with the other spine regions. Its kyphotic curvature is commonly measured between T4 and T12, and ranges in healthy subjects between 34 degrees and 44 degrees (Vialle et al., 2005). However, high intersubject variability has been reported. The lumbar spine provides a substantial part of the trunk mobility and is subjected to the highest spinal loads (White and Panjabi, 1990). Its lordotic curvature is conventionally measured between L1 and S1 and varies between 43 degrees and 63 degrees in adult asymptomatic subjects (Roussouly et al., 2005).

REFERENCES

Almécija, S., Tallman, M., Alba, D.M., Pina, M., Moyà-Solà, S., Jungers, W.L., 2013. The femur of Orrorin tugenensis exhibits morphometric affinities with both Miocene apes and later hominins. Nat. Commun. 4, 2888.

Benton, M.J., 2015. Vertebrate Paleontology, fourth ed. Wiley Blackwell, Hoboken.

Berge, C., 1998. Heterochronic processes in human evolution: An ontogenetic analysis of the hominid pelvis. Am. J. Phys. Anthropol. 105 (4), 441–459.

Bone, Q., Moore, R., 2007. Biology of Fishes, third ed. Taylor and Francis, London.

Boos, N., Aebi, M., 2008. Spinal Disorders—Fundamentals of Diagnosis and Treatment. Springer, Heidelberg.

Gore, D.R., Sepic, S.B., Gardner, G.M., 1986. Roentgenographic findings of the cervical spine in asymptomatic people. Spine 11 (6), 521–524.

Gruss, L.T., Gruss, R., Schmitt, D., 2017. Pelvic breadth and locomotor kinematics in human evolution. Anat. Rec. 300 (4), 739–751.

Harcourt-Smith, W.E., Aiello, L.C., 2004. Fossils, feet and the evolution of human bipedal locomotion. J. Anat. 204 (5), 403–416.

Kettler, A., Liakos, L., Haegele, B., Wilke, H.J., 2007. Are the spines of calf, pig and sheep suitable models for pre-clinical implant tests? Eur. Spine J. 16 (12), 2186–2192.

Lafage, V., Schwab, F., Skalli, W., Hawkinson, N., Gagey, P.M., Ondra, S., Farcy, J.P., 2008. Standing balance and sagittal plane spinal deformity: analysis of spinopelvic and gravity line parameters. Spine 33 (14), 1572–1578.

Le Huec, J.C., Saddiki, R., Franke, J., Rigal, J., Aunoble, S., 2011. Equilibrium of the human body and the gravity line: the basics. Eur. Spine J. 20 (Suppl. 5), 558–563.

Masao, F.T., Ichumbaki, E.B., Cherin, M., Barili, A., Boschian, G., Iurino, D.A., Menconero, S., Moggi-Cecchi, J., Manzi, G., 2016. New footprints from Laetoli (Tanzania) provide evidence for marked body size variation in early hominins. elife 5, e19568.

Nelson, J.S., 2006. Fishes of the World. John Wiley & Sons, Hoboken.

Pattappa, G., Li, Z., Peroglio, M., Wismer, N., Alini, M., Grad, S., 2012. Diversity of intervertebral disc cells: phenotype and function. J. Anat. 221 (6), 480–496.

Pierce, S.E., Clack, J.A., 2012. Three-dimensional limb joint mobility in the early tetrapod Ichthyostega. Nature 486, 523–526.

Pierce, S.E., Ahlberg, P.E., Hutchinson, J.R., Molnar, J.L., Sanchez, S., Taffroreau, P., Clack, J.A., 2013. Vertebral architecture in the earliest stem tetrapods. Nature 494, 226–229.

Richmond, B.G., Jungers, W.L., 2008. Orrorin tugenensis femoral morphology and the evolution of hominin bipedalism. Science 319 (5870), 1662–1665.

Rockwell, H., Evans, F.G., Pheasant, H.C., 1938. The comparative morphology of the vertebrate spinal column. Its form as related to function. J. Morphol. 63 (1), 87–117.

Romer, A.S., Parsons, T.S., 1977. The Vertebrate Body. Holt-Saunders International, Philadelphia.

Roussouly, P., Gollogly, S., Berthonnaud, E., Dimnet, J., 2005. Classification of the normal variation in the sagittal alignment of the human lumbar spine and pelvis in the standing position. Spine 30 (3), 346–353.

Schindler, T., Heidtke, U.H.J., 2013. The origin of the tetrapod limbs: from expeditions to enhancers. Trends Genet. 29, 419–426.

Schoch, R.R., 2014. Amphibian Evolution: The Life of Early Tetrapods. John Wiley & Sons, Chichester.

Shu, D.G., 2003. A paleontological perspective of vertebrate origin. Chin. Sci. Bull. 48, 725–735.

Vialle, R., Levassor, N., Rillardon, L., Templier, A., Skalli, W., Guigui, P., 2005. Radiographic analysis of the sagittal alignment and balance of the spine in asymptomatic subjects. J. Bone Joint Surg. Am. 87 (2), 260–267.

White, A.A., Panjabi, M.M., 1990. Clinical Biomechanics of the Spine, second ed. Lippincott Williams & Wilkins, Philadelphia, PA.

Yukawa, Y., Kato, F., Suda, K., Yamagata, M., Ueta, T., 2012. Age-related changes in osseous anatomy, alignment, and range of motion of the cervical spine Part I: radiographic data from over 1,200 asymptomatic subjects. Eur. Spine J. 21 (8), 1492–1498.

FURTHER READING

Ahlberg, P.E., Clack, J.A., Blom, H., 2005. The axial skeleton of the Devonian tetrapod Ichthyostega. Nature 437, 137–140.

Gibbons, A., 2009. A new kind of ancestor: Ardipithecus unveiled. Science 326 (5949), 36–40.

CHAPTER

THE CERVICAL SPINE

2

René Jonas, Hans-Joachim Wilke
Institute of Orthopaedic Research and Biomechanics, Ulm University, Ulm, Germany

The cervical spine is the most superior part of the human spine and consists of seven vertebrae. These vertebrae differ greatly from the vertebrae of the thoracic spine (midback) and lumbar spine (lower back). Those of the cervical spine even differ. The general functions of the cervical spine are mobile bearing of the head and protection of the spinal cord. Compared to the thoracic and lumbar spines, the cervical spine is the most agile part of the human spine. Each vertebra is named with a capital "C" followed by an index, starting with one. Thus the vertebra that is connected to the occiput is called C1 and the last vertebra of the cervical spine, which is connected to the first thoracic vertebra, is C7.

ANATOMY/PHYSIOLOGY

All cervical vertebrae are aligned along a lordotic curve. The curvature can be characterized by using either the "T1-tilt" (or "T1-slope") or the "sagittal vertical axis" (SVA). The T1-tilt represents the angle between the surface of the first thoracic vertebra's endplate and a horizontal line in the sagittal plane in the upright position. Regarding a study by Knott et al. (2010), a T1-tilt between 13 degree and 25 degree is considered to be a healthy lordosis. The sagittal vertical axis is the vertical axis passing through the center of mass of the first thoracic vertebral body. The parameter that is given with the sagittal vertical axis is the distance between a particular landmark of a superior vertebra and the axis itself (Fig. 1). Common landmarks are the center of mass of vertebra C2 or the tip of the dens. The mean distance for the center of mass of C2 is about 14.89 mm (±10.97) according to Harrison et al. (2000). Other methods for the characterization of the cervical lordosis are the "Cobb-Method" and the "Harrison Posterior Tangent Method," which can also be used for segmental categorization within the whole spine.

The mass and accordingly the volume of each cervical vertebra increase in caudal direction, except for C2, because of its unique anatomy, which will be presented in the following paragraphs. Compared with the lumbar spine, the cervical vertebrae are almost half the size of the lumbar vertebrae (Gilad and Nissan, 1986).

The first cervical vertebra is the "Atlas," which is also known as "C1." The Atlas is basically an arc-like vertebra. In some rare cases, the vertebral arc of C1 never closes during the growth phase. There are five articular surfaces on the Atlas. Two of them are oriented in superior direction. One faces the inner space of the arc, and the last two point in inferior direction. On the lateral sides of the vertebra, two *foramen transversarium* protect the arteries and nerves, one on each side (Fig. 2).

Biomechanics of the Spine. https://doi.org/10.1016/B978-0-12-812851-0.00002-1

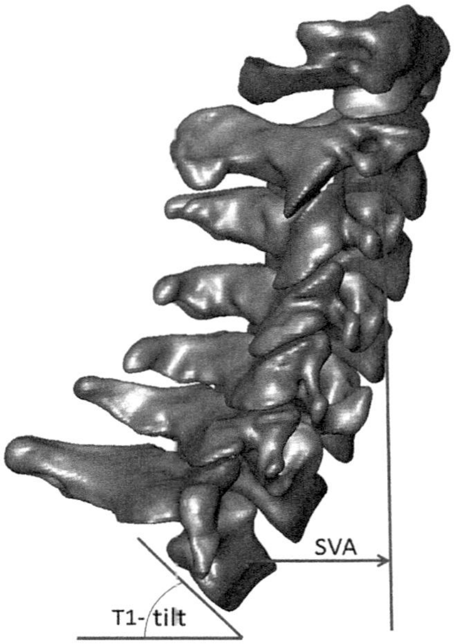

FIG. 1

Cervical lordosis, demonstrating T1-tilt and SVA (using center of mass of C2—vertebral body).

The second cervical vertebra is the "Axis," also known as C2. The Axis's most characteristic feature is the *dens axis (also: odontoid)*, which is usually referred to as only the dens. The dens is oriented in superior direction and consists of two articular surfaces located at the anterior and posterior sides of the dens. Apart from the dens, the Axis has four additional articular surfaces. Two of them are oriented in superior direction almost parallel to the transversal plane. The other two are located more posteriorly at the lateral sides of the *arcus vertebrae*. They target in anterior-inferior direction in an angle of about 45 degree. In addition they are slightly turned outwards in lateral direction. The *processus spinosus* of C2 is rather short as compared with the other cervical vertebrae. The inferior part of the vertebral body, which is connected to the first intervertebral disc, has a saddle-joint-like shape (Fig. 3).

The vertebrae C3 to C6 are very similar to each other. The superior as well as the inferior surface of each vertebral body have a saddle-joint-like shape, slightly different from the inferior surface of C2. One very important anatomical detail, which supports the saddle-joint characteristics of the superior surface, is the *processus uncinatus.* These bony formations are located at the lateral edges of the upper vertebral body, and their development starts around the age of 9 (Töndury and Theiler, 1958). Another very important detail is the *processus transversus* containing the foramen transversarium, which has a gutter-like shape and therefore supports the respective nerve root. The facet joints (also, zygapophysial joints) of the vertebrae C3 to C6 are also a very determinant anatomical feature. They are oriented at an angle of 45 degree toward the transversal plane, which has a great influence on their biomechanics (Fig. 4).

The last cervical vertebra, C7, is also very unique in its anatomical characteristics. The slope of the facet joints increase, the extent of the *processus uncinatus* decreases, and the *processus transversus* is wider. C7 is also well known for its processus spinosus, which is the longest along the whole spine (Fig. 5).

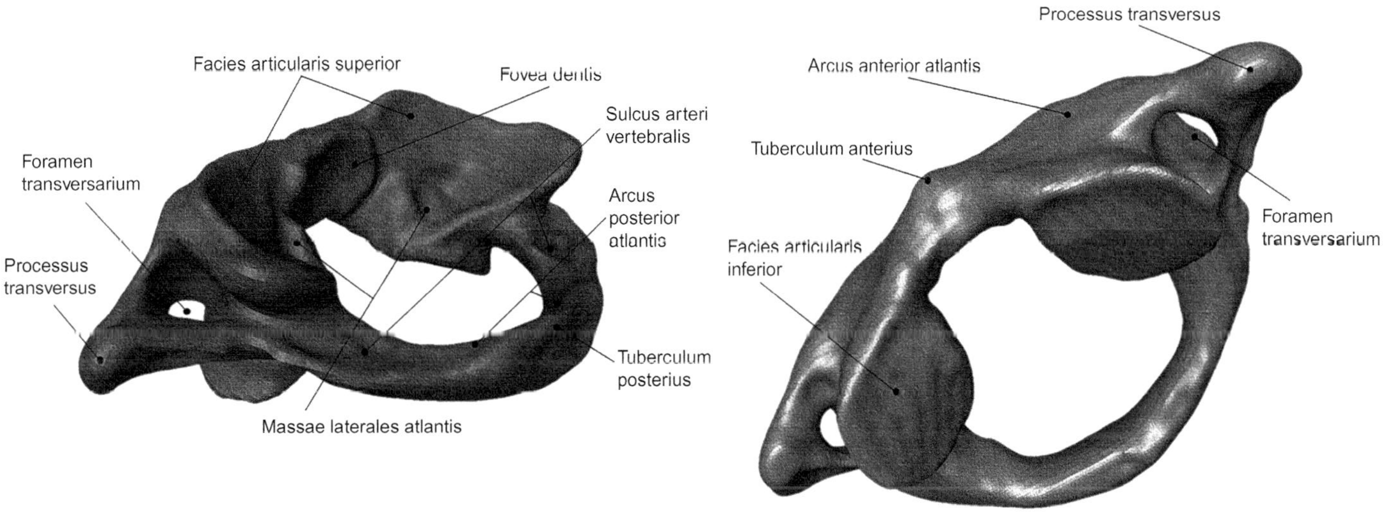

FIG. 2

Atlas (C1): *left*, view from superior, dorsal, and lateral; *right*, view from inferior, anterior, and lateral.

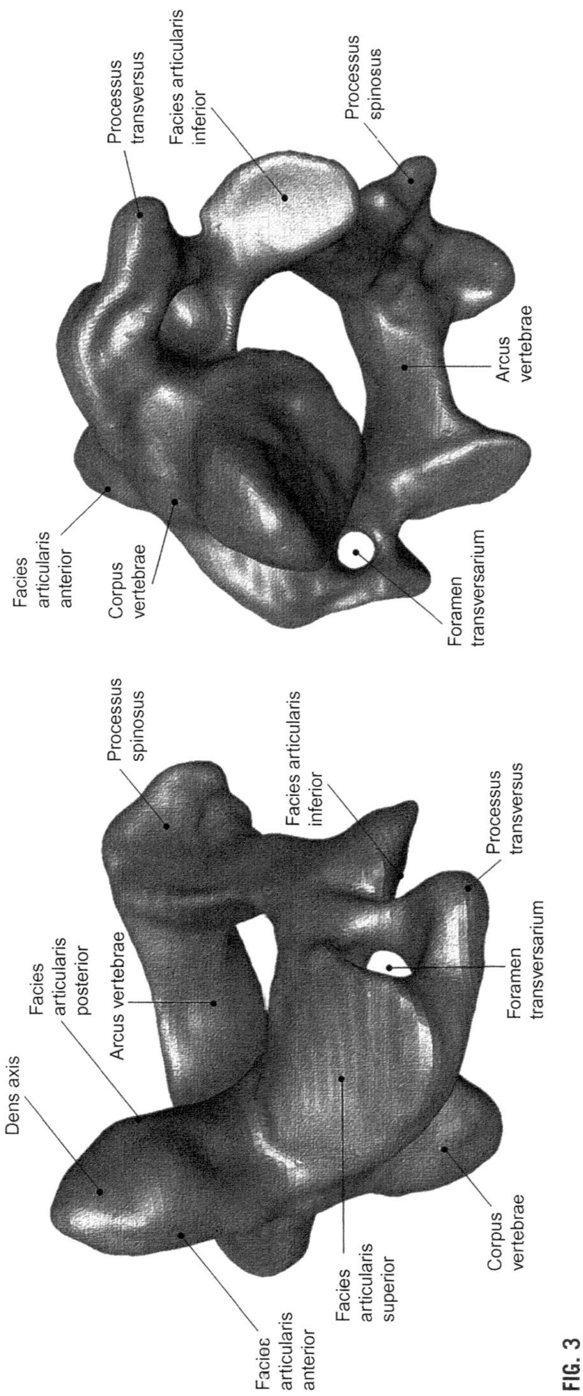

FIG. 3

Axis (C2): *left*, view from superior, anterior, and lateral; *right*, view from inferior, anterior, and lateral.

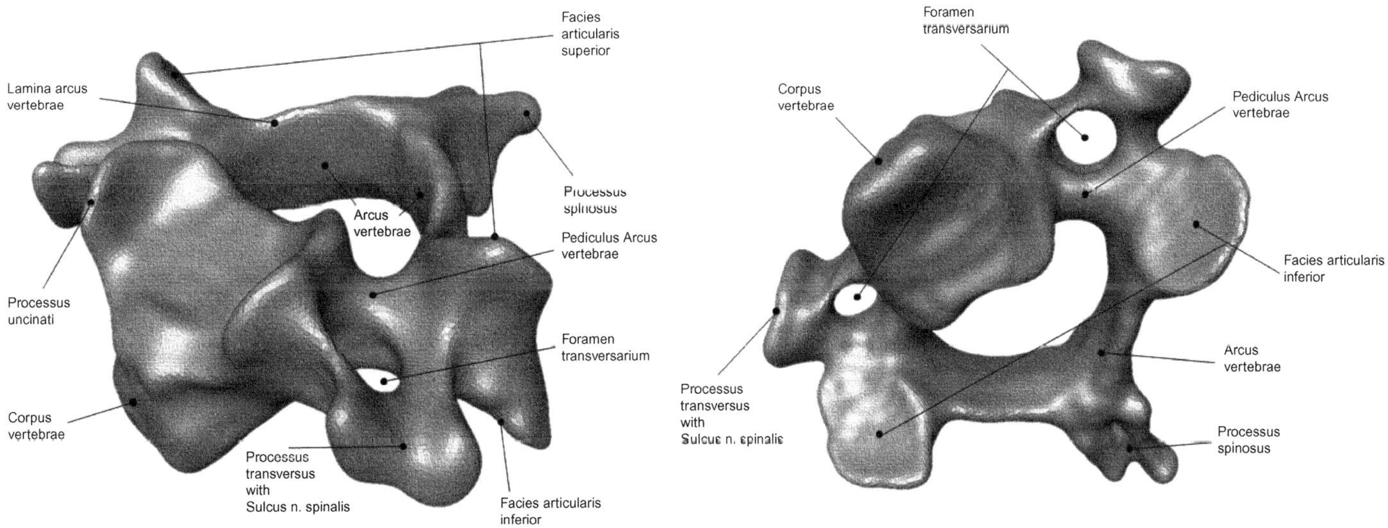

FIG. 4

Vertebra (C4): *left*, view from superior, anterior, and lateral; *right*, view from inferior, anterior, and lateral.

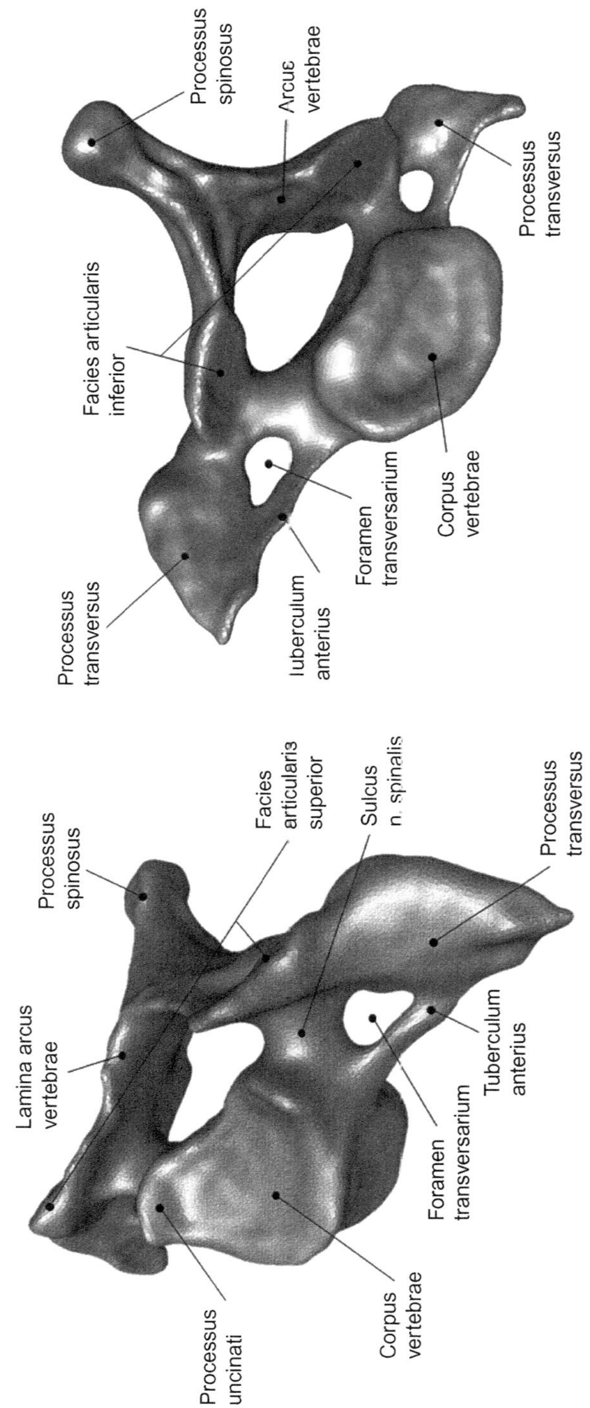

FIG. 5

Vertebra (C7): *left*, view from superior, anterior, and lateral; *right*, view from inferior and lateral.

ATLANTO-OCCIPITAL UNIT

The cervical spine can be divided into four functional units. The first unit is the "atlanto-occipital unit." This unit consists of the occipital bone, which is an inferior-posterior part of the skull, and the Atlas. The bond between the occiput and the Atlas is strong. Only small motions in flexion and extension are physiological, that is, rolling and gliding toward the anterior and posterior wall of the joint. In the case of lateral bending and axial rotation of the cervical spine, the occiput and the Atlas mainly move as one unit. Nonphysiological motions are restricted by the bony walls, which surround the superior articular surface of the Atlas. An impaction during flexion is mainly prevented by tension of the posterior neck muscles and compression of the submandibular tissue, but in extreme cases by an impingement of the bony parts. The limit for extension is given by the compression of the suboccipital muscles (Goel et al., 1988; Bogduk and Mercer, 2000).

Important ligaments that connect the occipital with either the Atlas or the Axis are the "*Lig. atlanto occipitale laterale,*" the "*Lig. apices dentis,*" the "*Lig. alaria,*" and the "*Lig. nuchae.*" The *Lig. atlantooccipitale laterale* surrounds the articular surfaces between the occiput and the Atlas and therefore forms the facet capsule (capsula articularis). The *Lig. apices dentis* connects the anterior part of the occiput (*Pars basilaris*) with the apex of the dens, whereas the *Lig. alaria* connects the lateral surface of the dens with the lateral rim of the *Foramen magnum.* The *Lig. nuchae* is located at the mid-sagittal plane and connects the posterior occiput with the *Arcus atlantis* of the Atlas as well as the *Proc. spinosus* of the Axis.

CRANIOCERVICAL UNIT

The "craniocervical unit" is composed of the Atlas and the Axis. It is the motion segment with the widest range for axial rotation. During axial rotation of the head, the Atlas rotates around the dens of the Axis (C2). Meanwhile, because of their horizontal alignment, the articular surfaces of each facet joint slide in the respective opposite direction. According to the literature, axial rotations in one direction up to 46 degree are possible (Penning and Wilmink, 1987; Bogduk and Mercer, 2000). Besides the facet capsule, the only ligaments that connect the Atlas with the Axis are the "*Lig. transversum*" and the "*Lig. nuchae.*" The *Lig. transversum* is attached to the anterior-lateral part of the Atlas and glides along the posterior articular surface of the dens. Together with the "*Fasciculi longitudinales,*" it forms the "*Lig. cruciform atlantis.*"

THE ROOT

The motion segment C2–C3 is also known as "the root" (Bogduk and Mercer, 2000). Its unique anatomy has a strong influence on the biomechanics of the cervical spine, even though they might be not prominent. The most relevant features are the facet joints. In contrast to the lower motion segments, the articular surfaces of the inferior vertebra are also pointing medially in an angle of about 40 degree (see Fig. 6), whereas the articular surfaces of the superior vertebra points in the opposite direction. They are also tilted in the sagittal plane as it can be observed within the lower segments. It can also be noticed that the superior articular surfaces of C3 lie more inferior with respect to its vertebral body. Together with the vertebral body of C2, which reaches more inferior than others do, the whole segment appears as root. The influences on the segment's biomechanical behavior due to these anatomical differences are described by Bogduk and Mercer (2000). They pointed out that there are differences in the coupling

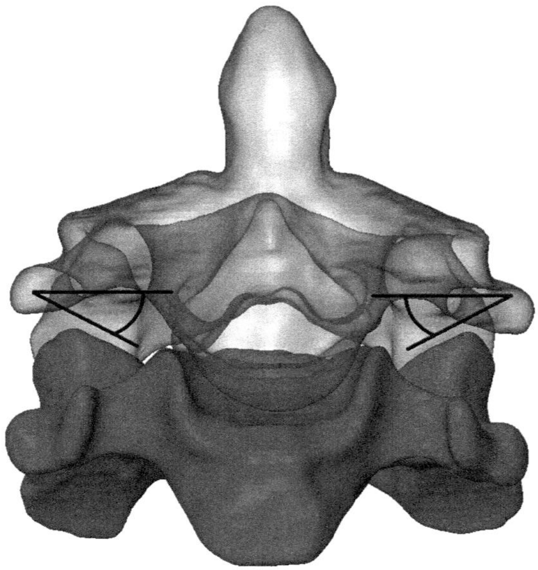

FIG. 6

The root (C2/3), highlighting the angulation in coronal view.

behavior during axial rotation of the head. In contrast to the lower cervical vertebrae C2 is rotating in the opposite direction of the head (Bogduk and Mercer, 2000).

THE COLUMN

"The column" basically includes all remaining motion segments: C3 to C7. However, it may also include C2 in some cases because motion segment C2–C3 shares the same soft tissue architecture. The basic ligaments of the column are *Lig. nuchae*, *Lig. interspinale*, *Lig. flava*, *Lig. longitudinale anterius*, *Lig. longitudinale posterius*, *Lig. transversarium*, and the capsula articularis. The *Lig. nuchae* and the *Lig. interspinale* connect the processus spinosus of the adjacent vertebrae. The *Lig. flava* is located in between each Arcus vertebrae of the respective motion segment. The *Lig. longitudinale* cover the anterior and posterior surface of the intervertebral discs as well as the vertebrae throughout the whole column. Each *Lig. transversarium* is attached to the processus transversus of the respective superior and inferior vertebra.

CERVICAL INTERVERTEBRAL DISC

The intervertebral disc of the cervical spine is unique in its anatomy. In fact, some anatomic details are not fully verified to this day. In general, the disc consists of a nucleus pulpous, an annulus fibrosus, and the Luschka's joints—also known as uncovertebral joints—which will be described later in more detail. In contrast to the intervertebral discs of the thoracic and lumbar spine, the nucleus is located more posteriorly within the whole disc. Published data about the anatomy of the annulus fibrosus is very inconsistent. Based on the study by the well-respected authors Mercer and Bogduk (1999), it can be assumed that the annulus does not fully surround the nucleus. Therefore the annulus is most prominent at the anterior part of the disc. The orientation of its fibers changes from longitudinal close to the surface to cross-linked within the deeper layers. At the posterior end of the disc, only a small layer of longitudinal fibers covers the nucleus pulposus. According to Mercer et al., no annulus fibers are at the lateral-posterior sides of the disc; instead small fissures—the uncovertebral joints—can be detected. Another study by Tonetti et al. in 2004 confirmed the data of Mercer et al. and described the annulus fibrosus as inconsistent.

The tissue of the nucleus pulposus is described as a fibrocartilaginous core, which is also unique compared to the thoracic and lumbar spine (Tonetti et al., 2005). Therefore Tonetti et al. confirms the findings of earlier studies. One study by Töndury (1972) stated that the cervical nucleus pulposus is at first gelatinous like the nucleus from the lumbar disc and then becomes dehydrated when an individual reaches the age of 30 (Töndury, 1972). Oda et al. also observed a gelatinous nucleus pulposus until the midteens, which is then replaced by fibrocartilage and fiber components (Oda et al., 1988; Mercer and Jull, 1996).

Nevertheless, not all literature concurs with these findings and instead describes the cervical intervertebral disc as a smaller version of the lumbar disc. However, because of the continuous development of fissures and the progression of dehydration throughout an entire lifetime, it is very difficult to clearly determine the basic anatomy (Figs. 7 and 8).

UNCOVERTEBRAL JOINTS

The Luschka's joints, also known as uncovertebral joints or uncovertebral clefts, were described first by Luschka in 1858. During the past decades, their anatomical detail has been widely investigated, but some aspects remain not fully clarified. The uncovertebral joints develop with age. When an individual reaches the age of 9, these joints start to form concurrent with the formation of the uncinate processes. They continuously progress until they reach the nucleus pulposus (Töndury and Theiler, 1958; Ecklin, 1960; Penning, 1988). Their status as a synovial joint has been extensively discussed ever since the discovery of their existence. For example, studies conducted by Hirsch et al. (1967) and Tonetti et al. (2005) were not able to detect any characteristics of a synovial joint. However, a study by Brismee et al. was able to observe synoviocytes and chondrocytes within the cleft. Therefore some investigators recommended to reconsider the status as joint (Brismee et al., 2009).

Another widely discussed topic of the uncovertebral clefts is their function or purpose. Because of the anatomy of the facet joints and the vertebral bodies, axial rotation and lateral bending occur in combination as a form of coupled motion. In addition, the vertebral bodies glide forward and backward as they rotate in flexion and extension. Those motions are assumed to provoke the formation of the uncovertebral clefts. These clefts are biggest in the upper cervical spine as is the amount of axial rotation, which leads to the assumption that coupled motions are primarily responsible for these clefts' existence. Consequently, it is commonly accepted that the intervertebral clefts improve the overall range of motion (ROM) (von Luschka, 1858; Töndury and Theiler, 1958; Ecklin, 1960; Penning, 1988; Sizer et al., 2001).

Nevertheless, because of its natural course, some investigators have assumed that the uncovertebral joint may be a trigger for neck pain, whereas others deny any correlation between its progression and symptoms for neck pain. Compared to a common fissure within the lumbar spine, the uncovertebral cleft progresses from the outside to the inside until it reaches the nucleus pulposus, which may

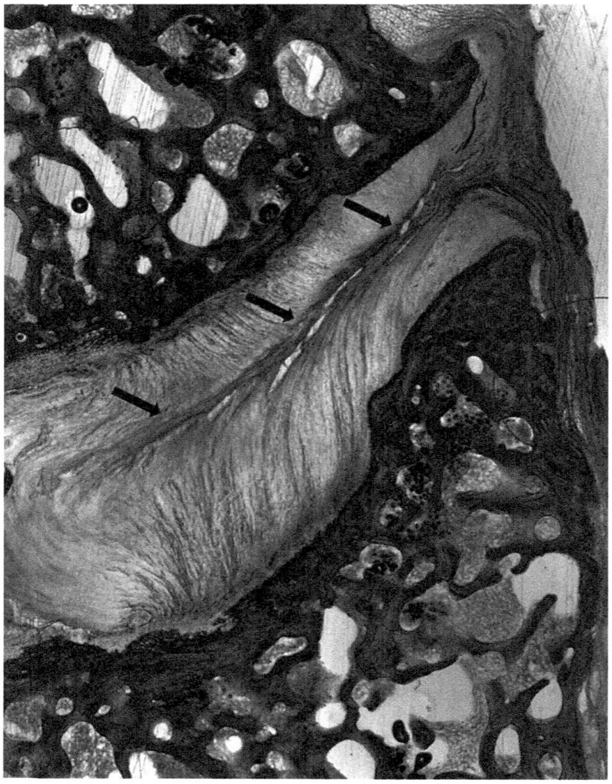

FIG. 7

Coronal histology cut through the lateral part of a cervical intervertebral disc, highlighting the uncovertebral cleft.

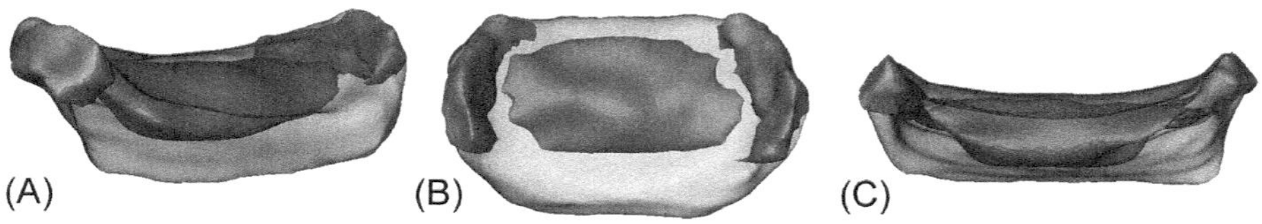

FIG. 8

Three-dimensional model of a cervical intervertebral disc, including nucleus pulposus (*in the middle*) and uncovertebral clefts (*on the lateral extremities*) reconstruction from ultra-high field MRI (11.7T). (A) coronal-sagittal-view, (B) transverse view, and (C) coronal view.

eventually lead to a migration of nucleus material into the spinal canal (Sizer et al., 2001). However Töndury and Theiler (1958) and Dvorak (1998) claimed that by the time extrusions may occur the nucleus pulposus is more or less dried out and therefore an acute protrusion cannot be expected. This finding may explain further studies that were not able to find any correlation between cervical disc degeneration and neck pain (Peterson et al., 2003). Consequently, the influence of the uncovertebral joint on the development of neck pain is still not fully understood.

FACET JOINTS

The facet joints (also, zygapophysial joints) of the cervical spine consist of two articular surfaces of cartilaginous tissue, the meniscoid, or synovial fold, and the capsular ligament (see Fig. 9). They are directly attached to the vertebral arch and transfer a considerable amount of compressive load to the inferior segments. Together with the intervertebral disc, they form a flexible connection of the cervical vertebrae and at the same time prevent the cervical spine from making extensive movements that could potentially damage the spinal cord. The facet joints also have an enormous influence on the biomechanical behavior of the cervical spine because of their unique orientation. All facet joints of the cervical spine, with a minor exception for the motion segment C2–C3, exhibit an inclination of 20 degree to 78 degree versus the axial plane and 70 degree to 96 degree versus the mid-sagittal plane (Jaumard et al., 2011).

The capsular ligament and the meniscoid of the facet joint lubricate and nourish the articular cartilage with synovial fluid. In addition, the synovial fluid contains macrophages for the nursing and maintenance of the cavity (Iwanaga et al., 2000; Farrell et al., 2017). The meniscoids are most prominent in the anterior and posterior regions of the articular surface, whereas circumferential extensions along the rim occur only partially (Jaumard et al., 2011). Concerning the anatomy of the articular cartilage, according to Yoganandan et al. (2003), the cartilage is the thickest in the middle of the articular surface. The thickness of the cartilage is assumed to be greater in the upper than in the lower cervical spine. Given values for the thickness range from 0.47 mm in lower regions to 0.73 mm in the upper regions. It is also mentioned that the cartilage in the upper cervical spine covers approximately 90% of the facet process but in the lower cervical spine only 68%, even though the facet joints in the upper cervical spine are wider than in the lower cervical spine (Yoganandan et al., 2001).

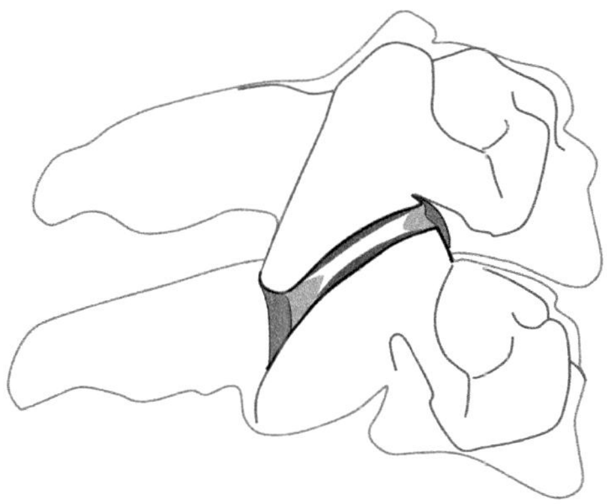

FIG. 9

Motion segment highlighting the facet joint, including capsule, menisci, and articular surfaces.

MUSCLES

A complex system of muscles in the neck and the back allow the neck a wide ROM. The anatomy of these muscles is well defined in several anatomy books. Most neck muscles are located laterally or posteriorly to the vertebral column (Moroney et al., 1988). There are short muscles, running from one vertebra to an adjacent one (e.g., *Mm. interspinales cervicis*), and there are longer muscles that, for example, connect the occipital bone with C3 to C6 (*M. longus capitis*) or the vertebrae C3 to C7 with the first and second rip (*Mm. scaleni*). These muscles are just some examples of a sophisticated system of muscles. In addition, Winkelstein et al. (2001) found that the muscles of the cervical spine even insert into the facet capsules. However, their purpose and relationship to neck pain remain unclear.

Morphometric studies of the neck musculature have shown that most muscles in men are bigger and therefore stronger than in women (Kamibayashi and Richmond, 1998; Vasavada et al., 2001). These studies, which are based on examination of cadaveric specimens or MR imaging, are also used to determine the cross-sectional area of each muscle in order to further estimate their contraction forces and their contribution to neck movements. The activation of each muscle depends not only on the motion direction but also on the current posture of the head and neck. Therefore each movement is a complex combination of multiple muscles, where each muscle acts depending on its current lever arm (Table 1).

Table 1 Involvement of Muscles for Each Motion Direction According to Vasavada et al. (1998)

Motion Direction	Contributing Muscles
Flexion	• Sternocleidomastoid
	• Longous Capitis/Colli
	• Scalenus Anterior
Extension	• Semispinalis
	• Splenius
	• Trapezius
	• Levator Scapulae
	• Erector Spinae
	• Suboccipital
Lateral bending	• Sternocleidomastoid
	• Trapezius
	• Semispinalis
	• Splenius
	• Suboccipital
Axial rotation	• Sternocleidomastoid
	• Scalenes
	• Trapezius
	• Semispinalis
	• Splenius
	• Levator Scapulae
	• Erector Spinae

NEURAL FORAMEN AND FORAMEN TRANSVERSARIUM

The neural foramen are located between the vertebral body and the facet joints. They obtain a functional connection between the intradural region and the extraspinal region. These foramen contain not only arteries, veins, and nerve roots but also fatty tissue. Nerve rootlets that exit through the neural foramen split mainly into a ventral and a dorsal bundle. The mean neural foraminal areas increase mainly from 33.7 mm^2 at the superior levels to 54.2 mm^2 at the more inferior motion segments, except for one motion segment: C2/3. This particular motion segment, which is also the first containing a neural foramen, obtains a mean cross-sectional area of about 55.5 mm^2 according to Smith et al. (2016). Based on an in vivo study by Anderst et al. (2015), it is also believed that the cross-sectional areas of the human neural foramen decrease in extension and increase in flexion, which has a big impact on physiotherapeutic treatment. The study by Smith et al. showed that an increased T1-tilt decreases the neural foraminal areas of the mid- and lower-cervical motion segments. On the contrary, an increase of the sagittal vertical alignment leads to an increase of the mid-to-lower neural foramen. The size of each neural foraminal area has a great influence on the neuronal interface. A loss of disc height and the formation of osteophytes might decrease the cross-sectional area of the neural foramen and could potentially cause a radiculopathy.

The foramen transversarium are a unique feature of the cervical vertebrae. They protect the vertebral vascular bundle and the sympathetic plexus. Their special anatomy is associated with the fusion of the true transverse process with vestigial costal elements. According to Zibis et al. (2016), their average diameter ranges from around 5.74 mm to 6.49 mm. The vertebral artery inserts mostly at the foramen transversarium of C6 and runs upward to the Atlas, whereas the foramen transversarium of C7 is usually occupied by veins and nerve branches along with areolar and adipose tissue. The huge variation in size and number on each side of the vertebra is associated with the presence and course of the vertebral vessels (Taitz et al., 1978).

BIOMECHANICS

For several decades, scientists around the world have conducted extensive study and research on the biomechanics of the cervical spine. One major motivation has been the neck pain experienced so frequently in modern society (Vos et al., 2015). Another reason may be the growing demand for safety in the automobile industry. Moreover, implants such as cages and disc replacements have become more and more important because of the increasing life expectancy.

The kinematics of the cervical spine are very complex and sophisticated. Because of the saddle-joint-like shape of the motion segments C2/3 to C6/7 and the unique orientation of the articular surfaces of the facet joints, lateral motions and axial rotation of the vertebrae C2 to C7 occur only as a combination of flexion/extension, lateral bending, and axial rotation. According to Bogduk and Mercer (2000), all motions of the head are performed using only two different axes of rotation. The first one, which is almost solely used during flexion and extension, is located in the inferior vertebra of one motion segment. Its orientation is perpendicular to the sagittal plane. The second axis is oriented perpendicular to the articular surfaces of the facet joints; then it goes through the respective vertebra and penetrates it close to its anterior, inferior tip (Fig. 10).

RANGE OF MOTION

One of the most relevant parameters in cervical biomechanics is the ROM. Many in vivo as well as in vitro studies about cervical biomechanics include this parameter as a basic entry into cervical

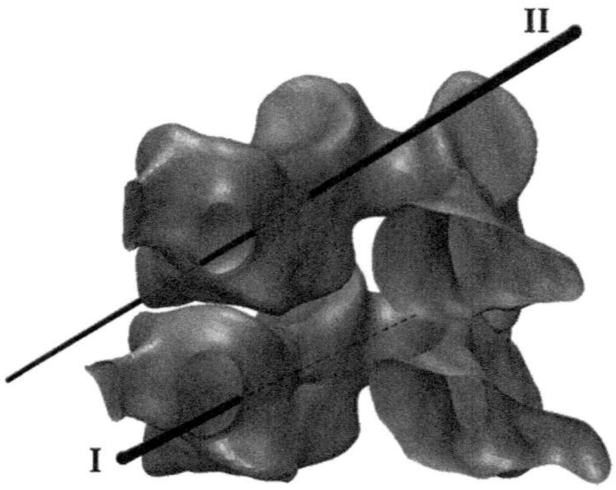

FIG. 10

General axes of rotation for motion segments C2/3 to C6/7 prcposed by Bogduk and Mercer (2000).

kinematics. In the development and testing of disc replacements, for instance, the ROM is used to describe general kinematic characteristics of the cervical spine in order to answer questions which address the capability of restoring the physiological ROM. Regarding the problematics of neck pain, the ROM was used to access the severity of degeneration or to quantify the amount of pain a patient had. There are various applications for the ROM; therefore the ROM has become a universal parameter in spine biomechanics. What makes this parameter so important? On one hand, the ROM is easy to understand, and on the other hand, it can tell us a lot about one specimen, especially in cervical kinematics regarding coupled motions.

During the past decades, scientists have produced ample data on the ROM in the cervical spine, which constitutes a valuable foundation for future research. Figs. 11–13 show some results of ROM experiments (in vivo and in vitro).

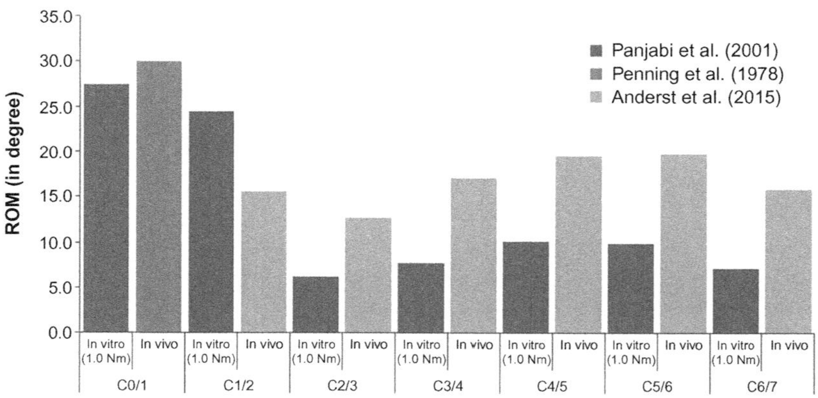

FIG. 11

ROM for flexion and extension.

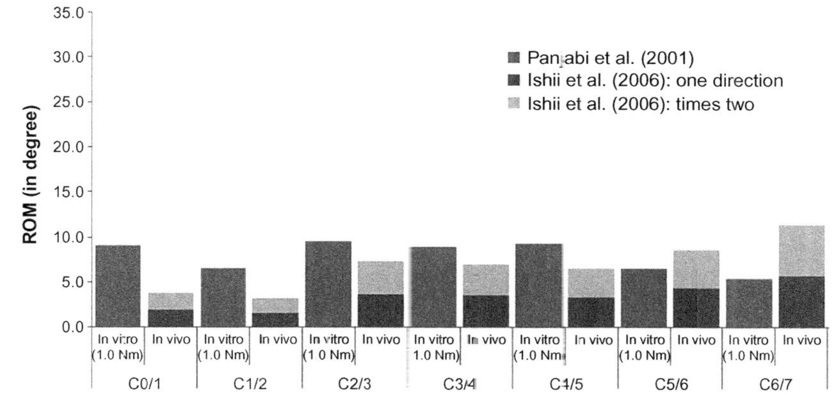

FIG. 12

ROM for lateral bending.

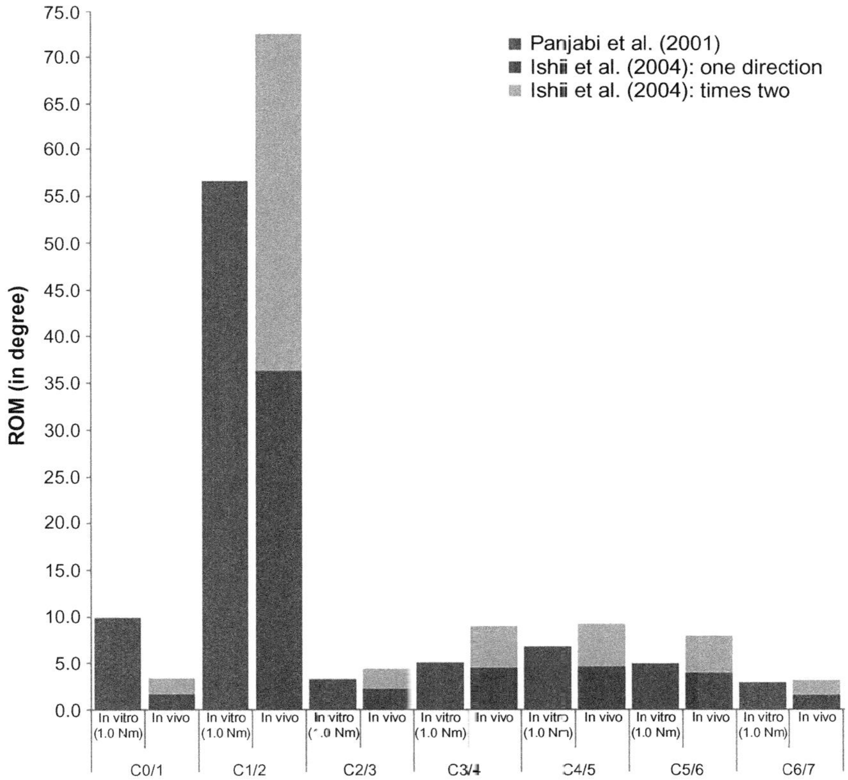

FIG. 13

ROM for axial rotation.

A further application for the ROM is the investigation of coupled motions as previously described. Coupled motions occur mainly during lateral bending and axial rotation, which means that the movement of the head in a specific direction is accomplished by a combination of a different motion direction within each motion segment. In order to measure coupled motions, subjects or scientists actively rotated the head or a vertebra in one specific direction (flexion, extension, lateral direction, or axial direction), while the ROMs within each motion segment for all three directions were measured. Measurements for lateral bending and axial rotation are shown in Figs. 14 and 15.

HELICAL AXES

A more qualitative analysis of cervical motion can be performed by using (instantaneous) helical axes (IHAs). Most investigations on IHA of the cervical spine are based on in vivo experiments. The very first studies dealing with this topic used X-ray images of the cervical spine. In these experiments, the subjects were asked to move their head in a specific direction while their shoulders and chest remained fixed. Meanwhile, at least two different X-ray images with different postures of the neck were taken. Afterward a graphical reconstruction of the center of rotation was easily performed by overlaying the X-ray images (Fig. 16). The center of rotation or instantaneous center of rotation (ICR) for flexion and extension can also be interpreted as an IHA, which is orientated perpendicular to the sagittal plane. In a further step, Penning and Wilmink (1987) used CT data to determine the ICRs in multiple, parallel planes and connected them. With this method, Penning was able to construct the IHA for lateral bending.

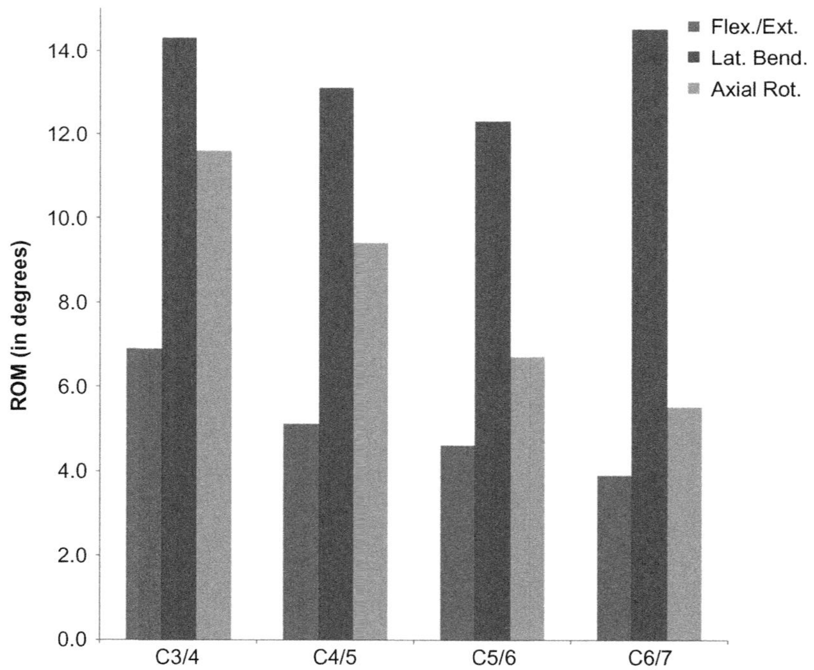

FIG. 14

Coupled motions (ROM) for lateral bending (active) by Anderst et al. (2015).

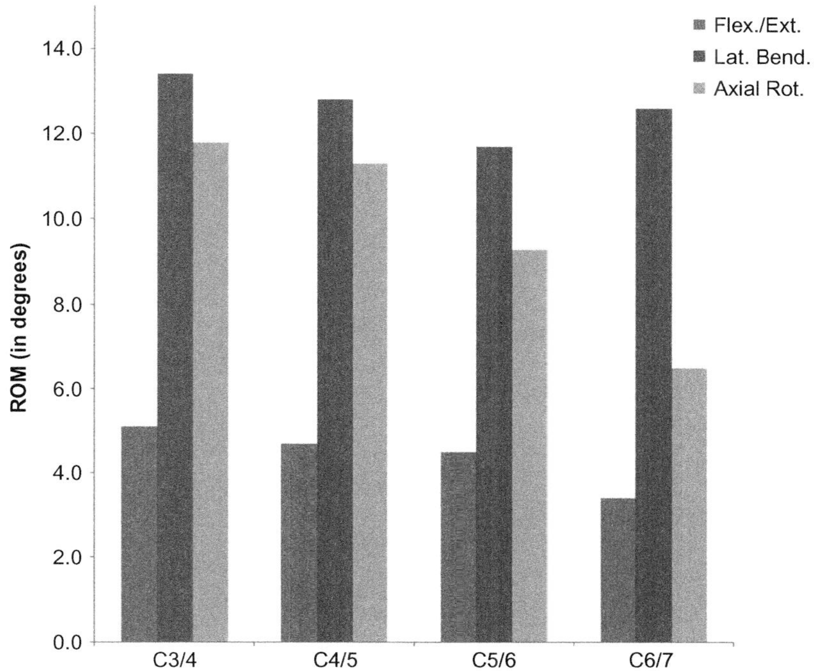

FIG. 15

Coupled motions (ROM) for axial rotation (active) by Anderst et al. (2015).

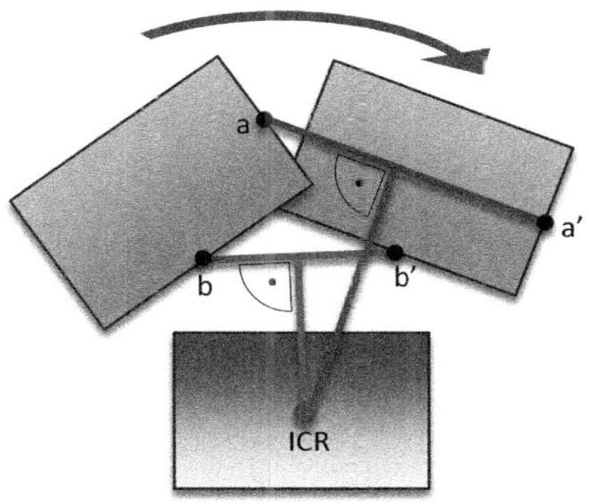

FIG. 16

Schematic reconstruction of the ICR for one increment.

A more recent in vivo study by Anderst et al. (2015) was able to show for the first time three-dimensional IHA in respect to a 3D reconstruction of the cervical spine. In order to do so, Anderst et al. used a biplanar X-ray device to record the three-dimensional motion and CT data for the 3D reconstruction of the subject's cervical spine. The data from this study confirmed the results of earlier studies based on X-ray images (Penning, 1978, 1988; Penning and Wilmink, 1987; Amevo et al., 1991; Van Mameren et al., 1992) as well as the axes of rotation proposed by Bogduk and Mercer (2000) (see Fig. 10).

Although a substantial amount of in vivo data about the IHA of the cervical spine have been published, data of IHA during in vitro testing are still very rare. To the author's knowledge, there are only two studies reporting about the IHA of the cervical spine during in vitro experiments. Dugailly et al. published an in vitro study in 2010 that dealt with the IHA of the upper cervical spine. They were able to visualize three-dimensional IHA for Flexion and Extension in C0/1 and axial rotation in C1/2. The IHA for Flexion and Extension were almost horizontally aligned. They were located within C0 and superior to the dens. The IHA for axial rotation were almost vertically oriented, going through the dens (Dugailly et al., 2010), as expected. The second in vitro study which was dealing with the IHA of the cervical spine was published by Jonas et al. (2018). This study was investigating the three-dimensional IHA of the motion segments C2-C3, C4-C5, and C6-C7 for all three types of rotation. Furthermore they compared their results from in vitro testing with IHA gained from in vivo experiments. The comparison could show that in vitro experiments can produce realistic IHA when using pure moments.

INTRADISCAL PRESSURE

Data about the intradiscal pressure (IDP) of the cervical intervertebral disc are very limited. Understanding of the IDP grew from the time of its first application in the lumbar spine. Very useful information could be obtained for this part of the spine in vitro and even in vivo. Lumbar intervertebral discs are larger than the cervical intervertebral discs because the former must carry a much higher load. Therefore it is easier to insert instruments for pressure measurements. In addition, as already mentioned, the tissue of the nucleus changes in character to a more fibrocartilaginous tissue when an individual reaches the age of 50 years. In contrast, measurement of IDP requires a hydrostatic pressure within the disc, which consequently limits the number of reliable subjects. However, Pospiech et al. (1999) were able to determine the IDP of the intervertebral discs C3/4 and C5/6 in an in vitro experiment, and even an in vivo study was reported by Hattori et al. (1981). The in vivo study included 80 discs from different motion segments (C3/4 to C6/7) in which the pressure was measured in sitting and supine positions (Fig. 17).

LOAD SHARING

The physiological load applied to the cervical spine because of the weight of the head and the activity of the muscles is shared by the anterior and posterior columns. The anterior column, which contains the vertebral bodies as well as the intervertebral discs, transmits an estimated load of about 36% to the thoracic spine. The rest is transmitted through the facet joints that belong to the posterior column. Additionally, the posterior column can be subdivided into left and right facet joints. Therefore the load that is applied to the cervical spine is carried by three pillars, with each carrying an almost equal amount of the total load: 36%/32%/32% (see Fig. 18) (Pal and Sherk, 1988).

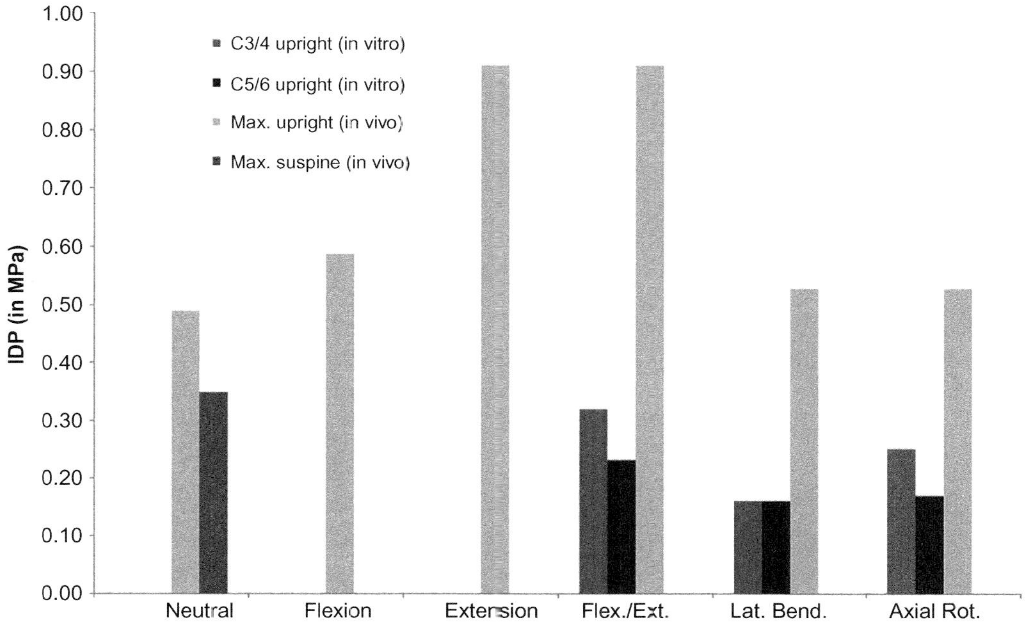

FIG. 17

Comparison between in vivo and in vitro measurements of the IDP in cervical intervertebral discs; in vitro data from Pospiech et al. (1999); in vivo from Hattori et al. (1981).

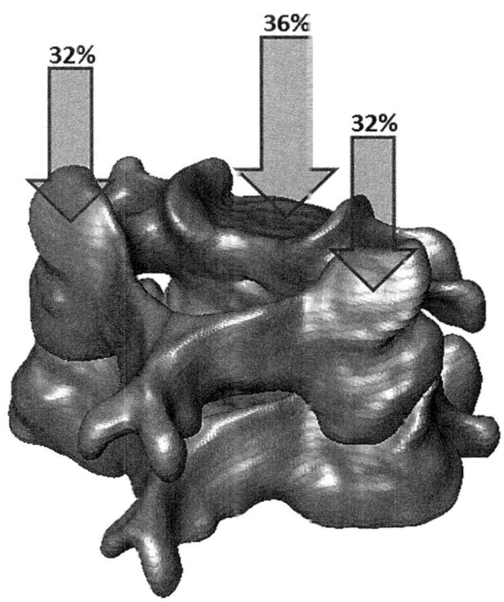

FIG. 18

Load distribution on cervical motion segment according to Pal and Sherk (1988).

NUMERICAL SIMULATION

Numerical models such as finite element (FE) models have heavily influenced spinal research and therefore the investigation of cervical biomechanics. Regarding cervical spine, FE models are used to estimate stresses, strains, displacements, and other important mechanical parameters also can be used to support in vivo as well as in vitro findings. They also come into consideration for areas that are difficult to investigate by the use of the appropriate measuring devices. But most importantly, they are of great value to the spinal research for ethical reasons. In vivo and in vitro investigations are controlled by ethical committees, whereas numerical simulations are not. Therefore many experiments that might potentially harm a subject or waste a valuable cadaveric specimen can be run in advance via numerical simulations. To some extent in vitro as well as in vivo experiments can even be replaced by finite element simulations. However, each model needs to be appropriately validated with in vivo or at least in vivo data. The outcome of the validation process depends heavily on the material parameters as well as on the approximations that were made for the anatomical characteristics. FE models for the cervical spine could include all vertebrae as well as the skull, but because of computational capacities, they are often restricted to specific motion segments.

With respect to the load application, it is possible to apply pure moments only to the uppermost vertebra or to use muscle forces interacting with the respective vertebrae. If using pure moments, the simultaneous application of a follower compressive load may be considered. The follower load applies the load from the superior part of the spine to the inferior part along the lordotic curve of the cervical spine.

SPECIMEN-SPECIFIC MODELS

The most common type of cervical FE model is the specimen-specific model. Specimen-specific models are based mostly on CT data. Although they need a substantial amount of time for generation and

Table 2 Overview of a Selection for Specimen-Specific FE Models

Author	Segments	Meshing
Yoganandan et al. (1996)	C4 to C6	8-node
Clausen et al. (1997)	C5 to C6	8-node
Wheeldon et al. (1999)	C2 to T1	8-node
Brolin and Halldin (2004)	Oc to C3	8-node
Panzer (2006)	C1 to T1	8-node
del Palomar et al. (2008)	C1 to C7	4- and 8-node
Galbusera et al. (2008)	C4 to C7	8-node
Tchako and Sadegh (2009)	C1 to C7	4-node
Kallemeyn et al. (2010)	C2 to C7	8-node
Toosizadeh and Haghpanahi (2011)	Oc to C7	10-node
Zhang et al. (2011)	C1 to C7	8-node
Bredbenner et al. (2014)	C3 to T1	4- and 8-node
Mustafy et al. (2014)	C2 to C3	4- and 8-node
Khuyagbaatar et al. (2017)	C2 to T1	8-node

Table 3 Selection of Parametric FE Models of the Cervical Spine

Author	Segments
Nitsche et al. (1996)	C1 to T1
Maurel et al. (1997)	C3 to C7 (C4–C5 validated)
Haghpanahi and Javadi (2012)	C3 to C7 (motion segments were validated separately)

validation, each model represents only one particular subject and, therefore, the use of such models may not be optimal in order to draw conclusions for the general population. On the other hand, specimen-specific models are very accurate in anatomical details that may have a huge impact on the results. In some cases, these models may even be validated with the help of the specimen (subject) that has also obtained the respective data for geometry. Table 2 shows a selection of specimen-specific FE models.

PARAMETRIC MODELS

Parametric models are usually based on correlations, which may concern either the anatomical parameters within one vertebra or those between the vertebrae themselves. The geometrical parameters are used to create a simplified FE model of the cervical spine. In that respect, simple geometries such as cylinders and hexahedrons are used to form vertebrae, intervertebral discs, and ligaments with little effort for the operator. Besides, these models provide an easy route to perform automated sensitivity analysis aimed to explore the biomechanical relevance of the various anatomical parameters. On the other hand, parametric models are less detailed than specimen-specific models in terms of their anatomical accuracy (Table 3).

MATERIAL MODELS/PARAMETERS

Once the geometry of a model is finished, it is necessary to assign material parameters to each part of the model. Although the material parameters for the cortical and cancellous bone are relatively relatively easy to measure, their determination is more challenging regarding the soft tissues. In the case of the ligaments, it is difficult to calculate the Young's modulus because of the difficulties involved in measuring their cross-sectional areas. Therefore force-displacement curves have been determined during in vitro tests (Moroney et al., 1988a, 1988b; Panjabi et al., 1998; Yoganandan et al., 2001; Ivancic et al., 2007). With respect to the complex structures of the intervertebral discs, different methodologies or material models have been proposed over the past decades. One approach is a material model developed by Holzapfel et al. (2005) that was originally proposed for the lumbar annulus fibrosus. This material model consists of a continuum formulation containing a ground substance in which nonlinear fibers are embedded by providing a contribution to the strain energy. Other researchers explicitly modeled the collagen fibers of the annulus fibrosus by means of unidimensional elements such as springs or trusses (Galbusera et al., 2006; Toosizadeh and Haghpanahi, 2011) or with two-dimensional elements in which fiber elements were embedded (Panzer, 2006). The nucleus pulposus, which is typically represented as an incompressible hydrostatic fluid-like substance following previous models of the lumbar spine, has been modeled either as a flexible solid with a high Poisson's ratio (Galbusera et al., 2008) or as a fluid-filled cavity. A comprehensive model including a detailed description of the fiber architecture and of the transition zone between the nucleus and annulus is still lacking. Similarly, the vast majority of available models neglected the presence of uncovertebral joints.

REFERENCES

Amevo, B., Worth, D., Bogduk, N., 1991. Instantaneous axes of rotation of the typical cervical motion segments: a study in normal volunteers. Clin. Biomech. 6 (2), 111–117.

Anderst, W.J., Donaldson, W.F., Lee, J.Y., Kang, J.D., 2015. Three-dimensional intervertebral kinematics in the healthy young adult cervical spine during dynamic functional loading. J. Biomech. 48 (7), 1286–1293.

Bogduk, N., Mercer, S., 2000. Biomechanics of the cervical spine. I. Normal kinematics. Clin. Biomech. 15 (9), 633–648.

Bredbenner, T.L., Eliason, T.D., Loren Francis, W., McFarland, J.M., Merkle, A.C., Nicolella, D.P., 2014. Development and validation of a statistical shape modeling-based finite element model of the cervical spine under low-level multiple direction loading conditions. Front. Bioeng. Biotechnol. 2, .

Brismee, J., Sizer Jr., P.S., Dedrick, G.S., Sawyer, B.G., Smith, M.P., 2009. Immunohistochemical and histological study of human uncovertebral joints—a preliminary investigation. Spine 34 (12), 1257–1263.

Brolin, K., Halldin, P., 2004. Development of a finite element model of the upper cervical spine and a parameter study of ligament characteristics. Spine 29 (4), 376–385.

Clausen, J.D., Goel, V.K., Traynelis, V.C., Scifert, J., 1997. Uncinate processes and luschka joints influence the biomechanics of the cervical spine: quantification using a finite element model of the c5-c6 segment. J. Orthop. Res. 15 (3), 342–347.

del Palomar, A.P., Calvo, B., Doblare, M., 2008. An accurate finite element model of the cervical spine under quasi-static loading. J. Biomech. 41 (3), 523–531.

Dugailly, P., Sobczak, S., Sholukha, V., Jan, S.V.S., Salvia, P., Feipel, V., Rooze, M., 2010. In vitro 3d-kinematics of the upper cervical spine: helical axis and simulation for axial rotation and flexion extension. Surg. Radiol. Anat. 32 (2), 141–151.

Dvorak, J., 1998. Epidemiology, physical examination, and neurodiagnostics. Spine 23 (24), 2663–2672.

Ecklin, U., 1960. Die Altersveränderungen der Halswirbelsäule. Springer Verlag, Berlin-GottingenHeidelberg.

Farrell, S.F., Osmotherly, P.G., Cornwall, J., Sterling, M., Rivett, D.A., 2017. Cervical spine meniscoids: an update on their morphological characteristics and potential clinical significance. Eur. Spine J. 9 (1), 1–9.

Galbusera, F., Fantigrossi, A., Raimondi, M.T., Assietti, R., Sassi, M., Fornari, M., 2006. Biomechanics of the c5-c6 spinal unit before and after placement of a disc prosthesis. Biomech. Model. Mechanobiol. 5 (4), 253–261.

Galbusera, F., Bellini, C.M., Raimondi, M.T., Fornari, M., Assietti, R., 2008. Cervical spine biomechanics following implantation of a disc prosthesis. Med. Eng. Phys. 30 (9), 1127–1133.

Gilad, I., Nissan, M., 1986. A study of vertebra and disc geometric relations of the human cervical and lumbar spine. Spine 11 (2), 154–157.

Goel, V.K., Clark, C.R., Gallaes, K., King Liu, Y., 1988. Moment-rotation relationships of the ligamentous occipito-atlanto-axial complex. J. Biomech. 21 (8), 673677–675680.

Haghpanahi, M., Javadi, M., 2012. A three dimensional parametric model of whole lower cervical spine c3-c7 under flexion, extension, torsion and lateral bending. Scientia Iranica 19 (1), 142–150.

Harrison, D.E., Harrison, D.D., Cailliet, R., Troyanovich, S.J., Janik, T.J., Holland, B., 2000. Cobb method or harrison posterior tangent method—which to choose for lateral cervical radiographic analysis. Spine 25 (16), 2072–2078.

Hattori, S., Oda, H., Kawai, S., 1981. Cervical intradiscal pressure in movements and traction of the cervical spine. Z. Orthop. 119, 568–569.

Hirsch, C., Schajowicz, F., Galante, J., 1967. Structural changes in the cervical spine—a study on autopsy specimens in different age groups. Acta Orthop. Scand. 38 (109), 1–77.

Holzapfel, G.A., Schulze-Bauer, C., Feigl, G., Regitnig, P., 2005. Single lamellar mechanics of the human lumbar anulus fibrosus. Biomech. Model. Mechanobiol. 3 (3), 125–140.

Ishii, T., Mukai, Y., Hosono, N., Sakaura, H., Fujii, R., Nakajima, Y., Tamura, S., Sugamoto, K., Yoshikawa, H., 2004a. Kinematics of the subaxial cervical spine in rotation in vivo three-dimensional analysis. Spine 29 (24), 2826–2831.

Ishii, T., Mukai, Y., Hosono, N., Sakaura, H., Nakajima, Y., Sato, Y., Sugamoto, K., Yoshikawa, H., 2004b. Kinematics of the upper cervical spine in rotation—in vivo three-dimensional analysis. Spine 29 (7), 139–144.

Ishii, T., Mukai, Y., Hosono, N., Sakaura, H., Fujii, R., Nakajima, Y., Tamura, S., Iwasaki, M., Yoshikawa, H., Sugamoto, K., 2006. Kinematics of the cervical spine in lateral bending—in vivo three-dimensional analysis. Spine 31 (2), 155–160.

Ivancic, P.C., Coe, M.P., Ndu, A.B., Tominaga, Y., Carlson, E.J., Rubin, W., Panjabi, M.M., 2007. Dynamic mechanical properties of intact human cervical spine ligaments. Spine J. 7 (6), 659–665.

Iwanaga, T., Shikichi, M., Kitamura, H., Yanase, H., Nozawa-Inoue, K., 2000. Morphology and functional roles of synoviocytes in the joint. Arch. Histol. Cytol. 63 (1), 17–31.

Jaumard, N.V., Welch, W.C., Winkelstein, B.A., 2011. Spinal facet joint biomechanics and mechanotransduction in normal, injury and degenerative conditions. J. Biomech. Eng. 133 (7), 7101001–7101031.

Jonas, R., Demmelmaier, R., Hacker, S.P., Wilke, H.J., 2018. Comparison of three-dimensional helical axes of the cervical spine between in vitro and in vivo testing. Spine J. (in press).

Kallemeyn, N., Gandhi, A., Kode, S., Shivanna, K., Smucker, J., Grosland, N., 2010. Validation of a c2-c7 cervical spine finite element model using specimen-specific flexibility data. Med. Eng. Phys. 32 (5), 482–489.

Kamibayashi, L.K., Richmond, F.J.R., 1998. Morphometry of human neck muscles. Spine 23 (12), 1314–1323.

Khuyagbaatar, B., Kim, K., Park, W.M., Kim, Y.H., 2017. Biomechanical investigation of post-operative c5 palsy due to ossification of the posterior longitudinal ligament in different types of cervical spinal alignment. J. Biomech. 57, 54–61.

Knott, P.T., Mardjetko, S.M., Techy, F., 2010. The use of the t1 sagittal angle in predicting overall sagittal balance of the spine. Spine J. 10 (11), 994–998.

Maurel, N., Lavaste, F., Skalli, W., 1997. A three-dimensional parameterized finite element model of the lower cervical spine, study of the influence of the posterior articular facets. J. Biomech. 30 (9), 921–931.

Mercer, S., Bogduk, N., 1999. The ligaments and anulus fibrosus of human adult cervical intervertebral discs. Spine 24 (7), 619–626.

Mercer, S.R., Jull, G.A., 1996. Morphology of the cervical intervertebral disc—implications for mckenzies model of the disc derangement syndrome. Man. Ther. 1 (2), 76–81.

Moroney, S.P., Schultz, A.B., Miller, J.A.A., Andersson, G.B.J., 1988. Load-displacement properties of lower cervical spine motion segments. J. Biomech. 21 (9), 769–779.

Moroney, S.P., Schultz, A.B., Miller, J.A.A., 1988a. Analysis and measurement of neck loads. J. Orthop. Res. 6 (5), 713–720.

Moroney, S.P., Schultz, A.B., Miller, J.A.A., Andersson, G.B.J., 1988b. Load-displacement properties of lower cervical spine motion segments. J. Biomech. 21 (9), 769–779.

Mustafy, T., El-Rich, M., Mesfar, W., Moglo, K., 2014. Investigation of impact loading rate effects on the ligamentous cervical spinal load-partitioning using finite element model of functional spinal unit c2-c3. J. Biomech. 47 (12), 2891–2903.

Nitsche, S., Krabbel, G., Appel, H., Haug, E., 1996. Validation of a finite-element-model of the human neck. J. Hum. Evol. 34 (3), 249–302.

Oda, J., Tanaka, H., Tsuzuki, N., 1988. Intervertebral disc changes with aging of human cervical vertebra: from the neonate to the eighties. Spine 13 (11), 1205–1211.

Pal, G.P., Sherk, H.H., 1988. The vertical stability of the cervical spine. Spine 13 (5), 447–449.

Panjabi, M.M., Crisco, J.J., Lydon, C., Dvorak, J., 1998. The mechanical properties of human alar and transverse ligaments at slow and fast extension rates. Clin. Biomech. 13 (2), 112–120.

Panjabi, M.M., Crisco, J.J., Vasavada, A., Oda, T., Cholewicki, J., Nibu, K., Shin, E., 2001. Mechanical properties of the human cervical spine as shown by three-dimensional load-displacement curves. Spine 26 (24), 2692–2700.

Panzer, M., (2006). Numerical Modelling of the Human Cervical Spine in Frontal Impact (Master-Thesis).

Penning, L., 1978. Normal movements of the cervical spine. Am. J. Roentgenol. 130 (2), 317–326.

Penning, L., 1988. Differences in anatomy, motion, development and aging of the upper and lower cervical disk segments. Clin. Biomech. 3 (1), 37–47.

Penning, L., Wilmink, J.T., 1987. Rotation of the cervical spine—a ct study in normal subjects. Spine 12 (8), 732–738.

Peterson, C., Bolton, J., Kim Humphreys, B., 2003. A cross-sectional study correlating degeneration of the cervical spine with disability and pain in united kingdom patients. Spine 28 (2), 129–133.

Pospiech, J., Stolke, D., Wilke, H.J., Claes, L.E., 1999. Intradiscal pressure recordings in the cervical spine. Neurosurgery 44 (2), 379–384.

Sizer, P.S., Phelps, V., Brismee, J.M., 2001. Differential diagnosis of local cervical syndrome versus cervical brachial syndrome. Pain Pract. 1 (1), 21–35.

Smith, Z.A., Khayatzadeh, S., Bakhsheshian, J., Harvey, M., Havey, R.M., Voronov, L.I., Muriuki, M.G., Patwardhan, A.G., 2016. Dimensions of the cervical neural foramen in conditions of spinal deformity: an ex vivo biomechanical investigation using specimen-specific ct imaging. Eur. Spine J. 25 (7), 2155–2165.

Taitz, C., Nathan, H., Arensburg, B., 1978. Anatomical observations of the foramina transversaria. J. Neurol. Neurosurg. Psychiatry 41 (2), 170–176.

Tchako, A., Sadegh, A., 2009. A cervical spine model to predict injury scenarios and clinical instability. Sports Biomech. 8 (1), 78–95.

Töndury, G., 1972. The behaviour of the cervical discs during life. In: Cervical Pain. Pergamon Press, Oxford, pp. 59–66.

Töndury, G., Theiler, K., 1958. Entwicklungsgeschichte und fehlbildungen der wirbelsaule. Hippokrates, Stuttgart.

Tonetti, J., Potton, L., Ribound, R., Peoch, M., Passagia, J.G., Chirossel, J.P., 2005. Morphological cervical disc analysis applied to traumatic and degenerative lesions. Surg. Radiol. Anat. 27 (3), 192–200.

Toosizadeh, N., Haghpanahi, M., 2011. Generating a finite element model of the cervical spine: estimating muscle forces and internal loads. Scientia Iranica 18 (6), 1237–1245.

Van Mameren, H., Sanches, H., Beursgens, J., Drukker, J., 1992. Cervical spine motion in the sagittal plane ii. Position of segmental averaged instantaneous centers of rotation-a cineradiographic study. Spine 17 (5), 467–474.

Vasavada, A.N., Li, S., Delp, S.L., 1998. Influence of muscle morphometry and moment arms on the moment-generating capacity of human neck muscles. Spine 23 (4), 412–422.

Vasavada, A.N., Li, S., Delp, S.L., 2001. Three-dimensional isometric strength of neck muscles in humans. Spine 26 (17), 1904–1909.

von Luschka, H., 1858. Die halbgelenke des menschlichen koerpers, eine monographie. Reimer, Berlin.

Vos, T., Barber, R.M., Bell, B., Bertozzi-Villa, A., Biryukov, S., Bolliger, I., Charlson, F., Davis, A., Degenhardt, L., Dicker, D., 2015. Global, regional, and national incidence, prevalence, and years lived with disability for 301 acute and chronic diseases and injuries in 188 countries, 1990–2013, a systematic analysis for the global burden of disease study 2013. Lancet 386 (9995), 743–800.

Wheeldon, J., Khouphongsy, P., Kumaresan, S., Yoganandan, N., Pintar, F.A., 1999. Finite element model of human cervical spinal column. Biomed. Sci. Instrum. 36, 337–342.

Winkelstein, B.A., McLENDON, R.E., Barbir, A., Myers, B.S., 2001. An anatomical investigation of the human cervical facet capsule, quantifying muscle insertion area. J. Anat. 198 (4), 455–461.

Yoganandan, N., Kumaresan, S.C., Voo, L., Pintar, F.A., Larson, S.J., 1996. Finite element modeling of the c4-c6 cervical spine unit. Med. Eng. Phys. 18 (7), 569–574.

Yoganandan, N., Kumaresan, S., Pintar, F.A., 2001. Biomechanics of the cervical spine part 2. Cervical spine soft tissue responses and biomechanical modeling. Clin. Biomech. 16 (1), 1–27.

Yoganandan, N., Knowles, S.A., Maiman, D.J., Pintar, F.A., 2003. Anatomic study of the morphology of human cervical facet joint. Spine 28 (20), 2317–2323.

Zhang, J., Wang, F., Zhou, R., Xue, Q., 2011. A three-dimensional finite element model of the cervical spine—an investigation of whiplash injury. Med. Biol. Eng. Comput. 49 (2), 193–201.

Zibis, A.H., Mitrousias, V., Baxevanidou, K., Hantes, M., Karachalios, T., Arvanitis, D., 2016. Anatomical variations of the foramen transversarium in cervical vertebrae: findings, review of the literature, and clinical significance during cervical spine surgery. Eur. Spine J. 25 (12), 4132–4139.

FURTHER READING

Anderst, W.J., 2012. Automated measurement of neural foramen cross-sectional area during in vivo functional movement. Comput. Methods Biomech. Biomed. Engin. 15 (12), 1313–1321.

CHAPTER

BASIC BIOMECHANICS OF THE THORACIC SPINE AND RIB CAGE

3

Christian Liebsch, Hans-Joachim Wilke

Institute of Orthopaedic Research and Biomechanics, Ulm University, Ulm, Germany

As its middle section, the thoracic spine represents a special case in the biomechanics of the spinal complex. Connecting the cervical and lumbar spinal sections, the thoracic spine must ensure high stability for the muscle-controlled sagittal balance of the spine through the erector spinae and abdominal muscles, for an optimum force transmission from the upper body to the lower spinal sections, and for flexibility adequate for performing three-dimensional motions. Whereas the intersegmental stability of the cervical and lumbar spine is determined mainly by the adjacent musculature, the thoracic spinal segments are stabilized primarily by the additional bony and ligamentous structures of the rib cage. Along with its functions as protector of the inner organs, supporter of the trunk muscle activity by providing a strong framework, and enhancer of the respiration by expanding and contracting, the rib cage is also seen as a fourth spinal column in the thoracic section (Berg, 1993). Consequently, the rib cage will be taken into account throughout the following discussion.

FUNCTIONAL ANATOMY

THE THORACIC SPINAL COMPLEX

The thoracic spine consists of 12 vertebrae (T1 to T12) that form a concave curvature to the ventral direction in the sagittal plane, also referred to as kyphosis (Fig. 1). The anterior concavity can be attributed to human phylogenetic development. That is, although the entire embryologic spinal curve is C-shaped with anterior concavity, the cervical spine and lumbar spine develop their anterior convexity after birth by muscle activation of the paraspinal muscles and the psoas muscle, respectively, to ensure an upright posture (Panjabi et al., 1991); therefore the thoracic curve can also be referred to as the primary curve (Maiman and Pintar, 1992).

The angle of the kyphosis, which is about 45 degree and varies between 20 degree and 70 degree in the asymptomatic spine (Roussouly et al., 2005), usually increases with age because of progressive fibrosis of the posterior muscles, leading to hyperkyphosis in old age, whereas the mean apex of the kyphosis is generally close to the T6 vertebra and independent of age and gender (Boyle et al., 2002). However, kyphosis starts increasing at a young age and continually progresses, causing a shift of the cervicothoracic inflection point from T3 to C7-T1 during a lifetime (Boyle et al., 2002). The thoracolumbar inflection point is located on average in the center of L1, but can generally be positioned in the lower thoracic or upper lumbar spine, whereas the angle of the thoracolumbar inflection is

Biomechanics of the Spine. https://doi.org/10.1016/B978-0-12-812851-0.00003-3

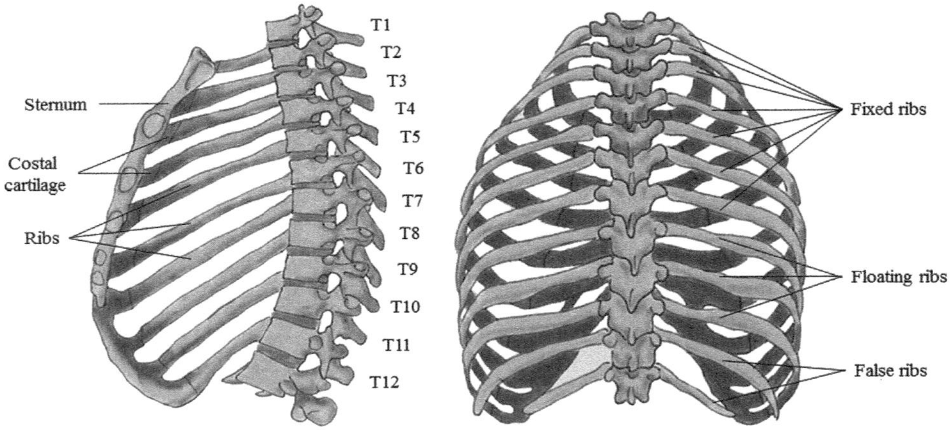

FIG. 1

The bony and cartilaginous thoracic spine and rib cage in sagittal view without the left ribs *(left)* and in back view *(right)*.

about 20 degree to the horizontal, varying between 7 degree and 35 degree in the asymptomatic spine (Roussouly et al., 2005). Indeed, the thoracic kyphosis is dependent on the lumbar lordosis angle and the positioning of the lower cervical vertebrae: Flat lumbar lordosis leads to a decreased lower thoracic curve, whereas a more curved lumbar lordosis induces a distinct thoracic kyphosis, each leading to an amplification of the effect in order to keep the lower cervical vertebrae in a balanced position (Roussouly and Pinheiro-Franco, 2011).

Because of the anteriorly located center of gravity in the case of kyphosis, which creates a natural lever arm, the thoracic spine is generally bent forward, making continuous posterior muscle forces necessary to maintain an upright spinal position, especially by the longissimus dorsi muscle. This effect is dependent on the initial spinal curvature and the body weight. With increasing axial load, the thoracic curvature straightens for low initial angles and increases for larger angles (Meakin et al., 2008). The kyphotic nature of the spine further leads to a primary compressive load distribution in anterior direction toward the vertebral body, whereas the posterior elements mainly have to resist tensile loads.

On both sides, 12 ribs are attached to the thoracic spine (Fig. 1). The upper seven pairs of ribs are rigidly connected to the sternum via costal cartilage and are therefore termed "fixed ribs." The 8th to 10th rib pairs are not directly attached to the sternum but to their adjacent superior ribs, also via costal cartilage, which is why they are referred to as "floating ribs." The "false ribs" are the lower two pairs of ribs, which do not have a bony or cartilaginous connection with other ribs. Because of the increasing integrity of the rib cage in superior direction, the stabilizing effect of the rib cage on the spine is higher in the upper thoracic spine. The articular surfaces of each rib-vertebra connection are on two sites. The costovertebral joint links the rib head with either one vertebra (T1, T11-T12) or two adjacent vertebrae (T1–T10), affecting the intervertebral flexibility in the upper 10 thoracic spinal segments in the latter case; whereas the costotransverse joint links the rib tubercles of the first 10 ribs with each of the single transverse processes of the upper 10 vertebrae (Fig. 1).

QUANTITATIVE ANATOMY OF THE THORACIC VERTEBRAE

The thoracic spinal curvature can be seen as a function of vertebral morphology, which is attributed to the inclined configuration of the thoracic vertebral bodies, having a higher effect on thoracic kyphosis than disc morphology (Goh et al., 1999). The average wedge angle is about 4 degree (Panjabi et al., 1991), which shows a trend for a more pronounced wedge configuration of the midthoracic vertebral bodies (Goh et al., 1999). In the case of osteoporosis, the force progression through the anterior spinal column can lead to wedge fractures, which are typically detected in the midthoracic spine. In the frontal plane, the vertebral bodies tend toward left lateral wedging in the asymptomatic spine, leading to a slight right-sided spinal convexity, especially in the upper thoracic vertebrae T1 to T3 and T5 to T7 (Masharawi et al., 2008). The thoracic vertebral bodies increase in size in inferior direction by approximately 50% (Zindrick et al., 1987; Panjabi et al., 1991; Kunkel et al., 2011), indicating an adaption to the increasing load caused by the body weight (Fig. 2). The laminae are generally broad and strongly overlap each other, similar to roof tiles, in order to protect the spinal cord posteriorly during flexion motions (Fig. 1). The lumen of the spinal canal varies throughout its length, but its narrowest segments are in the thoracic spine with an area of about 200 mm^2 from T1 to T10 (Panjabi et al., 1991), indicating lower deformation of the spinal cord and therefore lower segmental range of motion than the other spinal regions.

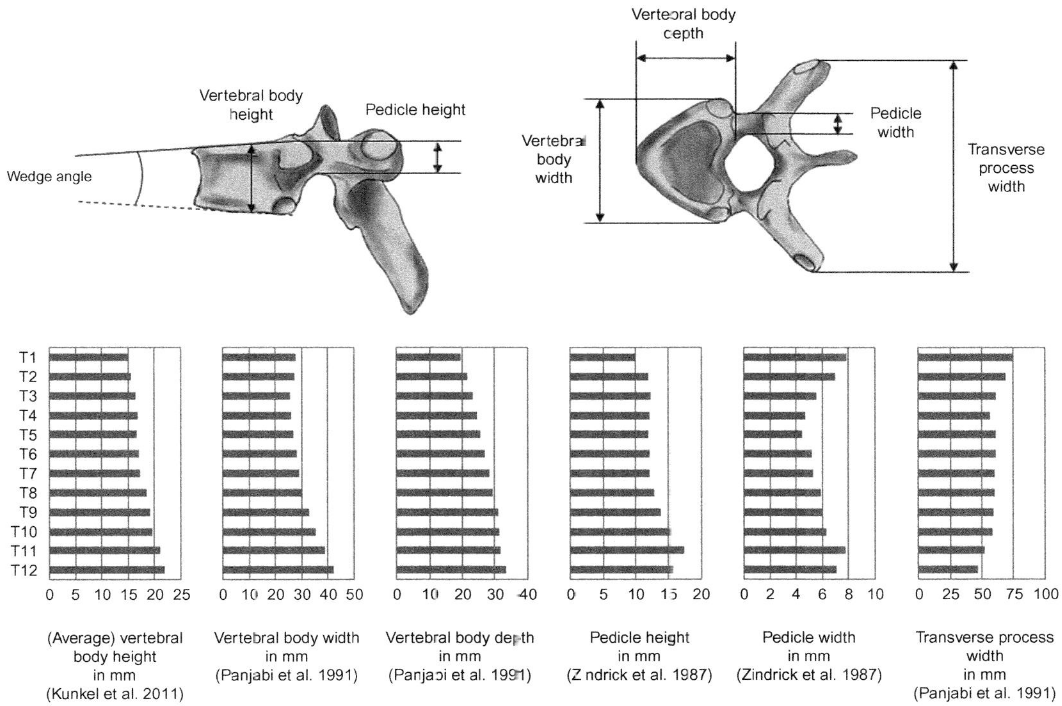

FIG. 2

Morphological parameters of thoracic vertebrae.

The thoracic pedicles are relatively thin and short, especially in the midthoracic spine. Because the pedicle height increases in inferior direction, the pedicles are usually narrowest in the T5 vertebra. Pedicle cross-sections, which are important for the development of screw implants, also depend on the thoracic section; in the upper thoracic spine, pedicle shape changes from a teardrop to a kidney postero-anteriorly, whereas in the middle and lower thoracic spine, pedicle shape changes from a tall, narrow teardrop to an inverted teardrop (Panjabi et al., 1997). The thoracic spinous processes are long and thin, pointing inferiorly and becoming horizontal in inferior direction, whereas the transverse processes diminish from superior to inferior direction (Panjabi et al., 1991).

THORACIC FACET ORIENTATIONS

Because of their specific spatial orientation, the articular facets of the thoracic spine substantially influence their motion characteristics by limiting the flexion/extension and axial rotation range of motion as well as by altering the instantaneous rotational axis in the transverse plane (White and Hirsch, 1971). The tilt angles of the facets increase gradually in inferior direction toward the lumbar spine by converging to the frontal plane, especially in the sagittal plane and slightly in the transversal plane (Panjabi et al., 1993; Masharawi et al., 2004), whereas the facets on the right side are oriented more vertically and more parallel to the sagittal plane than on the left side (Fig. 3). Because all of the thoracic facets are primary oriented parallel to the frontal plane, they provide a distinct resistance to anteroposterior

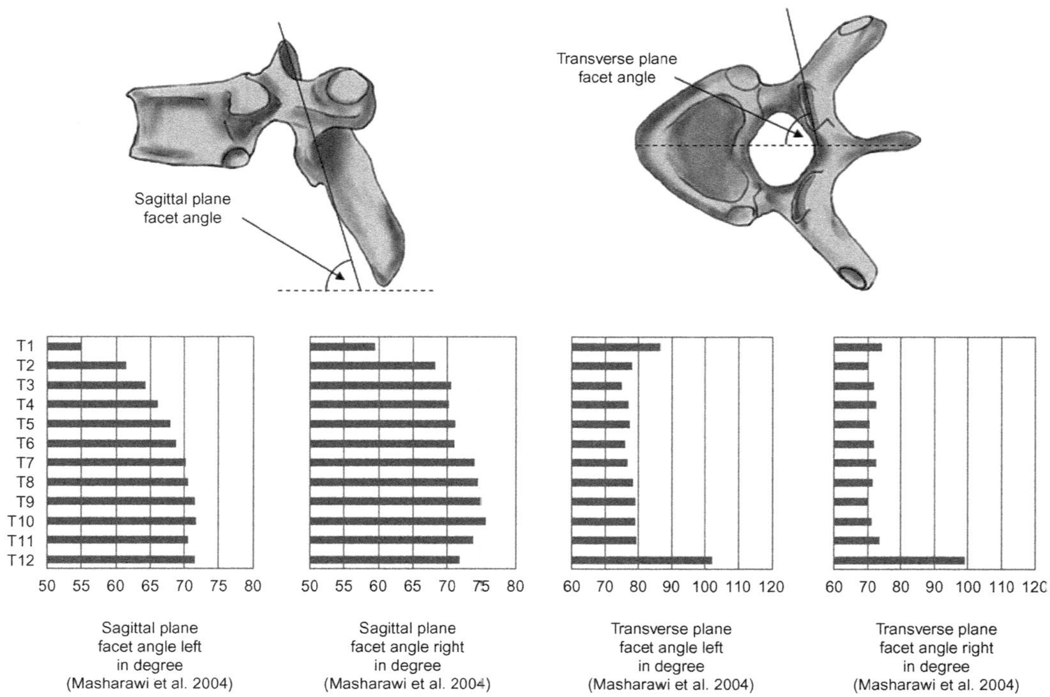

FIG. 3

Facet joint angles of the thoracic spinal motion segments.

translation and, to a lesser extent, to axial compression loads. In the thoracolumbar transition zone (T11 to L1), there is usually an abrupt or segmental change of the facet orientation in the transverse plane from frontal toward sagittal plane orientation of the articular surfaces (Singer et al., 1988); in the cervicothoracic transition zone, the sagittal plane angle tends to decrease and the transverse angle tends to increase in superior direction.

MORPHOLOGY OF THE THORACIC INTERVERTEBRAL DISCS

The intervertebral disc, especially the anulus fibrosus, substantially affects the load absorbing and motion characteristics of the thoracic spine. Because of the kyphotic curvature of the thoracic spine and the increasing bending moment in upright position, the anterior portion of the thoracic disc is compressed by the body weight, whereas the disc shows a wedge shape in uncompressed condition. This wedge shape, resulting from a higher anterior disc height, is especially present in the lower thoracic spine (T7–T12) by approximately 20%, whereas in the upper thoracic spine, the discs are either quite flat (T1–T4, T5–T7) or exhibit a larger posterior disc height (T4–T5) (Fig. 4). Because of the higher compression loads in the anterior portions of the midthoracic discs, the upper areas are more likely affected by bony

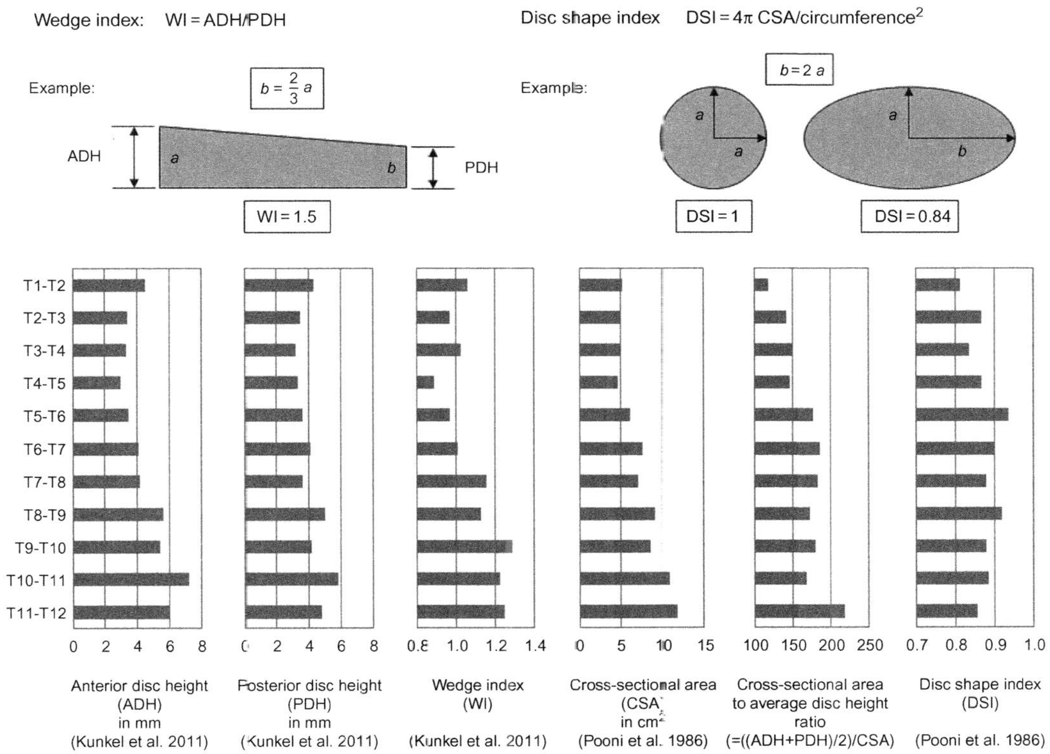

FIG. 4

Thoracic intervertebral disc dimensions without axial preload.

changes such as osteophytes, whereas degenerative disc changes are mainly in the lower thoracic spine (Goh et al., 2000). The average disc height decreases in inferior direction to a minimum at T4-T5 before increasing to reach a maximum height at T10-T11 (Fig. 4). Compared with the cervical or lumbar spine, the disc height is generally decreased, but the anulus fibrosus is thicker and stronger, especially in the posterior section (Goh et al., 1999), leading to limited rotation motions, whereas the nucleus is relatively small, indicating that axial compression loads are absorbed by other structures, such as the facet joints or the rib cage structures. Because the thoracic discs are relatively thin, herniations are quite rare and mostly asymptomatic, occurring mainly in the lower thoracic segments (Girard et al., 2004).

In the transverse plane, the cross-sectional areas of the discs of the upper thoracic spine are relatively smaller than those in the lower thoracic spine and increase inferiorly by more than 100% from T1-T2 to T11-T12 (Fig. 4). Because the ratio between the cross-sectional area and the average disc height is increasing in inferior direction (Fig. 4), a decreasing range of motion, especially in axial rotation, from the upper to the lower thoracic spinal motion segments can be expected because of the lower area moment of inertia of the discs. The transverse shape of the thoracic disc distinctly differs from those in the cervical and lumbar spine, being more circular than elliptic (Pooni et al., 1986). The disc shape index, equaling 1 for an ideal circular shape, increases from T1-T2 to T5-T6 and then decreases in inferior direction toward T11-T12 (Fig. 4).

THORACIC SPINAL LIGAMENTS

Each single ligament of the thoracic spine (Fig. 5) exhibits different stabilizing functions depending on the direction of spinal bending. In general, the ligaments get stronger and are more deformable in inferior direction of the thoracic spine, while showing high variations between the single segments (Myklebust et al., 1988). The anterior longitudinal ligament has a higher average failure load compared to the other ligaments (Table 1), mainly caused by the higher cross-sectional area. Because of limiting excessive extension motions, the anterior longitudinal ligament is about three times stronger in the upper than in the lower thoracic spine (Myklebust et al., 1988). The posterior longitudinal ligament, which limits primarily flexion motions exhibits a much lower failure load, but is stronger in the midthoracic region, indicating a tendency toward higher strength on the convex side, depending on the distance to the thoracic spinal center of gravity (Myklebust et al., 1988). Due to their functions of preventing hyperflexion and -extension, the anterior and posterior longitudinal ligaments are generally thicker in the thoracic spine than in the other sections (Panjabi et al., 1981). The ligamentum flavum, despite being one of the most elastic tissues in the human body, exhibits a high tensile failure load of about 300N in the lower thoracic levels (Table 1) because of its high cross-sectional area of about 100 mm^2 (Maiman and Pintar, 1992). One specific feature of the ligamentum flavum is that this ligament is in pretension with resting forces of about 10N when the spine is in neutral position. The thoracic facet joint capsules are short and tight, weakest in the upper thoracic spine, and located perpendicular to the facet plane (Maiman and Pintar, 1992); therefore these ligaments serve primarily to limit joint distraction and, to a lesser degree, to prevent articular surface deflection. Together with the facets, these ligaments limit flexion/extension and axial rotation motions, whereas their stabilizing effect decreases in inferior direction (White and Hirsch, 1971). The intertransverse ligaments, despite not being real ligaments (Jiang et al., 1994), are seen only in the thoracic and upper lumbar spine and have a restricting effect on the spinal flexibility in lateral bending and axial rotation because of their long lever arms, because they pass between the transverse processes (Panjabi et al., 1981). The interspinous

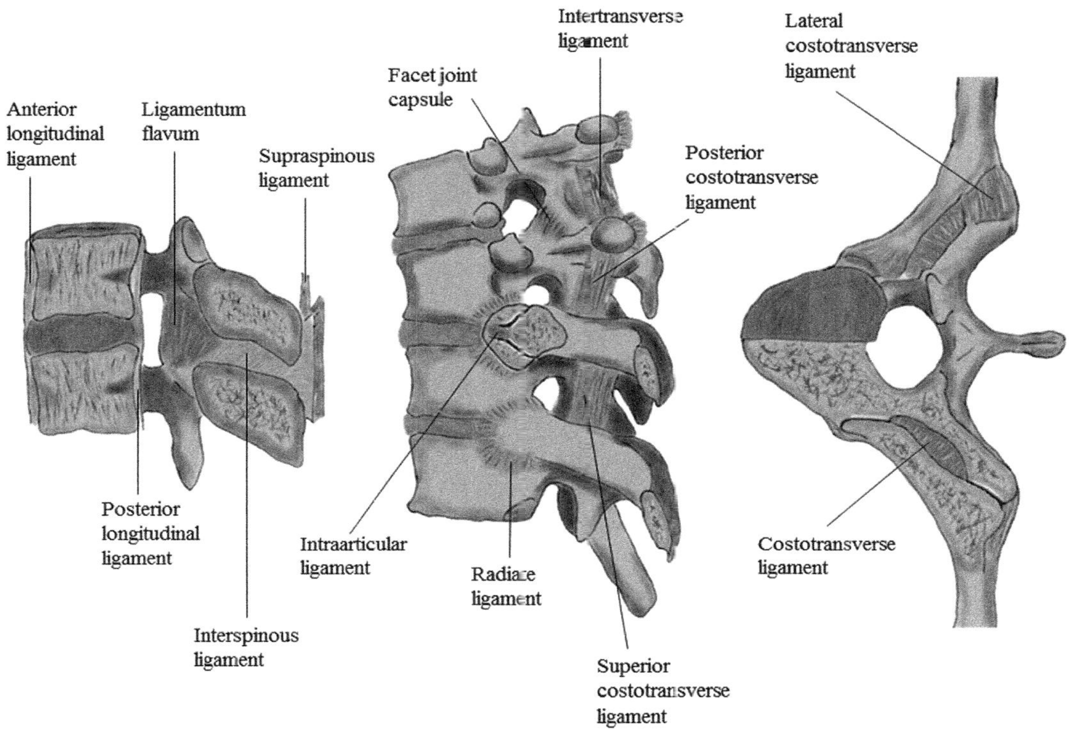

FIG. 5

Ligaments of the thoracic spine.

Table 1 Thoracic Spinal Ligament Failure Loads and Deflections at Failure

	Tensile Failure Load in N	Deflection at Failure in mm
Anterior longitudinal ligament	120–470	6–18
Posterior longitudinal ligament	70–140	3–7
Ligamentum flavum	130–280	6–11
Facet joint capsules (bilateral)	60–270	4–12
Interspinous ligament	20–170	4–10
Supraspinous ligament	90–540	7–19

Derived from Myklebust, J. B., Pintar, F., Yoganandan, N., Cusick, J. F., Maiman, D., Myers, T. J., Sances, A., 1988. Tensile strength of spinal ligaments. Spine 13, 526–531.

and supraspinous ligaments, connecting the adjacent spinous processes, serve to limit hyperflexion; thus they undergo maximum tension in full flexion. While becoming progressively stronger in inferior direction, the supraspinous ligament is much stronger and deflectable than the interspinous ligament because of the higher lever arm during flexion motion (Table 1). The most important ligament for spinal stabilization provided by the rib cage is the superior costotransverse ligament. The lateral and posterior costotransverse ligaments mainly assist in the integrity of the costotransverse joint (Jiang et al., 1994), whereas the radiate, intraarticular, and costotransverse ligaments have an indirect stabilizing effect by strongly binding the ribs to the intervertebral spaces.

COSTOVERTEBRAL AND COSTOTRANSVERSAL JOINTS

The rib-vertebra articulations play an active role in stabilization, load bearing, and protection of the thoracic spine, but also in chest wall movement. The range of motion is by far greatest around the longitudinal axis of the rib-vertebra connection between rib neck and transverse process, whereas in the other directions, there is more stiffness because of ligament attachments (Schultz et al., 1974). This ligament configuration facilitates respiratory motions and increases thoracic spinal stability, whereas the articular stiffness decreases in inferior direction (Duprey et al., 2010). During inspiration, the ribs are elevated by muscular forces to increase the intrathoracic volume. Because of the decreasing transverse process width in inferior direction (Fig. 2), the angular orientation of the rotational axis within the rib-vertebra joint changes (Fig. 6), while the average angle is about 45 degree (Wilson et al., 2001).

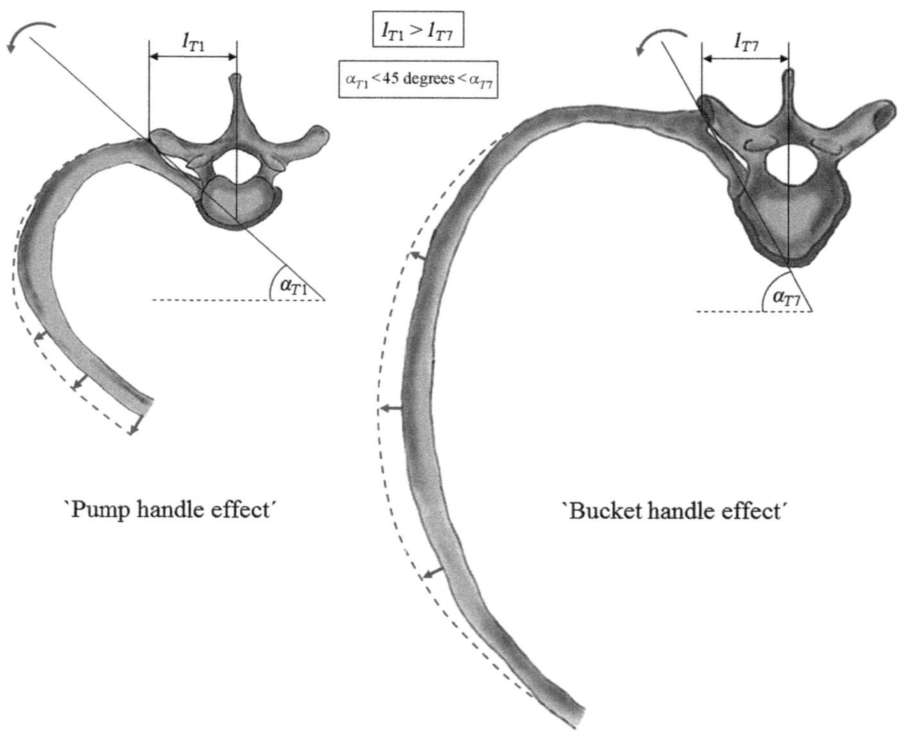

FIG. 6

Rotational axis orientations of the costovertebral and costotransversal joints for upper (T1, *left*) and lower (T7, *right*) thoracic vertebrae.

Therefore the sternal end of the upper ribs is elevated during inspiration, also referred to as "pump handle movement," which increases the anteroposterior rib cage diameter; whereas in the lower ribs, the lateral part is elevated, also called "bucket handle movement," which increases the transverse rib cage diameter (Wilson et al., 1987). However, because the rib-vertebra joint is a plane joint, a common type of synovial joint, it allows gliding movements under physiological conditions (Wilson et al., 2001).

MORPHOLOGY OF THE RIB CAGE

The bony rib cage significantly affects the biomechanics of the thoracic spine (and vice versa) by forming a strong external framework and thus increasing the thoracic spinal area moment of inertia, especially in the transverse plane. The morphology of the rib cage changes during lifetime (Grivas et al., 1991; Weaver et al., 2014), especially during infancy and adolescence and primarily in the sagittal plane with the ribs showing progressively downward, as well as in the transverse plane with increasing transverse diameter (Fig. 7). Consequently, the rib cage's morphology influences its stabilizing properties and affects the thoracic spine's three-dimensional motion characteristics.

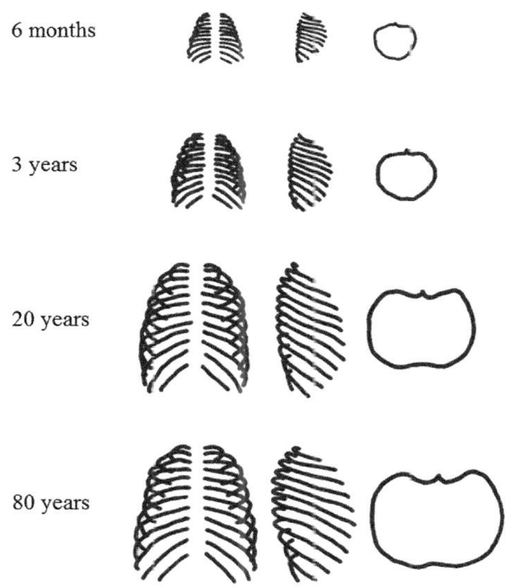

FIG. 7

Rib cage morphology in frontal *(left)* and lateral *(middle)* view, as well as transversal cut view *(right)*, depending on age.

Derived from Grivas, T. B., Burwell, R. G., Purdue, M., Webb, J. K., Moulton, A., 1991. A segmental analysis of thoracic shape in chest radiographs of children. Changes related to spinal level, age, sex, side and significance for lung growth and scoliosis. J. Anat. 178, 21–38 and Weaver, A. A., Schoell, S. L., Stitzel, J. D., 2014. Morphometric analysis of variation in the ribs with age and sex. J. Anat. 225, 246–261.

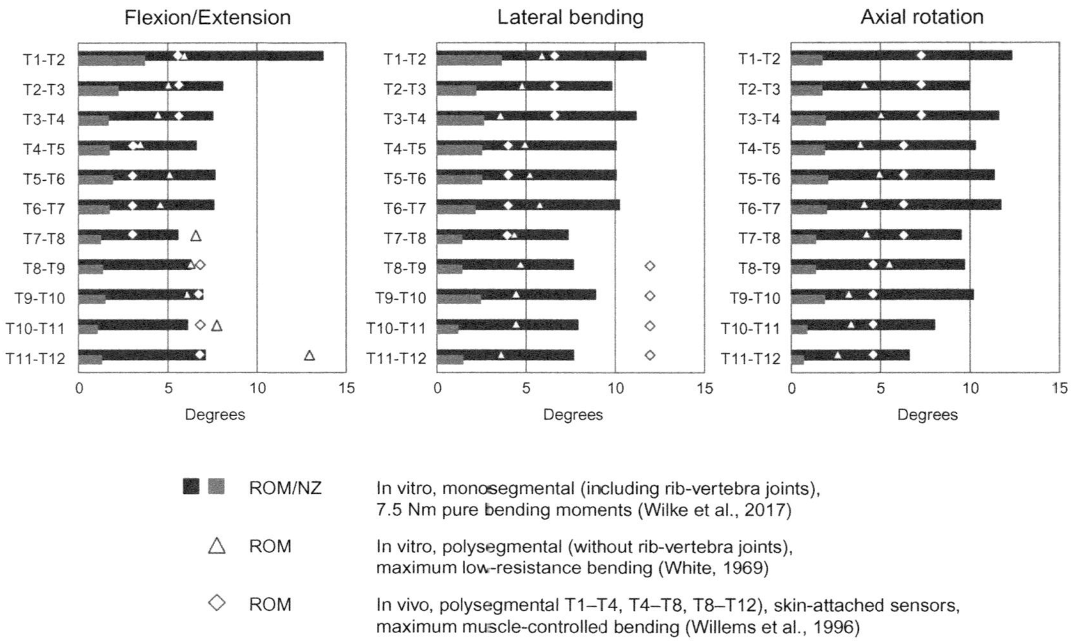

FIG. 8

ROM and NZ of the thoracic spinal motion segments in the three motion planes.

THORACIC SPINAL RANGE OF MOTION AND NEUTRAL ZONE

The mean segmental range of motion of the thoracic spine is distinctly smaller than in the lumbar or cervical spine, and its motion pattern in the different motion planes is highly determined by the rib cage and the orientation of the facets. In general, the range of motion and neutral zone were found to decrease in inferior direction for equal bending moments (Wilke et al., 2017), whereas in vivo, higher ranges of motion were found in the lower segments (Willems et al., 1996) (Fig. 8), indicating that either the intersegmental muscle forces are higher in the lower compared to the upper thoracic spine or that the floating and false ribs of the lower rib cage have a smaller effect on motion restriction (White, 1969). Overall, the presence of the rib cage was found to significantly decrease the range of motion in all motion planes, especially in axial rotation by a factor of about two (Liebsch et al., 2017a), whereas the intervertebral disc was the main determinant for the thoracic spinal range of motion (Oda et al., 2002).

KINEMATICS OF THE THORACIC SPINAL MOTION SEGMENT

The locations and orientations of the instantaneous rotational axes of the thoracic spinal motion segments depend on several factors, such as the type of loading, including the loading direction, additional shear forces, or additional compression or distraction loads (Panjabi et al., 1984), and also on the presence of the rib cage, which was found to shift the position of the rotational axis posteriorly in the sagittal plane during flexion/extension motions (Brasiliense et al., 2011). The instantaneous rotational

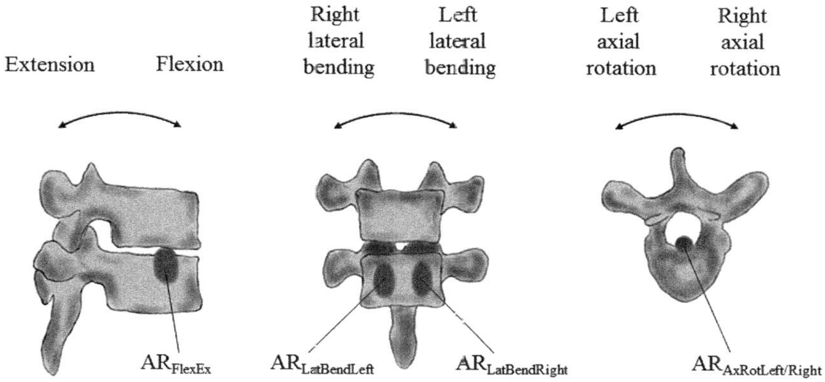

FIG. 9

Average locations of the rotational axes of the thoracic spinal motion segment in the three motion planes

Derived from White, A. A., Panjabi, M. M., 1975. Spinal kinematics. The research status of spinal manipulative therapy. NINCDS Monograph 15, 93 and Molnár, S., Manó, S., Kiss, L. Csernátony, Z., 2006. ex vivo and in vitro determination of the axial rotational axis of the human thoracic spine. Spine 31, E984–E991.

axes of the thoracic spinal motion segment run through the lower vertebra during flexion/extension and lateral bending (Fig. 9). In axial rotation, the rotational axis is located in the spinal canal close to the posterior part of the vertebral body to reduce the deformation of the intervertebral disc on the one hand, because of its close proximity to the center of the disc and to prevent the spinal cord being sheared off by the rotated vertebra (also referred to as "cigar cutting effect") and, on the other hand, because of its close proximity to the center of the spinal cord (White and Panjabi, 1975; Molnár et al., 2006).

COUPLED MOTIONS OF THE THORACIC SPINE

The pattern of coupled motions of the single motion segments is inconsistent throughout the entire thoracic spine because of thoracic kyphosis, varying disc morphologies and three-dimensional facet joint orientations, the presence of segmentally differently shaped ribs, as well as the specific ligamentous structures. Although the upper thoracic spine between T1 and T4 shows a segmental coupling behavior comparable with the cervical spine (White, 1969), showing associated ipsilateral axial rotation during primary lateral bending, no distinct coupling pattern is detectable for the lower single thoracic spinal motion segments. The overall coupling behavior from T1 to T12, including the effect of the rib cage, which is therefore more representative, exhibits a strong coupled ipsilateral axial rotation during primary lateral bending as well as a small coupled contralateral lateral bending during primary axial rotation when loaded with pure bending moments, whereas during flexion/extension movement, there is no considerable coupling behavior. In vivo studies, however, showed that coupled motions of the thoracic spine can vary considerably between individuals (Willems et al., 1996).

LOADS ACTING ON THE THORACIC SPINE

Because of the posteriorly convex curvature of the thoracic spine and therefore the anteriorly located gravitational force line in neutral position, the anterior sections of the motion segments are subjected to high compressive loads. About 75% of these loads are transferred through the vertebral bodies and intervertebral discs, respectively, in the upper thoracic region (Pal and Routal, 1986), whereas the T1 vertebra carries less than 10% of the body weight and the T12 already more than 35% in static upright position (Duval-Beaupère and Robain, 1987). The eccentric applied loads lead to an asymmetric pressure distribution below the compressed disc, similar to degenerated lumbar discs (Horst and Brinckmann, 1981). Together with the decreased disc height, the thoracic discs therefore exhibit more solid material properties compared with the more fluid material behavior of healthy lumbar discs.

STABILITY OF THE THORACIC SPINE

The physiologic stability of the thoracic spinal motion segments is predominantly maintained by the ligamentum flavum, the facet joint capsules, and the anterior and posterior ligaments, whereas the rib cage is the major structure resisting rotational motions in the entire thoracic spine (White and Hirsch, 1971; Panjabi et al., 1981). Therefore, when considering thoracic spinal stability, distinction must be made between overall, segmental, and vertebral stability. The overall stability of the thoracic spine is mainly a combination of the stiffness and strength properties of the differently shaped vertebral bodies and intervertebral discs; the entire thoracic spine, disregarding the rib cage, has a compressive failure force of about 2 kN, a compressive stiffness of about 300 N/mm, and an energy-absorbing capacity in compression of about 10 Nm (Yoganandan et al., 1993). In the case of high flexion-compression loading, posterior ligament injury is produced prior to vertebral body compression, leading to additional loads on the vertebral body, which itself exhibits a compressive failure load of about 10 kN, a compressive stiffness of about 1000 N/mm, and a compressive strength of about 4 N/mm^2 (Maiman and Pintar, 1992). The vertebral body compressive strength thereby highly correlates with the bone mineral density, which increases in inferior direction (Singer et al., 1995). However, isolated vertebral body fractures most likely do not cause spinal instability, while combined injuries, for example of the vertebral body and the posterior elements, very likely produce clinically relevant instability.

The thoracic spinal motion segment is stiffest in axial compression direction while allowing moderate shear (Panjabi et al., 1976). Because of its specific three-dimensional structure, it is also more flexible in flexion and axial rotation compared to extension and lateral bending (Table 2), while the axial rotation torsional stiffness is additionally increased by up to 50% by both compression and tension forces (Goodwin et al., 1994).

Table 2 Average Stiffness Coefficients of Thoracic Spinal Motion Segments

	Axial Stiffness in N/mm		Torsional Stiffness in Nm/degree
Tension	780	Flexion	2.7
Compression	1240	Extension	3.2
Anterior/posterior shear	110	Lateral bending	3.0
Lateral shear	110	Axial rotation	2.6

Derived from Panjabi, M. M., Brand, R. A., White, A. A., 1976. Three-dimensional flexibility and stiffness properties of the human thoracic spine. J. Biomech. 9, 185–192.

EFFECT OF THE RIB CAGE ON THE THORACIC SPINE

The single rib cage structures play an important role in the stability of the thoracic spine. The posterior rib cage elements, especially the rib-vertebra articulations, contribute to spinal stability by restricting motions and thereby increasing the intersegmental stiffness, whereas the anterior rib cage elements, especially the sternal complex, provide strength and energy-absorbing capacity to increase the overall stiffness. Consequently, there is less stabilizing musculature in the thoracic spine compared with the cervical or lumbar spine but the same relative load bearing properties. In fact, all rib cage structures contribute to thoracic spinal stability, although the integrity of the anterior rib cage is the most important stabilizing factor (Liebsch et al., 2017a). Moreover, the intact rib cage distinctly increases the neutral and elastic zone stiffness in all motion directions (Mannen et al., 2015) as well as the compressive stiffness (Watkins IV et al., 2005) of the thoracic spine. The biggest effect of the rib cage on spinal stability can be observed in axial rotation (Liebsch et al., 2017a), because it increases the area moment of inertia in the transverse plane and thus the stiffness of the spine during rotational loading. Because the rib cage is essentially a cylindrical tube with low wall thickness, any discontinuity of its annular cross-section will increase its maximum shear stress by a factor of 3 wall thickness/tube radius and its torsional deformation by a factor of 3 (wall thickness/tube radius)^2, whereas bending loads will have a lesser effect on its stiffness, because the torsional section modulus of a tube is twice its axial bending section modulus (Fig. 10). This corresponds to the finding that median longitudinal sternotomy, which is mostly performed during cardiac surgeries, decreases the stability mainly in axial rotation and less in flexion/extension or lateral bending (Liebsch et al., 2017b).

NUMERICAL MODELS OF THE THORACIC SPINE AND RIB CAGE

In 1974, Andriacchi and colleagues presented the first numerical model to investigate the interaction of the thoracic spine and rib cage (Andriacchi et al., 1974). Using a static multibody model that included rigid body elements for all bony structures and deformable elements for the discs, ligaments, and costal cartilages, they found, after validating the model by in vitro experiments, that the costovertebral joint geometry strongly influenced their mechanical response and that the compression tolerance of the thoracic spine was distinctly increased by including the rib cage. In 2005, Sham and colleagues developed a CT volumetric data-based elastostatic finite element model of the thoracic spine to analyze the stabilizing effect of the rib cage during spinal bending (Sham et al., 2005). In accordance with Andriacchi et al., they concluded that the overall flexibility of the thoracic spine is determined by the rib cage in all loading planes. In 2016, Ignasiak et al. performed a three-dimensional inverse dynamic multibody simulation to calculate muscle forces and spinal loading patterns during daily living activities (Ignasiak et al., 2016). They found that the rib cage predominantly had an influence on thoracic spinal compressive forces and muscle activation during spinal motion tasks, while reducing and distributing the compressive forces compared to a model without a rib cage in mild flexion and hyperextension. Finally, Wang and colleagues presented a parametric elastostatic thoracic spine and rib cage finite element model to determine the effects of age, sex, height, and BMI on the model geometry (Wang et al., 2016). They concluded that height and age-sex interaction were the dominant factors for the size of their model, which was designed to analyze traumatic injuries to the thorax.

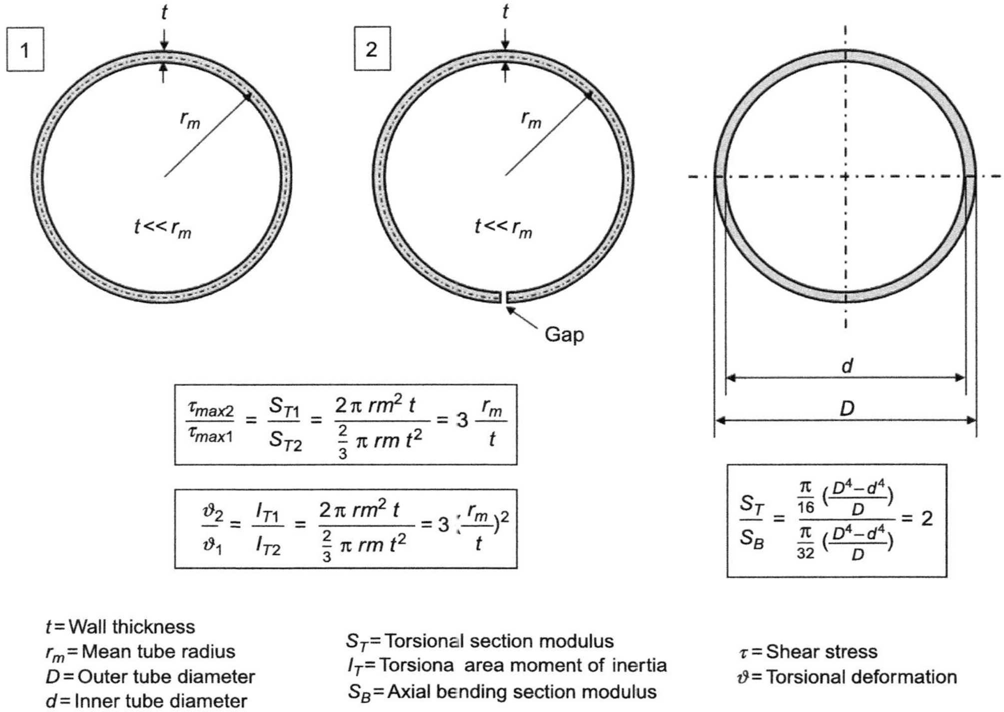

FIG. 10

Idealized mechanical model of the rib cage transverse cross-section as thin walled ring, illustrating the effect of torsional resistance.

REFERENCES

Andriacchi, T., Schultz, A., Belytschko, T., Galante, J., 1974. A model for studies of mechanical interactions between the human spine and rib cage. J. Biomech. 7, 497–507.

Berg, E.E., 1993. The sternal-rib complex: a possible fourth column in thoracic spine fractures. Spine 18, 1916–1919.

Boyle, J.J., Milne, N., Singer, K.P., 2002. Influence of age on cervicothoracic spinal curvature: an ex vivo radiographic survey. Clin. Biomech. 17, 361–367.

Brasiliense, L.B., Lazaro, B.C., Reyes, P.M., Dogan, S., Theodore, N., Crawford, N.R., 2011. Biomechanical contribution of the rib cage to thoracic stability. Spine 36, E1686–E1693.

Duprey, S., Subit, D., Guillemot, H., Kent, R.W., 2010. Biomechanical properties of the costovertebral joint. Med. Eng. Phys. 32, 222–227.

Duval-Beaupère, G., Robain, G., 1987. Visualization on full spine radiographs of the anatomical connections of the centres of the segmental body mass supported by each vertebra and measured in vivo. Int. Orthop. 11, 261–269.

Girard, C.J., Schweitzer, M.E., Morrison, W.B., Parellada, J.A., Carrino, J.A., 2004. Thoracic spine disc-related abnormalities: longitudinal MR imaging assessment. Skelet. Radiol. 33, 216–222.

Goh, S., Price, R.I., Leedman, P.J., Singer, K.P., 1999. The relative influence of vertebral body and intervertebral disc shape on thoracic kyphosis. Clin. Biomech. 14, 439–448.

Goh, S., Tan, C., Price, R.I., Edmondston, S.J., Song, S., Davis, S., Singer, K.P., 2000. Influence of age and gender on thoracic vertebral body shape and disc degeneration: an MR investigation of 169 cases. J. Anat. 197, 647–657.

Goodwin, R.R., James, K.S., Daniels, A.U., Dunn, H.K., 1994. Distraction and compression loads enhance spine torsional stiffness. J. Biomech. 27, 1049–1057.

Grivas, T.B., Burwell, R.G., Purdue, M., Webb, J.K., Moulton, A., 1991. A segmental analysis of thoracic shape in chest radiographs of children. Changes related to spinal level, age, sex, side and significance for lung growth and scoliosis. J. Anat. 178, 21–38.

Horst, M., Brinckmann, P., 1981. Measurement of the distribution of axial stress on the end-plate of the vertebral body. Spine 6, 217–232.

Ignasiak, D., Dendorfer, S., Ferguson, S.J., 2016. Thoracolumbar spine model with articulated ribcage for the prediction of dynamic spinal loading. J. Biomech. 49, 959–966.

Jiang, H., Raso, J.V., Moreau, M.J., Russell, G., Hill, D.L., Bagnall, K.M., 1994. Quantitative morphology of the lateral ligaments of the spine Assessment of their importance in maintaining lateral stability. Spine 19, 2676–2682.

Kunkel, M.E., Herkommer, A., Reinehr, M., Bockers, T.M., Wilke, H.J., 2011. Morphometric analysis of the relationships between intervertebral disc and vertebral body heights: an anatomical and radiographic study of the human thoracic spine. J. Anat. 219, 375–387.

Liebsch, C., Graf, N., Appelt, K., Wilke, H.J., 2017a. The rib cage stabilizes the human thoracic spine: an in vitro study using stepwise reduction of rib cage structures. PLoS One 12, e0178733.

Liebsch, C., Graf, N., Wilke, H.J., 2017b. EUROSPINE 2016 FULL PAPER AWARD: wire cerclage can restore the stability of the thoracic spine after median sternotomy: an in vitro study with entire rib cage specimens. Eur. Spine J. 26, 1401–1407.

Maiman, D.J., Pintar, F.A., 1992. Anatomy and clinical biomechanics of the thoracic spine. Clin. Neurosurg. 38, 296–324.

Mannen, E.M., Anderson, J.T., Arnold, P.M., Friis, E.A., 2015. Mechanical contribution of the rib cage in the human cadaveric thoracic spine. Spine 40, E760–E766.

Masharawi, Y., Rothschild, B., Dar, G., Peleg, S., Robinson, D., Been, E., Hershkovitz, I., 2004. Facet orientation in the thoracolumbar spine: three-dimensional anatomic and biomechanical analysis. Spine 29, 1755–1763.

Masharawi, Y., Salame, K., Mirovsky, Y., Peleg, S., Dar, G., Steinberg, N., Hershkovitz, I., 2008. Vertebral body shape variation in the thoracic and lumbar spine: characterization of its asymmetry and wedging. Clin. Anat. 21, 46–54.

Meakin, J.R., Smith, F.W., Gilbert, F.J., Aspden, R.M., 2008. The effect of axial load on the sagittal plane curvature of the upright human spine in vivo. J. Biomech. 41, 2850–2854.

Molnár, S., Manó, S., Kiss, L., Csernátony, Z., 2006. Ex vivo and in vitro determination of the axial rotational axis of the human thoracic spine. Spine 31, E984–E991.

Myklebust, J.B., Pintar, F., Yoganandan, N., Cusick, J.F., Maiman, D., Myers, T.J., Sances, A., 1988. Tensile strength of spinal ligaments. Spine 13, 526–531.

Oda, I., Abumi, K., Cunningham, B.W., Kaneda, K., McAfee, P.C., 2002. An in vitro human cadaveric study investigating the biomechanical properties of the thoracic spine. Spine 27, E64–E70.

Pal, G.P., Routal, R.V., 1986. A study of weight transmission through the cervical and upper thoracic regions of the vertebral column in man. J. Anat. 148, 245–261.

Panjabi, M.M., Brand, R.A., White, A.A., 1976. Three-dimensional flexibility and stiffness properties of the human thoracic spine. J. Biomech. 9, 185–192.

Panjabi, M.M., Hausfeld, J.N., White, A.A., 1981. A biomechanical study of the ligamentous stability of the thoracic spine in man. Acta Orthop. Scand. 52, 315–326.

Panjabi, M.M., Krag, M.H., Dimnet, J.C., Walter, S.D., Brand, R.A., 1984. Thoracic spine centers of rotation in the sagittal plane. J. Orthop. Res. 1, 387–394.

Panjabi, M.M., Takata, K., Goel, V., Federico, D., Oxland, T., Duranceau, J., Krag, M., 1991. Thoracic human vertebrae quantitative three-dimensional anatomy. Spine 16, 888–901.

Panjabi, M.M., Oxland, T., Takata, K., Goel, V., Duranceau, J., Krag, M., 1993. Articular facets of the human spine quantitative three-dimensional anatomy. Spine 18, 1298–1310.

Panjabi, M.M., O'Holleran, J.D., Crisco, J.J., Kothe, R., 1997. Complexity of the thoracic spine pedicle anatomy. Eur. Spine J. 6, 19–24.

Pooni, J.S., Hukins, D.W.L., Harris, P.F., Hilton, R.C., Davies, K.E., 1986. Comparison of the structure of human intervertebral discs in the cervical, thoracic and lumbar regions of the spine. Surg. Radiol. Anat. 8, 175–182.

Roussouly, P., Pinheiro-Franco, J.L., 2011. Sagittal parameters of the spine: biomechanical approach. Eur. Spine J. 20 (Suppl. 5), 578–585.

Roussouly, P., Gollogly, S., Berthonnaud, E., Dimnet, J., 2005. Classification of the normal variation in the sagittal alignment of the human lumbar spine and pelvis in the standing position. Spine 30, 346–353.

Schultz, A.B., Benson, D.R., Hirsch, C., 1974. Force-deformation properties of human costo-sternal and costo-vertebral articulations. J. Biomech. 7, 311–318.

Sham, M.L., Zander, T., Rohlmann, A., Bergmann, G., 2005. Effects of the rib cage on thoracic spine flexibility. Biomed. Tech. 50, 361–365.

Singer, K.P., Breidahl, P.D., Day, R.E., 1988. Variations in zygapophyseal joint orientation and level of transition at the thoracolumbar junction. Preliminary survey using computed tomography. Surg. Radiol. Anat. 10, 291–295.

Singer, K., Edmondston, S., Day, R., Breidahl, P., Price, R., 1995. Prediction of thoracic and lumbar vertebral body compressive strength: correlations with bone mineral density and vertebral region. Bone 17, 167–174.

Wang, Y., Cao, L., Bai, Z., Reed, M.P., Rupp, J.D., Hoff, C.N., Hu, J., 2016. A parametric ribcage geometry model accounting for variations among the adult population. J. Biomech. 49, 2791–2798.

Watkins IV, R., Watkins III, R., Williams, L., Ahlbrand, S., Garcia, R., Karamanian, A., Sharp, L., Vo, C., Hedman, T., 2005. Stability provided by the sternum and rib cage in the thoracic spine. Spine 30, 1283–1286.

Weaver, A.A., Schoell, S.L., Stitzel, J.D., 2014. Morphometric analysis of variation in the ribs with age and sex. J. Anat. 225, 246–261.

White, A.A., 1969. Analysis of the mechanics of the thoracic spine in man. An experimental study of autopsy specimens. Acta Orthop. Scand. Suppl 127, 1–105.

White, A.A., Hirsch, C., 1971. The significance of the vertebral posterior elements in the mechanics of the thoracic spine. Clin. Orthop. Relat. Res. 81, 2–14.

White, A.A., Panjabi, M.M., 1975. Spinal kinematics. The research status of spinal manipulative therapy. NINCDS Monograph 15, 93.

Wilke, H.J., Herkommer, A., Werner, K., Liebsch, C., 2017. In vitro analysis of the segmental flexibility of the thoracic spine. PLoS One 12, e0177823.

Willems, J.M., Jull, G.A., Ng, J. K. F., 1996. An in vivo study of the primary and coupled rotations of the thoracic spine. Clin. Biomech. 11, 311–316.

Wilson, T.A., Rehder, K., Krayer, S., Hoffman, E.A., Whitney, C.G., Rodarte, J.R., 1987. Geometry and respiratory displacement of human ribs. J. Appl. Physiol. 62, 1872–1877.

Wilson, T.A., Legrand, A., Gevenois, P.A., Troyer, A., 2001. Respiratory effects of the external and internal intercostal muscles in humans. J. Physiol. 530, 319–330.

Yoganandan, N., Maiman, D.J., Pintar, F.A., Bennett, G.J., Larson, S.J., 1993. Biomechanical effects of laminectomy on thoracic spine stability. Neurosurgery 32, 604–610.

Zindrick, M.R., Wiltse, L.L., Doornik, A., Widell, E.H., Knight, G.W., Patwardhan, A.G., Thomas, J.C., Rothman, S.L., Fields, B.T., 1987. Analysis of the morphometric characteristics of the thoracic and lumbar pedicles. Spine 12, 160–166.

CHAPTER

BASIC BIOMECHANICS OF THE LUMBAR SPINE

4

Hans-Joachim Wilke, David Volkheimer
Institute of Orthopaedic Research and Biomechanics, Ulm University, Ulm, Germany

ANATOMY AND PHYSIOLOGY

The lumbar spinal column consists of five vertebrae (L1 to L5). Neighboring vertebrae are connected by a three-joint complex consisting of the fibrocartilaginous intervertebral disc and two synovial zygapophysial joints that are stabilized by various ligaments.

BONY STRUCTURES

A lumbar vertebra can be divided into three functional elements: vertebral body, posterior elements, and pedicles connecting the vertebral body with the posterior elements. The posterior elements are formed by the two laminae with their respective processes: two transverse, four articular, and one spinous process. Whereas the shape of the two upper (L1 and L2) and lower (L4 and L5) lumbar segments is transitional toward their neighboring spinal regions (the thoracic and sacral spine, respectively), L3 can be regarded as a "typical" lumbar vertebra (Panjabi et al., 1992).

The vertebral body is the major load-bearing structure of the lumbar spine (Adams and Hutton, 1980). It is frequently described as being drum-shaped with a transverse cross-section resembling the shape of a kidney. The cranial and caudal surfaces of the vertebral bodies, termed bony vertebral endplates, are slightly concave and provide attachment for the intervertebral discs. To support the high loads acting on the lumbar vertebrae, they are comparatively large with their size increasing toward the sacrum. While the two uppermost lumbar vertebral bodies are moderately lower anteriorly than posteriorly (between about 4% at L2 and 12% at L1 at the sagittal midsection), the endplates of the L3 vertebra are almost parallel. This relation reverses for the L4 and L5 vertebrae, where the anterior height is greater than the posterior height (between approximately 7% at L4 and 14% at L5 at the sagittal midsection) (Parent et al., 2004). In contrast, all lumbar intervertebral discs (L1/L2 to L5/S1) are at least twice as high anteriorly than posteriorly (Amonoo-Kuofi, 1991), meaning that the lumbar lordosis of approximately 60 degrees (Roussouly et al., 2005) is mainly constituted by the intervertebral discs rather than vertebrae.

The internal architecture of the vertebral bodies is ideally designed to sustain the high compressive forces acting on the lumbar spine. This architecture consists of a shell of compact bone, termed corticalis, which has a thickness of less than 0.4 mm at the ventral and dorsal shells of the L5 vertebral body (Ritzel et al., 1997). To prevent the thin shell from collapsing at physiologic loads, it is reinforced by thin vertical struts that are connected by horizontal struts, called the vertical and horizontal trabeculae.

Biomechanics of the Spine. https://doi.org/10.1016/B978-0-12-812851-0.00004-5

The horizontal trabeculae serve to prevent bending or buckling of the vertical trabeculae (Euler's buckling), thus reinforcing the whole structure. The composition of the vertebral bodies resembles an extreme weight-saving design, which additionally allows shocks to be absorbed more effectively than a solid block of bone does. The cavities formed by the trabeculae also allow blood vessels to grow into the vertebral body, thus helping to provide nutrients to the intervertebral discs. The posterior wall of the vertebral bodies is pierced by one or more large holes, termed nutrient foramina, which transmit the nutrient arteries and basivertebral veins. Similar, but much smaller foramina can be found at the anterior shell of the vertebral bodies. The major portion of the bony vertebral endplates is perforated to promote the nutrient supply to the intervertebral discs. An exception is represented by a narrow ring of less perforated bone (named ring apophysis) at the periphery of the caudal and cranial surfaces of the vertebral bodies. The posterior structures of the vertebrae are connected to the vertebral bodies via two short pillars that extend from the posterior wall of the vertebral bodies. These structures have a cranial offset with respect to the transversal mid-plane of the vertebral bodies. The pedicles as well as most posterior structures consist of spongious bone covered by a shell of compact bone. The oval-shaped pedicles of the lumbar spine are larger than those in the other spinal regions. Depending on the subject's height and general constitution, the average pedicle height ranges from 14.4mm at L3 to 15.9mm at L1 in the upper and middle lumbar spine and increases to 19.6mm at L5. The pedicle width of the upper lumbar spine ranges from 8.2mm (L2) to 8.6mm (L1). From L2 to L5, the width increases by about 3.5mm per vertebral level and reaches an average of 18.9mm at L5 (Panjabi et al., 1992). The pedicles provide attachment for the posterior spinal structures, which are formed by the laminae with the corresponding processes. The processes can be differentiated by their mechanical function. While the transverse, accessory, mammillary, and spinous processes serve as attachment for the musculature, the articular processes constitute synovial joints, which strongly modify the kinematics of the lumbar spine and contribute to the load sharing between the anterior and posterior spinal columns. The larger processes (transverse and spinous processes) form substantial levers, thus amplifying the action of the attached muscles. The spinal cord passes through a space, called vertebral foramen, which is formed by the posterior surface of the vertebral bodies and the neural arch (both pedicles and the laminae). The transverse cross-sectional area of the vertebral foramen decreases from approximately 320mm^2 at L1 to 280–290mm^2 at L2 to L4, and increases to 330mm^2 at L5 (Panjabi et al., 1993). The nerve branches leave the vertebral foramen via intervertebral foramina, which are formed by the notches below the pedicles of the cranial articulating vertebra and the notches above the pedicles of the caudal vertebra. The joints formed by the inferior articular processes of the cranial and superior articular processes of the caudal adjacent vertebra are correctly termed "zygapophysial joint." However, they are also frequently referred to as "apophyseal joint" or "facet joint." However, these terms are comparably imprecise in their meaning. The articulating areas, which are the medial surface of the superior and the lateral surface of the inferior articular processes, are covered by a thin layer of articular cartilage. In the lumbar spine, the joint surfaces are cylindrically formed. The sagittal orientation of the lumbar facet joints is distinct from other spinal regions. While these joints form a sagittal angle between 60 degree and 90 degree in the cervical and thoracic spine (i.e., they are more frontally oriented), they have a sagittal orientation of 120 degree to 150 degree in the lumbar region (i.e., they are more sagittally oriented). The steepest transition of facet joint orientation between the thoracic and lumbar spine occurs at the thoracolumbar junction (Panjabi et al., 1993). These characteristics play an important role in the biomechanics of the lumbar spine because they determine the load sharing between the intervertebral disc and the facet joints, as well as the resistance against translational and rotational movement.

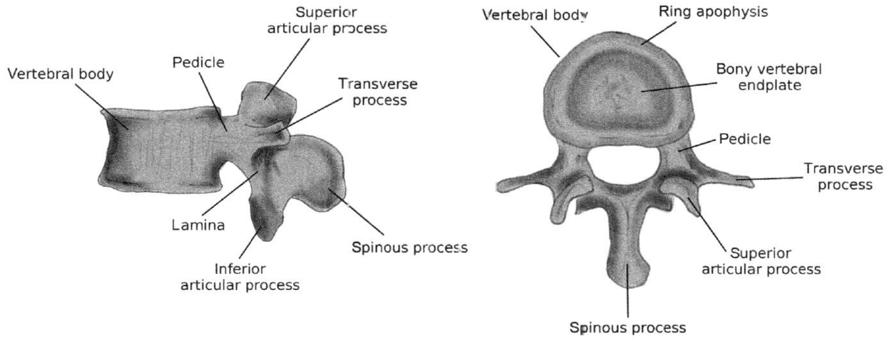

FIG. 1

Bony structures of the lumbar vertebrae depicted in latero-lateral *(left)* and cranio-caudal *(right)* direction.

Adapted from http://www.tana-rose.com/pic-of-lumbar-spine.

By axially loading spinal motion segments resected from different levels of the lumbar spine, Adams and Hutton found that the facet joints resisted about 16% of the compressive force in upright standing, whereas they did not resist the compressive loads in slight flexion, which resembles the lumbar profile during sitting (Adams and Hutton, 1980). By sequentially dissecting spinal structures, Heuer et al. investigated the influence of the facet joints on the qualitative and quantitative motion of L4/L5 motion segments (Heuer et al., 2007). The authors reported a mild to moderate increase in flexion/extension range of motion (ROM) with dissection of the facet joints. The influence of the facet joints was highest in axial rotation, whereas they had a negligible contribution to lateral bending ROM. Additionally, it is assumed that the facet joints are mainly responsible for evoking coupled motions in the lumbar spine, which means that the spine also moves in motion planes other than the primary plane of motion. Skrzypiec et al. evaluated the importance of the lumbar facet joints (L2/L3 and L4/L5) in resisting anterior and posterior shear loads by testing either single intervertebral discs or complete motion segments (Skrzypiec et al., 2013). The authors reported that the removal of the facet joints caused the shear stiffness to decrease by 66% and 43% for anterior and posterior shear directions, respectively (Fig. 1).

INTERVERTEBRAL DISC

Neighboring vertebral bodies are connected by flexible fibrocartilaginous joints, termed intervertebral discs. The intervertebral discs consist of the central nucleus pulposus, which is surrounded by the concentric annulus fibrosus. There is no clear distinction between the amorphous nucleus pulposus and the highly oriented structure of the annulus fibrosus. Nowadays, this zone is referred to as "transition zone." The nucleus pulposus and the inner annulus fibrosus are separated from the bony vertebral endplate by a thin layer of hyaline cartilage. Mainly consisting of proteoglycans and collagen fibers, the fibrocartilaginous endplates' chemical composition is similar to the other components of the intervertebral disc, meaning that, in a healthy state, they do not constitute a further barrier for diffusion of nutrients through the endplates, which is the key pathway for nutrient support to the intervertebral discs. The supply of nutrients via diffusion, and to a minor extent by bulk transport, is of crucial importance for the intervertebral disc, because a healthy adult disc has no or only a very limited number of blood vessels, which are restricted to the outer annulus.

The central nucleus pulposus mainly consists of proteoglycans (about 50% of the dry weight of the nucleus of a child), which represent long chains of glycosaminoglycans (GAGs) linked to proteins. Proteoglycans are highly negatively charged, which allows them to attract positively charged cations from the interstitial fluid. Therefore they have the capacity to attract and imbibe water (Urban and Mcmullin, 1988), which amounts to about 80% of the total weight of the nucleus pulposus in young individuals. With slightly less than 20% of the dry weight of the nucleus pulposus, collagen type II is the cardinal collagen of the nucleus pulposus and serves to hold the proteoglycans together. Furthermore, the nucleus pulposus contains small amounts of collagen type I, elastin fibers, and other non-collagenous proteins.

The nucleus pulposus is surrounded by the highly ordered circular annulus fibrosus, which consists of up to 20 layers of circular collagen type I-rich sheets, called lamellae. The lamellae occasionally are interconnected by radial collagen bundles (Schollum et al., 2008), but most of the space is filled with proteoglycans. Within a lamella, the collagen fibers are arranged at an approximate angle of 30 degree with respect to the vertebral endplates. This angle is reversed between adjacent lamellae. The annulus fibrosus is greater and thicker anteriorly than posteriorly in the lumbar spine, which is why the nucleus pulposus lies posteriorly to the geometric center of the intervertebral disc. The fibers in the outer zone of the annulus fibrosus attach directly to the bony apophyseal ring via Sharpey's fibers, whereas the inner annulus fibers join the cartilaginous vertebral endplate. Compared to the nucleus pulposus, the annulus fibrosus of a healthy disc contains significantly fewer proteoglycans and consequently approximately 10% less water, but more than double the amount of collagens. In contrast with the nucleus pulposus, where collagen type II is predominant, the main collagen of the annulus fibrosus is type I, which is more suitable to transmit and sustain tensile forces.

The low permeability of the endplates and the surrounding annulus fibrosus, in conjunction with the highly water-attracting nucleus pulposus, causes the nucleus pulposus of healthy specimens to exhibit an intrinsic pressure ranging from 0.5 to 1.5 MPa in a neutral position (i.e., no flexion or extension). Thus the annulus fibrosus becomes pre-tensed, which allows the intervertebral disc to sustain high compressive loads. The intradiscal pressure generally increases with deflection of the motion segments. Loading of the intervertebral disc in motion planes other than by flexion/extension and lateral bending, such as shear movements or axial rotations, is primarily resisted by the unique lamellar structure of the annulus fibrosus, where resistance is primarily provided by tensioning the collagen fibers, similar to the fiber reinforcement of a tire.

LIGAMENTS

The osteoligamentous spine is further stabilized by passive ligaments, with each ligament primarily resisting a specific motion direction. Ventrally, the spinal column is covered by the anterior longitudinal ligament. It represents a broad and narrow ligament, which runs over the vertebral bodies and intervertebral discs. In the lumbar spine, the anterior longitudinal ligament has a cross-sectional area of approximately 32 mm^2 (Pintar et al., 1992). Because of this ligament's ventral position, it mainly serves to resist extension movement of the lumbar spine, but has only a small influence in the other motion directions (Heuer et al., 2007). Dorsally, the anterior spinal column is covered by the posterior longitudinal ligament, which runs down the vertebral foramen on the posterior surface of the vertebral bodies and intervertebral discs. It is narrower than the anterior longitudinal ligament, but is also thin and thus possesses by far the smallest cross-sectional area of the lumbar ligaments (Pintar et al., 1992).

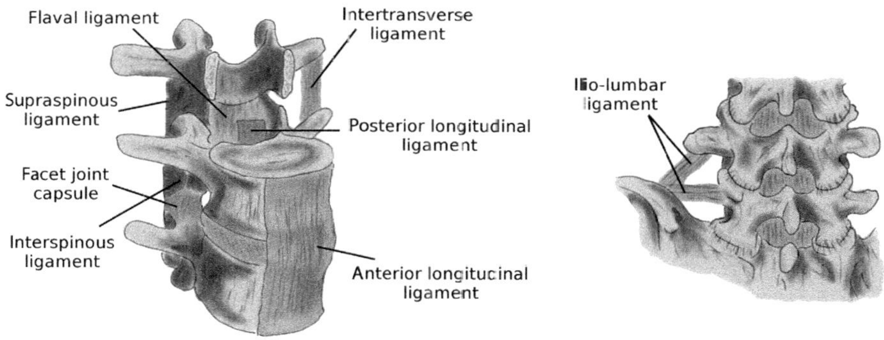

FIG. 2

Lumbar ligaments depicted in oblique *(left)* and posterior *(right)* direction.

Adapted from http://emedicine.medscape.com/article/1899031-overview and http://www.backpain-guide.com/Chapter_Fig_folders/Ch01_Spine_Folder/1LumbarAnat.html

Positioned behind the intervertebral disc, the posterior longitudinal ligament mainly resists flexion movements of the spine. The dorsal surface of the spinal canal is covered by the yellow ligamentum flavum, which connects the laminae of adjacent vertebrae. Its high level of thickness leads to a cross-sectional area that by far exceeds the cross-sectional areas of the other lumbar ligaments. In contrast to the remaining ligaments, which mainly consist of interlacing collagen fibrils, the ligamentum flavum mainly comprises elastic fibers, producing a pre-stress of this ligament. The elastic behavior of this ligament prevents it from buckling into the spinal canal during extension of the spine, which would cause compression of the spinal canal. Like most posterior ligaments, the ligamentum flavum mainly resists flexion movement. The transverse processes are connected by intertransverse ligaments, which are very thin and membranous in the lumbar spine. Their lateral position predestines them to resist lateral bending movements of the spine. Two blended ligaments connect the spinous processes. While the supraspinous ligament attaches to the tips of the spinous processes, the interspinous ligaments connect the spinous processes. The supraspinous ligament is thicker and broader in the lumbar spine than in the thoracic spine and blends with the thoracolumbar fascia. The interspinous ligaments are generally thin and membranous, but thicker and broader in the lumbar spine. This ligament complex has the longest posterior lever arm and mainly resists spinal flexion (Heuer et al., 2007). Like all synovial joints, the zygapophysial joint is surrounded by a joint capsule. Besides its function as an integral part of the synovial joint, the facet joint capsule resists flexion of the spine (Heuer et al., 2007). The lumbosacral junction is additionally stabilized by the iliolumbar ligament. It connects the L5 transverse processes with the top of the iliac crest. Occasionally, an additional weak connection to the L4 vertebra is found. It is assumed to play an important role in maintaining stability of the lumbosacral junction (Fig. 2).

LUMBAR SPINAL MOTION

The osteoligamentous components of the spine and their influence on its biomechanical behavior were the focus of the previous section. The following discussion describes the motion of the lumbar spine resulting from an interaction of the single structures, both in vivo and in vitro.

The lumbar spine allows movements in all 6 degrees of motion, which can be divided into three rotations around and three translations along the principal axes of a coordinate system. Rotations in the sagittal plane are termed flexion/extension, whereas rotations in the frontal plane are called lateral bending. Axial rotation represents a rotation around the vertical axis of the spine. The motion segments also experience translations in the anterior, posterior, and lateral directions as well as compression and decompression. The anatomy of the lumbar spine evokes a coupling of the motion directions, meaning that motion in one principal motion plane is accompanied by movements in other motion planes. In contrast to translations, which need to be defined with respect to a reference point on a rigid moving body, rotations are independent of the point of location within a moving rigid body because all points experience the same rotation. Thus the measurement of rotations from combined rotatory and translatory movements is generally simpler than the calculation of translations. In vivo, the kinematic parameters, are usually derived from either healthy subjects or patients suffering a spinal disease. Most frequently, functional flexion/extension radiographs are performed for this purpose. The intersegmental kinematics are commonly determined from medical imaging techniques such as single- or dual-plane radiography and fluoroscopy, roentgen stereophotogrammetric analysis (RSA), computed tomography (CT), and magnetic resonance imaging (MRI). Because of its wide availability and cost-effectiveness, plane radiography is still the standard method for determining intersegmental motion, even though this technique strongly suffers from out-of-plane motion. Thus, low levels of accuracy and precision, ranging from approximately 1 degree to 5 degree, are reported in the literature. Multi-planar imaging methods such as dual-plane radiography generally offer a higher accuracy (about 0.5 mm and 0.5 degree) and a higher precision (about 0.5 mm and 0.5 degree) by compensating for out-of-plane motion. In contrast to static imaging methods such as radiography, in which the patient must hold a certain position and from which only the endpoints of motion (i.e., the quantitative motion) are usually detected, videofluroscopy captures the dynamic motion behavior of the spine, from which qualitative and time-dependent kinematic parameters (such as the lumbo-pelvic rhythm) can also be determined. RSA offers the highest accuracy (about 0.1 mm and 0.2 degree) and precision (about 0.1 mm and 0.1 degree) of all medical imaging methods. However, this invasive technique requires the implantation of small tantalum markers as landmarks for motion tracking, which impedes its widespread use. With the very limited ability of the patient to move within the scanner being a common drawback of CT and MRI, these imaging techniques have not gained widespread popularity for capturing intervertebral motion. However, upright or even open scanners, which do not limit or to a much smaller extent limit the subjects' motion, might promote the use of these three-dimensional imaging methods in the near future.

Compared to the in vivo assessment of lumbar spine motion, in vitro studies allow a better standardization, because the loads applied to the spine are known and motivation- and pain-related influences can be ruled out. Furthermore, the intersegmental motion can be directly measured from markers or other measuring devices attached to the individual vertebrae, which generally provide a higher accuracy than the medical imaging methods (except for RSA) without the need for radiation exposure.

ROTATORY PARAMETERS

The kinematic behavior of the spine is nonlinear, meaning that the resistance against further deflection depends on the present deflection of the motion segment. To express the qualitative and quantitative kinematic behavior, certain phases of the load-deformation curves are distinguished. The neutral zone

(NZ) resembles the range over which a spine specimen can be moved without the application of appreciable loads. The ROM represents the maximal deformation of the spine obtained by application of physiologic loads. While the ROM can be easily determined both in vivo and in vitro, NZ is generally measured only in vitro tests because the loads, which are acting on the spine, cannot be assessed in vivo, rendering an evaluation impossible.

RANGE OF MOTION

Most in vivo studies included in this overview on the most relevant kinematic parameters employed single-plane radiography for determining lumbar motion (Hayes et al., 1989; Boden and Wiesel, 1990; Yoshioka et al., 1990; Dvorak et al., 1991). Other in vivo studies used dual-plane radiography (Pearcy et al., 1984; Pearcy, 1985), a combination of electrogoniometry with single-plane videofluroscopy (Lee et al., 2002), CT (Ochia et al., 2006), and open MRI (Mcgregor et al., 2001). The subjects actively or passively performed submaximal or maximal movements while sitting on a chair, standing, or lying prone with and without constraints of the pelvis. Therefore the deviation in ROM of the lumbar spine segments was generally large (Fig. 3, left column). In flexion/extension, no general trend toward an increase or decrease between the different lumbar levels could be found among the individual studies. The deviation in ROM determined for the lumbar levels was smaller in the upper lumbar levels than in the lower lumbar segments. From the two studies that measured the ROM in lateral bending, it can be seen that it is about the same for the levels L1/L2 to L3/L4 and then decreases for L4/L5 and L5/S1. The ROM in axial rotation was smaller than for the aforementioned motion directions, with no marked difference between the lumbar levels.

In this overview, only in vitro studies that measured the ROM of each motion segment of the lumbar spine obtained under application of pure bending moments are included. Yamamoto et al. (1989) and Panjabi et al. (1994) both applied pure bending moments of ± 10 Nm in a stepwise fashion, which was supplemented by a compressive preload of 100 N in the study published by the latter author group. In the unpublished study by Volkheimer, pure bending moments of ± 7.5 Nm were applied to 10 T12-S1 specimens in a continuous fashion. Generally, good agreement between these three studies was found (Fig. 3, right column). In contrast to the in vivo studies, the in vitro studies showed an increase in flexion/extension ROM for the lumbar levels L4/L5 and L5/S1, which might be attributed to the resection of certain stabilizing ligaments (e.g., the iliolumbar ligament) or the missing muscular activity. The same difference between the in vivo and in vitro data can be seen for lateral bending, where a decreasing ROM was found for L4/L5 and L5/S1 in the in vivo studies, whereas all lumbar levels had about the same ROM in the in vitro studies. In agreement with the in vivo studies, the ROM in axial rotation was smallest among the three principal spinal rotations, with no marked difference between the lumbar levels.

NEUTRAL ZONE

Two in vitro studies were identified that quantified the neutral zone of the lumbar levels under application of pure bending moments (Fig. 4). Generally, a good agreement between these two studies can be seen. Both studies found a higher NZ for the L4/L5 and L5/S1 motion segments in flexion/extension, whereas it was almost constant among the lumbar levels in lateral bending and axial rotation. As for ROM, NZ was smallest in axial rotation.

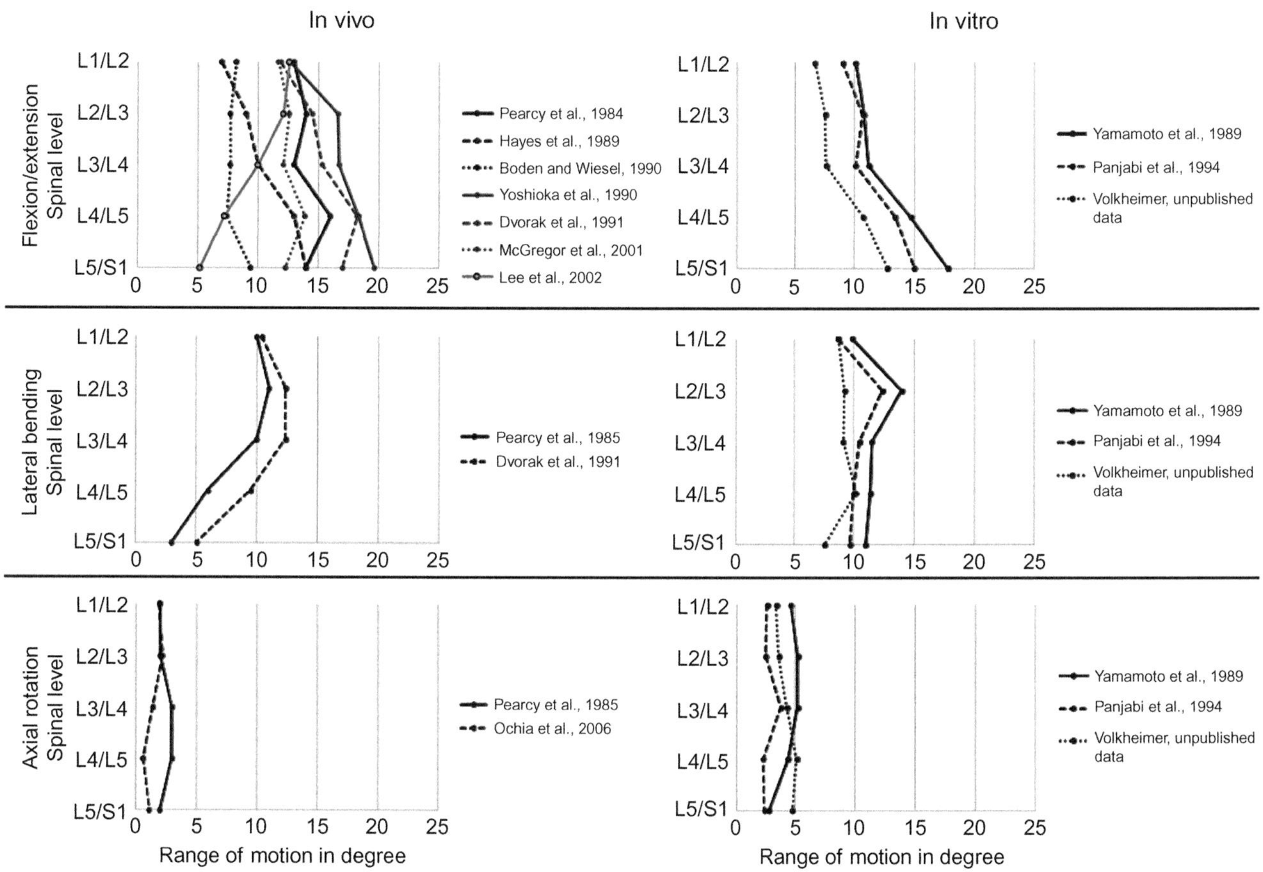

FIG. 3

Range of motion (ROM) obtained from in vivo (*left* column) and in vitro (*right* column) studies. The rows represent the three principal motion directions, flexion/extension, lateral bending, and axial rotation. Only studies that measured the ROM of each motion segment of the lumbar spine were included.

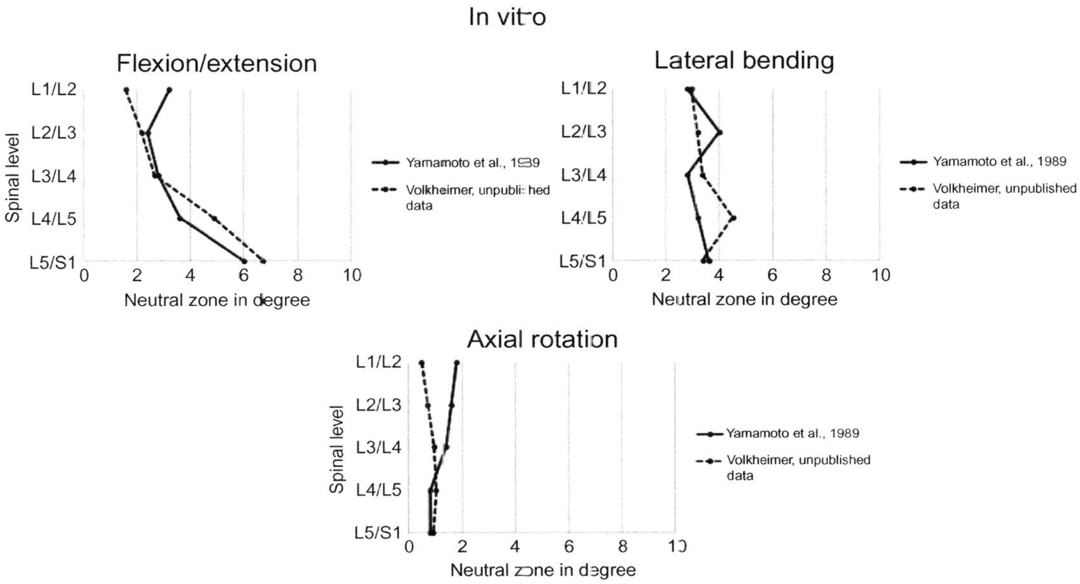

FIG. 4

Neutral zone (NZ) obtained from in vitro studies. Only studies that measured the ROM of each motion segment of the lumbar spine under application of pure bending moments were included.

SHEAR TRANSLATION

Nachemson et al. (Nachemson et al., 1979) measured the shear translation of the lumbar segments resulting from a shear force of 145 N applied in anterior, posterior, and lateral direction while compressing the motion segments with a preload of 400 N (Fig. 5). In all motion directions, the shear translations ranged between 0.06 mm and 0.13 mm.

HEIGHT DECREASE FROM COMPRESSION

In the same study, Nachemson et al. (1979) quantified the compressibility of lumbar motion segments when applying a compressive force of 400 N (Fig. 6). The authors reported lumbar spine axial compressions ranging from 0.04 mm to 0.07 mm.

INSTANTANEOUS CENTER OF ROTATION/CENTRODES

The instantaneous center of rotation is a common method used to describe the relative movement of one vertebra with respect to another vertebra. The instantaneous center of rotation represents an axis perpendicular to the plane of rotation, which does not move during rotation of two rigid bodies. Therefore rotations can be described by the location of the instantaneous axis of rotation and the magnitude of

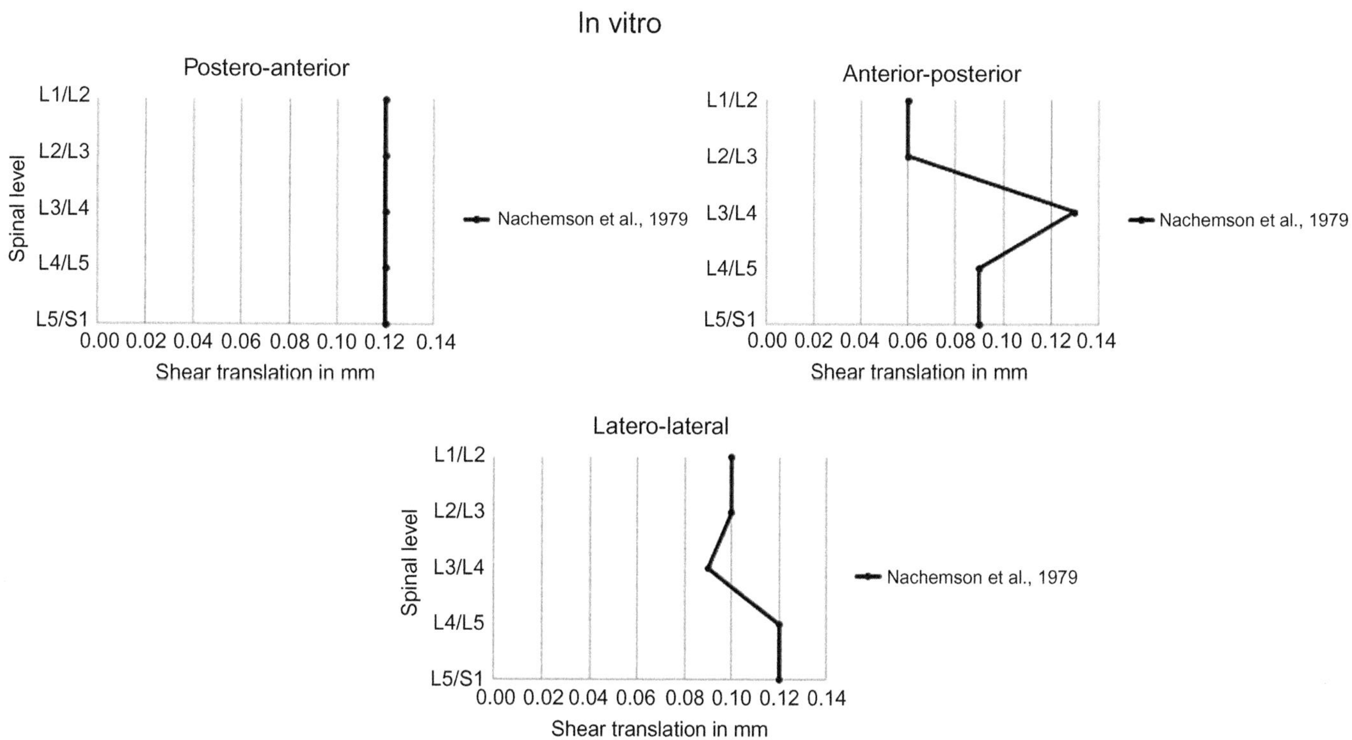

FIG. 5

Shear translations obtained by Nachemson et al. in 1979 under application of 145 N shear and a compressive preload of 400 N (Nachemson et al., 1979). Only studies that measured the shear translation of all lumbar levels were included.

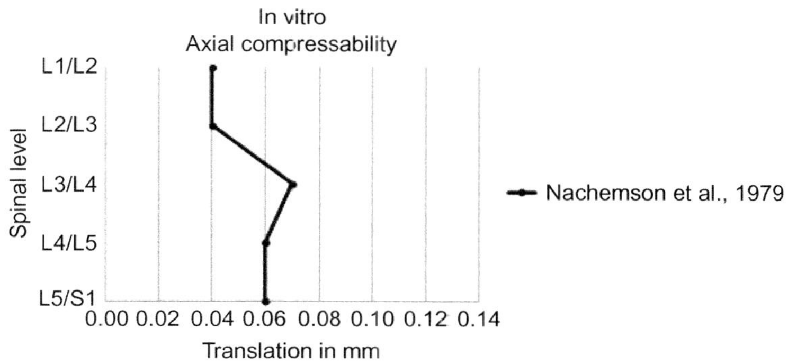

FIG. 6

Axial compression obtained by Nachemson et al. in 1979 under application of a compressive load of 400 N (Nachemson et al., 1979). Only studies that measured the axial compression of all lumbar levels were included.

rotation about it. Centrodes are obtained when the timely course of the instantaneous centers of rotation is plotted, meaning that they represent the pathway of the instantaneous centers of rotation. The centrode length is zero for pure rotations, whereas its length increases with increasing translations.

The centrodes of the lumbar spine have been assessed both in vivo (Ogston et al., 1986, Pennal et al., 1972, Pearcy, 1985, Yoshioka et al., 1990) and in vitro (Gertzbein et al., 1985, Gertzbein et al., 1986, Cossette et al., 1971) by using medical imaging methods. For healthy spines, consensus is forming that the flexion/extension centers of rotation are located in the posterior region of the lumbar vertebral bodies, which migrate anteriorly with flexion of the spine from an extended position. Generally, the centers of rotation fall within a small centrode area, meaning that flexion/extension is accompanied by only a small amount of shear translation. Because the determination of the location of the center of rotation is highly prone to errors for small joint movements, care must be taken when the centrodes are determined from landmarks on radiographs. Thus centrodes obtained from incremental movements of less than 5 degree must be interpreted cautiously (Pearcy, 1985). Gertzbein et al. noted that the centrode length is an indicator for spinal degeneration, because large and unpredictable centrodes were obtained from specimens with minimal radiographic evidence of degeneration, whereas those of normal specimens fell within a markedly smaller area, leading the authors to conclude that minimal degeneration causes erratic rather than excessive motion (Gertzbein et al., 1985). Cossette et al. in vitro assessed the center of rotation of the third to fifth lumbar vertebrae during axial rotation movement (Cossette et al., 1971). The authors reported that the instantaneous center of rotation was located anterior to the facet joints in the region of the posterior part of the nucleus and tended to move toward the side to which the spine rotated.

HELICAL AXES

A major drawback of expressing lumbar spine kinematics by calculating centers of rotation is that out-of-plane motions that accompany motion in each principal motion direction by means of coupled spinal motion cannot be assessed with this method. To overcome this issue, helical axes represent an extension of the two-dimensional center of rotation into the three-dimensional space in the form of a screw motion. The screw motion represents a superposition of translations along and rotations around the helical axis at each instant of time. Thus helical axes provide a comprehensive description and visualization of the three-dimensional spinal movement.

With their in vitro experiment, Kettler et al. in 2004 investigated the position and orientation of eight L4/L5 motion segments under application of pure bending moments in all principal motion directions (Kettler et al., 2004). For each motion direction, the authors identified two different helical axes patterns. In flexion/extension, the helical axes were oriented parallel to each other and the frontal plane. They were either running through the discs or through the inferior vertebra. In lateral bending, the helical axes were either parallel to each other and the sagittal plane and running through the intervertebral disc or running through the inferior vertebra. For the second pattern, the helical axes were spread out in the horizontal plane with the point of intersection lying in the anterior third of the intervertebral disc space. In axial rotation, the helical axes had a vertical orientation and were running close to the facet joints. For the first pattern, they had a strict cranio-caudal orientation, whereas for the second pattern, they were tilted slightly anteriorly. Recently, Ellingson and Nuckley investigated in vitro how the lateral bending helical axes of L4/L5 are affected by degeneration of the intervening intervertebral disc (Ellingson and Nuckley, 2015). The authors reported that the orientation of the helical axes became less aligned in the anterior direction with worsening degeneration, indicating that there was significantly more out-of-plane motion in the degenerated motion segments. This finding indicates that the different motion patterns described by Kettler et al. might be attributed either to intrinsic anatomic differences between the specimens or to varying grades of degeneration, which were not assessed in this experiment (Kettler et al., 2004).

LOADING OF THE LUMBAR SPINE

By providing sufficient stability, the complex structure of the human lumbar spine is optimized to carry physiologic loads without damage. Trauma, degeneration, or other musculoskeletal disorders can lead to a loss in stability, potentially causing back pain. Knowledge about the loads on the spine is necessary in order to optimally treat patients and to provide adequate advice.

Unfortunately, it is not possible to quantify spinal loading directly in vivo because force transducers usually cannot be introduced into the spine of living humans. Therefore, many different methods have been used to indirectly quantify the loads acting on the spine. The following paragraphs present and discuss the findings of studies in which the intradiscal pressure was measured in vivo.

The most reliable in vivo data about the loading of the spine are derived from intradiscal pressure measurements as first reported by Nachemson and colleagues (Nachemson, 1966, Nachemson and Elfstrom, 1970, Nachemson and Morris, 1964, Nachemson et al., 1986). These researchers inserted a pressure-sensitive needle that was connected to an amplifier into the intervertebral disc. This group usually measured the intradiscal pressure in the L3/L4 intervertebral disc, which was then normalized to the intradiscal pressure in each subject while standing. These measurements showed a 50% increase of intradiscal pressure when the subjects bent forward. When they also held weights in their hands, the pressure rose further by 70% to 220% as compared to the reference value.

The authors also found similar results when performing the same exercises in a sitting position, where the intradiscal pressure rose when the subjects bent forward and even when they held additional weights in their hands.

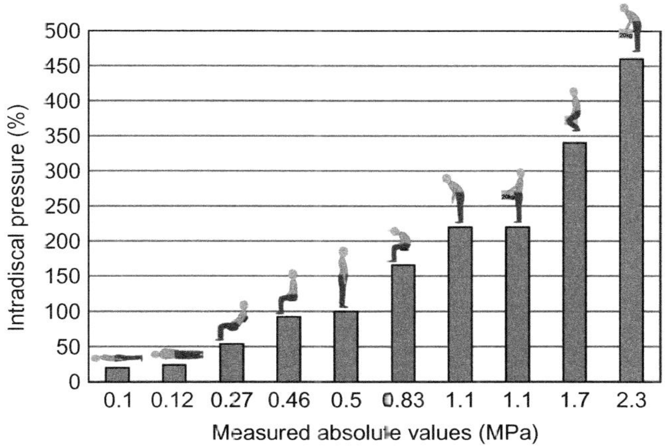

FIG. 7

Intradiscal pressure was measured in vivo in the disc L4–5 for different postures and activities with a special transducer.

Nachemson's measurements were replicated and complemented in more recent studies in which additional positions and exercises of the spine in the lower lumbar motion segment L4-L5 were done by means of an optimized measuring technique (Wilke et al., 1999, Wilke and Rohlmann, 2012) (Fig. 7, Table 1). The pressure transducer was integrated into a 7-mm-long steel tip with a diameter of 1.5 mm, which was connected to a flexible silicone cable.

The first measurements were performed while the subject was still on the operating table. With an intradiscal pressure of 0.08 MPa, the lowest pressure values were recorded when the subject was lying supine in a relaxed way with slightly flexed legs. When the legs were extended, the pressure rose to 0.11 MPa; rotation into the lateral position raised the pressure only slightly to 0.12 MPa. With 0.11 MPa, the intradiscal pressure in the prone position was intermediate between those measured in the supine and lateral positions. The turning maneuver alone (e.g., from supine to lateral) produced pressure peaks of up to 0.8 MPa in magnitude. Coughing and sneezing in a supine position produced pressures as high as 0.38 MPa, but heartily laughing raised the pressure only up to 0.15 MPa.

During relaxed standing, the pressure was measured reproducibly between 0.48 and 0.5 MPa. Abdominal pressing raised it to 0.92 MPa and increased with forward bending to 1.1 MPa. In sitting, the pressure varied strongly depending on the degree of support from the backrest or armrests. Sitting on a chair with a normal, straight back produced a pressure of 0.45 to 0.5 MPa. Active extension of the lumbar spine raised the pressure to 0.55 MPa. Pressure in the sitting position could be strongly reduced by leaning back comfortably in an armchair. Relaxed reclining brought the pressure down to 0.27 MPa.

The highest pressure values (2.3 MPa) were measured during lifting of a full crate of beer weighing 19.8 kg with a rounded back and straight knees. Lifting the same crate keeping the back straight and bending at the hip and knee, as recommended by back school, reduced the peak pressure to 1.7 MPa. When the crate was held up against the body at chest height, the pressure was 1.1 MPa; when it was held at the same height but at arm's length (approximately 60 cm) from the upper body, the pressure was 1.8 MPa.

Table 1 Absolute Values of Intradiscal Pressure for Different Postures and Exercises, Normalized to Relaxed Standing (Chosen Arbitrarily as 100%) and Converted Into Representative Mean Force Values

Position	Pressure (MPa)	Percentage	Force (N) With Factor 1:3[a]	Force (N) With Factor 1:8[b]
Lying supine	**0.1**	**20**	**140**	**100**
Side-lying	**0.12**	**24**	**165**	**120**
Lying prone	0.11	22	150	110
Lying prone, extended back, Supporting elbows	0.25	50	345	250
Laughing heartly, lying laterally	0.15	30	210	150
Sneezing, lying laterally	0.38	76	525	380
Peaks by turning around	0.7–0.8	140–160	970–1110	700–800
Relaxed standing	**0.5**	**100**	**690**	**500**
Standing, performing Valsalva maneuver	0.92	184	1275	920
Standing, bent forward	**1.1**	**220**	**1520**	**1100**
Sitting relaxed, without back rest	**0.46**	**92**	**635**	**460**
Sitting actively straightening the back	**0.55**	**110**	**760**	**550**
Sitting with maximum flexion	**0.83**	**166**	**1150**	**830**
Sitting bent forward with thigh supporting the elbows	0.43	86	595	430
Sitting slouched into the chair	0.27	54	375	270
Standing up from the chair	1.1	220	1525	1100
Walking barefoot	0.53–0.65	106–130	735–900	530–650
Walking with tennis shoes	0.53–0.65	106–130	735–900	530–650
Jogging with hard street shoes	0.35–0.95	70–190	485–1315	350–950
Jogging with tennis shoes	0.35–0.95	70–190	485–1315	350–950
Climbing stairs, one at a time	0.5–0.7	100–140	690–970	500–700
Climbing stairs, two at a time	0.3–1.2	60–240	415–1660	300–1200
Walking down stairs, one at a time	0.38–0.6	76–120	525–830	380–600
Walking down stairs, two at a time	0.3–0.9	60–180	415–1250	300–900
Lifting 20 kg, bent over with round back	**2.3**	**460**	**3185**	**2300**
Lifting 20 kg as taught in back school	**1.7**	**340**	**2350**	**1700**
Holding 20 kg close to the body	**1.1**	**220**	**1480**	**1100**
Holding 20 kg, 60 cm away from the chest	1.8	360	2490	1800
Carrying 20 kg (in the left hand or in the right hand)	1.0	200	1385	1000
Carrying 40 kg (20 kg left and 20 kg right)	0.9	180	1250	900
Pressure increase during the night rest (over a period of 7 h)	0.1–0.24	20–48	140–330	100–240

The activities in **bold** *are represented in Fig. 7*
[a]*Minimum factor to convert pressure into force values (determined by Brinckmann and Grootenboer, 1991).*
[b]*Maximum factor to convert pressure into force values (determined by Brinckmann and Grootenboer, 1991)*
Adapted from Wilke, H.J., Rohlmann, A., 2012. Forces on the spine. In Ambrosio, L., Tanner, E. (Eds.) Biomaterials for Spinal Surgury. Woodhead Publishing Limited.

The findings of these more recent studies are in agreement with the results of Nachemson and Morris (Nachemson and Morris, 1964) with respect to the pressure values during standing, lying, lifting, and carrying. In particular, the pressure differences between lifting with a straight back and flexed knees versus a rounded back and extended knees were reproduced with the more recent studies. Lifting a heavy weight, such as a 20-kg crate of beer, with rounded back and straight knees raised the intradiscal pressure to 450% of its initial value. Such high pressure values were not reached under any other activity, even during jumping on the trampoline.

Important differences between the findings of the new pressure measurements and the data provided by Nachemson et al. are indicated (Nachemson, 1966, Nachemson and Morris, 1964) and need to be emphasized here. The pressures found in standing versus sitting, or for lying in different positions, were not confirmed in the more recent studies. It could not be confirmed, for example, that comfortable (relaxed) sitting actually raises the intradiscal pressure by 40%. The current studies show instead that sitting in this position lowers the pressure by 10%, which was demonstrated with both high-precision stadiometric measurements (Althoff et al., 1992) and instrumented internal fixators (Rohlmann et al., 2001). Dreischarf et al. showed that the loads strongly depend on the arm position and the kyphosis angle in the thoracic spine (Dreischarf et al., 2010). The earlier finding that the intradiscal pressure was three times as high in the lateral decubitus position than in the supine position also could not be replicated; the current studies show only a small difference when subjects are in those different positions.

The new findings on the relaxed sitting position are of great importance. For example, relaxed sitting with diagonal positioning of the back and "deposition" of the upper body on a backrest markedly lowered the intradiscal pressure, and the subject also felt that this position was comfortable. As soon as the subject rose from the relaxed sitting position to the upright position, pressure in the lumbar disc increased markedly. Therefore, even when seated, an individual can improve diffusion, and thus the nutritive process within the intervertebral discs by regularly alternating between upright and relaxed sitting positions.

The differences between the newer findings and those of Nachemson et al. (Nachemson, 1966, Nachemson and Morris, 1964) can probably be explained by the use of different pressure transducers. Nachemson et al. integrated their measuring devices into a stiff cannula, which reached from the surface of the back through the muscle layers into the disc (Nachemson, 1966, Nachemson and Morris, 1964). When a person changes between sitting and standing, a tilt of the pelvis and thus a shift of the muscle layers relative to the spinal structures can be expected. Such a situation might have bent the long cannula and thus created false pressure values in the studies by Nachemson et al. The pressure gauges that were used in the newer study by Wilke et al. (1999) could not have been bent because the only stiff part was the 7-mm-long metal tip, which contained the sensor and was located completely within the nucleus pulposus, where it was exposed to hydrostatic and therefore equally distributed pressure. The rest of the transducer was a flexible cable with the same diameter as the tip. Therefore it can be assumed that the newer data are free of artifacts. The newer findings were also confirmed by their excellent agreement with measurements made on external fixator implants (Wilke et al., 1992) and stadiometric measurements (Van Deursen et al., 2005).

Researchers often are interested in calculating spinal loading from these pressures. Unfortunately, it is not as expedient to calculate forces from the pressure because it cannot be assumed that the pressure can simply be multiplied by the cross-sectional area of the disc. This correlation between force and pressure was investigated thoroughly by Brinckmann and Grootenboer (1991), who found that the result must be divided by a factor of 1.3 to 1.8 in order to account for the complex structure of the disc, where pressure is not only transferred by the nucleus pulposus, but also by the inner zone of the annulus fibrosus.

REFERENCES

Adams, M.A., Hutton, W.C., 1980. The effect of posture on the role of the apophysical joints in resisting intervertebral compressive forces. J. Bone Joint Surg. (Br.) 62, 358–362.

Althoff, I., Brinckmann, P., Frobin, W., Sandover, J., Burton, K., 1992. An improved method of stature measurement for quantitative determination of spinal loading. Application to sitting postures and whole body vibration. Spine (Phila Pa 1976) 17, 682–693.

Amonoo-Kuofi, H.S., 1991. Morphometric changes in the heights and anteroposterior diameters of the lumbar intervertebral discs with age. J. Anat. 175, 159–168.

Boden, S.D., Wiesel, S.W., 1990. Lumbosacral segmental motion in normal individuals. Have we been measuring instability properly? Spine (Phila Pa 1976) 15, 571–576.

Brinckmann, P., Grootenboer, H., 1991. Change of disc height, radial disc bulge, and intradiscal pressure from discectomy. An in vitro investigation on human lumbar discs. Spine (Phila Pa 1976) 16, 641–646.

Cossette, J.W., Farfan, H.F., Robertson, G.H., Wells, R.V., 1971. The instantaneous center of rotation of the third lumbar intervertebral joint. J. Biomech. 4, 149–153.

Dreischarf, M., Bergmann, G., Wilke, H.J., Rohlmann, A., 2010. Different arm positions and the shape of the thoracic spine can explain contradictory results in the literature about spinal loads for sitting and standing. Spine (Phila Pa 1976) 35, 2015–2021.

Dvorak, J., Panjabi, M.M., Chang, D.G., Theiler, R., Grob, D., 1991. Functional radiographic diagnosis of the lumbar spine. Flexion-extension and lateral bending. Spine (Phila Pa 1976) 16, 562–571.

Ellingson, A.M., Nuckley, D.J., 2015. Altered helical axis patterns of the lumbar spine indicate increased instability with disc degeneration. J. Biomech. 48, 361–369.

Gertzbein, S.D., Seligman, J., Holtby, R., Chan, K.H., Kapasouri, A., Tile, M., Cruickshank, B., 1985. Centrode patterns and segmental instability in degenerative disc disease. Spine (Phila Pa 1976) 10, 257–261.

Gertzbein, S.D., Seligman, J., Holtby, R., Chan, K.W., Ogston, N., Kapasouri, A., Tile, M., 1986. Centrode characteristics of the lumbar spine as a function of segmental instability. Clin. Orthop. Relat. Res. 48–51.

Hayes, M.A., Howard, T.C., Gruel, C.R., Kopta, J.A., 1989. Roentgenographic evaluation of lumbar spine flexion-extension in asymptomatic individuals. Spine (Phila Pa 1976) 14, 327–331.

Heuer, F., Schmidt, H., Klezl, Z., Claes, L., Wilke, H.J., 2007. Stepwise reduction of functional spinal structures increase range of motion and change lordosis angle. J. Biomech. 40, 271–280.

Kettler, A., Marin, F., Sattelmayer, G., Mohr, M., Mannel, H., Durselen, L., Claes, L., Wilke, H.J., 2004. Finite helical axes of motion are a useful tool to describe the three-dimensional in vitro kinematics of the intact, injured and stabilised spine. Eur. Spine J. 13, 553–559.

Lee, S.W., Wong, K.W., Chan, M.K., Yeung, H.M., Chiu, J.L., Leong, J.C., 2002. Development and validation of a new technique for assessing lumbar spine motion. Spine (Phila Pa 1976) 27, E215–20.

Mcgregor, A.H., Anderton, L., Gedroyc, W.M., Johnson, J., Hughes, S.P., 2001. Assessment of spinal kinematics using open interventional magnetic resonance imaging. Clin. Orthop. Relat. Res. 341–348.

Nachemson, A., 1966. The load on lumbar disks in different positions of the body. Clin. Orthop. Relat. Res. 45, 107-22.

Nachemson, A., Elfstrom, G., 1970. Intravital dynamic pressure measurements in lumbar discs. A study of common movements, maneuvers and exercises. Scand. J. Rehabil. Med. Suppl. 1, 1–40.

Nachemson, A., Morris, J.M., 1964. In vivo measurements of intradiscal pressure discometry, a method for the determination of pressure in the lower lumbar discs. J. Bone Joint Surg. Am. 46, 1077–1092.

Nachemson, A.L., Schultz, A.B., Berkson, M.H., 1979. Mechanical properties of human lumbar spine motion segments. Influence of age, sex, disc level, and degeneration. Spine (Phila Pa 1976) 4, 1–8.

Nachemson, A.L., Andersson, B.J., Schultz, A.B., 1986. Valsalva maneuver biomechanics. Effects on lumbar trunk loads of elevated intraabdominal pressures. Spine (Phila Pa 1976) 11, 476–479.

Ochia, R.S., Inoue, N., Renner, S.M., Lorenz, E.P., Lim, T.H., Andersson, G.B., An, H.S., 2006. Three-dimensional in vivo measurement of lumbar spine segmental motion. Spine (Phila Pa 1976) 31, 2073–2078.

Ogston, N.G., King, G.J., Gertzbein, S.D., Tile, M., Kapasouri, A., Rubenstein, J.D., 1986. Centrode patterns in the lumbar spine. Baseline studies in normal subjects. Spine (Phila Pa 1976) 11, 591–595.

Panjabi, M.M., Goel, V., Oxland, T., Takata, K., Duranceau, J., Krag, M., Price, M., 1992. Human lumbar vertebrae. Quantitative three-dimensional anatomy. Spine (Phila Pa 1976) 17, 299–306.

Panjabi, M.M., Oxland, T., Takata, K., Goel, V., Duranceau, J., Krag, M., 1993. Articular facets of the human spine. Quantitative three-dimensional anatomy. Spine (Phila Pa 1976) 18, 1298–1310.

Panjabi, M.M., Oxland, T.R., Yamamoto, I., Crisco, J.J., 1994. Mechanical behavior of the human lumbar and lumbosacral spine as shown by three-dimensional load-displacement curves. J. Bone Joint Surg. Am. 76, 413–424.

Parent, S., Labelle, H., Skalli, W., De Guise, J., 2004. Vertebral wedging characteristic changes in scoliotic spines. Spine (Phila Pa 1976) 29, E455–62.

Pearcy, M.J., 1985. Stereo radiography of lumbar spine motion. Acta Orthop. Scand. Suppl. 212, 1–45.

Pearcy, M., Portek, I., Shepherd, J., 1984. Three-dimensional x-ray analysis of normal movement in the lumbar spine. Spine (Phila Pa 1976) 9, 294–297.

Pennal, G.F., Conn, G.S., Mcdonald, G., Dale, G., Garside, H., 1972. Motion studies of the lumbar spine: a preliminary report. J. Bone Joint Surg. (Br.) 54, 442–452.

Pintar, F.A., Yoganandan, N., Myers, T., Elhagediab, A., Sances, A., Jr., 1992. Biomechanical properties of human lumbar spine ligaments. J. Biomech., 25, 1351–6.

Ritzel, H., Amling, M., Posl, M., Hahn, M., Delling, G., 1997. The thickness of human vertebral cortical bone and its changes in aging and osteoporosis: a histomorphometric analysis of the complete spinal column from thirty-seven autopsy specimens. J. Bone Miner. Res. 12, 89–95.

Rohlmann, A., Arntz, U., Graichen, F., Bergmann, G., 2001. Loads on an internal spinal fixation device during sitting. J. Biomech. 34, 989–993.

Roussouly, P., Gollogly, S., Berthonnaud, E., Dimnet, J., 2005. Classification of the normal variation in the sagittal alignment of the human lumbar spine and pelvis in the standing position. Spine (Phila Pa 1976) 30, 346–353.

Schollum, M.L., Robertson, P.A., Broom, N.D., 2008. ISSLS prize winner: microstructure and mechanical disruption of the lumbar disc annulus: part I: a microscopic investigation of the translamellar bridging network. Spine (Phila Pa 1976) 33, 2702–2710.

Skrzypiec, D.M., Bishop, N.E., Klein, A., Puschel, K., Morlock, M.M., Huber, G., 2013. Estimation of shear load sharing in moderately degenerated human lumbar spine. J. Biomech. 46, 651–657.

Urban, J.P., Mcmullin, J.F., 1988. Swelling pressure of the lumbar intervertebral discs: influence of age, spinal level, composition, and degeneration. Spine (Phila Pa 1976) 13, 179–187.

Van Deursen, L.L., Van Deursen, D.L., Snijders, C.J., Wilke, H.J., 2005. Relationship between everyday activities and spinal shrinkage. Clin. Biomech. (Bristol, Avon) 20, 547–550.

Wilke, H.J., Rohlmann, A., 2012. Forces on the spine. In: Ambrosio, L., Tanner, E. (Eds.), Biomaterials for Spinal Surgury. Woodhead Publishing Limited, Cambridge, UK.

Wilke, H.J., Durselen, L., Claes, L., Worsdorfer, O., 1992. A method for in vivo measurements of implants used in the reconstruction of spinal fractures. Biomed Tech (Berl) 37, 78–85.

Wilke, H.J., Neef, P., Caimi, M., Hoogland, T., Claes, L.E., 1999. New in vivo measurements of pressures in the intervertebral disc in daily life. Spine (Phila Pa 1976) 24, 755–762.

Yamamoto, I., Panjabi, M.M., Crisco, T., Oxland, T., 1989. Three-dimensional movements of the whole lumbar spine and lumbosacral joint. Spine (Phila Pa 1976) 14, 1256–1260.

Yoshioka, T., Tsuji, H., Hirano, N., Sainoh, S., 1990. Motion characteristic of the normal lumbar spine in young adults: instantaneous axis of rotation and vertebral center motion analyses. J. Spinal Disord. 3, 103–113.

SECTION

BIOMECHANICS OF SPINAL TISSUES

2

CHAPTER

5

THE VERTEBRAL BONE

Mirit Sharabi*, Hans-Joachim Wilke†, Rami Haj-Ali*

Tel-Aviv University, Tel-Aviv, Israel Institute of Orthopaedic Research and Biomechanics, Ulm University, Ulm, Germany†*

THE HIERARCHICAL STRUCTURE OF THE VERTEBRAL BONE

Bone provides mechanical structural support for the human body, enables movements, protects vital organs, and stores minerals. In general, the bone's structure makes it lightweight yet capable of load bearing. Its structure reduces stress at the joints in the human body, increases the absorption of energy, and enables the human body to bear and resist high mechanical loads associated with many modes of operation. Therefore bone has unique mechanical material properties: high level of stiffness, strength, and fracture toughness. Generally, biological bone is a material with structural hierarchy. Its nanostructure consists apatite minerals (inorganic phase) (33%–43% bone volume), organic phase (mainly collagen type I) (32%–44% bone volume), and water (15%–25% bone volume) (Currey, 1990; Olszta et al., 2007).

The vertebral structure can be examined at several scale levels. Fig. 1 shows its nested hierarchical nature: macroscale (vertebra), mesoscale (cortical and trabecular bone), microscale (osteon and trabeculae), sub-microscale (lamella built by mineralized collagen fibril), and nanoscale (type I collagen and hydroxyapatite) (Gibson et al., 2010).

THE MACROSTRUCTURE OF THE VERTEBRAE

At the macroscale, the vertebrae can be divided into three functional parts: the vertebral body, the pedicles, and the posterior elements consisting of the laminae and their processes. These parts have different mechanical functions, and their integration contributes to the overall biomechanical behavior. In general, the vertebral bodies are designed for weight bearing. The posterior elements are designed to withstand different forces, such as resisting sliding and twisting and providing stability to the vertebral body in the horizontal plane. The pedicles can transmit both tension and bending forces to the vertebral bodies (Bogduk, 2005).

At the mesoscale, the vertebral body has a unique internal structure that is responsible for its load-bearing ability. The vertebral body is constructed as an assembly of a trabecular core surrounded by a cortical bone shell (Fig. 1E–F). The microstructure of the trabecular bone (also called cancellous or spongy bone) consists of plates or struts called "trabeculae," each about 200 μm thick. Cortical bone (also called compact bone) is a highly dense bone, which is responsible for its high mechanical stiffness.

Biomechanics of the Spine. https://doi.org/10.1016/B978-0-12-812851-0.00005-7

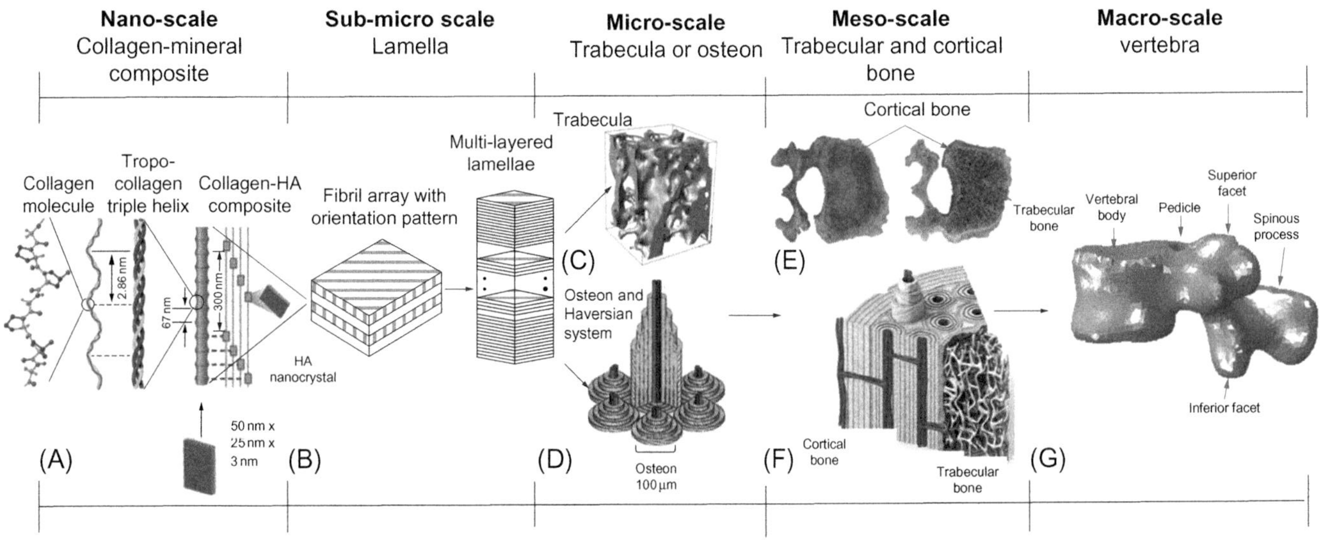

FIG. 1

The hierarchical structure of the vertebral bone: Nanoscale to macroscale. (A) Nanoscale: collagen fiber-HA nanoplatelets composite. (B) Sub-microscale: oriented lamellae unit cell and periodicity. (C–D) Microscale: (C) Trabecula. (D) Osteon and Haversian system. (E–F) Mesoscale: cortical and trabecular bone. (G) Macroscale—the vertebrae divided to the vertebral body, the pedicles and the posterior elements.

Subparts (A) and (D) are adapted from Wegst, U.G., Bai, H., Saiz, E., Tomsia, A.P., Ritchie, R.O., 2015. Bioinspired structural materials. Nat. Mater. 14, 23–36 with permission.

This structure has several advantages: The cancellous bone is lighter in weight than a solid block of cortical bone is and also is more favorable for dynamic load bearing, because the crystalline structure of solid blocks tends to fracture along cleavage planes under sudden force due to its inability to dissipate and absorb energy.

The outer cortical shell cannot withstand longitudinal compression; therefore an inner reinforcement of trabeculae prevents the cortical shell from collapsing. The trabeculae give the vertebral body resilience and weight-bearing strength. Thus, when a load is applied, the vertical trabeculae carry it, and the horizontal trabeculae restrain the vertical trabeculae from bowing by creating transverse tension (lateral support). The load is supported by a combination of vertical pressure and transverse tension in the trabeculae, which provides the unique resilience of the vertebrae. The trabecular structure of the vertebral body extends through the pedicles into the posterior elements and reinforces these processes (Bogduk, 2005).

THE MICROSTRUCTURE OF THE CORTICAL BONE

At the microscale, the cortical bone is a dense bone with an approximate porosity of 6%, and its thickness ranges between tenths of a millimeter in the vertebrae, to even a few centimeters in the mid-shaft of long bones (Fratzl and Weinkamer, 2007). Blood vessels in the cortical bone are surrounded by three to eight concentric lamellae. This structure is called "osteon" or "Haversian system" (Fig. 1D). The arrangement of the lamellar bone is similar to a rotated plywood structure, where the fibers are parallel inside a thin sublayer. The fiber direction rotates around an axis that is vertical to the layers (Fratzl and Weinkamer, 2007).

The cylinder-like osteons are roughly parallel to the long axis of the bone. An osteon is structured from a 3–7-μm-wide lamellae consisting of parallel mineralized collagen fibers arranged in a planar orientation. The cortical bone can also be arranged in less ordered structures called "woven bones." Lamellar bone is formed when the lamellae, together with woven bone tissue, form stacked thicker layers (150–300 mm) around the entire circumference of the bone (Rho et al., 1998).

Mineralized fibrils are arranged into fibril arrays (or fibers) on a scale of 1–10 μm. Lamellae can be found in a repetitive stacked arrangement in both trabecular (cancellous) and cortical (compact) bone (Fratzl and Weinkamer, 2007).

THE MICROSTRUCTURE OF THE TRABECULAR BONE

Trabecular bone is usually found in bones that are under compressive stress. It is found at the ends of all long bones, often within flat and irregular bones, such as the sternum, pelvis, and spine. Trabecular bone is structured as a foam-like interconnecting network of trabeculae in a number of combinations: rod-rod, rod-plate, or plate-plate (Fratzl and Weinkamer, 2007). This tissue is made of trabeculae, which are lamellar structures formed at the microscale level. At the sub-micronscale level, a single lamella is built up by mineral collagen fibrils (MCFs) and some cavities, called lacunae (Gibson et al., 2010).

The typical thickness of the trabeculae struts is about 50–300 μm, with an optimized orientation that depends on the load distribution in the bone (Rho et al., 1998). The trabeculae are surrounded by bone marrow that contains blood nutrients needed by the osteocytes inside the bone material (Fratzl and Weinkamer, 2007). The trabecular microstructure is anisotropic, such that there is one direction in which the mechanical stiffness and strength are the largest. Therefore trabecular bone is classified as a composite, anisotropic, open porous cellular solid (Keaveny et al., 2001). Thus the effective mechanical elastic material of the trabecular bone is orthotropic and often transversely isotropic.

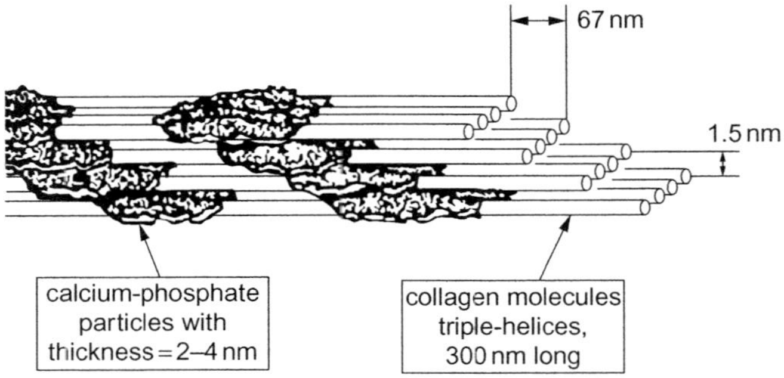

FIG. 2

The nanostructure of the bone. The mineral crystals are parallel to each other and to the collagen fibers, in a repetitive periodic pattern. The staggered structure is a result of mineral nucleation in the gap zone of the collagen fibrils.

Reproduced from Fratzl, P., Weinkamer, R., 2007. Nature's hierarchical materials. Prog. Mater. Sci. 52, 1263–1334 with permission.

THE NANOSTRUCTURE OF THE BONE

The basic building block of the bone material is nano-composite mineral nanoparticles made of carbonated hydroxyapatite nano-platelets and collagen fibrils of about 100 nm in diameter, as shown in Figs. 1A and 2. The fibrils are assembled from 300 nm long and 1.5 nm thick organic collagen molecules, and 2–4 nm inorganic apatite crystals staggered in the collagen fibrils. The collagen in the fibrils limits the possible primary growth of the crystals, forcing them to be discrete and discontinuous. The apatite crystals are arranged in an approximate repeated distance of 67 nm, corresponding to the distance between adjacent collagen molecules (Eliaz and Metoki, 2017; Fratzl et al., 2004; Fratzl and Weinkamer, 2007).

BONE REMODELING

Bone is a highly dynamic tissue with an evolving microstructure. It continuously adjusts itself to the physiologic and mechanical environment by changes in its composition and microscopic architecture. Living bone is continually undergoing processes of growth, reinforcement, and resorption. Remodeling processes in living bone are governed by mechanisms in which the bone adapts its macro and micro structure to changes in enduring loads (Cowin and Hegedus, 1976). The processes of bone formation and resorption are continuous throughout life and are necessary for the maintenance of a normal functional skeleton. Resorption occurs when osteoclasts digest old bone, and bone formation occurs when osteoblasts lay down new bone until the resorbed bone is completely replaced. These two processes are tightly coupled (Daci et al., 2003). It is commonly agreed that bone is able to adapt itself to a variety of mechanical loadings.

An important principle of skeletal physiology is that bones are able to sense the mechanical loads they bear and modify their structure to suit changes in the loads. This principle is called Wolff's law (Cowin, 1986; Wolff, 1892, 1986), which states that the structure of bone tissue can adapt and have an optimal microstructural configuration under the influence of mechanical loads. The optimal structure of the bone tissue is expressed in the minimal mass needed to bear repeated loading (Jovanović and Jovanović, 2004). Wolff concluded that trabecular morphology matches the (perpendicular) principal stress trajectories.

One of the first fundamental theories of bone remodeling was proposed by Cowin and Hegedus, who developed a comprehensive phenomenological theory called "Adaptive Elasticity" (Cowin and Hegedus, 1976; Cowin and Van Buskirk, 1979). This theory is based on linear elasticity with an additional constitutive equation that allows for changes in the density, hence the stiffness of bone, and also changes in external shapes of bone. Based on loads, material properties, and boundary conditions, the mechanical remodeling equilibrium can be represented by the strain state. Then a new strain state is referenced to the equilibrium initial strain, which is produced by altering the local geometry and loading of the bone. Cowin and associates separated internal and surface (or external) remodeling. Cowin and Van Buskirk (1979) performed external remodeling by adding or removing material. First-order bone remodeling rate equations were also proposed to specify the change in bone density and/or shape (Hart et al., 2001).

A phenomenological approach similar to Cowin's adaptive elasticity theory for shape changes was developed by Huiskes et al. (1987) with several differences. Weinans et al. (1992) expanded the computational work to simulate adaptive bone remodeling, using a time-dependent rule for regulation of density. Mullender et al. (1994) proposed a mechanistic model for the functional adaptation of trabecular bone. The basis of the model was the hypothesis that a local mechanical signal is appraised by a sensor (osteocyte), which in turn produces a stimulus for bone mass regulation to actors in its environment. Changes of bone mass were translated to change in the mechanical properties, thus also affecting the mechanical signals.

The trabecular bone has a higher remodeling level than the cortical bone. The trabecular 3D microarchitecture is directly influenced by the applied mechanical strains. During childhood, trabecular bone is made of a dense isotropic network of plates. However, in adults, the trabecular plates are arranged in a preferential orientation along the direction of the strains, which are exerted on the bone (Chappard et al., 2008).

MECHANICAL BEHAVIOR OF THE VERTEBRAL BONE

THE MATERIAL PROPERTIES OF BONE

Understanding the mechanical response of bone is important for assessing the risk of fracture. The material properties of trabecular bone are important for characterizing bone pathologies and the remodeled bone adjacent to joint implants because trabecular bone is affected by disease earlier than cortical bone (Rho et al., 1998). Trabecular bone, like many biological materials, demonstrates time-dependent behavior and susceptibility to damage during cyclic loading. It presents substantial heterogeneity that stems from variations in volume fraction, the three-dimensional architecture of the individual trabeculae, and other associated tissue properties. The overall mechanical properties are strongly dependent on age in humans. For example, ultimate stress is reduced by almost 11% per decade for the human spine in populations above the age of 20 (Mosekilde and Mosekilde, 1986; Mosekilde et al., 1987). Therefore factors such as age, gender, and location need to be declared when discussing the mechanical properties. Different ranges in the mechanical properties are summarized in Table 1.

Table 1 Selected Mechanical Properties of the Vertebral Trabecular Bone

Mechanical Property	Loading Mode	Range	Reference
Ultimate strength	Compressive	0.038–10.6 MPa	Hans et al. (1999), Hou et al. (1998)
	Tensile	1.33–3.53 MPa	Kopperdahl and Keaveny (1998)
Yield strength	Compressive	0.1–24 MPa	Kopperdahl and Keaveny (1998), Kopperdahl et al. (2002), Nazarian et al. (2008)
	Tensile	0.77–2.75 MPa	Kopperdahl and Keaveny (1998)
	Shear	1.4–7.8 MPa	Cendre et al. (1999)
Modulus	Compressive	1–976 MPa	Hans et al. (1999), Hou et al. (1998)
	Tensile	139–620 MPa	Kopperdahl and Keaveny (1998), Wolfram et al. (2010)
	Shear	40–120 MPa	Wolfram et al. (2010)
Ultimate strain	Compressive	0.96%–2.30%	Kopperdahl and Keaveny (1998)
	Tensile	1.09%–2.51%	Kopperdahl and Keaveny (1998)
Yield strain	Compressive	0.65%–0.87%	Kopperdahl et al. (2002)
	Tensile	0.71%–0.88%	Kopperdahl and Keaveny (1998)

The anisotropy of trabecular bone is pronounced in both the modulus (Mosekilde et al., 1987) and strength (Galante et al., 1970; Mosekilde et al., 1987). The anisotropy is mild compared with fiber-reinforced composites, but its biomechanical influence on whole bone strength or bone-implant performance is still unknown (Keaveny et al., 2001).

The compressive strength becomes more anisotropic when the porosity of trabecular bone increases. This is consistent with the concepts of Wolff's law (Cowin, 1986; Wolff, 1892, 1986), according to which bone is placed where it is most needed (Keaveny et al., 2001).

Trabecular bone demonstrates orthotropic symmetry (Zysset et al., 1998) and, in some cases, presents transverse isotropy (Odgaard et al., 1997). For example, in the human vertebral bone, the mean values of strength and modulus in the superior-inferior direction are higher than those in the transverse direction by factors of 2.8 and 3.4, respectively. Therefore mechanical testing should be performed along the principal directions of the trabecular architecture. Otherwise, the resulting measurements are difficult to explain because of the influence of multiple material properties (Keaveny et al., 2001).

The elastic modulus of trabecular bone depends primarily on the density. However, the exact form of this relationship is still under dispute. It seems that the anatomic sight, loading direction, and testing method have a significant effect on this relationship. The differences between the linear and power law relationships are negligible because of the limited range of density found within a single anatomic site (Helgason et al., 2008; Keaveny et al., 2001).

The strength also depends on the density and can vary by an order of magnitude across sites. Moreover, the strength behavior is lower in tension than in compression (Keaveny et al., 1994a), and the shear mode also presents a lower strength (Ford and Keaveny, 1996). The relative differences in these strengths increase with increasing modulus. Therefore the strength of trabecular bone is heterogeneous (changes with age and location), anisotropic (depends on loading direction), and asymmetric (different in tension, compression, and shear) (Keaveny et al., 2001). However, the strain-based description of failure is relatively simple. Failure strains for trabecular bone are relatively constant with a minor dependence on density. Moreover, the yield strains for human trabecular bone are higher in compression than tension, and can differ across sites, but this yield strain may be considered uniform within a given site.

Experimental studies of irreversible loading (overloading and damage) of the trabecular bone showed a reduction of elastic moduli and strength and the accumulation of residual deformations in the trabecular bone (Keaveny et al., 1994b; Keaveny et al., 1999; Zysset, 1994). In a study about the mechanism of failure in human vertebral bone compressed by a large deformation (Fyhrie and Schaffler, 1994), the authors found that the dominant mechanism of failure was microscopic cracking, as a result of shear stresses in the trabeculae. The complete fractures of trabeculae occurred transversely to the loading direction.

The full load–displacement curve of the trabecular bone includes a sequence of several responses: (1) an initial linear undamaged region; (2) a material nonlinear response, including strain-softening caused by damage accumulation; (3) ultimate load peak region; (4) post-peak unloading with increased strain and displacement; and (5) a region where, with increased displacements, the load is increased with very small values (Fyhrie and Schaffler, 1994; Gibson and Ashby, 1999; Hayes and Carter, 1976).

MECHANICAL TESTING

The trabecular bone carries a substantial proportion of the compressive loadings applied on the vertebral body (Rockoff et al., 1969). The hierarchical nature of its microstructure renders its overall mechanical properties. It is strongly dependent on the geometrical assembly and the material properties of the individual trabeculae (bone-wall). Different experimental methods have been used to measure in vitro and in vivo mechanical properties of bone tissues. Classical mechanical tests of bone have been conducted using classical uniaxial or pure shear methods. In these tests, a specimen of trabecular bone was cut out of a bone and loaded in a material testing machine. The mechanical properties were calculated from the load–displacement curve. Different testing setups were developed, such as the platen technique, where the bone specimen was placed between two blocks that compress the specimen. The measurement of strain was obtained from the displacements of the platens. Platen compression, for example, includes substantial systematic underestimation of the end effect because of the interface between the platens and the specimens, which leads to random errors of approximately 20%–40% (Helgason et al., 2008; Keaveny et al., 1997). Another remedy is to employ an extensometer attached to the specimen (Keaveny et al., 1997). In the end-cap technique, the specimen ends were embedded in a layer of polymethylmethacrylate (Keaveny et al., 1997; Odgaard et al., 1997).

It was found that in order to satisfy the continuum assumption, the size of the specimens should be greater than 4–5 mm and contain at least five trabeculae (Harrigan et al., 1988). Cylinder and cube specimen geometry demonstrated similar mechanical properties. However, structural artifacts were decreased with specimen size and length-to-width ratio. A more precise method was to use the end-caps technique with a 2:1 cylindrical geometry (Keaveny et al., 1997; Linde et al., 1992).

In addition to the preceding classical contact mechanical testing methods, other techniques such as ultrasonic and nanoindentation were applied in order to obtain bone properties (Rho et al., 1993; Rho et al., 1997; Yoon and Katz, 1976; Zysset et al., 1999). These techniques have been recommended because of their advantages of using small volume samples, targeting a wider range of elastic constants, and repeated nondestructive tests over different length scales (Peterson and Bronzino, 2008). Nanoindentation experiments have demonstrated that substantial variations in tissue modulus can exist both within and across specimens (Rho et al., 1997; Zysset et al., 1999). These studies proposed that the elastic Young's modulus in the longitudinal direction is greater than in the transverse direction (Rho et al., 1998; Rho et al., 1997). In recent studies, in addition to elastic constants, strength values were identified by means of nanoindentation (Schwiedrzik et al., 2014).

DENSITY AND MICROARCHITECTURE MEASUREMENTS

The mechanical properties of bone, including fracture behavior of the trabecular bone, are strongly dependent on its mineral density (BMD) and microarchitecture (Ulrich et al., 1999). Additional factors that affect bone quality are the remodeling level, the size of hydroxyapatite crystals, and the quality of collagen (Chappard et al., 2008). The BMD represents the mass of mineral for bone volume. It is usually measured by dual-energy X-ray absorptiometry (DXA), a two-dimensional, projection-based radiographic technique that measures integral BMD of both cortical and trabecular bone. Additional methods, such as high-resolution peripheral quantitative computed tomography (HR-pQCT), are also being employed to measure the BMD and the microarchitecture of the bone. An imaging tool that has widespread use in the laboratory is the Microtomography (μCT); it provides a high resolution of up to few microns; rapid, nondestructive imaging of bone specimens; and noninvasive imaging in animal models (Boehm et al., 2003; Boutroy et al., 2005; Burghardt et al., 2011; Chappard et al., 2008; Fuller et al., 2015; Krug et al., 2010). The vertebral trabecular bone demonstrates a wide range of densities, depending on age, pathology, and location. Studies have reported densities ranging from 0.09 g/cm^3 (Keaveny et al., 1997) to 0.35 g/cm^3 (Morgan et al., 2001). The density and trabecular structure has been reported as varying between anatomical sites (Morgan and Keaveny, 2001) and within the same site (Chen et al., 2008).

It has been shown that the BMD alone can give a good estimation of the average stiffness and strength for a large range of densities. However, BMD measurements alone do not take into account the trabecular microstructure and that the mineral may account for only 60% to 80% of the biomechanical strength (Boehm et al., 2003). The most classical histomorphometric parameters currently used for the description of trabecular bone microarchitecture are based on Parfitt's principles (Parfitt et al., 1987)—for example, trabecular thickness (Tb.Th, in microns), trabecular number (or more exactly trabecular density) (Tb.N, per millimeter), and the trabecular separation (Tb.Sp, in microns) (Chappard et al., 2008). Moreover, it has been found that, in cases in which the specimens could not be identified as either osteoporotic or normal, the BMD alone is insufficient for accurate prediction of the mechanical properties. The addition of these microstructural measurements, such as BV/TV (volume of trabeculae/Total volume), Tb.Sp (trabecular separation), and MIL (mean intercept length), increased the correlation of the mechanical properties up to 92% (Ulrich et al., 1999).

COMPUTATIONAL MODELS OF THE VERTEBRAL BONE

Computational and analytical models have been performed to predict the mechanical elastic properties of a trabecular bone. Results of early studies, for example, (Gibson, 1985), suggest that the mechanical behavior of cancellous bone is similar to that of cellular materials. This similarity is manifested by the dependence on the relative density, cell geometrical configurations, and cell-wall properties. The Gibson's model (Gibson, 1985) employs open- and closed-cell configurations to predict bone properties with low and high relative densities, respectively. The former case exists when the trabecular bone is made up of a network of rod-like elements; the latter is more valid when the microstructure consists of plate-like elements. In addition, Maquer et al. (2015) stated that the volume fraction and the bone anisotropy parameters are prime factors affecting trabecular bone stiffness. Therefore, to provide a more reliable evaluation of the mechanical material properties of a trabecular bone, a high-resolution quantitative CT and other imaging devices should be used in order to improve the accuracy of calculation for the bone volume fraction and more importantly the bone anisotropic properties.

The finite element (FE) computational method has also been utilized to investigate the mechanical behavior of the trabecular bone. van Rietbergen et al. (1995) demonstrated a relatively refined FE model by analyzing large regions of the trabecular bone. The authors used three-dimensional serial reconstruction techniques to build a large scale FE model by directly converting voxels to elements. They offered a new methodology in investigating the trabecular bone by taking into consideration its native architectural form. More recently, Wolfram et al. (2010) introduced micro-FE models of vertebral bone, constructed from μCT scans along with averaged tissue mechanical properties obtained from the wet nanoindentation technique for the bone-wall material properties.

Integrated multiscale mechanical methods have been advocated recently and applied to bone materials in order to evaluate the effective elastic properties. For example, McDonald et al. (2010) developed a multiscale FE model of an osteoporotic lumbar vertebral body to investigate how this pathology would affect the vertebral stiffness and strength in the lumbar spine. Hamed et al. (2012), considering the hierarchical structure of the trabecular bone, presented a multiscale computational model that used five scales, from the nanoscale scale up to the mesoscale level. In their analysis, continuum micromechanics, the lamination theory, the Mori-Tanaka (MT) scheme, and the FE method were integrated and used. More recently, Colloca et al. (2014) introduced an analytical micromechanical multiscale framework to study the influence on bone macroscale structural properties at the nanoscale level. The bone composition at each scale level was assumed to be heterogeneous and to comprise two phases. A typical representative volume element (RVE) tissue level, representing the trabecular surface, was considered with both the bone matrix and cylindrical voids acting as the constituents of the RVE. At the bone matrix lower level, the composite material was built from hydroxyapatite mineral foam, reinforced with collagen fibrils in the longitudinal direction. Their solution was based on the strain-energy density. It is interesting to note that a wide range of values for the elastic properties of the single trabecula (bone wall) were used in the abovementioned computational and analytical models. Haj-Ali et al. (2016) presented a 3D nested modeling analysis framework, adept at predicting the overall linear anisotropic mechanical properties of the highly porous microstructure (i.e., vertebral trabecular bone), as illustrated in Fig. 3. They used the HFGMC-3D (Haj-Ali and Aboudi, 2009, 2013, 2016) micromechanical method twice at both the nanoscale and the microscale combined with the 3D sublaminate mesoscale and intermediate model to compute the effective elastic properties of different trabecular assemblies. The trabecular micromodel (i.e., the VTBs' bone wall) consists of a unit cell with a repeated stack of multilayered lamellae. This 3D sublaminate model combines the stress and strain fields within each lamina in an effective anisotropic continuum for laminated material in order to generate the overall stiffness matrix. The sublaminate approach is generally applied to multilayered repeated sublaminate in which the stacking sequence effects are not recognized. The three-dimensional (3D) lamination theory is used assuming a perfect bond between the laminae, as described in (Pecknold and Haj-Ali, 1993). The overall modeling framework predicts the effective elastic properties of the human bone, as shown in Fig. 4B, for a wide range of volume fractions that reflects different ages and the related bone microstructures.

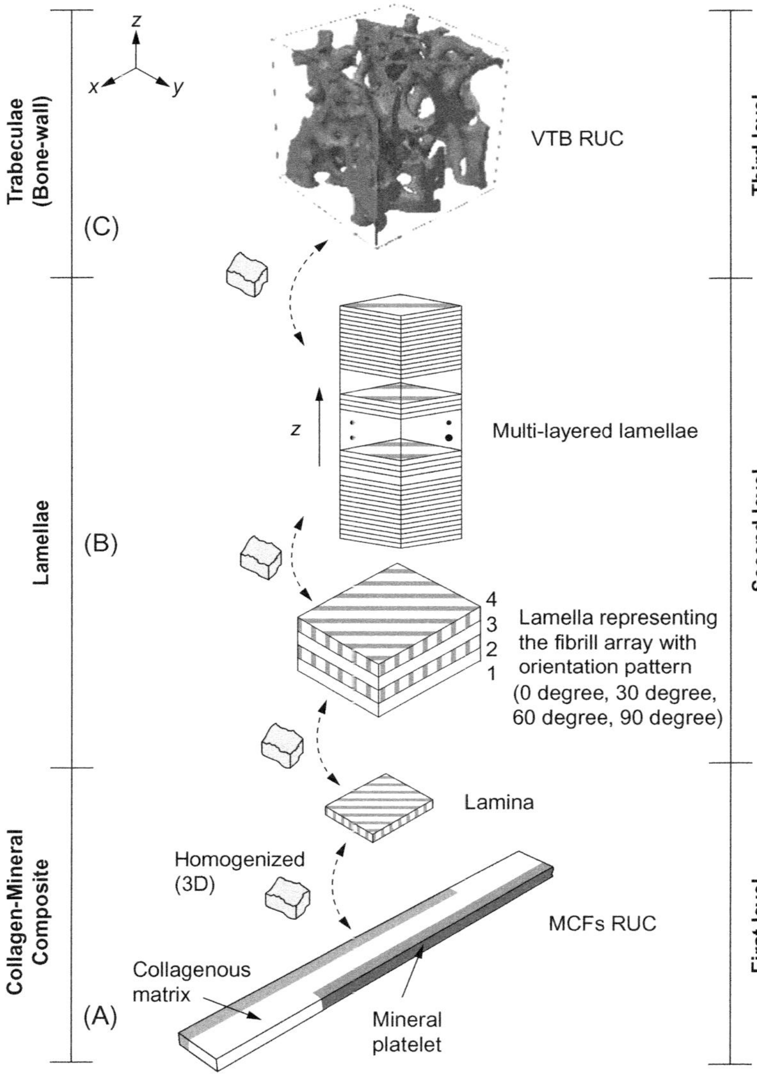

FIG. 3

A schematic representation of the three-level multiscale *modeling* framework of the vertebral trabecular bone (VTB) using two different combined HFGMC and sublaminate micromodels at the three scale levels. (A) Nano scale of the collagen-mineral comopsite, (B) sub-micron scale of the lamellae and their multi-layered lamellae, and (C) the micron scale of the trabeculae.

Reproduced from Haj-Ali, R., Massarwa, E., Aboudi, J., Galbusera, F., Wolfram, U., Wilke, H.-J., 2016. A new multiscale micromechanical model of vertebral trabecular bones. Biomech. Model. Mechanobiol., 1–14 with permission.

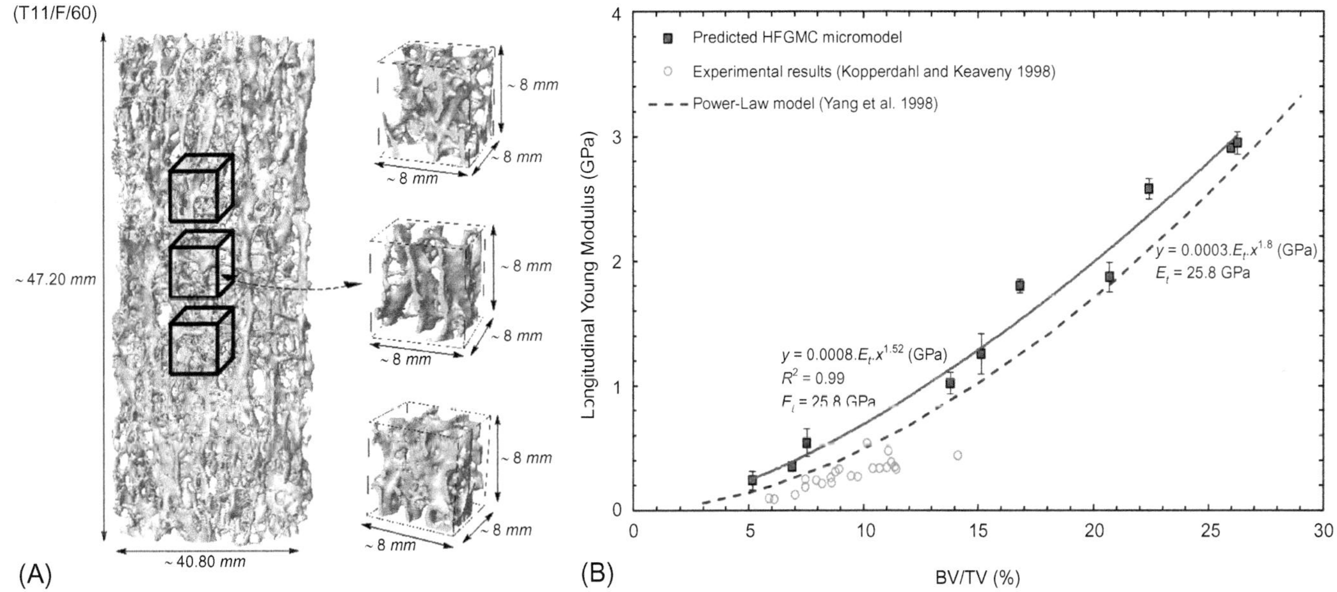

FIG. 4

(A) Representative volume elements (RVE, n=3) obtained (~8.0×8.0×8.0 mm) from µCT scan of T11 vertebra from 60-year-old female (diameter=~40 0.8 mm, total height=~47.2 mm); (B) Predicted longitudinal Young's modulus for human VTB samples compared with experimental data and power law models available in the literature.

Reproduced from Haj-Ali, R., Massarwa, E., Aboudi, J., Galbusera, F., Wolfram, U., Wilke, H.-J., 2016. A new multiscale micromechanical model of vertebral trabecular bones. Biomech. Model. Mechanobiol., 1–14 with permission.

A nonlinear and damage bone model was recently introduced (Hamed and Jasiuk, 2013). The authors presented a multiscale computational finite element model, with dominant damage mechanisms to predict stiffness and bone strength at different structural scales by using cohesive interface elements. Harrison et al. (2013) also developed a 3D high-resolution model with fracture criteria and a large-deformation computational model, predicting the full range of response, including the pre- and post-ultimate stress states (ultimate strength). This analysis model was successful in predicting ultimate strength following softening of the trabecular bone. To analyze human trabecular bone samples under compressive loading, an isotropic μCT finite element model coupled with a damage law with element deletion method was developed (Hambli, 2013).

In a series of papers, Goda et al. (Goda et al., 2012; Goda and Ganghoffer, 2015; Goda et al., 2016) presented a micromechanical method to investigate the microstructural effects of the macroscopic mechanical properties of trabecular bones. An anisotropic continuum model was modeled from the discrete homogenization of a quasi-periodical lattice model for the microstructural trabecular bone (Goda et al., 2012). The discrete homogenization technique was applied to construct the plastic yield surfaces of 2D and 3D bending-dominated periodic lattices, which represented the topology of trabecular bones and established criteria for brittle and ductile collapse based on micromechanical methods (Goda and Ganghoffer, 2015). Moreover, the authors developed a numerical model to predict the effective properties of porous polymeric biomaterials. Finite element models, based on an accurate three-dimensional solid model from μCT data, were generated to quantify the effects of the porosity. In the latter study, the second gradient of deformation, which is expressed as in coupled stress continuum theories, was used to address the size dependency in porous structures (Goda et al., 2016).

The linear framework model (Haj-Ali et al., 2016) was extended to the damage behavior of the vertebral bone (Massarwa et al., 2017). The latter used the 3D parametric HFGMC with progressive

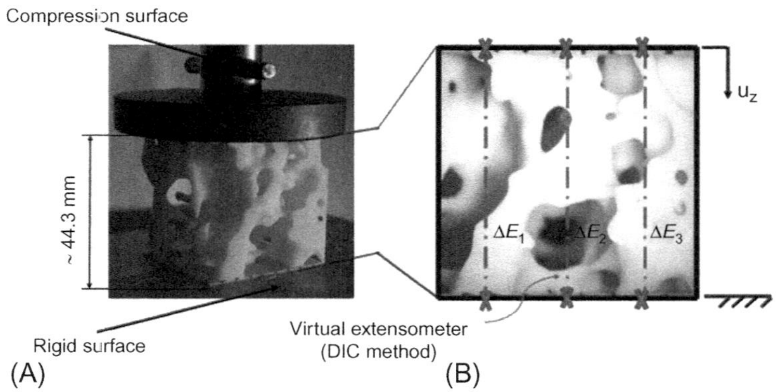

FIG. 5

Compression testing and modeling of 3D-printed representative volume element (RVE). The RVE is from vertebral trabecular bone micro-CT scan scaled (80×) in order to obtain bone micro-structure manufactured using 3D printer from polymeric material. (A) A compressed polymeric VTB-RUC highly porous microstructure. (B) An axial deformation gauge.

Reproduced from Massarwa, E., Aboudi, J., Galbusera, F., Wilke, H.-J., Haj-Ali, R., 2017. A nonlinear micromechanical model for progressive damage of vertebral trabecular bones. J. Mech. Mater. Struct. 12, 407–424 with permission.

damage in order to predict the overall nonlinear mechanical response with evolving damage of the highly porous vertebral trabecular bone. Verification of the model was done by comparing the nonlinear stress-strain predicted curve to one obtained from compression tests performed on a printed polymeric unit cell of trabecula bone microstructure (Fig. 5), as well as curves reported in the literature for native vertebral trabecular bone (Fig. 6). It was shown that the 3D nonlinear micromechanical modeling approach has the ability to predict the compression response of a tested porous polymeric material system, manufactured using a 3D printer based on a μCT scan geometry. Moreover, the proposed HFGMC progressive damage micromodel was used to simulate compression failure of native vertebral trabecular bone. Good predictive ability has been demonstrated for the compression yield stresses and strains compared to results taken from the literature.

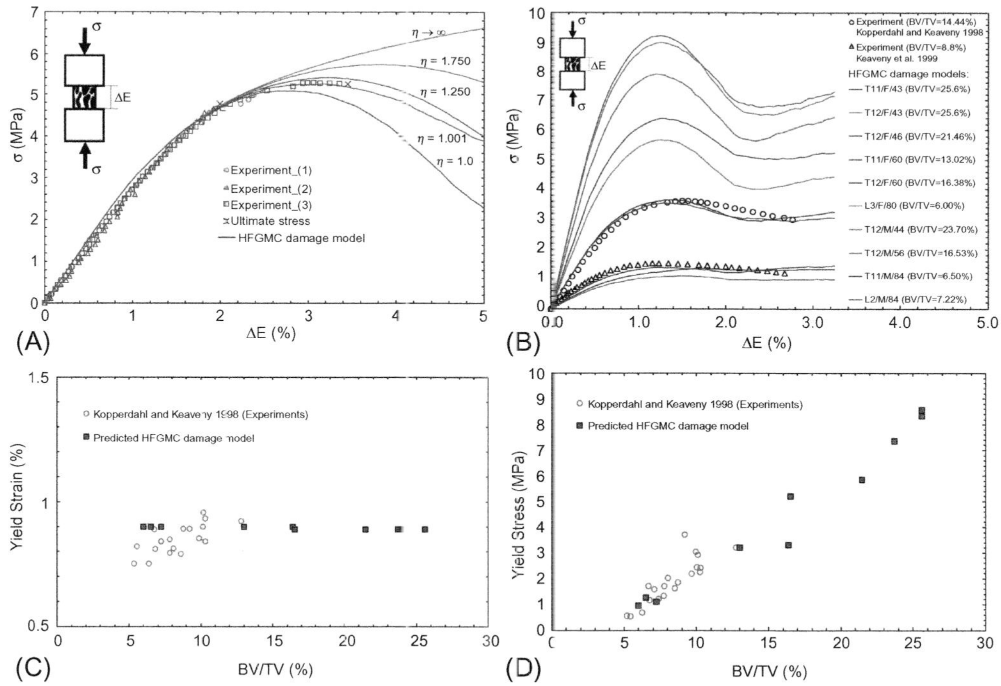

FIG. 6

Results from the nonlinear HFGMC analysis of the compression test in the form of stress versus compression strain of the printed polymeric VTB-like compression test. (A) The predicted nonlinear HFGMC response included post-peak strain-softening degradation behaviors. This degradation was related to a strain-softening parameter that controlled the energy dissipation rate. (B) Predicted HFGMC mechanical responses versus compression overloading tests of native VTBs available in the literature. (C) Predicted HFGMC yield strain versus bone volume fraction of native VTBs. (D) Predicted HFGMC yield stress versus bone volume fraction of native VTBs.

Reproduced from Massarwa, E., Aboudi, J., Galbusera, F., Wilke, H.-J., Haj-Ali, R., 2017. A nonlinear micromechanical model for progressive damage of vertebral trabecular bones. J. Mech. Mater. Struct. 12, 407–424 with permission.

Acknowledgments

This research was supported by the Israeli Ministry of Science, Technology and Space under Grant No. 3-12961. The last author gratefully acknowledges the support of the Nathan Cummings Chair of Mechanics.

REFERENCES

Boehm, H., Raeth, C., Monetti, R., Mueller, D., Newitt, D., Majumdar, S., Rummeny, E., Morfill, G., Link, T., 2003. Local 3D scaling properties for the analysis of trabecular bone extracted from high-resolution magnetic resonance imaging of human trabecular bone: comparison with bone mineral density in the prediction of biomechanical strength in vitro. Investig. Radiol. 38, 269–280.

Bogduk, N., 2005. Clinical Anatomy of the Lumbar Spine and Sacrum, 4th ed. Elsevier Health Sciences, Edinburgh, Churchill Livingstone.

Boutroy, S., Bouxsein, M.L. Munoz, F., Delmas, P.D., 2005. In vivo assessment of trabecular bone microarchitecture by high-resolution peripheral quantitative computed tomography. J. Clin. Endocrinol. Metab. 90, 6508–6515.

Burghardt, A.J., Link, T.M., Majumdar, S., 2011. High-resolution computed tomography for clinical imaging of bone microarchitecture. Clin. Orthop. Relat. Res. 469, 2179–2193.

Cendre, E., Mitton, D., Roux, J.P., Arlot, M.E., Duboeuf, F., Burt-Pichat, B., Rumelhart, C., Peix, G., Meunier, P.J., 1999. High-resolution computed tomography for architectural characterization of human lumbar cancellous bone: relationships with histomorphometry and biomechanics. Osteoporos. Int. 10, 353–360.

Chappard, D., Baslé, M.-F., Legrand, E., Audran, M., 2008. Trabecular bone microarchitecture: a review. Morphologie 92, 162–170.

Chen, H., Shoumura, S., Emura, S., Bunai, Y., 2008. Regional variations of vertebral trabecular bone microstructure with age and gender. Osteoporos. Int. 19, 1473–1483.

Colloca, M., Blanchard, R., Hellmich, C., Ito, K., van Rietbergen, B., 2014. A multiscale analytical approach for bone remodeling simulations: linking scales from collagen to trabeculae. Bone 64, 303–313.

Cowin, S.C., 1986. Wolff's law of trabecular architecture at remodeling equilibrium. J. Biomech. Eng. 108, 83–88.

Cowin, S., Hegedus, D., 1976. Bone remodeling I: theory of adaptive elasticity. J. Elast. 6, 313–326.

Cowin, S.C., Van Buskirk, W., 1979. Surface bone remodeling induced by a medullary pin. J. Biomech. 12, 269–276.

Currey, J.D., 1990. Biomechanics of mineralized skeletons. In: Skeletal Biomineralization: Patterns, Processes and Evolutionary Trends. Van Nostrand Reinhold, New York.

Daci, E., Everts, V., Torrekens, S., Van Herck, E., Tigchelaar-Gutterr, W., Bouillon, R., Carmeliet, G., 2003. Increased bone formation in mice lacking plasminogen activators. J. Bone Miner. Res. 18, 1167–1176.

Eliaz, N., Metoki, N., 2017. Calcium phosphate bioceramics: a review of their history, structure, properties, coating technologies and biomedical applications. Materials 10, 334.

Ford, C.M., Keaveny, T.M , 1996. The dependence of shear failure properties of trabecular bone on apparent density and trabecular orientation. J. Biomech. 29, 1309–1317.

Fratzl, P., Weinkamer, R., 2007. Nature's hierarchical materials. Prog. Mater. Sci. 52, 1263–1334.

Fratzl, P., Gupta, H., Paschalis, E., Roschger, P., 2004. Structure and mechanical quality of the collagen–mineral nano-composite in bone. J. Mater. Chem. 14, 2115–2123.

Fuller, H., Fuller, R., Pereira, R.M.R., 2015. High resolution peripheral quantitative computed tomography for the assessment of morphological and mechanical bone parameters. Rev. Bras. Reumatol. (English Edition) 55, 352–362.

Fyhrie, D.P., Schaffler, M.B., 1994. Failure mechanisms in human vertebral cancellous bone. Bone 15, 105–109.

Galante, J., Rostoker, W., Ray, R., 1970. Physical properties of trabecular bone. Calcif. Tissue Res. 5, 236–246.

Gibson, L.J., 1985. The mechanical-behavior of cancellous bone. J. Biomech. 18, 317.

Gibson, L.J., Ashby, M.F., 1999. Cellular Solids: Structure and Properties, second ed. Cambridge University Press, Cambridge, UK.

Gibson, L.J., Ashby, M.F., Harley, B.A., 2010. Cellular Materials in Nature and Medicine. Cambridge University Press, Cambridge, UK.

Goda, I., Ganghoffer, J.-F., 2015. 3D plastic collapse and brittle fracture surface models of trabecular bone from asymptotic homogenization method. Int. J. Eng. Sci. 87, 58–82.

Goda, I., Assidi, M., Belouettar, S., Ganghoffer, J.F., 2012. A micropolar anisotropic constitutive model of cancellous bone from discrete homogenization. J. Mech. Behav. Biomed. Mater. 16, 87–108.

Goda, I., Rahouadj, R., Ganghoffer, J.-F., Kerdjoudj, H., Siad, L., 2016. 3D couple-stress moduli of porous polymeric biomaterials using μCT image stack and FE characterization. Int. J. Eng. Sci. 100, 25–44.

Haj-Ali, R., Aboudi, J., 2009. Nonlinear micromechanical formulation of the high fidelity generalized method of cells. Int. J. Solids Struct. 46, 2577–2592.

Haj-Ali, R., Aboudi, J., 2013. A new and general formulation of the parametric HFGMC micromechanical method for two and three-dimensional multi-phase composites. Int. J. Solids Struct. 50, 907–919.

Haj-Ali, R., Aboudi, J., 2016. Integrated microplane model with the HFGMC micromechanics for nonlinear analysis of composite materials with evolving damage. Int. J. Solids Struct. 90, 129–143.

Haj-Ali, R., Massarwa, E., Aboudi, J., Galbusera, F., Wolfram, U., Wilke, H.-J., 2016. A new multiscale micromechanical model of vertebral trabecular bones. Biomech. Model. Mechanobiol. 1–14.

Hambli, R., 2013. Micro-CT finite element model and experimental validation of trabecular bone damage and fracture. Bone 56, 363–374.

Hamed, E., Jasiuk, I., 2013. Multiscale damage and strength of lamellar bone modeled by cohesive finite elements. J. Mech. Behav. Biomed. Mater. 28, 94–110.

Hamed, E., Jasiuk, I., Yoo, A., Lee, Y., Liszka, T., 2012. Multi-scale modelling of elastic moduli of trabecular bone. J. R. Soc. Interface 9, 1654–1673.

Hans, D., Wu, C., Njeh, C.F., Zhao, S., Augat, P., Newitt, D., Link, T., Lu, Y., Majumdar, S., Genant, H.K., 1999. Ultrasound velocity of trabecular cubes reflects mainly bone density and elasticity. Calcif. Tissue Int. 64, 18–23.

Harrigan, T.P., Jasty, M., Mann, R.W., Harris, W.H., 1988. Limitations of the continuum assumption in cancellous bone. J. Biomech. 21, 269–275.

Harrison, N.M., McDonnell, P., Mullins, L., Wilson, N., O'Mahoney, D., McHugh, P.E., 2013. Failure modelling of trabecular bone using a non-linear combined damage and fracture voxel finite element approach. Biomech. Model. Mechanobiol. 12, 225–241.

Hart, K., Shaw, J., Vajda, E., Hegsted, M., Miller, S.C., 2001. Swim-trained rats have greater bone mass, density, strength, and dynamics. J. Appl. Physiol. 91, 1663–1668.

Hayes, W.C., Carter, D.R., 1976. Postyield behavior of subchondral trabecular bone. J. Biomed. Mater. Res. 10, 537–544.

Helgason, B., Perilli, E., Schileo, E., Taddei, F., Brynjólfsson, S., Viceconti, M., 2008. Mathematical relationships between bone density and mechanical properties: a literature review. Clin. Biomech. 23, 135–146.

Hou, F.J., Lang, S.M., Hoshaw, S.J., Reimann, D.A., Fyhrie, D.P., 1998. Human vertebral body apparent and hard tissue stiffness. J. Biomech. 31, 1009–1015.

Huiskes, R., Weinans, H., Grootenboer, H., Dalstra, M., Fudala, B., Slooff, T., 1987. Adaptive bone-remodeling theory applied to prosthetic-design analysis. J. Biomech. 20, 1135–1150.

Jovanović, J.D., Jovanović, M.L., 2004. Biomechanical model of vertebra based on bone remodeling. La Medicina Biologica 11, 35–39.

Keaveny, T.M., Wachtel, E.F., Ford, C.M., Hayes, W.C., 1994a. Differences between the tensile and compressive strengths of bovine tibial trabecular bone depend on modulus. J. Biomech. 27, 1137–1146.

Keaveny, T.M., Wachtel, E.F., Guo, X.E., Hayes, W.C., 1994b. Mechanical behavior of damaged trabecular bone. J. Biomech. 27, 1309–1318.

Keaveny, T.M., Pinilla, T.P., Crawford, R.P., Kopperdahl, D.L., Lou, A., 1997. Systematic and random errors in compression testing of trabecular bone. J. Orthop. Res. 15, 101–110.

Keaveny, T.M., Wachtel, E.F., Kopperdahl, D.L., 1999. Mechanical behavior of human trabecular bone after overloading. J. Orthop. Res. 17, 346–353.

Keaveny, T.M., Morgan, E.F., Niebur, G.L., Yeh, O.C., 2001. Biomechanics of trabecular bone. Annu. Rev. Biomed. Eng. 3, 307–333.

Kopperdahl, D.L., Keaveny, T.M., 1998. Yield strain behavior of trabecular bone. J. Biomech. 31, 601–608.

Kopperdahl, D.L., Morgan, E.F., Keaveny, T.M., 2002. Quantitative computed tomography estimates of the mechanical properties of human vertebral trabecular bone. J. Orthop. Res. 20, 801–805.

Krug, R., Burghardt, A.J., Majumdar, S., Link, T.M., 2010. High-resolution imaging techniques for the assessment of osteoporosis. Radiol. Clin. N. Am. 48, 601–621.

Linde, F., Hvid, I., Madsen, F., 1992. The effect of specimen geometry on the mechanical-behavior of trabecular bone specimens. J. Biomech. 25, 359–368.

Maquer, G., Musy, S.N., Wandel, J., Gross, T., Zysset, P.K., 2015. Bone volume fraction and fabric anisotropy are better determinants of trabecular bone stiffness than other morphological variables. J. Bone Miner. Res. 30, 1000–1008.

Massarwa, E., Aboudi, J., Galbusera, F., Wilke, H.-J., Haj-Ali, R., 2017. A nonlinear micromechanical model for progressive damage of vertebral trabecular bones. J. Mech. Mater. Struct. 12, 407–424.

McDonald, K., Little, J., Pearcy, M., Adam, C., 2010. Development of a multi-scale finite element model of the osteoporotic lumbar vertebral body for the investigation of apparent level vertebra mechanics and micro-level trabecular mechanics. Med. Eng. Phys. 32, 653–661.

Morgan, E.F., Keaveny, T.M., 2001. Dependence of yield strain of human trabecular bone on anatomic site. J. Biomech. 34, 569–577.

Morgan, E.F., Yeh, O.C., Chang, W.C., Keaveny, T.W., 2001. Nonlinear behavior of trabecular bone at small strains. J. Biomech. Eng.-Trans. ASME 123, 1–9.

Mosekilde, L., Mosekilde, L., 1986. Normal vertebral body size and compressive strength: relations to age and to vertebral and iliac trabecular bone compressive strength. Bone 7, 207–212.

Mosekilde, L., Mosekilde, L., Danielsen, C., 1987. Biomechanical competence of vertebral trabecular bone in relation to ash density and age in normal individuals. Bone 8, 79–85.

Mullender, M., Huiskes, R., Weinans, H., 1994. A physiological approach to the simulation of bone remodeling as a self-organizational control process. J. Biomech. 27, 1389–1394.

Nazarian, A., von Stechow, D., Zurakowski, D., Muller, R., Snyder, B.D., 2008. Bone volume fraction explains the variation in strength and stiffness of cancellous bone affected by metastatic cancer and osteoporosis. Calcif. Tissue Int. 83, 368–379.

Odgaard, A., Kabel, J., van Rietbergen, B., Dalstra, M., Huiskes, R., 1997. Fabric and elastic principal directions of cancellous bone are closely related. J. Biomech. 30, 487–495.

Olszta, M.J., Cheng, X., Jee, S.S., Kumar, R., Kim, Y.-Y., Kaufman, M.J., Douglas, E.P., Gower, L.B., 2007. Bone structure and formation: a new perspective. Mater. Sci. Eng. R. Rep. 58, 77–116.

Parfitt, A.M., Drezner, M.K., Glorieux, F.H., Kanis, J.A., Malluche, H., Meunier, P.J., Ott, S.M., Recker, R.R., 1987. Bone histomorphometry: standardization of nomenclature, symbols, and units: report of the ASBMR Histomorphometry Nomenclature Committee. J Bone Miner. Res. 2, 595–610.

Pecknold, D.A., Haj-Ali, R., 1993. Integrated micromechanical/structural analysis of laminated composites. In: Hyer, M.W. (Ed.), Mechanics of Composite Materials—Nonlinear Effects. Proceedings of the 1st Joint Mechanics Meeting of SES/ASME/ASCE, June,Charlottesville, VA. AMD-vol. 159. pp. 197–206.

Peterson, D., Bronzino, J., 2008. Mechanics of hard tissue. In: Biomechanics: Principles and Applications. CRS Press, Boca Raton.

Rho, J.Y., Ashman, R.B., Turner, C.H., 1993. Youngs modulus of trabecular and cortical bone material—ultrasonic and microtensile measurements. J. Biomech. 26. 111–119.

Rho, J.Y., Tsui, T.Y., Pharr, G.M., 1997. Elastic properties of human cortical and trabecular lamellar bone measured by nanoindentation. Biomaterials 18, 1325–1330.

Rho, J.-Y., Kuhn-Spearing, L. Zioupos, P., 1998. Mechanical properties and the hierarchical structure of bone. Med. Eng. Phys. 20, 92–102.

Rockoff, S.D., Sweet, E., Bleustein, J., 1969. The relative contribution of trabecular and cortical bone to the strength of human lumbar vertebrae. Calcif. Tissue Int. 3, 163–175.

Schwiedrzik, J., Raghavan, R., Burki, A., LeNader, V., Wolfram, U., Michler, J., Zysset, P., 2014. In situ micropillar compression reveals superior strength and ductility but an absence of damage in lamellar bone. Nat. Mater. 13, 740–747.

Ulrich, D., Van Rietbergen, B., Laib, A., Ruegsegger, P., 1999. The ability of three-dimensional structural indices to reflect mechanical aspects of trabecular bone. Bone 25, 55–60.

van Rietbergen, B., Weinans, H., Huiskes, R., Odgaard, A., 1995. A new method to determine trabecular bone elastic properties and loading using micromechanical finite-element models. J. Biomech. 28, 69.

Weinans, H., Huiskes, R., Grootenboer, H., 1992. The behavior of adaptive bone-remodeling simulation models. J. Biomech. 25, 1425–1441.

Wolff, J., 1892. Das gesetz der transformation der knochen. A Hirshwald 1, 1–152.

Wolff, J., 1986. The Law of Bone Remodeling (Translation of the German 1892 edition). Springer, Berlin, Heildelberg, New York.

Wolfram, U., Wilke, H.J., Zysset, P.K., 2010. Valid mu finite element models of vertebral trabecular bone can be obtained using tissue properties measured with nanoindentation under wet conditions. J. Biomech. 43, 1731–1737.

Yang, G., Kabel, J., Van Rietbergen, B., Odgaard, A., Huiskes, R.I.K., Cown, S.C., 1998. The anisotropic Hooke's law for cancellous bone and wood. J. Elast. 53 (2), 125–146.

Yoon, H.S., Katz, J.L., 1976. Ultrasonic wave-propagation in human cortical bone .2. Measurements of elastic properties and microhardness. J. Biomech. 9, 459.

Zysset, P., 1994. A Constitutive Law for Trabecular Bone.

Zysset, P., Goulet, R., Hollister, S., 1998. A global relationship between trabecular bone morphology and homogenized elastic properties. Ann Arbor 1050, 43109.

Zysset, P.K., Guo, X.E., Hoffler, C.E., Moore, K.E., Goldstein, S.A., 1999. Elastic modulus and hardness of cortical and trabecular bone lamellae measured by nanoindentation in the human femur. J. Biomech. 32, 1005–1012.

FURTHER READING

Wegst, U.G., Bai, H., Saiz, E., Tomsia, A.P., Ritchie, R.O., 2015. Bioinspired structural materials. Nat. Mater. 14, 23–36.

CHAPTER

INTERVERTEBRAL DISC

6

Andrea Malandrino[a]
Institute for Bioengineering of Catalonia, Barcelona, Spain

INTRODUCTION

A uniquely large avascular organ in the human body, the intervertebral disc consists of two histologically different cartilaginous tissues (Fig. 1A and B): annulus fibrosus (AF) and nucleus pulposus (NP). In addition, a distinctive tissue called cartilage endplate (CEP) lies at the disc-vertebral bone interface and is often considered as a functional region of the disc. The annulus can be further subdivided into posterior (PAF) and anterior (AAF) (Fig. 1B) regions because of their often distinct biomechanical roles. Together, these components can be studied as a structural unit capable of supporting and distributing a significant load magnitude, while passively controlling the relative motion of the vertebrae. In healthy conditions, discs resist biomechanical loading of a predominantly compressive nature (Fig. 1D), especially in the lumbar portion of the human spine. The classically recognized view is that these forces produce an increase in hydrostatic pressure within the disc's central part—the gelatinous and highly hydrated NP—with this pressure balanced at the AF by tensile stresses retained in the specially organized collagen fibers, located within the concentric lamellae and coexisting with the radial lamellar cross bridge network (Fig. 1A and E).

This chapter focuses on the intervertebral disc structural biomechanics based on a cross-talk to both the disc's tissue compositional organization and the nutrient constraints for disc cellular environment. The disc structure and cellular maintenance are thought to be a prerequisite for many strategies for restoring spinal biomechanics and remedying back pain issues.

BIOLOGICAL AND MOLECULAR CHARACTERISTICS OF THE INTERVERTEBRAL DISC

CELLS OF THE INTERVERTEBRAL DISCS AND DEVELOPMENTAL ASPECTS

The IVD is populated by (i) chondrocytes in the CEPs, (ii) vacuolated notochordal cells in immature NP, and/or chondrocyte-like cells in the mature NP, and (iii) fibroblast-like cells in the AF (Fig. 2A). Cells of the AF and CEPs originate during development from the mesenchyme of the sclerotome, synthesizing throughout their lifecycle proteins like types I and II collagen and aggregating proteoglycans (Lawson and Harfe, 2017), characteristics of fibroblasts and chondrocytes. In human NPs, cells

[a]Present address: European Molecular Biology Laboratory (EMBL), Barcelona, Spain.

Biomechanics of the Spine. https://doi.org/10.1016/B978-0-12-812851-0.00006-9

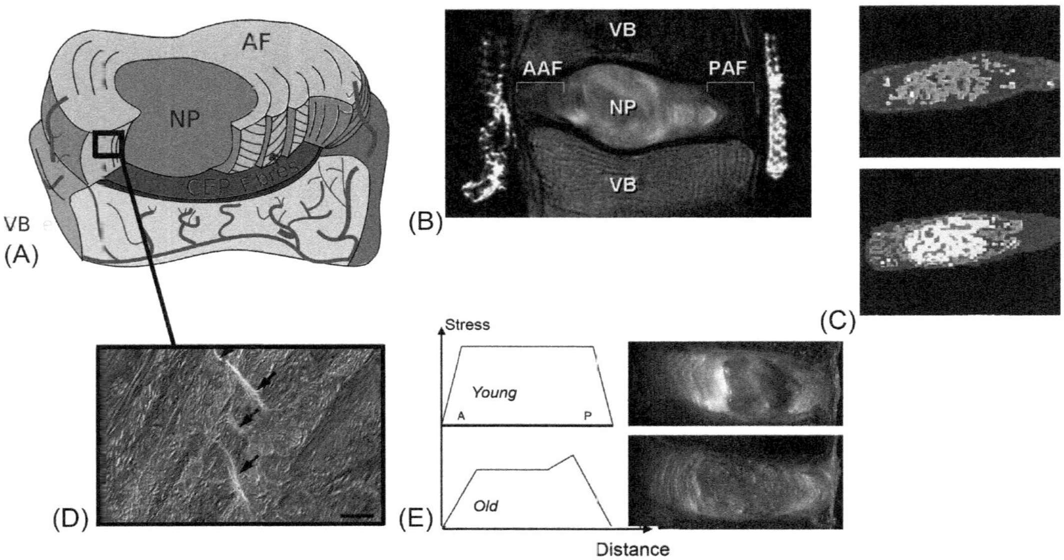

FIG. 1

The human intervertebral disc structural features. (A) Schematics of the disc tissue organization. In the representation, the disc is sectioned with the NP, AF, CEP, and vertebral bone (VB) visible. Collagen fibers organized in the concentric lamellae of the AF are shown, as well as an example of the surrounding vasculature. (B) Magnetic resonance image (MRI) of disc tissues. (C) Pixel maps of T2-relaxation time from MRI estimating water content (from *light green* and *blue* [low] to *red* and *dark green* [high]) computed for human lumbar intervertebral disc in the evening *(upper)* and morning *(lower)*. (D) Differential interference contrast (DIC) imaging of human annulus fibrosus radial sections (bars = 200 μm) showing the lamellar organization and trans-lamellar bridges (black arrows). (E) Photographs of mid-sagittal sections flanked with schemes of stress profiles for a human's young disc *(upper)* and old disc *(middle)*; P = posterior, A = anterior.

Part (B) adapted from Nerurkar, N.L., Elliott, D.M., Mauck, R.L., 2010. Mechanical design criteria for intervertebral disc tissue engineering. J. Biomech. 43, 1017–1030 with permissions; (C) adapted from Roberts, N., Hogg, D., Whitehouse, G.H., Dangerfield, P., 1998. Quantitative analysis of diurnal variation in volume and water content of lumbar intervertebral discs. Clin. Anat. 11, 1–8 with permissions; (D) Adapted from Smith, L.J., Elliott, D.M., 2011. Formation of lamellar cross bridges in the annulus fibrosus of the intervertebral disc is a consequence of vascular regression. Matrix Biol. 30, 267–274 with permissions. (E) Adapted from Adams, M.A., Dolan, P., McNally, D.S., 2009. The internal mechanical functioning of intervertebral discs and articular cartilage, and its relevance to matrix biology. Matrix Biol. 28, 384–389 with permissions.

are originally derived from the notochord, which is dominated by chondrocyte-like cells as the IVD matures. The adult NP of the IVD develops as hyaline cartilage-like tissue, with synthesis of collagen II, collagen IX, and aggrecans, typical signature of chondrocytes. The characteristic lamellar structure of the AF, with alternate orientations of parallel collagen bundles from one lamella to another, seems to correlate well with initial AF fibroblast elongation. Such elongation, present in the embryonic stage and removed shortly after birth, involves alignment of intracellular stress fibers—the contractile acto-myosin cell machinery—and deposition of oriented extracellular proteins, such as fibronectin and the characteristic oriented AF collagen fibers (Hayes et al., 1999).

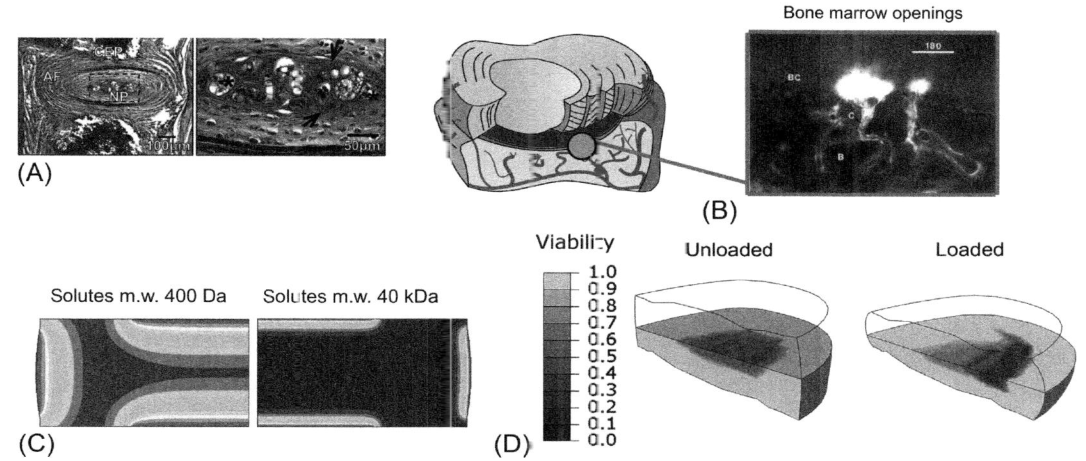

FIG. 2

Cellular and nutritional characteristics of the intervertebral disc. (A) Histological sections of mouse IVD, with red color in the NP proportional to proteoglycan content. In the right view, the arrows indicate aggregations of vacuolated notochord cells, and the asterisks indicate cartilage-like cells. (B) Details of the VB-CEP-NP interface in which staining solution was pumped for 15 min, demonstrating fluid flows from the marrow contact channels (labeled "C") in the VB (labeled "B") through the marrow openings at the bone-cartilage interface (labeled "BC"). Scale bar is 1 µm. (C) Computational model of solute diffusion after 24 h as a function of their molecular weight. Results are normalized to the concentration of the solute supply at the boundaries. (D) Computational model of solute diffusion coupled with mechanics and cell viability. Results after preconditioning for diffusion solver show a beneficial effect for cell viability in the central NP because of shortening of diffusion distances due to deformation, and a detrimental effect in the AF due to tissue compaction and drop in bulk diffusivity.

Part (A) adapted from McCann, M.R., Bacher, C.A., Séguin, C.A., 2011. Exploiting notochord cells for stem cell-based regeneration of the intervertebral disc, J. Cell Commun. Signal. 5, 39–43; (B) Modified from Ayotte, D.C., Ito, K., Tepic, S., 2001. Direction-dependent resistance to flow in the endplate of the intervertebral disc: an ex vivo study. J. Orthop. Res. 19, 1073–1077; (C) Modified from Ferguson, S.J., Ito, K., Nolte, L.-P., 2004. Fluid flow and convective transport of solutes within the intervertebral disc. J. Biomech. 37, 213–221; (D) Modified from Malandrino, A., Lacroix, D., Hellmich, C., Ito, K., Ferguson, S.J., Noailly, J., 2014a. The role of endplate poromechanical properties on the nutrient availability in the intervertebral disc. Osteoarthr. Cartil. 22, 1053–1060 and Malandrino, A., Noailly, J., Lacroix, D., 2014b. Numerical exploration of the combined effect of nutrient supply, tissue condition and deformation in the intervertebral disc. J. Biomech. 47, 1520–1525.

NUTRIENTS AND CELLULARITY

The intervertebral disc is the largest avascular tissue in the human body. According to the cartilaginous environment, the disc shows cell densities as low as ~4000 cells/mm^3 in the central IVD, increasing to ~9000 cells/mm^3 at the annulus edge (Maroudas et al., 1975). Essential cell nutrients such as oxygen and glucose are supplied by the blood capillaries that reach to the disc's periphery (Fig. 3): the external annulus edge and the thin CEP region. Nutrients released from these peripheral routes diffuse through the dense extracellular matrix of the disc and become available for cells, despite the large transport distance

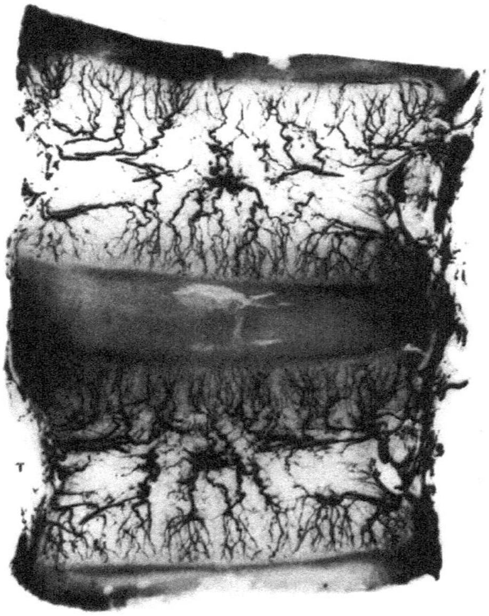

FIG. 3

Vascular supply in the lumbar disc. Photograph of a coronal section of a lumbar motion segment from a 65-year-old male, showing distribution of arteries and capillaries around the disc boundaries.

Reproduced from Crock, H. V, Yoshizawa, H., 1976. The blood supply of the lumbar vertebral column. Clin. Orthop. Relat. Res., 6–21 with permissions.

posed by the disc geometry. Conversely, metabolic waste molecules, such as lactate, are released by cells and cleared from the disc tissue to the periphery. The disc shows finely tuned pumping dynamics, with compressive forces and osmotic swelling inducing circadian variations in water content (Fig. 1C) driven by interstitial fluid flows that are direction-dependent thanks to the endplate's composition that allows passage of fluids through the contact openings at the bone-cartilage interface (Ayotte et al., 2001) (Fig. 2B). Nonetheless, computational modeling has confirmed that interstitial fluid flows have a negligible effect on transporting low molecular weight solutes—such as relevant IVD nutrients—in and waste products out of the disc (Malandrino et al., 2011; Urban et al., 2004), although the role of flows (advection) might become more important for higher molecular weight solutes (Ferguson et al., 2004), possibly enhancing the limited penetration of the latter within the disc (Fig. 2C). Loading and consequent large soft tissue strains also substantially modify the diffusion distances dynamics, providing an explanation for the mechanics-dependent diffusion of small molecules such as glucose (Malandrino et al., 2015a). Therefore, because of the load-driven changes in tissue compaction and bulk diffusivity, cell survival, proportional to the glucose availability, can be affected by a prolonged loading scenario (Fig. 2D). The nutritional demand of cells varies throughout the disc and depends locally on cell density and the local concentration. However, simple and universal rules seem to exist, as the cell density directly correlates to the nutritional landscape, the number of cells being higher close to the nutrient supply and steeply falling with distance. Also, average cell density through the nucleus is higher in small animals where diffusional distances are small and is exponentially lower when the disc height is greater.

Energy Metabolism Within the Disc

Carbohydrate assimilation in the intervertebral disc is dominated by the conversion of glucose to lactate via glycolytic pathways, a relatively fast process in air-saturated media, with a theoretical lactate/glucose mole ratio of 2:1. Because of this rapid process concomitant to low oxygen consumption (Bibby et al., 2001), it has been estimated that glycolysis is the main energy source in the form of adenosine triphosphate (ATP). Glycolysis generates 2 mol of ATP per mole of glucose and lactic acid as resulting waste metabolite:

$$\text{glucose} + 2\text{NAD}^+ + 2\text{ADP} + 2\text{Pi} \rightarrow 2\ \text{pyruvate} + 2\text{NADH} + 2\text{H}^+ + 2\text{ATP} + 2\text{H2O}$$

Here NAD^+ stands for nicotinamide adenine dinucleotide, a coenzyme found in all living cells that carries electrons from other molecules reducing to NADH. ADP is the adenosine diphosphate, product of ATP dephosphorylation. Pi is a phosphate group. If glycolysis were to continue indefinitely, all of the NAD^+ would be used up, and glycolysis would stop. To allow glycolysis to cycle, organisms must be able to oxidize NADH back to NAD^+. One way for this to happen is to have the pyruvate triggering such oxidation. The pyruvate is thus converted to lactate, the conjugate base of lactic acid, in a process called lactic acid fermentation:

$$\text{pyruvate} + \text{NADH} + \text{H}^+ \rightarrow \text{lactate} + \text{NAD}^+$$

Lactic acid fermentation happens in anaerobic conditions. However, the cellular aerobic respiration includes additional energy-conversion and metabolic pathways: Krebs cycle (or citric acid cycle) and oxidative phosphorylation, which happen following glycolysis. In oxidative phosphorylation, the electrons removed from molecules are transferred to oxygen, and the energy released is used to produce ATP much more efficiently than glycolysis alone can. Nonetheless, oxidative phosphorylation in the IVD appears to contribute up to only 15% of the total ATP formed (Bibby et al., 2001). In the IVD, very low oxygen concentrations result in low oxygen consumption, so the rate of glycolysis should accelerate to maintain the energy supply. This metabolic environment often appears to be inefficient because the dramatic oxygen deficiency has clear detrimental effects on disc cells, such as a marked decrease of proteoglycan and protein synthesis (Ishihara and Urban 1999).

FROM TISSUE COMPOSITION TO BIOMECHANICAL PROPERTIES

NUCLEUS PULPOSUS

In a healthy disc, NP gelatinous extracellular matter is mainly water, which comprises about 80% of the overall weight of the NP, with the remaining amount consisting of proteoglycans and collagen—primarily type II. Proteoglycans account for 65% of the healthy NP dry weight, and collagen accounts for 25% of the dry weight (Eyre, 1979). Proteoglycans are macromolecules assembled from negatively charged polysaccharide chains called glycosaminoglycans (GAGs), which possess a high molecular weight. Proteoglycans are involved in binding cations (e.g., sodium, potassium, calcium), thus retaining water through local osmotic effects (Fig. 4D). Because of the dependence on the net fixed-charge concentration in a given volume, osmotic effects in such volume are dynamic according to the tissue loading history (Malandrino et al., 2015a). Overall, the NP tissue organization provides fluidity and viscoelasticity to the IVD structure, acting as a biomechanical shock absorber and cushioning the spine against complex loading conditions.

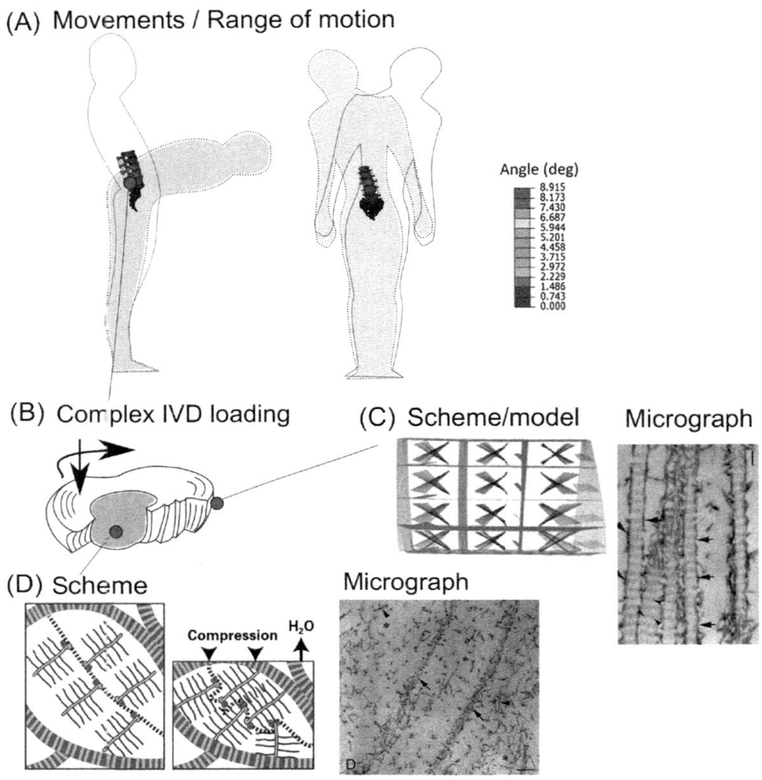

FIG. 4

Multiscale view of AF and NP tissue features in relation to spine biomechanics. (A) Schematic example of the entire body-level spinal movements (from left to right, flexion/extension, lateral bending) with angular displacements modified from Malandrino et al. (2015b). (B) At the IVD level, a drawing represents how spinal movements are translated to complex loading, such as combined compressive, shear, and torsional loads. At the tissue level, such IVD loads are resisted by specialized organization. (C) In the AF, with lamellar structure with crisscrossed patterns of collagen bundles (left scheme/model adapted from Noailly et al., 2011 with permissions). On the right, the electron micrograph of the AF of a rhesus monkey IVD, with longitudinally sectioned collagen fibrils with intrafibrillar precipitates (arrowheads) covered by proteoglycan precipitates (arrows). Bar = 100 nm, modified from Stoeckelhuber et al. (2005) with permissions. (D) In the NP, swelling pressure contribution to mechanical resistance is schematized on the left at the level of proteoglycan aggregates entrapped by collagen fibrils (reproduced from Sivan et al., 2014, with permissions): under compression, water is expelled and the aggrecan molecules are brought into closer proximity increasing the effective concentration of aggrecans at the site of compression, and consequently the swelling potential. In a relaxed state, the aggrecan expands, with water flowing back into the tissue, restoring the initial equilibrium. On the right, an electron micrograph of the central NP of a rhesus monkey IVD, with wide areas (asterisks), collagen fibrils (arrows), and PG precipitates (arrowheads). Bar = 200 nm. Modified from Stoeckelhuber et al. (2005) with permissions.

ANNULUS FIBROSUS

In healthy discs, there is a relatively clear transition between the NP and the AF. After a reduced zone of wispy fibrocartilage encompasses the NP, a compositional radial gradient of the annular ring from the NP to the external disc boundary can be observed, with a possible distinction between an inner AF and an outer AF. On average, the human AF consists of water (65% of the total weight), proteoglycans (20% of dry weight), and collagen (66% of dry weight) (Eyre, 1979). Type I and II collagen fibers (36% and 64% of total collagen content, respectively (Eyre, 1979)) are arranged primarily in concentric layers called lamellae; the direction of the fibers with respect to a circumferential plane alternates from one lamella at angles of ~30 degree to the successive lamella at angles of ~60 degree (Figs. 1 and 4C). At the periphery, some annular fibers extend beyond the CEP to enter the bone as Sharpey's fibers. The presence of translamellar elastin bridges has also been observed (Schollum et al., 2008) (Fig. 1D). Elastin bridges are thought to have a non-negligible mechanical role, although the elastin content in nondegenerate tissue, constant throughout the IVD, has been reported as low as 2% of the dry weight (Cloyd and Elliott, 2007). The anisotropic microstructure of the annulus fibrosus is certainly a key responsibility of the efficient IVD mechanical properties, which are dominated by the orientation of the collagen fibers embedded in the hydrated matrix of proteoglycans.

EXPERIMENTAL MECHANICS OF THE INTERVERTEBRAL DISC

Like all tissue in the human body, a highly hydrated soft matter comprises the IVD, the percentage of fluid phase being particularly high as with other fibrocartilagineous tissues. Thus the fluid phase drives the mechanical properties of healthy intervertebral discs, and degradation of fluid-related properties is regarded as fundamental in disc pathophysiology (Malandrino et al., 2015a). Grouping all remaining non-fluid proteins in the disc as solid phase, one can theoretically analyze and experimentally characterize the mechanics of the disc soft matter as biphasic.

A fundamental facet of the biphasic nature of the IVD is that the mechanical response to compressive stress involves a controlled exudation of water and generates relatively high, hydrostatic-like intradiscal pressures. Nachemson and coworkers recognized this aspect early on and, using a pressure-sensitive needle, developed a technique to measure the pressures experienced within the IVD. In a series of papers, they reported the lumbar intradiscal pressures resulting from external loads on the disc, both in vitro (Nachemson, 1960) and in vivo (Nachemson and Morris, 1964). They showed that, except for severely degenerated tissues, the stress in the central portion of the disc is hydrostatic and the pressure is almost always 1.3 to 1.5 times the applied pressures. Wilke et al. (1999) later found positive correlation with Nachemson's data; however, in these later studies, the intradiscal pressure was lower when a subject was sitting than when standing erect, which could be due to a single-subject study design or to the use of different transducers. These results, when seen in the context of biphasic material characterization, were used to argue that the nucleus pulposus could be treated theoretically as a fluid experiencing hydrostatic pressures.

Further insight on the biphasic characterization of IVD subtissue was provided by the stress profilometry technique, developed by McNally and Adams (1992). The technique consists of pulling a pressure transducer through disc specimens along the mid-sagittal diameter while the specimens are subjected to sustained mechanical loads. Verification of the technique (McNally et al., 1992) showed that the values of compressive "stress" measured corresponded approximately to the average axial compressive stress acting perpendicular to the transducer membrane. Depending on the transducer orientation, both superior-inferior (i.e., crano-caudal) and transversal (to the IVD) stress measurements could be measured. Interestingly, results revealed that the stress in the central region did not vary with either location or transducer orientation, suggesting again that the NP behaves like a pressurized fluid. Fluid-like behavior was also observed by profiling the inner and middle AF in young subjects, whereas steep gradients of compressive stress were measured in the outer AF (Fig. 1E). In degenerated subjects, the stress profilometry technique has shown exaggerated changes: NP was decompressed and large stress concentrations appeared in the AF, usually posterior to the NP (Adams et al., 2009). It is thus feasible that a distinction in the mechanical behavior of the subtissue within the IVD will depend on the amount and relative proportions of biochemical constituents, such as water, proteoglycans, and collagen, which vary spatially (NP, AF, CEP) and temporally (with age and degeneration).

A significant amount of experimental work and other analyses based on the biphasic mixture theory has been done. Such work depicted biphasic mechanical properties as a function of the distinct IVD tissues and studied the relative changes with age or degeneration. For instance, Best and coworkers analyzed the compressive properties (compressive modulus, hydraulic permeability, swelling pressure) of human AF through the linear biphasic theory and correlated regional variations within the AF subtissue with the local biochemical composition (Best et al., 1994). Similar techniques focused later on the fluid kinetics and hydraulic permeability (Gu et al., 1999). Iatridis and coworkers in 1996 studied experimentally the human NP and stated that the nucleus exhibited the properties of both solid and fluid material phases (Iatridis et al., 1996). They found that the NP is relatively viscous and capable of sustaining significant shear stresses, suggesting that the assumption of the nucleus pulposus as an incompressible and inviscid fluid in models might be inappropriate, proposing instead a biphasic material description. Périé and coworkers investigated both linear and nonlinear biphasic properties of both the AF and the NP, suggesting the use of confined swelling pretest conditions and a nonlinear biphasic model as a proper combination to minimize the relative variance in the results under water contents representative of physiological conditions (Périé et al., 2005).

The emerging picture is that fluid-phase kinetics play a key role in the form, the function, and the pathologies of IVDs. Looking at circadian rhythms, around 25% of the disc fluid is daily expelled and re-imbibed, causing a loss and regain of around 1–2 cm in height (Tyrrell et al., 1985) (see also Fig. 1C). Loss of water and progressive solid-phase disruption, as happens in degenerative disc disease, would thus alter the biomechanics of the intervertebral disc (Iatridis et al., 1998; Johannessen and Elliott, 2005). Biphasic and poroelasticity constitutive material theories, mathematically equivalent under certain assumptions, have been used to analyze the fluid flow paths and the hydrostatic pressure distributions within the disc to draw valuable clinical conclusions (Argoubi and Shirazi-Adl, 1996; Ferguson et al., 2004; Malandrino et al., 2015a) for the treatment of disc degeneration, including the design of new regenerative strategies (Gan et al., 2000; Iatridis, 2009; Revell et al., 2007). Finite element analyses have been performed to determine the most sensitive structural parameters (Malandrino et al., 2009). Including fluid flow calculations in these studies is relevant for the biomechanics of spinal segments. For instance, an abundance of published literature, partially validated by experimental data, exists on poroelastic models of lumbar motion segments (or only IVD) (Ferguson et al., 2004; Argoubi and Shirazi-Adl, 1996; Laible et al., 1993; Williams et al., 2007; Ruiz et al., 2013; Wills et al., 2016). These models clearly show the link between fluid flow and mechanical response within spinal segments for clinically relevant investigations.

BIOMECHANICAL RELEVANCE OF DISTINCT PHYSICOCHEMICAL COMPOSITIONS OF THE IVD TISSUES

Swelling Pressure in the Nucleus

Because of a high concentration of fixed charges, the osmotic pressure largely governs the compressive behavior of the NP. The loss of osmotic pressure has clinically relevant consequences, such as a negative effect on the activity of disc cells (Neidlinger-Wilke et al., 2009) or on disc herniation (Wognum et al., 2006).

Biphasic experiments have measured swelling pressures in the IVD—defined as the external pressure at which the tissue neither loses nor gains water—ranging from 0.1 MPa to 0.25 MPa, with lower values in the AF and higher values in the NP (Best et al., 1994; Iatridis et al., 1998; Périé et al., 2005). To incorporate osmotic phenomena, triphasic and quadriphasic mathematical models have been implemented by including the diffusion of ions trough the interstitial fluid and the interaction with the fixed charges. Simpler approaches exist to partially simulate the swelling phenomenon in structural analyses with commercial finite element software, for instance by introducing a fixed pore pressure at the external boundaries of the disc (Argoubi and Shirazi-Adl, 1996; Ferguson et al., 2004) or prescribing the osmotic pressure as a fixed contribution to the spherical part of the stress tensor (Olsen and Oloyede, 2002).

Incorporating osmotic effects through rather simplified models offers an opportunity to obtain fast and reliable solutions for complex FE simulations in spine biomechanics. Under this line, the different swelling algorithms to simulate osmotic pressure in IVD subtissues were assessed by Galbusera and coworkers. They demonstrated that results predicted with the fixed pore pressure boundary and the fixed osmotic pressure in the whole disc models were nearly identical, with the only difference being that the boundary pore pressure model could not simulate differential osmotic pressures in disc regions (Galbusera et al., 2011). The swelling model offered the most potential to provide accurate results, provided reliable values for the required coefficients and material properties are available. However, the other approaches were a good compromise between ease of implementation and reliability of results, especially when considering higher loads or when the focus is on global results such as spinal kinematics.

AF Collagen Fibers

A challenge for structural models of the AF is to capture the intricate structure of the tissue and relate this structure to the specific AF mechanical function. The concentric fiber-reinforced lamellae that build up the AF generate a strong anisotropy because of specific collagen fiber orientation patterns. At the same time, the annulus presents significant regional differences in mechanical behavior. For instance, a study showed that under a moderate circumferential strain of 5% to 6%, specimens in the anterior outer region of the AF experienced circumferential stresses about 10 times higher than those occurring in posterior inner specimens (Ebara et al., 1996). Collagen fibers, embedded in a crisscross pattern within the AF matrix are essential to mechanically stabilize the IVD and tune the functional biomechanics of the intersegmental joints. Interestingly, while this crisscross pattern seems common to human AF, particular regional variations in AF fiber crisscross angles have been anatomically observed (Cassidy et al., 1989; Holzapfel et al., 2005; Zhu et al., 2008). Such particularity could be further related consistently with the regional-dependent functional biomechanics of the AF tissue (Noailly et al., 2011). As a result of several fiber angle distributions in the literature, different fiber patterns could be used as input for models. For example, following the description from Marchand and Ahmed (Marchand and Ahmed, 1990), some studies did not model any particular changes of fiber angle throughout the AF, taking an average crisscross angle value of ±30 degree with respect to the AF circumferential direction (Argoubi and Shirazi-Adl, 1996). According to the findings of Cassidy et al. (1989), more transversal fibers in the outer AF than in the inner AF have been introduced in AF models (see Noailly et al., 2007, for instance). Eberlein et al. (2001) reported a model where experimentally based fiber angle distributions were in perfect agreement with the micro-dissection results further described by Holzapfel et al. (2005), who reported an increasing fiber angle with respect to the AF transverse plane, in the tangential direction, from the anterior to the posterior annulus. More recently, Zhu et al. (2008) investigated the lumbosacral AF microstructure and reported both radial and tangential crisscross angle changes. A circumferential and radial fiber angle variation was introduced in Schmidt et al. (2006) in a FE calibration study that considered comparisons against in vitro tests on motion segments. However, calibrated parameters involved stiffness of the AF fibers and of the ground substance, whereas fiber angle values were unvaried. Also, such a bundle configuration was qualitatively studied and led to biomechanical predictions different from those given by other AF collagen network definitions (Noailly et al., 2011). In most structural models of the AF, the collagen network is simulated by using fiber-reinforced finite elements, as well as through continuum constitutive theories that can take into account fiber orientations and fiber stiffness-related parameters. Using a continuum approach, it was shown how, by varying the fiber stiffness-related parameters, two different mechanical behaviors for posterior and anterior AF regions can be reproduced (Eberlein et al., 2001). The regional variations in fiber angles were studied only circumferentially, and the experimental stress-strain responses were calibrated by varying the fiber stiffness parameters. Moreover, anatomical data on radial gradients in AF collagen network could be important for the mechanical stabilization of the lumbar segments under specific motions (Noailly et al., 2011). Finally, a model calibration procedure through the adjustment of fiber orientations in four discretely different AF regions was proposed (Malandrino et al., 2013). For this, a continuum anisotropic poro-hyperelastic approach was chosen and regional AF stress-strain measurements were reproduced by locally calibrating fiber orientation and stiffness. Calibration results in terms of fiber angles were finally implemented in a complete FE IVD model for which the predicted behavior was compared to experimental data. It was concluded that annulus fiber variations (both radial and circumferential) might be an effective tool to calibrate lumbar spine IVD and segment models.

FUTURE DIRECTIONS AND CONCLUSIVE REMARKS

COUPLING NUTRITION AND COMPOSITION TO BIOMECHANICS

The Role of Fluid Phase

In the present chapter, the role of the finely tuned composition of the soft IVD material was presented as it relates to physical and chemical characteristics and, in turn, their relationship to the material's structural function. Tissue in the human body is constantly renewed and maintained by the resident cells, and nutritional aspects that drive structure and functioning of the biological disc machinery were highlighted. One relevant aspect that has emerged is the hydrated nature of the healthy disc material and all intertwined biophysical properties. By using poroelasticity and biphasic theories, fluid and solid phases within the IVD have been coupled. The importance of hydraulic effects for the local biomechanics as well as for the transport phenomena within the intervertebral disc has become clear. For instance, the hydraulic permeability of both the AF and the CEP associated with the fluid exchanges at the disc boundaries were shown to influence the local and global biomechanics of the disc. Many studies have highlighted that the CEP route is the most relevant hydraulic permeability involved, confirming its role for proper disc mechanics (Ayotte et al., 2001; Roberts et al., 1989; Urban et al., 2004). For instance, the local fluid dynamics at the bony part of the endplate was studied via human vertebral bone microtomographic images confirming that the permeability related to the vertebral bony endplate is much higher than the permeability of the cartilaginous endplate, also for very low bone porosities, which again points out the importance of the cartilage at the vertebra-disc interface (Malandrino et al., 2014a, b). Therefore pathologies related to the CEP, such as calcification could be important processes responsible for altering the exchange of nutrients with the disc boundaries, by acting on the fluid content and thus reducing the diffusion of metabolites.

The Effective Stiffness of the Disc Material and Its Maintenance

When analyzing the biomechanics of spinal segment one fundamental parameter is the effective stiffness of the disc as a whole (Yang et al., 2016). Therefore we reviewed the contribution of the distinct tissues' stiffness in the IVD. Many parametric studies have highlighted that the stiffness of the AF is the major contributor to an effective intervertebral modulus (Cortes et al., 2013; Malandrino et al., 2009; Ng et al., 2004; Rao and Dumas, 1991). The AF stiffness in turn highly depends on the oriented network of collagen fibers. Collagen fibers' orientations are thus being targeted in many structural studies (Malandrino et al., 2013; Marchand and Ahmed, 1990; Mengoni et al., 2017; Noailly et al., 2011). Other determinants of the IVDs' effective mechanical resistance are the disc shape, the fluid content, and the loss of swelling pressure (Malandrino et al., 2011; Noailly et al., 2007; Schroeder et al., 2006). Interestingly, these factors, together with the cell density, were also found to be relevant to the diffusion of solutes in a multiscale modeling study that coupled the transport of relevant biomolecule with the local continuum poromechanics (Malandrino et al., 2011). Thus a feedback loop of composition, cell nutrition, transport of metabolites, and biochemical properties must be considered (Barthelemy et al., 2016; Malandrino et al., 2015b; Travascio et al., 2014; Wills et al., 2016), a task that requires advanced modeling strategies for complex biophysical couplings. These models ultimately allow relating the local changes in metabolites concentrations with the application of external forces in the spine, for instance relating it to muscle activity and daily tasks (Toumanidou and Noailly, 2015), an ambitious aim in the field of spine biomechanics and mechanobiology.

Toward a Multiscale and Multiphysics Comprehension of the Disc Biomechanics

In this chapter, we discussed the intricate connections between disc biology, composition, biomechanics, and transport of important solutes related to the disc cell pathways. Studies pursuing this systems' biology-integrated vision will open new strategies for designing analytical procedures to associate the local disc mechanics with the cellular environment. These studies can finally relate spine mechanics to the stimuli sensed by the avascular disc cells (e.g., via solid deformation and fluid shear stresses) and open an unprecedented path in the field of mechanobiological models of soft tissues. These mechanochemical models are being combined with medical images and applied to patient-specific geometries and image-derived fluid content and diffusivities to gain a real-life understanding of the mechanical and geometrical effects on metabolic transport and to support clinical decision-making processes related to disc pathologies and regenerative strategies.

REFERENCES

Adams, M.A., Dolan, P., McNally, D.S., 2009. The internal mechanical functioning of intervertebral discs and articular cartilage, and its relevance to matrix biology. Matrix Biol. 28, 384–389. https://doi.org/10.1016/j.matbio.2009.06.004.

Argoubi, M., Shirazi-Adl, A., 1996. Poroelastic creep response analysis of a lumbar motion segment in compression. J. Biomech. 29, 1331–1339.

Ayotte, D.C., Ito, K., Tepic, S., 2001. Direction-dependent resistance to flow in the endplate of the intervertebral disc: an ex vivo study. J. Orthop. Res. 19, 1073–1077. https://doi.org/10.1016/S0736-0266(01)00038-9.

Barthelemy, V.M.P., van Rijsbergen, M.M., Wilson, W., Huyghe, J.M., van Rietbergen, B., Ito, K., 2016. A computational spinal motion segment model incorporating a matrix composition-based model of the intervertebral disc. J. Mech. Behav. Biomed. Mater. 54, 194–204. https://doi.org/10.1016/j.jmbbm.2015.09.028.

Best, B.A., Guilak, F., Setton, L.A., Zhu, W., Saed-Nejad, F., Ratcliffe, A., Weidenbaum, M., Mow, V.C., 1994. Compressive mechanical properties of the human anulus fibrosus and their relationship to biochemical composition. Spine (Phila Pa 1976) 19, 212–221.

Bibby, S.R.S., Jones, D.A., Lee, R.B., Yu, J., Urban, J.P.G., 2001. Pathophysiology of the intervertebral disc. Joint Bone Spine 68, 537–542. https://doi.org/10.1002/jmri.20874.

Cassidy, J.J., Hiltner, A., Baer, E., 1989. Hierarchical structure of the intervertebral disc. Connect. Tissue Res. 23, 75–88. https://doi.org/10.3109/03008208909103905.

Cloyd, J.M., Elliott, D.M., 2007. Elastin content correlates with human disc degeneration in the anulus fibrosus and nucleus pulposus. Spine (Phila Pa 1976) 32, 1826–1831. https://doi.org/10.1097/BRS.0b013e3181132a9d.

Cortes, D.H., Jacobs, N.T., Delucca, J.F., Elliott, D.M., 2013. Elastic, permeability and swelling properties of human intervertebral disc tissues: a benchmark for tissue engineering. J. Biomech. 1–7, https://doi.org/10.1016/j.jbiomech.2013.12.021.

Ebara, S., Iatridis, J.C., Setton, L.A., Foster, R.J., Mow, V.C., Weidenbaum, M., 1996. Tensile properties of nondegenerate human lumbar anulus fibrosus. Spine (Phila Pa 1976) 21, 452–461.

Eberlein, R., Holzapfel, G.A., Schulze-Bauer, C., 2001. An anisotropic model for annulus tissue and enhanced finite element analyses of intact lumbar disc bodies. Comput. Methods Biomech. Biomed. Engin. 4, 209–229. https://doi.org/10.1080/10255840108908005.

Eyre, D.R., 1979. Biochemistry of the intervertebral disc. Int. Rev. Connect. Tissue Res. 8, 227–291.

Ferguson, S.J., Ito, K., Nolte, L.-P., 2004. Fluid flow and convective transport of solutes within the intervertebral disc. J. Biomech. 37, 213–221. https://doi.org/10.1016/S0021-9290(03)00250-1.

Galbusera, F., Schmidt, H., Noailly, J., Malandrino, A., Lacroix, D., Wilke, H.J., Shirazi-Adl, A., 2011. Comparison of four methods to simulate swelling in poroelastic finite element models of intervertebral discs. J. Mech. Behav. Biomed. Mater. 4, 1234–1241. https://doi.org/10.1016/j.jmbbm.2011.04.008.

Gan, J.C., Ducheyne, P., Vresilovic, E., Shapiro, I.M., 2000. Bioactive glass serves as a substrate for maintenance of phenotype of nucleus pulposus cells of the intervertebral disc. J. Biomed. Mater. Res. 51, 596–604.

Gu, W.Y., Mao, X.G., Foster, R.J., Weidenbaum, M., Mow, V.C., Rawlins, B.A., 1999. The anisotropic hydraulic permeability of human lumbar anulus fibrosus. Influence of age, degeneration, direction, and water content. Spine (Phila Pa 1976) 24, 2449–2455.

Hayes, A.J., Benjamin, M., Ralphs, J.R., 1999. Role of actin stress fibres in the development of the intervertebral disc: cytoskeletal control of extracellular matrix assembly. Dev. Dyn. 215, 179–189. https://doi.org/10.1002/(SICI)1097-0177(199907)215:3<179::AID-AJA1>3.0.CO;2-Q.

Holzapfel, G.A., Schulze-Bauer, C.A.J., Feigl, G., Regitnig, P., 2005. Single lamellar mechanics of the human lumbar anulus fibrosus. Biomech. Model. Mechanobiol. 3, 125–140. https://doi.org/10.1007/s10237-004-0053-8.

Iatridis, J.C., 2009. Tissue engineering: function follows form. Nat. Mater. 8, 923–924. https://doi.org/10.1038/nmat2577.

Iatridis, J.C., Weidenbaum, M., Setton, L.A., Mow, V.C. 1996. Is the nucleus pulposus a solid or a fluid? Mechanical behaviors of the nucleus pulposus of the human intervertebral disc. Spine (Phila Pa 1976) 21, 1174–1184.

Iatridis, J.C., Setton, L.A., Foster, R.J., Rawlins, B.A., Weidenbaum, M., Mow, V.C., 1998. Degeneration affects the anisotropic and nonlinear behaviors of human anulus fibrosus in compression. J. Biomech. 31, 535–544.

Ishihara, H., Urban, J.P., 1999. Effects of low oxygen concentrations and metabolic inhibitors on proteoglycan and protein synthesis rates in the intervertebral disc. J. Orthop. Res. 17, 829–835.

Johannessen, W., Elliott, D.M., 2005. Effects of degeneration on the biphasic material properties of human nucleus pulposus in confined compression. Spine (Phila Pa 1976) 30, E724–9.

Laible, J.P., Pflaster, D.S., Krag, M.H., Simon, B.R., Haugh, L.D., 1993. A poroelastic-swelling finite element model with application to the intervertebral disc. Spine (Phila Pa 1976) 18, 659–670.

Lawson, L.Y., Harfe, B.D., 2017. Developmental mechanisms of intervertebral disc and vertebral column formation. Wires Dev Biol 1–14. https://doi.org/10.1002/wdev.283.

Malandrino, A., Planell, J.A., Lacroix, D., 2009. Statistical factorial analysis on the poroelastic material properties sensitivity of the lumbar intervertebral disc under compression, flexion and axial rotation. J. Biomech. 42, 2780–2788. https://doi.org/10.1016/j.jbiomech.2009.07.039.

Malandrino, A., Noailly, J., Lacroix, D., 2011. The effect of sustained compression on oxygen metabolic transport in the intervertebral disc decreases with degenerative changes. PLoS Comput. Biol. 7, e1002112https://doi.org/10.1371/journal.pcbi.1002112.

Malandrino, A., Noailly, J., Lacroix, D., 2013. Regional annulus fibre orientations used as a tool for the calibration of lumbar intervertebral disc finite element models. Comput. Methods Biomech. Biomed. Engin. 5842, 37–41. https://doi.org/10.1080/10255842.2011.644539.

Malandrino, A., Lacroix, D., Hellmich, C., Ito, K., Ferguson, S.J., Noailly, J., 2014a. The role of endplate poromechanical properties on the nutrient availability in the intervertebral disc. Osteoarthr. Cartil. 22, 1053–1060. https://doi.org/10.1016/j.joca.2014.05.005.

Malandrino, A., Noailly, J., Lacroix, D., 2014b. Numerical exploration of the combined effect of nutrient supply, tissue condition and deformation in the intervertebral disc. J. Biomech. 47, 1520–1525. https://doi.org/10.1016/j.jbiomech.2014.02.004.

Malandrino, A., Jackson, A R., Huyghe, J.M., Noailly, J., 2015a. Poroelastic modeling of the intervertebral disc: a path toward integrated studies of tissue biophysics and organ degeneration. MRS Bull. 40, 324–332. https://doi.org/10.1557/mrs.2015.58.

Malandrino, A., Pozo, J.M., Castro-Mateos, I., Frangi, A.F., van Rijsbergen, M.M., Ito, K., Wilke, H.-J., Dao, T.T., Ho Ba Tho, M.-C., Noailly, J., 2015b. On the relative relevance of subject-specific geometries and degeneration-specific mechanical properties for the study of cell death in human intervertebral disk models. Front. Bioeng. Biotechnol. 3, 1–15. https://doi.org/10.3389/fbioe.2015.00005.

Marchand, F., Ahmed, A.M., 1990. Investigation of the laminate structure of lumbar disc anulus fibrosus. Spine (Phila Pa 1976) 15, 402–410.

Maroudas, A., Stockwell, R.A , Nachemson, A., Urban, J., 1975. Factors involved in the nutrition of the human lumbar intervertebral disc: cellularity and diffusion of glucose in vitro. J. Anat. 120, 113–130.

McNally, D.S., Adams, M., 1992. Internal intervertebral disc mechanics as revealed by stress profilometry. Spine (Phila Pa 1976) 17, 66–73.

McNally, D.S., Adams, M.A., Goodship, A E., 1992. Development and validation of a new transducer for intradiscal pressure measurement. J. Biomed. Eng. 14, 495–498.

Mengoni, M., Kayode, O., Sikora, S.N.F., Zapata-Cornelio, F.Y., Gregory, D.E., Wilcox, R.K., 2017. Annulus fibrosus functional extrafibrillar and fibrous mechanical behaviour: experimental and computational characterisation. R. Soc Open Sci. 4, .

Nachemson, A., 1960. Lumbar intradiscal pressure Experimental studies on post-mortem material. Acta Orthop. Scand. Suppl. 43, 1–104.

Nachemson, A., Morris, J., 1964. In vivo measurements of Intradiscal pressure. J. Bone Joint Surg. 46, 1077–1092.

Neidlinger-Wilke, C., Liedert, A., Wuertz, K., Buser, Z.C., 2009. Mechanical stimulation alters pleiotrophin and aggrecan expression by human intervertebral disc cells and influences their capacity to stimulate endothelial cell. Spine (Phila Pa 1976) 34, 663–669.

Ng, H.W., Teo, E.C., Lee, V.S., 2004. Statistical factorial analysis on the material property sensitivity of the mechanical responses of the C4-C6 under compression, anterior and posterior shear. J. Biomech. 37, 771–777. https://doi.org/10.1016/j.jbiomech.2003.09.025.

Noailly, J., Wilke, H.-J., Planell, J.A., Lacroix, D., 2007. How does the geometry affect the internal biomechanics of a lumbar spine bi-segment finite element model? Consequences on the validation process. J. Biomech. 40, 2414–2425. https://doi.org/10.1016/j.jbiomech.2006.11.021.

Noailly, J., Planell, J.A., Lacroix, D., 2011. On the collagen criss-cross angles in the annuli fibrosi of lumbar spine finite element models. Biomech. Model. Mechanobiol. 10, 203–219. https://doi.org/10.1007/s10237-010-0227-5.

Olsen, S., Oloyede, A., 2002. A finite element analysis methodology for representing the articular cartilage functional structure. Comput. Methods Biomech. Biomed. Engin. 5, 377–386.

Périé, D., Korda, D., Iatridis, J.C., 2005. Confined compression experiments on bovine nucleus pulposus and annulus fibrosus: sensitivity of the experiment in the determination of compressive modulus and hydraulic permeability. J. Biomech. 38, 2164–2171. https://doi.org/10.1016/j.jbiomech.2004.10.002.

Rao, A.A., Dumas, G.A., 1991. Influence of material properties on the mechanical behaviour of the L5-S1 intervertebral disc in compression: a nonlinear finite element study. J. Biomed. Eng. 13, 139–151.

Revell, P.A., Damien, E., Di Silvio, L., Gurav, N., Longinotti, C., Ambrosio, L., 2007. Tissue engineered intervertebral disc repair in the pig using injectable polymers. J. Mater. Sci. Mater. Med. 18, 303–308. https://doi.org/10.1007/s10856-006-0693-6.

Roberts, S., Menage, J., Urban, J.P., 1989. Biochemical and structural properties of the cartilage end-plate and its relation to the intervertebral disc. Spine (Phila Pa 1976) 14, 166–174.

Ruiz, C., Noailly, J., Lacroix, D., 2013. Material property discontinuities in intervertebral disc porohyperelastic finite element models generate numerical instabilities due to volumetric strain variations. J. Mech. Behav. Biomed. Mater. 26, 1–10.

Schmidt, H., Heuer, F., Simon, U., Kettler, A., Rohlmann, A., Claes, L., Wilke, H.-J., 2006. Application of a new calibration method for a three-dimensional finite element model of a human lumbar annulus fibrosus. Clin. Biomech. (Bristol, Avon) 21, 337–344. https://doi.org/10.1016/j.clinbiomech.2005.12.001.

Schollum, M.L., Robertson, P.A., Broom, N.D., 2008. ISSLS prize winner: microstructure and mechanical disruption of the lumbar disc annulus: part I: a microscopic investigation of the translamellar bridging network. Spine (Phila Pa 1976) 33, 2702–2710. https://doi.org/10.1097/BRS.0b013e31817bb92c.

Schroeder, Y., Wilson, W., Huyghe, J.M., Baaijens, F.P.T., 2006. Osmoviscoelastic finite element model of the intervertebral disc. Eur. Spine J. 15 (Suppl. 3), S361–71. https://doi.org/10.1007/s00586-006-0110-3.

Sivan, S.S., Wachtel, E., Roughley, P., 2014. Structure, function, aging and turnover of aggrecan in the intervertebral disc. Biochim. Biophys. Acta https://doi.org/10.1016/j.bbagen.2014.07.013.

Stoeckelhuber, M., Brueckner, S., Spohr, G., Welsch, U., 2005. Proteoglycans and collagen in the intervertebral disc of the rhesus monkey (Macaca Mulatta). Ann. Anat. 187, 35–42. https://doi.org/10.1016/j.aanat.2004.08.007.

Toumanidou, T., Noailly, J., 2015. Musculoskeletal modeling of the lumbar spine to explore functional interactions between back muscle loads and intervertebral disk Multiphysics. Front. Bioeng. Biotechnol. 3, 1–13. https://doi.org/10.3389/fbioe.2015.00111.

Travascio, F., Elmasry, S., Asfour, S., 2014. Modeling the role of IGF-1 on extracellular matrix biosynthesis and cellularity in intervertebral disc. J. Biomech. 47, 2269–2276. https://doi.org/10.1016/j.jbiomech.2014.04.046.

Tyrrell, A.R., Reilly, T., Troup, J.D., 1985. Circadian variation in stature and the effects of spinal loading. Spine (Phila Pa 1976) 10, 161–164.

Urban, J.P.G., Smith, S., Fairbank, J.C.T., 2004. Nutrition of the intervertebral disc. Spine (Phila Pa 1976) 29, 2700–2709.

Wilke, H.J., Neef, P., Caimi, M., Hoogland, T., Claes, L.E., 1999. New in vivo measurements of pressures in the intervertebral disc in daily life. Spine (Phila Pa 1976) 24, 755–762.

Williams, J.R., Natarajan, R.N., Andersson, G.B.J., 2007. Inclusion of regional poroelastic material properties better predicts biomechanical behavior of lumbar discs subjected to dynamic loading. J. Biomech. 40, 1981–1987. https://doi.org/10.1016/j.jbiomech.2006.09.022.

Wills, C.R., Malandrino, A., Van Rijsbergen, M., Lacroix, D., Ito, K., Noailly, J., 2016. Simulating the sensitivity of cell nutritive environment to composition changes within the intervertebral disc. J. Mech. Phys. Solids 90, 108–123. https://doi.org/10.1016/j.jmps.2016.02.003.

Wognum, S., Huyghe, J.M., Baaijens, F.P.T., 2006. Influence of osmotic pressure changes on the opening of existing cracks in 2 intervertebral disc models. Spine (Phila Pa 1976) 31, 1783–1788. https://doi.org/10.1097/01.brs.0000227267.42924.bb.

Yang, H., Jekir, M.G., Davis, M.W., Keaveny, T.M., 2016. Effective modulus of the human intervertebral disc and its effect on vertebral bone stress. J. Biomech. 49, 1134–1140. https://doi.org/10.1016/j.jbiomech.2016.02.045.

Zhu, D., Gu, G., Wu, W., Gong, H., Zhu, W., Jiang, T., Cao, Z., 2008. Micro-structure and mechanical properties of annulus fibrous of the L4-5 and L5-S1 intervertebral discs. Clin. Biomech. 23 (Suppl. 1), S74–S82.

FURTHER READING

Crock, H.V., Yoshizawa, H., 1976. The blood supply of the lumbar vertebral column. Clin. Orthop. Relat. Res. 6–21. https://doi.org/10.1097/00003086-197603000-00003.

McCann, M.R., Bacher, C.A., Séguin, C.A., 2011. Exploiting notochord cells for stem cell-based regeneration of the intervertebral disc, J. Cell Commun. Signal. Springer, 5, pp. 39–43. doi:https://doi.org/10.1007/s12079-010-0116-9

Nerurkar, N.L., Elliott, D.M., Mauck, R.L., 2010. Mechanical design criteria for intervertebral disc tissue engineering. J. Biomech. 43, 1017–1030. https://doi.org/10.1016/j.jbiomech.2009.12.001.

Roberts, N., Hogg, D., Whitehouse, G.H., Dangerfield, P., 1998. Quantitative analysis of diurnal variation in volume and water content of lumbar intervertebral discs. Clin. Anat. 11, 1–8. https://doi.org/10.1002/(SICI)1098-2353(1998)11:1<1::AID-CA1>3.0.CO;2-Z.

Smith, L.J., Elliott, D.M., 2011. Formation of lamellar cross bridges in the annulus fibrosus of the intervertebral disc is a consequence of vascular regression. Matrix Biol. 30, 267–274. https://doi.org/10.1016/j.matbio.2011.03.009.

CHAPTER

7

THE MECHANICAL ROLE OF COLLAGEN FIBERS IN THE INTERVERTEBRAL DISC

Mirit Sharabi*, Kelly Wade[†], Rami Haj-Ali*

Tel-Aviv University, Tel-Aviv, Israel[*] Institute of Orthopaedic Research and Biomechanics, Ulm University, Ulm, Germany[†]

COLLAGEN MULTISCALE SYSTEMS AND THEIR MECHANICAL FUNCTION IN THE INTERVERTEBRAL DISC

COLLAGEN IN GENERAL

Collagen is one of the most abundant proteins in mammals. It is a family comprising 29 proteins forming a triple helix of three polypeptide chains, although their size, function, and tissue distribution vary. The most abundant and widespread family of collagens, with about 90% of the total collagen, is represented by the fibril-forming collagens, among them collagen type I is the most prevalent. Collagen type I is the major constituent of the extracellular matrix of connective tissue; it is found in diverse tissues, such as skin, intervertebral disc, cornea, bone, and ligaments. It is the main load-bearing element in a wide variety of soft tissues and plays a very important role in human physiology. Collagen type I has a unique fibrous structure and cross-linking ability, providing support for the tissue and enabling high tensile strength (Parenteau-Bareil et al., 2010). In addition, it holds a variety of biological functions in cell-matrix interactions and serves as a binder for other proteins to promote essential cellular functions (Parenteau-Bareil et al., 2010; Tamayol et al., 2013).

Most of the collagens consist of three amino acids glycine (33%), which enhances the stability of the molecule; proline (15%); and hydroxyproline (15%) (Holzapfel, 2001). Collagen molecules are linked to each other by covalent bonds that build collagen fibrils. Collagen is cross-linked in vivo through lysine side chains by lysyl oxidase enzymes via an aldehyde reaction mechanism.

The cross-linking process determines the fibers' mechanical strength, degradation rate, and cell-fiber interactions (Eyre and Wu, 2005; Friess, 1998). Collagen fibrils are aggregates of collagen molecules (Fig. 1) (Parenteau-Bareil et al., 2010) and are assembled into varied multiscale and hierarchical structures with diverse mechanical behaviors and functions (Mauck et al., 2009). This structural hierarchy results in a special stress-strain behavior that is nonlinear, hyperelastic, and anisotropic (Fratzl, 2008; Fratzl and Weinkamer, 2007). The collagen basic building block is the fibril whose thickness ranges between 50 nm to a few hundred nanometers.

The mechanical behavior of collagen is nonlinear and hyperelastic. It depends on the specific structure and hierarchy of the fibrous tissue. For example, a tendon consists of a high content of collagen organized in a unidirectional arrangement (Holzapfel, 2001). Its mechanical behavior includes three main

Biomechanics of the Spine. https://doi.org/10.1016/B978-0-12-812851-0.00007-0

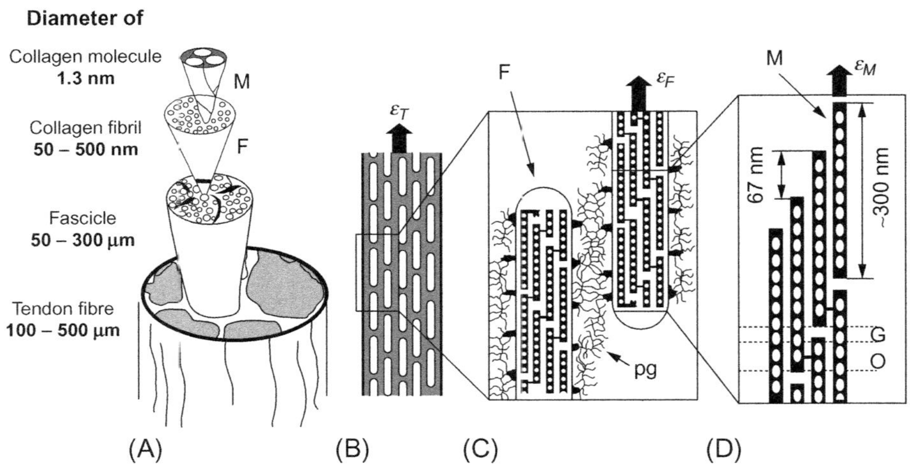

FIG. 1

(A) Collagen system and its multiscale structures in a tendon that is made of a number of parallel fascicles containing collagen fibrils (F); these are assemblies of parallel molecules (M). (B) The tendon fascicle can be viewed as a composite of collagen fibrils in a proteoglycan-rich matrix. (C) Fascicle-level strain is taken up in part by the deformation of the proteoglycan (pg) matrix. (D) Triple-helical collagen molecules (M) are packed within fibrils in a staggered fashion with an axial spacing of D=67 nm. The length of the molecules is 300 nm and consists of gap (G) and overlap (O) zones.

Reproduced from Fratzl, P., Weinkamer, R., 2007. Nature's hierarchical materials. Prog. Mater. Sci. 52, 1263–1334 with permission.

regions (Fig. 2): toe, heel, and linear regions. In the toe region (small strains), a very small stress is sufficient to elongate the tendon. Structurally, the macroscopic crimp of the fibrils is removed, and additional structural changes occur at the fibrillar level. This stage can be visualized using polarized light. Straightening of molecular kinks occurs in the heel region, whereas the linear region is associated with structural changes at the fibril level: stretching of the collagen triple-helices and gliding of neighboring molecules, creating a linear dependence between force and displacement. This stage can be captured by synchrotron diffraction. The physiological mechanical working environments of tendons are in the heel and linear regions (maximum strains of 8% to 10%) (Fratzl et al., 1997; Fratzl and Weinkamer, 2007).

COLLAGEN IN THE INTERVERTEBRAL DISC: GENERAL REMARKS

Intervertebral discs (IVDs) are the largest avascular organs in the body. Greater than 70% of their dry weight are collagens (Eyre et al., 2002), where they constitute 70% of the dry weight in the annulus fibrosus (AF) and 20% of the central nucleus pulposus (NP). Proteoglycans constitute a few percent of the dry weight of the AF, but 50% of the dry weight of the NP (Buckwalter, 1995). The IVD also consists of approximately 2% elastin (Roberts et al., 2006) that probably allows the collagen fibers to return to their undeformed state. The collagen enables the tissue's tensile properties and strength, whereas the proteoglycans, through water binding, gives the tissue stiffness and resilience in compression (Buckwalter, 1995). Therefore collagen is the most prominent constituent of the IVD tissue. Overall mechanical functions of fibrous collagen are summarized in Fig. 3 and described in the next sections.

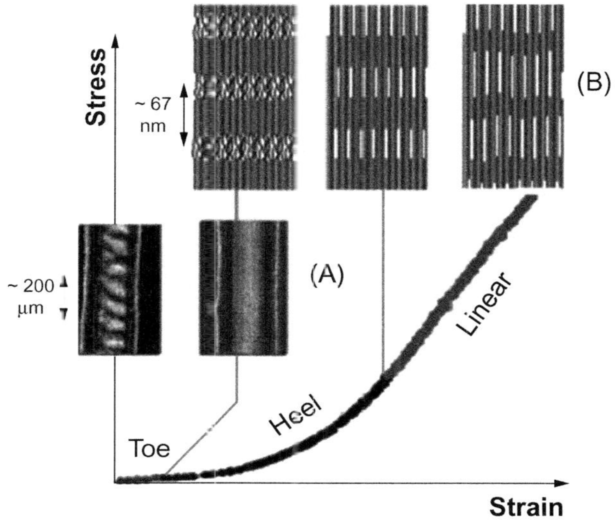

FIG. 2

Typical hyperelastic stress-strain curve of a rat tail tendon demonstrates three distinct regions: toe, heel and linear regions. In the toe region, a little force is needed for the extension. In the heel region, the stiffness increases with extension, followed by a linear stress-strain regime. (A) Removal of a macroscopic crimp of the fibrils can be visualized using polarized light. Additional structural changes occur at the fibrillar level, where the molecular kinks are straightening in the gap and there is a gliding of molecules. Moreover, the gap/overlap interface becomes fuzzy (as seen in synchrotron diffraction data) (B).

Reproduced from Fratzl, P., Weinkamer, R., 2007. Nature's hierarchical materials. Prog. Mater. Sci. 52, 1263–1334 with permission.

COLLAGEN IN THE ANNULUS FIBROSUS

The AF comprises a set of circumferential lamellae consisting of inclined collagen fibers, as illustrated in Fig. 3A. The end of these fibers are bound to the cartilaginous endplates in collagenous fibers called Sharpey's fibers that anchor the lamellae (Fig. 3C) (Inoue and Takeda, 1975; Pech and Haughton, 1985; Rodrigues et al., 2017; Rodrigues et al., 2012).

In the IVD, the stiffness of the outer AF region is greater than it is in the inner region, and the anterior region is stiffer than the posterior region (Holzapfel et al., 2005). The reason for the radial variation of the lamellar stiffness is mechanically beneficial: a heterogeneous thick-walled cylinder with internal soft layers will reduce stress gradients, in comparison with a homogenous inner thick-walled cylinder that yields a high gradient of circumferential stress (Holzapfel et al., 2005). These stiffness gradients are the result of a structural gradient of collagen I and II ratio (Eyre and Muir, 1974, 1976), the proteoglycans type (Adams et al., 1977), the water content (Skaggs et al., 1994), and the collagen fibril diameter (Han et al., 2012). The fibril diameter was measured and reported to be 87.8 ± 27.9 nm in the outer AF region and 71.0 ± 26.9 nm in the inner AF region (Han et al., 2012). Moreover, in the outer region of the AF, the lamellae were found to be about 120 μm, or one-half the thickness of the lamellae in the transitional zone (Cassidy et al., 1989).

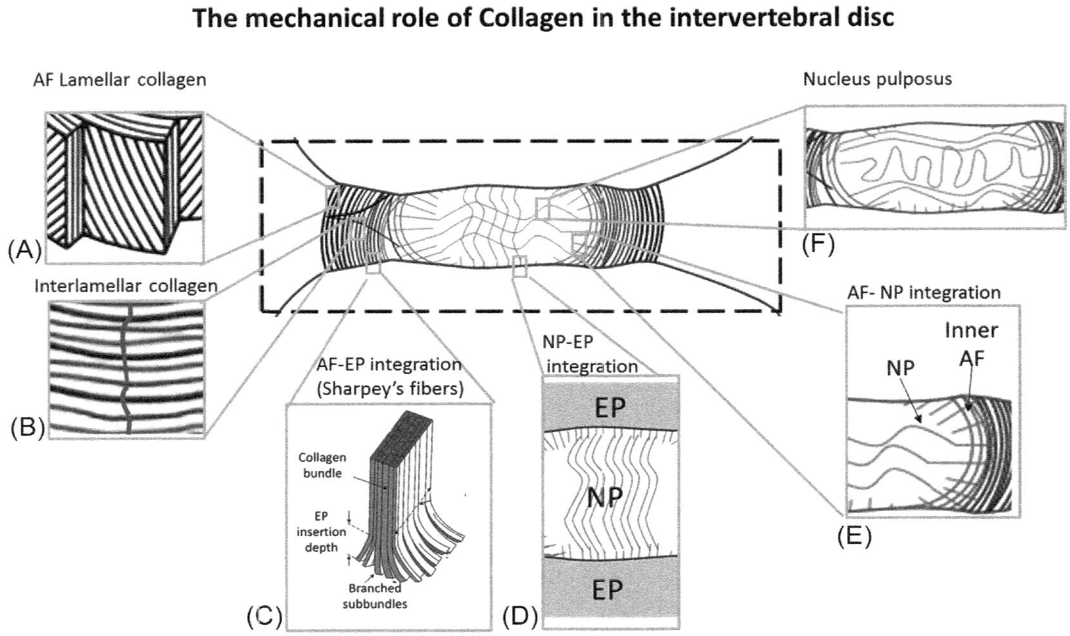

FIG. 3

Schematic representations for the mechanical roles of fibrous collagen in the IVD: (A) collagen fibers in AF lamella; (B) interlamellar cross-bridges consisting of collagen types VI and I; (C) an AF collagen bundle that penetrated to the EP; (D) the integration between NP and EP; (E) the integration between inner AF-NP fibrous in the transition zone; and (F) the fibrous structure of the NP.

Part (C) Reproduced from Rodrigues, S.A., Wade, K.R., Thambyah, A., Broom, N.D., 2012. Micromechanics of annulus-end plate integration in the intervertebral disc. Spine J. 12, 143–150 with permission.

It was proven that posterior bulging and protrusion is twice as common as anterior bulging (Schollmeier et al., 2000). The reasons for the variance between the circumferential regions are less clear. The lamellae thicknesses in the lateral and posterior regions are widely distributed through depth and radial location; however, the average lamellar thickness in the lateral regions is greater than in the posterior region (Cassidy et al., 1989). Moreover, posterior multilamellar samples were found to be less stiff than anterior-lateral multilamellar samples, even if the single lamellae are equally stiff (Holzapfel et al., 2005). Therefore the weakness and elevated risk of rupture in the posterior region is attributed to frequent interruptions and an elevated number of incomplete lamellae in that region, and not to significant circumferential variations of the tensile behavior (Holzapfel et al., 2005; Marchand and Ahmed, 1990; Skaggs et al., 1994; Tsuji et al., 1993). In addition, proteoglycan concentration is higher in the posterior region than in the anterior region (Tavakoli, 2017).

Single Lamella of the Annulus Fibrosus

In adjacent layers, the orientation alternates relative to the spinal column. The collagen fiber angle ranges between 46 degree and 60 degree with the loading direction in human discs (Cassidy et al., 1989; Guerin and Elliott, 2006; Hickey and Hukins, 1980; Marchand and Ahmed, 1990). In addition, elastin

fibers and microfibrils are found within the lamellae, aligned in the same direction as the collagen fibers (Yu et al., 2007), and their mechanical role is probably associated with restoring the collagen crimp under very small stresses. The positive and negative collagen fiber orientations of adjacent lamellae are similar, and also similar in different depth levels. However, fiber angles increase from the anterior to the posterior region. The larger angles for the posterior fibers restrict the maximal axial distance of endplates in flexion and minimize spinal cord overstretching (Holzapfel et al., 2005; Marchand and Ahmed, 1990).

Every lamella consists of collagen fibers arranged in parallel arrays. The fibers introduce a joint planar crimp waveform. The plane of the waveform is vertical to the lamellar plane and bent relative to the spinal axis by the interlamellar angle. Although the length of the crimp stays constant, there is a gradient in the crimp geometry: the crimp angle increases from 20 degree to 45 degree, whereas the crimp period decreases from 26 to 20 μm linearly from the outer region of the intervertebral disc to its inner region. This gradient influences the mechanical response to disc compression: when the IVD is compressed, axial loads are translated to radial and tangential tensile stresses in the AF lamellae. Therefore, the collagen fibers reorient in the radial direction (Klein and Hukins, 1982). Consequentially, there is a decrease in the crimp angle and increase in the crimp period. Therefore the known hyperelastic J-shape stress-strain behavior, typical for type I collagen tissues under tension, is seen (Fratzl et al., 1998), with low modulus crimp straightening and reorientation, a linear elastic region, which is a result of the fibers' extension and yield region, where the structure undergoes irreversible damage. Because of the spatial gradients in the structure, the outer lamellae are less deformable, stiffer, and have lower energy dissipation properties than the inner lamellae (Cassidy et al., 1989).

In the torsional mode, shear stresses are applied mainly on the lamellae that is in the direction of the applied torque angle (Hickey and Hukins, 1980). In lumbar motion segments without degeneration, annulus fibers restrict axial rotation before the facets begin to work (Krismer et al., 1996). Therefore positive fibers actively resist positive axial rotation.

Under a physiological level of strain (10%), the fibers reorient with an average decrease of 17 degree in the outer AF. Decreasing the fiber angle increases rotation under torsion, AF circumferential modulus, and compressive stiffness and reduces vertical displacement under compression and AF bulging. In general, fiber reorientation may be responsible for stabilization of motion segments under axial loading (Guerin and Elliott, 2006).

The tensile behavior of single lamella in the fiber direction is nonlinear and hyperelastic. The nonlinearity can be represented by a trilinear stress-strain curve. The modulus of a single lamella in the linear region (between 0.5 and 1 MPa) ranges between 28 and 78 MPa (Holzapfel et al., 2005). Skaggs et al. (1994) reported a tensile modulus of 59 to 136 MPa. The latter difference may be a result of the different spinal levels used in the two tests, the testing method, or even different definitions of the modulus. Stiffness of multilamellar specimens with axial orientation can range in the anterior region between 0.82 and 0.96 MPa (Elliott and Setton, 2001). The tensile moduli of the heel and linear regions are larger for the outer lamella than for the inner lamellae, but the change in the circumferential location is still under dispute (Holzapfel et al., 2005; Skaggs et al., 1994). In the toe region, a significant spatial change does not occur in the tensile modulus. Therefore the lamellar matrix can be considered homogenous because the toe region is a matrix-dominated region (Holzapfel et al., 2005).

Integration/Interlamellar Region

The interlamellar space between the AF lamellae is compartmentally structured and separated by fibrous interconnections that were identified by Gracovetsky and Farfan (1986) (Fig. 3B). These radial interconnections are made of proteoglycans and collagen types VI and I in ovine discs

(Melrose et al., 2008). A high degree of radial connectivity characterizes the fibrous interlamellar bridging. In ovine discs, the width of the cross-bridges is between 300 and 600 μm. The cross-bridges connect lamellae with the same orientation. The cross-bridges consist of obvious crimp and stream into the collagen fibers in the lamellae bend at 90 degree and merge with them (Schollum et al., 2008, 2009). Moreover, a network of microfibrils and elastin fibers is found in the interlamellar space (Yu et al., 2007).

The mechanical role of the cross-bridges is not fully known. However, in a "lap test" involving two adjacent lamellae, it was found that the interlamellar region plays a major role in preventing delamination between the layers (Gregory et al., 2011). Therefore their mechanical role is probably due to limited slipping between the lamellae, ultimately leading to decohesion and delamination. Moreover, under transverse shear, intralamellar skewing occurs, and not sliding between the lamellae. Therefore the lack of sliding indicates that interlamellar connections are governed by rigid fibrous cross-bridges (collagen and fibrilin). Elastin was found to stiffen the AF fiber bundles transversely, and it plays a role in orthotropic damping. Its absence yields an increased fiber rotation and phase angle between adjacent lamellae (Michalek et al., 2009). Shear strains were found to be larger in the interlamellar matrix than in the lamellae (Han et al., 2016).

Under uniaxial stretching of the AF, no sliding was observed between lamellae, and the linear strain was almost three times higher than that of the corresponding intralamellar, thus confirming a strong interlamellar cohesion. Moreover, inter-bundle shear and linear strains were higher than intra-bundle strains; therefore slipping occurs between the bundles. This inter-bundle slipping is responsible for the strain inside the lamellae (Vergari et al., 2016).

NUCLEUS PULPOSUS AND TRANSITION ZONE

The nucleus pulposus encloses a type II collagen fibrous network (Inoue and Takeda, 1975) (Fig. 3F). Most of the studies present the nucleus pulposus as a hydrostatically functioning entity that is separated from its confining boundaries and lacks any attachment or cohesion (Inoue, 1981; Inoue and Takeda, 1975). However, tensile loading of the AF-NP transverse section in ovine discs resulted in a peak load of 11.5 N and a "J-shape" load-extension curve, which characterizes collagen fiber mechanical behavior (Wade et al., 2012b). These results strongly suggest and provide evidence for the mechanical role of the organized collagenous network that interconnects the inner AF and NP through the transition zone (Fig. 3E). Moreover, endplate-NP integration was also shown in ovine discs. Tensile loading of vertebrae-NP-vertebra samples demonstrated a peak load of 20 N, together with the evident NP fibers inserting to the endplates (Wade et al., 2011, 2012a). Because both cases of integration relate to load-bearing fibers, collagen fibers are probably interconnecting the whole disc. So far, the authors have not found data in the literature concerning the nature of these interconnecting fibers. It is known that elastin fibers are found in these areas (Yu et al., 2015). However, elastin is not a load-bearing material.

COLLAGEN TYPES IN THE INTERVERTEBRAL DISC

COLLAGEN TYPES: DISTRIBUTION AND FUNCTION

The unique mechanical behavior of the intervertebral disc results from the combination between cartilage and ligaments and their different collagen types (Eyre et al., 2002). In general, mature AF consists of nine types of collagen)I, II, III, V, VI, IX, XI, XII, and XIV) (Kobielarz et al., 2016), and the NP consists of five types (I, II, VI, IX, and XI) (Roberts et al., 1991c).

The most common types are fibril-forming types I and II collagens, which constitute approximately 80% of the collagen content in the IVD (Eyre et al., 2002; Roberts et al., 1991c). Type II collagen is prevalent in the NP, whereas type I is prevalent in the AF. Actually, there is a smooth and interchanging gradation in the collagen type I and II ratio from the outermost region of the AF that contains exclusively collagen type I to the NP innermost region that contains exclusively collagen type II (Eyre and Muir, 1974, 1976). There is a decrease of approximately 35% in the level of collagen type I from the outer part to the inner part. On the other hand, the amount of type II collagen increases by almost 40% from the outer part to the inner part (Kobielarz et al., 2016).

The fibrous network restrains the swelling pressure of the closely packed proteoglycans in the matrix, creating resilient material with tensile and compressive strength. In the AF, type I and II collagens are arranged in a lamellar structure, where a minimum of 15 lamellae can be found in the posterior part of the AF and a maximum of 25 lamellae in the lateral part of the AF (Kobielarz et al., 2016).

In general, type I collagen is abundant in fibrous connective tissues, such as skin, ligaments, and tendons, whereas type II is predominate mainly in hyaline cartilage. Type II collagen forms a fibrillar network that allows cartilage to trap the proteoglycan aggregate along with providing the tissue tensile strength (Eyre and Muir, 1977).

The other types of collagens, usually existing in small amounts (1% to 3%), are believed to contribute to the overall functioning of the tissue, as described in Table 1. For example, type IX collagen is cross-linked to type II in articular cartilage and hence regulates its size. Types VI and III are found pericellularly in the disc and cartilage endplates. Type VI is particularly prominent in the nucleus pulposus and creates chondrons, fibrous capsules that encapsulate cells, as seen in the deep and middle zones of articular cartilage. Type III is believed to be involved in the organization of the pericellular environment and may link the chondron capsule to the interterritorial matrix (Roberts et al., 1991a, b, c).

CHANGES IN COLLAGEN CONTENT WITH GROWTH AND DEGENERATION

The collagenous structure is crucial for maintaining IVD biomechanical stability and overall functions. Therefore the composition of the collagen and its condition have an enormous influence on the maintenance of its biomechanical function. During degeneration, the IVD is subjected to microstructural disarrangement of the AF matrix, to changes in the composition of the IVD, to a decrease of proteoglycans and elastin, and to an increase of enzymatic activity (Kobielarz et al., 2016).

The AF lamellae initiate their life cycle as individual discrete bundles of collagen fibers. However, with aging the lamellar organization becomes increasingly complex with more bifurcations expressed in part by increased thickness of the lamellae (Roberts et al., 2006). The annulus fibers become disrupted with a growing number of clefts and fissures. These changes in the organization of the normal collagen contribute to a reduced biomechanical function of the AF (Gruber and Hanley Jr, 2002). Variations in the cross-section of the fibril diameter have also been reported (Gruber and Hanley Jr, 2002). Increased type II denaturation and type I collagen synthesis have also been observed with aging, along with widened collagen fibrils (with widths of approximately 150 nm) that are also identified in articular cartilage in degenerative joint disease. Their role is not yet understood, but these changes in the ultrastructure of the ECM may change the normal biomechanical properties (Gruber and Hanley Jr, 2002).

Table 1 Collagen Types Within the IVD and Their Properties

Collagen Type	Distribution	% of Total Collagen	Type	Special Features and Property	Reference
I	AF, translamellar cross-bridges	Up to 80%	Fibrillar	High to low gradient from outer to inner AF	Eyre and Muir (1976), Melrose et al. (2008), Roberts et al. (1991c), Schollmeier et al. (2000)
II	Inner AF, NP, cartilaginous EP	Up to 80%	Fibrillar	High to low gradient from NP to inner AF	Eyre and Muir (1976), Nerlich et al. (1998), Roberts et al. (1991c), Schollmeier et al. (2000)
III	AF, NP, pericellular, EP		Fibrillar	Mostly pericellular, perhaps linking the chondron capsule to the interterritorial matrix	Nerlich et al. (1998), Roberts et al. (1991b), Roberts et al. (1991c)
V	AF, NP, EP	~3%	Minor fibrillar	Often associate with type I, II	Nerlich et al. (1998), Roberts et al. (1991c)
VI	AF, NP, pericellular, translamellar cross-bridges, EP	10%–20% In bovine: AF 5% NP 30% In human: AF 10% NP >15%	110 nm microfibrils	Creates chondrons	Buckwalter (1995), Melrose et al. (2008), Nerlich et al. (1998), Roberts et al. (1991a, c), Wu et al. (1987)
IX	AF, NP, Pericellular, EP	1%–2%	Minor	Covalent bind to type II surface and may regulate it size and swelling	Buckwalter (1995), Nerlich et al. (1998), Roberts et al. (1991c), Wu et al. (1992)
XI	NP	~3%	Minor	Codistributed with type II?	Ayad et al. (1981), Ayad et al. (1982), Buckwalter (1995), Roberts et al. (1991c)
XII				Bound to fibril surface	Eyre et al. (2002)
XIV				Bound to fibril surface	Eyre et al. (2002)

Furthermore, the NP becomes smaller and decompressed with aging; therefore the inner AF is subjected to more compressive loads (this is the reason for the relatively high proteoglycan content at that region). However, aged AF consists of fewer proteoglycans and thus becomes stiffer and weaker (Adams and Roughley, 2006).

Additional changes with degeneration are manifested by a decrease in the fibers' reorientation. The fibers in healthy discs reorient toward the loading direction at an average of 17 degree, whereas in degenerated discs, the average reorientation has been measured to be 9.5 degree at a 10% strain level (Guerin and Elliott, 2006). Decreased fiber reorientation may contribute to increased bulging of the AF and decreased stability. The reason for this decrease may be attributed to the elevation in collagen cross-links, reduced water content, and disorganization of the lamellae that constrains the reorientation of the fibers (Guerin and Elliott, 2006).

The changes in collagen types and distribution are described in Table 2.

Type I collagen was seen in normal AF and in degenerated NP, indicating severe tissue disarrangement in the NP, but not within the EP (Nerlich et al., 1998; Nerlich et al., 1997). It is absent in juvenile/adolescent NPs and mainly exists in the outer annulus fibrosus.

Type II and IX collagens were observed in the EP, the inner AF, and the NP. They were prominent in areas with minor degenerative changes, but not in areas with advanced degeneration. Type II collagen is the main component of the NP but also exists in the inner AF; these results are consistent with the interchanging gradation in collagen type (Eyre and Muir, 1974, 1976). Type II collagen is also a principal constituent of the EP, and its loss in the degenerated endplate indicates major disarrangement (Nerlich et al., 1998). During aging, the expression of type II collagen is reduced, whereas type I collagen increases, especially in the NP region (Colombini et al., 2008).

Table 2 Changes in Collagen Types in Degeneration

	Fetal/Infantile			Adolescent/Young Adults			Adults/Elderly		
Ages	0–13			16–25			31–86		
Degeneration	–			(+)/+			(+)/++		
Collagen type	AF	NP	EP	AF	NP	EP	AF	NP	EP
I	+	–	–	+	–	–	++	+	–
II	+	++	+	+	+++	++	(+)	+/++	+
III	+	(+)	–	+/++	+++	+	+	+/+++	+/++
IV	–	–	–	(+)	++	–	–	–/+	–
V	+	+	–	+	+	+	+	+	+
VI	+	+	+	++	+++	++	+/++	+/+++	+/++
IX	+	++	+	+	+++	++	–/(+)	+/++	(+)
X	–	+	++	–	–	+/++	–	(+)/+	–

"–" absent, "(+)" weakly positive, "+/+++" positive with increased levels.

Adapted from Nerlich, A., Boos, N., Wiest, I., Aebi, M., 1998. Immunolocalization of major interstitial collagen types in human lumbar intervertebral discs of various ages. Virchows Arch. 432, 67–76 and Nerlich, A.G., Schleicher, E.D., Boos, N., 1997. 1997 Volvo award winner in basic science studies: immunohistologic markers for age-related changes of human lumbar intervertebral discs. Spine 22, 2781–2795.

Significant changes were seen in type III and VI collagens in all areas with any degree of degeneration (Colombini et al., 2008; Nerlich et al., 1998). These collagens are distributed with a typical pericellular arrangement in the AF, NP, and EP.

Type V collagen is mainly found in the AF and in the NP. It demonstrated only minor changes with degeneration (Nerlich et al., 1998). It is distributed in the same regions as collagen types I and II (Nerlich et al., 1998). Collagen X is associated with EP calcification and morphological signs of degeneration, such as fractures and clefts (Colombini et al., 2008).

DIFFERENCES IN COLLAGEN TYPES AND CONTENT BETWEEN DIFFERENT SPECIES

Collagen types and distribution vary between humans and different animal species (Roberts et al., 1991c); however, little is known about these differences. One interesting difference is that human AFs contain a higher proportion of collagen type II (50% to 65%, an approximate 1:1 ratio between type I and type II) than pigs (20%, an approximate 4:1 ratio between type I and type II). In the kangaroo, whose posture is intermediate between biped and quadruped, the proportion of collagen is midway between that of humans and pigs. The latter finding strengthens the assumption that this elevation in type II collagen is a consequence of mechanical posture (Eyre et al., 2002; Eyre and Muir, 1976, 1977).

STRUCTURAL AND MECHANICAL DIFFERENCES BETWEEN TYPES I AND II COLLAGEN

Because they constitute most of the IVD and because of their graded nature, here we attempt to explain the structural differences between the two fibrous collagen types I and II. A difference between these two collagen types is not visible under electron microscopy. Moreover, young NP collagen fibers are thinner than their counterparts in the AF. This is not the case for adult discs (Eyre and Muir, 1977).

Apart from well-known differences between these collagen types, such as their amino acid sequence and high glycosylated hydroxylysines content, it was found that in a wet environment under equatorial diffraction, the native type II collagen molecules are spaced farther apart laterally than they are in type I collagen (Grynpas et al., 1980). The average lateral spacing was found to be 14Å for an outer AF fiber (mostly collagen type I) and 16–17Å for the NP (collagen type II); this difference was not seen in dried specimens. Therefore, under physiological conditions, type II collagen fibrils contain 50% to 100% more water than type I fibrils (Grynpas et al., 1980). This larger water content is crucial for type II collagen's mechanical function of resisting compressive forces. Another difference is that type II is strongly entrapped with proteoglycans and is much richer in hydroxylysine-bound hexoses than type I collagen is. These sugar residues may mediate interactions with proteoglycans (Grynpas et al., 1980).

INTEGRATION BETWEEN DIFFERENT COLLAGEN TYPES

Because there is a graded interchange between collagen I and II ratios from the outer region of the IVD to its inner region, and the unique properties of the disc are strongly influenced by the copolymerization, and integration of the different fibrillar collagens, it is important to understand how the different collagen types are integrated. The exact mechanism is not yet fully understood (Eyre et al., 2002). Different fibril types may be fused and cross-linked. In other tissues, such as the knee meniscus, the proportion of type I and type II collagens is usually distributed toward one of the constituents, but not in an equally mixed fashion as in the human IVD. The other collagen types found in the IVD are usually associated with either collagen type I or collagen type II. For example, collagens IX and XI are *co*-polymerized with type II collagen as in cartilage, and collagen V is co-polymerized with type I collagen as in other collagen I-based tissues (Eyre et al., 2002). Coexistence of collagens I and II was found in different regions in the meniscus. It is possible that the two types may colonize the same fibril or bundle of fibrils (Kambic and McDevitt, 2005) In the AF, lateral interaction between collagen I and II was hypothesized in an apparent basic fibril type mixing but not in hybrid bundles.

Moreover, collagen type III can interact with the surface of type I and type II collagen fibrils, and it may form thin fibrils that cross-link and bind them laterally (Eyre et al., 2002).

The interactions between collagen and proteoglycans occur at the fibril and molecular levels, where each collagen molecule appears to have at least three binding sites such that one is weaker than the other two (Berthet-Colominas et al., 1982).

NUMERICAL MODELING OF THE COLLAGEN FIBERS IN THE IVD (HETEROGENEOUS MODELS)

Finite element (FE) models are widely used to investigate the biomechanics of the spine (Dreischarf et al., 2014; Moramarco et al., 2010; Niemeyer et al., 2012; Schmidt et al., 2012; Schmidt et al., 2006; Schmidt et al., 2007). The anisotropic nature of the intervertebral disc can be explored through two main approaches: (1) by homogenized models that illustrate the effective mechanical behavior (Caner et al., 2007; Eberlein et al., 2001; Hollingsworth and Wagner, 2011; Peng et al., 2006; Reutlinger et al., 2014), and (2) by structural models that describe the AF as a heterogeneous and anisotropic fiber-reinforced composite material (Goel et al., 1995; Schmidt et al., 2006; Shirazi-Adl et al., 1984). The FE structural models assist in examining the influence of mechanical loads on individual components and in understanding their inherent function. The mechanical behaviors of the AF matrix and the collagen fibers' constituents used in heterogeneous computational models are shown in Figs. 4 and 5,

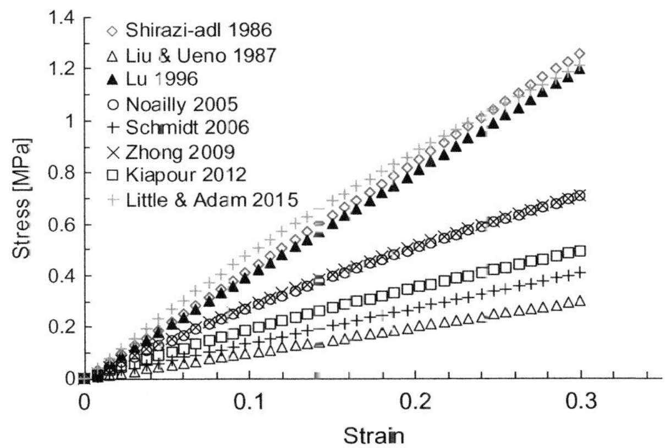

FIG. 4

The mechanical stress-strain response of different AF matrices used in different heterogeneous computational models.

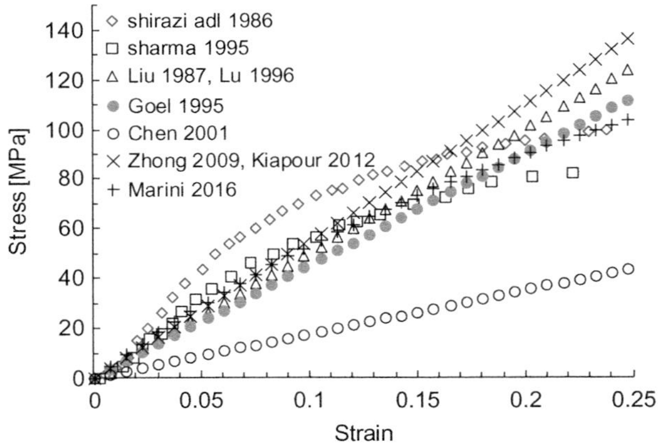

FIG. 5

The mechanical stress-strain response of AF collagen fibers used in different heterogeneous computational models.

respectively. In general, the collagen fibers' behavior is nonlinear but does not always demonstrate the "J-shaped" response characteristic of collagen. The matrix is almost linear and is less stiff than the collagen fibers on an approximate order of two magnitudes.

Translamellar cross-bridges can be embedded in the AF models using the homogenized methods (Hollingsworth and Wagner, 2011; Labus et al., 2014; Luxmoore et al., 2012; Luxmoore, 2013; Mengoni et al., 2015). An examination of the cross-bridges' influence on the mechanical behavior of regional parts of the IVD shows that the preceding models can be extended to take into account the heterogeneous nature of the translamellar cross-bridges or their noncontinuous distribution.

Structural models use fiber-reinforced materials, for example, 6 to 20 layers of bar or truss elements that represent the collagen fibers and reinforce the solid elements representing the annulus matrix. These models consist of crisscrossed fibers arranged at approximately ±30 degree angles (24–46 degree) and with average fiber volume fractions of 16% to 20%. The orientation changes gradually from the outer to the inner AF (Noailly et al., 2005; Noailly et al., 2007; Schmidt et al., 2012; Schmidt et al., 2006; Schmidt et al., 2007) and from ventral to dorsal (Marini et al., 2016; Schmidt et al., 2012; Schmidt et al., 2006; Schmidt et al., 2007). Some of the published models consist of a gradual decrease in the fiber volume fraction from the outermost region (23%) of the annulus to the innermost region (5%), representing the transition zone. A gradual decrease in both the mechanical properties and cross-section of the fibers has also been employed (Liu et al., 2011; Sharma et al., 1995; Shirazi-Adl et al., 1986; Shirazi-Adl et al., 1984; Zhong et al., 2009).

One of the advantages of these structural models lies in their ability to examine the local stresses and strains of the isolated components under different architectures. For example, Noailly et al. (2011) tested the influence of the inner AF fiber orientation changing by radial and circumferential locations under axial rotation and compression. They calculated the stresses in the fibers, load distribution, and matrix shear strains. They also took into account a gradual collagen I to II ratio. They found that given a different collagen network organization, fiber stresses could be locally changed by 100% or more and that average matrix shear strains could be changed by up to 50% (Noailly et al., 2011). These results provide a better understanding of the structure-function behavior of the IVD and may aid in the development of future IVD grafts.

Reference	Spinal Segment	FVF	No. of Layers	Mechanical Behavior of the AF Fibers	Mechanical Behavior of the AF Matrix	Fiber Orientation
Shirazi-Adl et al. (1984)	L2–L3	19%	8	$\sigma = 23000\varepsilon^{1.9}$ RCS-1, 0.78, 0.62,0.47 RMP-1, 0.9, 0.75, 0.65	E = 4.2 MPa G = 1.6 MPa $\nu = 0.45$	29 degree
Shirazi-Adl et al. (1986)	L2–L3	16% changed with radial location	8	Attached curve RCS-1, 0.78, 0.62, 0.47 RMP-1, 0.9, 0.75, 0.65	E = 4.2 MPa G = 1.6 MPa $\nu = 0.45$	29 degree
Liu and Ueno (1987)	L4–L5	Shirazi-Adl et al. (1984)	8	E = 500 MPa	E = 1 MPa $\nu = 0.48$	37.7 degree
Goel et al. (1995)	L3–L4	20%	8	E = 450 MPa $\nu = 0.3$	E = 4.2 MPa $\nu = 0.45$	30 degree
Sharma et al. (1995)	L3–L4	19%	8	Attached curve RCS-1, 0.78, 0.62, 0.47 RMP-1, 0.9, 0.75, 0.65	E = 4.2 MPa $\nu = 0.45$	30 degree
Lu et al. (1996)	L2–L3	Outer to inner gradient 23%–5%	6	E = 500 MPa $\nu = 0.35$	E = 4 MPa $\nu = 0.4$	29.6 degree 24.5–36.3 degree
Chen et al. (2001)	L1–L5			E = 175 MPa	E = 4.2 MPa $\nu = 0.45$	
Schmidt et al. (2012), Schmidt et al. (2006), Schmidt et al. (2007)	L4–L5 L1–L5	Outer to inner gradient 23%–5%	8	Modified Shirazi-Adl et al. (1986) $\lambda_{fiber} = 1.03$ RMP-1, 0.9, 0.75, 0.65	Moony Rivlin C1 = 0.18, C2 = 0.045 E = 1.36 MPa $\nu = 0.45$	Ventral to dorsal 24–46 degree Inner layer 45 degree
Noailly et al. (2005), Noailly et al. (2007)	L3–L5	16% changed with radial location	20	Sharma et al. (1995)	Neo-Hookean M = 0.5 MPa	Outer to inner 28–45 degree
Kiapour et al. (2012) based on Goel et al. (1995)	L3–S1		8	E = 357–550 MPa $\nu = 0.3$	Neo-Hookean C10 = 0.348, D1 = 0.3	

Continued

Reference	Spinal Segment	FVF	No. of Layers	Mechanical Behavior of the AF Fibers			Mechanical Behavior of the AF Matrix	Fiber Orientation
Liu et al. (2011), Zhong et al. (2009)	L1–L5	Shirazi-Adl et al. (1984)	12	E outer to inner 550, 495, 412.5357.5 MPa			Mooney Rivlin	Marchand and Ahmed (1990)
				RCS-1, 0.78, 0.62, 0.47			C10 = 0.42, C01 = 0.105	
				RMP-1, 0.9, 0.75, 0.65				
Little and Adam (2015)			8	E = 500 MPa			Mooney Rivlin	
				$\nu = 0.3$			C10 = 0.7, C01 = 0.2	
Marini et al. (2016)			10	Polynomial	μ	γ	Yeoh	$\theta = 23.2\psi + 0.130\phi$
				Ventro-lateral external	0.109	106.405	C10 = 38.793 KPa	
							C20 = 55.049 KPa	
				Ventro-lateral internal	0.029	45.449	$\kappa = 73.521$	
				Dorsal external	0.045	55.625		
				Dorsal internal	0.004	32.649		

ACKNOWLEDGMENTS

This research was supported by the Israeli Ministry of Science, Technology and Space under Grant No. 3-12961. The last author gratefully acknowledges the support of the Nathan Cummings Chair of Mechanics.

REFERENCES

Adams, M.A., Roughley, P.J., 2006. What is intervertebral disc degeneration, and what causes it? Spine 31, 2151–2161.

Adams, P., Eyre, D., Muir, H., 1977. Biochemical aspects of development and ageing of human lumbar intervertebral discs. Rheumatology 16, 22–29.

Ayad, S., Abedin, M., Grundy, S., Weiss, J.B., 1981. Isolation and characterisation of an unusual collagen from hyaline cartilage and intervertebral disc. FEBS Lett. 123, 195–199.

Ayad, S., Abedin, M., Weiss, J.B., Grundy, S., 1982. Characterisation of another short-chain disulphide-bonded collagen from cartilage, vitreous and intervertebral disc. FEBS Lett. 139, 300–304.

Berthet-Colominas, C., Miller, A., Herbage, D., Ronziere, M.-C., Tocchetti, D., 1982. Structural studies of collagen fibres from intervertebral disc. Biochim. Biophys. Acta Protein Struct. Mol. Enzymol. 706, 50–64.

Buckwalter, J.A., 1995. Aging and degeneration of the human intervertebral disc. Spine 20, 1307–1314.

Caner, F.C., Guo, Z., Moran, B., Bazant, Z.P., Carol, I., 2007. Hyperelastic anisotropic microplane constitutive model for annulus fibrosus. J. Biomech. Eng. 129, 632–641.

Cassidy, J., Hiltner, A., Baer, E., 1989. Hierarchical structure of the intervertebral disc. Connect. Tissue Res. 23, 75–88.

Chen, C.-S., Cheng, C.-K., Liu, C.-L., Lo, W.-H., 2001. Stress analysis of the disc adjacent to interbody fusion in lumbar spine. Med. Eng. Phys. 23, 485–493.

Colombini, A., Lombardi, G., Corsi, M.M., Banfi, G., 2008. Pathophysiology of the human intervertebral disc. Int. J. Biochem. Cell Biol. 40, 837–842.

Dreischarf, M., Zander, T., Shirazi-Adl, A., Puttlitz, C., Adam, C., Chen, C., Goel, V., Kiapour, A., Kim, Y., Labus, K., 2014. Comparison of eight published static finite element models of the intact lumbar spine: predictive power of models improves when combined together. J. Biomech. 47, 1757–1766.

Eberlein, R., Holzapfel, G.A., Schulze-Bauer, C.A., 2001. An anisotropic model for annulus tissue and enhanced finite element analyses of intact lumbar disc bodies. Comput. Methods Biomech. Biomed. Engin. 4, 209–229.

Elliott, D.M., Setton, L.A., 2001. Anisotropic and inhomogeneous tensile behavior of the human anulus fibrosus: experimental measurement and material model predictions. J. Biomech. Eng. 123, 256–263.

Eyre, D.R., Muir, H., 1974. Collagen polymorphism: two molecular species in pig intervertebral disc. FEBS Lett. 42, 192–196.

Eyre, D.R., Muir, H., 1976. Types I and II collagens in intervertebral disc. Interchanging radial distributions in annulus fibrosus. Biochem. J. 157, 267–270.

Eyre, D.R., Muir, H., 1977. Quantitative analysis of types I and II collagens in human intervertebral discs at various ages. Biochim. Biophys. Acta Protein Struct. 492, 29–42.

Eyre, D.R., Wu, J.J., 2005. Collagen cross-links. Top. Curr. Chem. 247, 207–229.

Eyre, D.R., Matsui, Y., Wu, J.-J., 2002. Collagen polymorphisms of the intervertebral disc. Biochem. Soc. Trans. 30, 844–848.

Fratzl, P., 2008. Collagen: Structure and Mechanics. Springer Science & Business Media, New York.

Fratzl, P., Weinkamer, R., 2007. Nature's hierarchical materials. Prog. Mater. Sci. 52, 1263–1334.

Fratzl, P., Misof, K., Zizak, I., 1997. Fibrillar structure and mechanical properties of collagen. J. Struct. Biol. 122, 119–122.

Fratzl, P., Misof, K., Zizak, I., Rapp, G., Amenitsch, H., Bernstorff, S., 1998. Fibrillar structure and mechanical properties of collagen. J. Struct. Biol. 122, 119–122.

Friess, W., 1998. Collagen—biomaterial for drug delivery. Eur. J. Pharm. Biopharm. 45, 113–136.

Goel, V., Monroe, B., Gilbertson, L., Brinckmann, P., 1995. Interlaminar shear stresses and laminae separation in a disc: finite element analysis of the L3-L4 motion segment subjected to axial compressive loads. Spine 20, 689–698.

Gracovetsky, S., Farfan, H., 1986. The optimum spine. Spine 11, 543–573.

Gregory, D.E., Veldhuis, J.H., Horst, C., Brodland, G.W., Callaghan, J.P., 2011. Novel lap test determines the mechanics of delamination between annular lamellae of the intervertebral disc. J. Biomech. 44, 97–102.

Gruber, H.E., Hanley Jr, E.N., 2002. Ultrastructure of the human intervertebral disc during aging and degeneration: comparison of surgical and control specimens. Spine 27, 798–805.

Grynpas, M.D., Eyre, D.R., Kirschner, D.A., 1980. Collagen type II differs from type I in native molecular packing. Biochim. Biophys. Acta Protein Struct. 626, 346–355.

Guerin, H.A.L., Elliott, D.M., 2006. Degeneration affects the fiber reorientation of human annulus fibrosus under tensile load. J. Biomech. 39, 1410–1418.

Han, W.M., Nerurkar, N.L., Smith, L.J., Jacobs, N.T., Mauck, R.L., Elliott, D.M., 2012. Multi-scale structural and tensile mechanical response of annulus fibrosus to osmotic loading. Ann. Biomed. Eng. 40, 1610–1621.

Han, S.K., Chen, C.-W., Labus, K.M., Puttlitz, C.M., Chen, Y., Hsieh, A.H., 2016. Optical coherence tomographic Elastography reveals mesoscale shear strain Inhomogeneities in the annulus Fibrosus. Spine 41, E770–777.

Hickey, D.S., Hukins, D.W., 1980. Relation between the structure of the annulus fibrosus and the function and failure of the intervertebral disc. Spine 5, 106–116.

Hollingsworth, N.T., Wagner, D.R., 2011. Modeling shear behavior of the annulus fibrosus. J. Mech. Behav. Biomed. Mater. 4, 1103–1114.

Holzapfel, G.A., 2001. Biomechanics of soft tissue. In: The Handbook of Materials Behavior Models. vol. 3. Academic Press, Boston, pp. 1049–1063.

Holzapfel, G.A., Schulze-Bauer, C., Feigl, G., Regitnig, P., 2005. Single lamellar mechanics of the human lumbar anulus fibrosus. Biomech. Model. Mechanobiol. 3, 125–140.

Inoue, H., 1981. Three-dimensional architecture of lumbar intervertebral discs. Spine 6, 139–146.

Inoue, H., Takeda, T., 1975. Three-dimensional observation of collagen framework of lumbar intervertebral discs. Acta Orthop. Scand. 46, 949–956.

Kambic, H.E., McDevitt, C.A., 2005. Spatial organization of types I and II collagen in the canine meniscus. J. Orthop. Res. 23, 142–149.

Kiapour, A., Ambati, D., Hoy, R.W., Goel, V.K., 2012. Effect of graded facetectomy on biomechanics of Dynesys dynamic stabilization system. Spine 37, E581–E589.

Klein, J.A., Hukins, D.W., 1982. X-ray diffraction demonstrates reorientation of collagen fibres in the annulus fibrosus during compression of the intervertebral disc. Biochim. Biophys. Acta Gen. Subj. 717, 61–64.

Kobielarz, M., Szotek, S., Głowacki, M., Dawidowicz, J., Pezowicz, C., 2016. Qualitative and quantitative assessment of collagen and elastin in annulus fibrosus of the physiologic and scoliotic intervertebral discs. J. Mech. Behav. Biomed. Mater. 62, 45–56.

Krismer, M., Haid, C., Rabl, W., 1996. The contribution of anulus fibers to torque resistance. Spine 21, 2551–2557.

Labus, K.M., Han, S.K., Hsieh, A.H., Puttlitz, C.M., 2014. A computational model to describe the regional interlamellar shear of the annulus fibrosus. J. Biomech. Eng. 136, 051009.

Little, J., Adam, C., 2015. Geometric sensitivity of patient-specific finite element models of the spine to variability in user-selected anatomical landmarks. Comput. Methods Biomech. Biomed. Engin. 18, 676–688.

Liu, Y., Ueno, K., 1987. A three-dimensional nonlinear finite element model of lumbar intervertebral joint in torsion. J. Biomech. Eng. 109, 200–209.

Liu, C.-L., Zhong, Z.-C., Hsu, H.-W., Shih, S.-L., Wang, S.-T., Hung, C., Chen, C.-S., 2011. Effect of the cord pretension of the Dynesys dynamic stabilisation system on the biomechanics of the lumbar spine: a finite element analysis. Eur. Spine J. 20, 1850–1858.

Lu, Y.M., Hutton, W.C., Gharpuray, V.M., 1996. Do bending, twisting, and diurnal fluid changes in the disc affect the propensity to prolapse? A viscoelastic finite element model. Spine 21, 2570–2579.

Luxmoore, B.J., 2013. Computational Simulation of the Intervertebral Disc. The University of Leeds, Leeds, UK.

Luxmoore, B., Wijayathunga, V., Rehman, S., Wade, K., Rodrigues, S., Broom, N., Wilcox, R., 2012. In: Investigating the mechanical role of cross-bridging in the annulus Fibrosus using finite element analysis. Transactions of the 58th Annual Meeting of Orthopaedic Research Society San Francisco. pp. 2164.

Marchand, F., Ahmed, A.M., 1990. Investigation of the laminate structure of lumbar disc anulus fibrosus. Spine 15, 402–410.

Marini, G., Studer, H., Huber, G., Püschel, K., Ferguson, S.J., 2016. Geometrical aspects of patient-specific modelling of the intervertebral disc: collagen fibre orientation and residual stress distribution. Biomech. Model. Mechanobiol. 15, 543–560.

Mauck, R.L., Baker, B.M., Nerurkar, N.L., Burdick, J.A., Li, W.-J., Tuan, R.S., Elliott, D.M., 2009. Engineering on the straight and narrow: the mechanics of nanofibrous assemblies for fiber-reinforced tissue regeneration. Tissue Eng. B Rev. 15, 171–193.

Melrose, J., Smith, S.M., Appleyard, R.C., Little, C B., 2008. Aggrecan, versican and type VI collagen are components of annular translamellar crossbridges in the intervertebral disc. Eur. Spine J. 17, 314–324.

Mengoni, M., Luxmoore, B.J., Wijayathunga, V.N., Jones, A.C., Broom, N.D., Wilcox, R.K., 2015. Derivation of inter-lamellar behaviour of the intervertebral disc annulus. J. Mech. Behav. Biomed. Mater. 48, 164–172.

Michalek, A.J., Buckley, M.R., Bonassar, L.J., Cohen, I., Iatridis, J.C., 2009. Measurement of local strains in intervertebral disc anulus fibrosus tissue under dynamic shear: contributions of matrix fiber orientation and elastin content. J. Biomech. 42, 2279–2285.

Moramarco, V., del Palomar, A.P., Pappalettere, C., Doblaré, M., 2010. An accurate validation of a computational model of a human lumbosacral segment. J. Biomech. 43, 334–342.

Nerlich, A.G., Schleicher, E.D., Boos, N., 1997. 1997 Volvo award winner in basic science studies: immunohistologic markers for age-related changes of human lumbar intervertebral discs. Spine 22, 2781–2795.

Nerlich, A., Boos, N., Wiest, I., Aebi, M., 1998. Immunolocalization of major interstitial collagen types in human lumbar intervertebral discs of various ages. Virchows Arch. 432, 67–76.

Niemeyer, F., Wilke, H.-J., Schmidt, H., 2012. Geometry strongly influences the response of numerical models of the lumbar spine—a probabilistic finite element analysis. J. Biomech. 45, 1414–1423.

Noailly, J., Lacroix, D., Planell, J.A., 2005. Finite element study of a novel intervertebral disc substitute. Spine 30, 2257–2264.

Noailly, J., Wilke, H.-J., Planell, J.A., Lacroix, D., 2007. How does the geometry affect the internal biomechanics of a lumbar spine bi-segment finite element model? Consequences on the validation process. J. Biomech. 40, 2414–2425.

Noailly, J., Planell, J.A., Lacroix, D., 2011. On the collagen criss-cross angles in the annuli fibrosi of lumbar spine finite element models. Biomech. Model. Mechanobiol. 10, 203–219.

Parenteau-Bareil, R., Gauvin, R., Berthod, F., 2010. Collagen-based biomaterials for tissue engineering applications. Materials 3, 1863–1887.

Pech, P., Haughton, V., 1985. Lumbar intervertebral disk: correlative MR and anatomic study. Radiology 156, 699–701.

Peng, X., Guo, Z., Moran, B., 2006. An anisotropic hyperelastic constitutive model with fiber-matrix shear interaction for the human annulus fibrosus. J. Appl. Mech. 73, 815–824.

Reutlinger, C., Bürki, A., Brandejsky, V., Ebert, L., Büchler, P., 2014. Specimen specific parameter identification of ovine lumbar intervertebral discs: on the influence of fibre–matrix and fibre–fibre shear interactions. J. Mech. Behav. Biomed. Mater. 30, 279–289.

Roberts, S., Ayad, S., Menage, P., 1991a. Immunolocalisation of type VI collagen in the intervertebral disc. Ann. Rheum. Dis. 50, 787–791.

Roberts, S., Menage, J., Duance, V., Wotton, S., 1991b. Type III collagen in the intervertebral disc. Histochem. J. 23, 503–508.

Roberts, S., Menage, J., Duance, V., Wotton, S., Ayad, S., 1991c. 1991 Volvo award in basic sciences: collagen types around the cells of the intervertebral disc and cartilage end plate: an immunolocalization study. Spine 16, 1030–1038.

Roberts, S., Evans, H., Trivedi, J., Menage, J., 2006. Histology and pathology of the human intervertebral disc. J. Bone Joint Surg. Am. 88, 10–14.

Rodrigues, S.A., Wade, K.R., Thambyah, A., Broom, N.D., 2012. Micromechanics of annulus-end plate integration in the intervertebral disc. Spine J. 12, 143–150.

Rodrigues, S.A., Thambyah, A., Broom, N.D., 2017. How maturity influences annulus-endplate integration in the ovine intervertebral disc: a micro-and ultra-structural study. J. Anat. 230, 152–164.

Schmidt, H., Heuer, F., Simon, U., Kettler, A., Rohlmann, A., Claes, L., Wilke, H.-J., 2006. Application of a new calibration method for a three-dimensional finite element model of a human lumbar annulus fibrosus. Clin. Biomech. 21, 337–344.

Schmidt, H., Kettler, A., Rohlmann, A., Claes, L., Wilke, H.-J., 2007. The risk of disc prolapses with complex loading in different degrees of disc degeneration—a finite element analysis. Clin. Biomech. 22, 988–998.

Schmidt, H., Galbusera, F., Rohlmann, A., Zander, T., Wilke, H.-J., 2012. Effect of multilevel lumbar disc arthroplasty on spine kinematics and facet joint loads in flexion and extension: a finite element analysis. Eur. Spine J. 21, 663–674.

Schollmeier, G., Lahr-Eigen, R., Lewandrowski, K.-U., 2000. Observations on fiber-forming collagens in the anulus fibrosus. Spine 25, 2736–2741.

Schollum, M.L., Robertson, P.A., Broom, N.D., 2008. ISSLS prize winner: microstructure and mechanical disruption of the lumbar disc annulus: part I: a microscopic investigation of the translamellar bridging network. Spine 33, 2702–2710.

Schollum, M.L., Robertson, P.A., Broom, N.D., 2009. A microstructural investigation of intervertebral disc lamellar connectivity: detailed analysis of the translamellar bridges. J. Anat. 214, 805–816.

Sharma, M., Langrana, N.A., Rodriguez, J., 1995. Role of ligaments and facets in lumbar spinal stability. Spine 20, 887–900.

Shirazi-Adl, S.A., Shrivastava, S.C., Ahmed, A.M., 1984. Stress analysis of the lumbar disc-body unit in compression. A three-dimensional nonlinear finite element study. Spine 9, 120.

Shirazi-Adl, A., Ahmed, A.M., Shrivastava, S.C., 1986. Mechanical response of a lumbar motion segment in axial torque alone and combined with compression. Spine 11, 914–927.

Skaggs, D., Weidenbaum, M., Ratcliffe, A., Mow, V., 1994. Regional variation in tensile properties and biochemical composition of the human lumbar anulus fibrosus. Spine 19, 1310–1319.

Tamayol, A., Akbari, M., Annabi, N., Paul, A., Khademhosseini, A., Juncker, D., 2013. Fiber-based tissue engineering: progress, challenges, and opportunities. Biotechnol. Adv. 31, 669–687.

Tavakoli, J., 2017. Region–media coupling in characterization and modelling of the disc annulus single lamella swelling. Med. Biol. Eng. Comput. 1–10.

Tsuji, H., Hirano, N., Ohshima, H., Ishihara, H., Terahata, N., Motoe, T., 1993. Structural variation of the anterior and posterior Anulus Fibrosus in the development of human lumbar intervertebral disc: a risk factor for intervertebral disc rupture. Spine 18, 204–210.

Vergari, C., Mansfield, J., Meakin, J.R., Winlove, P.C., 2016. Lamellar and fibre bundle mechanics of the annulus fibrosus in bovine intervertebral disc. Acta Biomater. 37, 14–20.

Wade, K.R., Robertson, P.A., Broom, N.D., 2011. A fresh look at the nucleus-endplate region: new evidence for significant structural integration. Eur. Spine J. 20, 1225–1232.

Wade, K.R., Robertson, P.A., Broom, N.D., 2012a. On how nucleus-endplate integration is achieved at the fibrillar level in the ovine lumbar disc. J. Anat. 221, 39–46.

Wade, K.R., Robertson, P.A., Broom, N.D., 2012b. On the extent and nature of nucleus-annulus integration. Spine 37, 1826–1833.

Wu, J.-J., Eyre, D.R., Slayter, H., 1987. Type VI collagen of the intervertebral disc. Biochemical and electron-microscopic characterization of the native protein. Biochem. J. 248, 373–381.

Wu, J.-J., Woods, P.E., Eyre, D.R., 1992. Identification of cross-linking sites in bovine cartilage type IX collagen reveals an antiparallel type II-type IX molecular relationship and type IX to type IX bonding. J. Biol. Chem. 267, 23007–23014.

Yu, J., Tirlapur, U., Fairbank, J., Handford, P., Roberts, S., Winlove, C.P., Cui, Z., Urban, J., 2007. Microfibrils, elastin fibres and collagen fibres in the human intervertebral disc and bovine tail disc. J. Anat. 210, 460–471.

Yu, J., Schollum, M.L., Wade, K.R., Broom, N.D., Urban, J.P., 2015. ISSLS prize winner: a detailed examination of the elastic network leads to a new understanding of annulus fibrosus organization. Spine 40, 1149–1157.

Zhong, Z., Chen, S., Hung, C., 2009. Load-and displacement-controlled finite element analyses on fusion and non-fusion spinal implants. Proc. Inst. Mech. Eng. H J. Eng. Med. 223, 143–157.

CHAPTER

8

VERTEBRAL ENDPLATES

Kelly Wade

Institute of Orthopaedic Research and Biomechanics, Ulm University, Ulm, Germany

ENDPLATES

The cartilaginous endplates enclose the disc superiorly and inferiorly and partially constrain the hydrostatic nucleus. They are the junction between the relatively hard and rigid vertebrae and the soft, flexible tissue of the disc, and are elegantly structured to enable the effective transmission of load. At this point, it is important to make a distinction between the vertebral endplate and the cartilaginous endplate. The vertebral endplate has an obvious role in supporting the loads acting on the disc while also allowing blood to reach the avascular disc tissue. Strictly speaking, the vertebral endplate is not considered to be part of the disc, but it is included in this section because it plays an important role in the health and structure of the disc (Moore, 2006). The vertebral endplate is also known to play an important role in back pain when it is disturbed by injury or degeneration (Lotz, 2013). Because of the complex combination of forces acting on the diverse tissues of the endplate, it is a common site of disc failure. In fact, experimental evidence indicates this is the case under many loading conditions. These loading conditions and their effects will be considered later in this chapter.

Fig. 1 presents a basic structural overview of the endplate region showing the different regions and terminology that will be used in this chapter. This schematic also highlights the structural features that will be described in the following sections. It is intended to highlight the gradual transitions between the regions of the disc and how the attachment of disc to endplate varies from annulus to inner nucleus.

The basic structure of the cartilaginous endplates consists of a hydrated proteoglycan gel reinforced by a network of collagen fibrils. This is hyaline cartilage, whose fibers run largely parallel to the vertebral (osseous) endplate and is much like the structure of articular cartilage, but without the stratified zonation that is present in this structure. However, the fibers of the cartilaginous endplates are not thought to be directly connected to the fibers of the vertebral endplates (Moore, 2006; Inoue, 1981; Francois, 1982). There is a layer of calcified cartilage adjacent to the vertebral endplate (Roberts et al., 1989) that, as will be seen, is thought to at least partially anchor the disc to the vertebrae and that plays an important role in the nutrition and eventual degeneration of the disc. In humans, the cartilaginous endplates are around 0.5 to 1.0 mm thick, but the thickness varies with the region of the disc—it is thinner in the center of the disc than it is in the outer regions (Laffosse et al., 2009; Roberts et al., 1989).

The vertebral endplates consist of a layer of subchondral bone overlaying the trabecular bone of the vertebral body. It is important to note that the vertebral structure of most animals is very different from that of humans. In humans, the outer epiphyseal ring (also known as the ring apophysis) of the vertebrae is covered in cartilage, which acts as a growth plate for the vertebrae during maturation (Francois, 1975;

Biomechanics of the Spine. https://doi.org/10.1016/B978-0-12-812851-0.00008-2

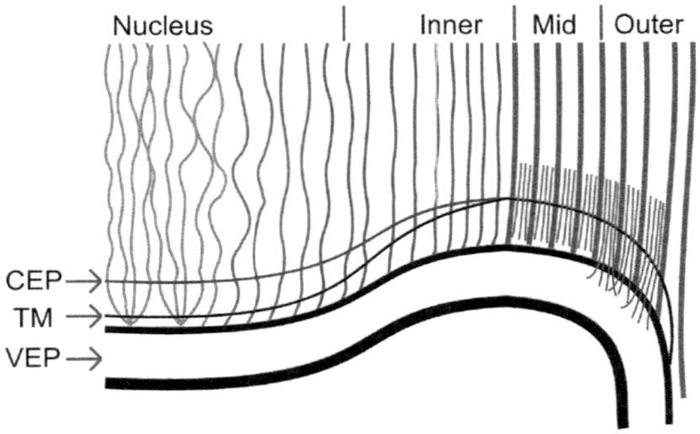

FIG. 1

Schematic representation of a central sagittal section of the posterior disc/endplate showing the relevant regions as will be referred to subsequently in this chapter. *CEP*, cartilaginous endplate; *TM*, tidemark (upper limit of calcified cartilaginous endplate); *VEP*, vertebral endplate. Note that the lamellae of the outer annulus are shown to branch into the vertebral endplate, in order to be consistent with recent evidence as shown in Figs. 3 and 4.

Francois and Dhem, 1974; Francois, 1982). In other animals, however, the vertebral endplate has a layer of cartilage between it and the bulk of the vertebra from which growth occurs (Laffosse et al., 2009). It should be noted that this is distinct from the recent observation that in humans the vertebral endplate can be double layered, which may offer advantages in terms of biomechanics and nutrition of the disc (Fields et al., 2012). Generally, the cranial endplate is thicker than the caudal endplate (Roberts et al., 1997; Laffosse et al., 2009; Francois, 1975; Francois and Dhem, 1974). The vertebral endplate is perforated with a network of arteries and veins that allow blood to contact the cartilaginous endplates and provide nutrition to the disc (Crock and Yoshizawa, 1976; Hirsch and Schajowicz, 1952). It has been established that the central zone of the endplate is more permeable than the outer regions (Accadbled et al., 2008).

In terms of biochemistry, the cartilaginous endplate consists largely of type II collagen, with types III, VI, and X present in the pericellular matrix (Roberts et al., 1991a, b). Type X collagen is of particular interest as it is thought to be related to hypertrophic chondrocytes and is produced by the cells of the endplate when they are involved in calcification of the endplate (Roberts et al., 1998; Moore, 2006). The proteoglycans present in the cartilaginous endplate are decorin and biglycan. These proteoglycans are likely to be different from those found in articular cartilage as they contain only chondroitin sulfate, whereas proteoglycans in articular cartilage also contain dermatan sulfate (Johnstone et al., 1993; Roberts et al., 1994).

Because the intervertebral disc is mainly avascular (Adams et al., 2002), the cells of the nucleus must receive their nutrition by diffusion through the endplates of the disc (Roberts et al., 1989). The ease by which this diffusion occurs through the endplates is governed by the proteoglycan content. Because proteoglycans have a negative charge, positively charged ions such as sodium and calcium easily diffuse into the disc, as do neutrally charged molecules such as glucose and oxygen. Large molecules are excluded simply because of their size (Urban et al., 1978; Urban et al., 1977). Another function of the endplate is thought to be to keep fragments of osmotically active proteoglycans in the disc, as it has been shown that removing proteoglycans from the endplate results in loss of proteoglycans from the nucleus (Roberts et al., 1993; Roberts et al., 1996).

The vertebral endplates are an important pathway through which nutrition can reach the disc. One possible cause of degeneration is the progressive calcification of these endplates with age, which may have the effect of reducing the nutrient supply to the cells of the disc, especially the nucleus. This hypothesis was tested by Hutton, who sealed the endplates of canine spines with bone cement and found that this procedure had very little effect on the appearance of the discs after a year (Hutton et al., 2004). However, other researchers have shown that the calcification present in degenerate and scoliotic disc endplates can significantly impede solute transfer (Urban et al., 2004; Roberts et al., 1996). Occlusion of the pores of the endplates by calcification has also been shown to have an effect on nutrion and correlates with increasing degeneration of the disc (Benneker et al., 2005).

FORCES ACTING ON THE ENDPLATES

Needle puncture experiments have shown that when the motion segment is compressed/loaded that hydrostatic pressure is generated within the nucleus, which is contained radially by the annulus, thus subjecting the annulus to tensile forces (McNally and Adams, 1992; Nachemson, 1963; Nachemson, 1981; Wilke et al., 1999; Sato et al., 1999). Therefore the endplates have multiple roles in life. In the central region, they must contain the pressure generated in the hydrostatic functional nucleus, and at the periphery, they anchor the lamellae of the annulus. In addition, they must be able to resist the shearing forces associated with the relative motion of the vertebrae. The endplates are intricately structured to achieve this feat, and breakdown of this structure is one of the key features of disc herniation and also degeneration. The factors leading to endplate damage will be discussed later in this chapter.

ENDPLATE STRUCTURE AND HOW IT IS INTEGRATED WITH ITS SURROUNDINGS

As briefly mentioned in the introduction to this chapter, the endplates are intricately structured and well integrated with the disc and vertebral bodies. The nature of this integration is dictated by the forces experienced by the endplates previously mentioned, which is described in the following sections. Unless otherwise noted, the following images are from ovine discs; analogous or identical structural features have been observed in human discs recently (Brown et al., 2016; Berg-Johansen et al., 2017).

CARTILAGINOUS ENDPLATE

Fibrils of the cartilaginous endplate are said to be parallel to the fibrils of the vertebral endplate (Hashizume, 1980; Inoue, 1981; Inoue and Takeda, 1975; Roberts et al., 1989). But what does this really mean? Does it mean that the fibrils of the cartilaginous endplate have a preferred orientation, or do they form a multidirectional woven fibrillar mat? To investigate this issue, sections of endplate adjacent to the nucleus were imaged in three different axial section planes: sagittal, 45 degree, and coronal. Fig. 2 shows scanning electron microscope images from the nucleus and cartilaginous endplate regions of these samples. In all three sets of images, the fibrils of the nucleus appear to be vertically aligned, as expected. Examination of the images from the cartilaginous endplate reveals that the fibrils in this region appear to have been cut mainly in cross-sections in all planes, thus indicating that they form a multidirectional (isotropic) fibrillar mat. As will be seen in the following sections, this densely woven mat of collagen fibrils anchors the fibrils of the annulus and nucleus in place. See Fig. 2.

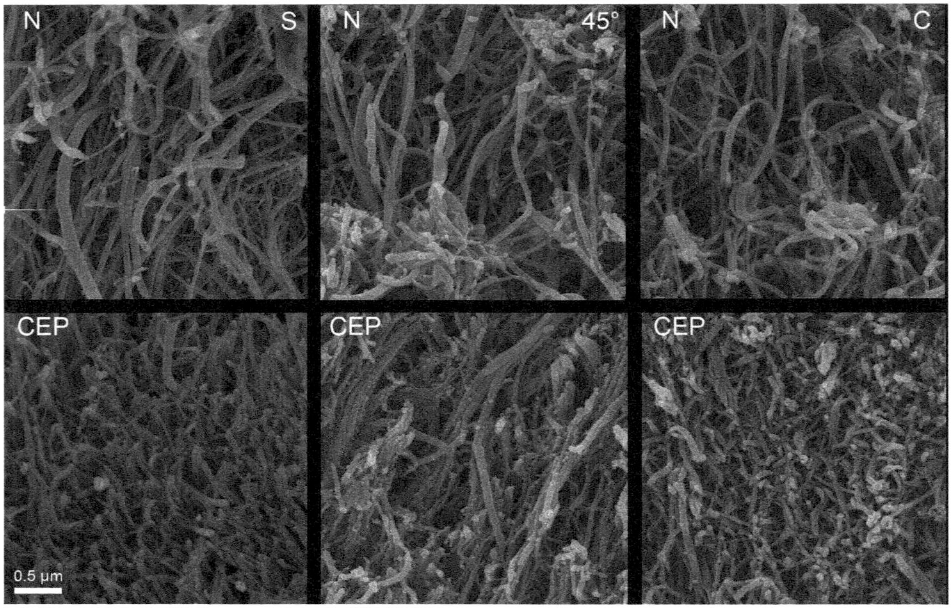

FIG. 2

Shows high magnification SEM views of the fibrils of the nucleus (N) and cartilaginous endplate (CEP). These images were taken from samples that were sectioned at three different angles: sagittal (S), 45 degree, and coronal (C). The differences between the fibril orientations in the nucleus and cartilaginous endplate are clearly visible.

ANNULUS-ENDPLATE JUNCTION

In general, the fibers of the annulus are inserted into the cartilaginous endplate. The depth of this insertion appears to vary from being moderately shallow in the inner annular region to deep in the outer annulus when viewed in histologic sections. The fibers of the outer annulus penetrate the vertebral bodies directly, and therefore anchor the disc to the vertebral bodies. However, this has been a contentious issue in the past, with one argument being that the outer layers should in fact be classified as ligaments (Francois, 1975; Francois and Dhem, 1974; Francois, 1982). It is believed that the layer of calcified cartilage in the cartilaginous endplate plays an important role in anchoring these annular fibers within the endplate along with interdigitation of the fiber bundles of the annulus and that the calcified cartilage has a "keying" effect, thus anchoring the disc to the underlying bone (Rodrigues et al., 2012). It was argued that there is no fibril-level anchoring structure between the calcified disc tissue (i.e., endplates and annular lamellae) and the vertebral bone (Hashizume, 1980; Inoue, 1981); however, this argument was subsequently shown to be incorrect (Rodrigues et al., 2015). This topic is discussed further in this section. Usually, the entire depth of the cartilaginous endplate is calcified in the mid and outer annulus (Francois and Dhem, 1974; Francois, 1975; Roberts et al., 1989), with the depth reducing toward the nucleus, as the schematic overview in Fig. 1 shows. Note that the extent of calcification increases with age, and this is thought to be related to the nutrition of the disc, as the increased calcification likely impairs the diffusion of nutrients into the disc (Roberts et al., 1996; Urban et al., 2004), as previously mentioned.

Elastic fibers run from the bulk of the annulus into the endplate. It is thought that these play some role in anchoring the annulus to the endplate and vertebral bodies (Johnson et al., 1984; Johnson et al.,

1982; Yu et al., 2002; Yu, 2002). Because these fibers are elastin, they may provide a restoring force on the disc, helping it to recover its shape after being deformed by the motion of the spinal column.

Recently, it was shown that the bundles of the lamellae divide into sub-bundles upon entering the endplate (Rodrigues et al., 2015; Rodrigues et al., 2017; Rodrigues et al., 2012). Each separate sub-bundle, or leaf, is then separated by a cartilaginous endplate matrix. Fig. 3 shows an example of this

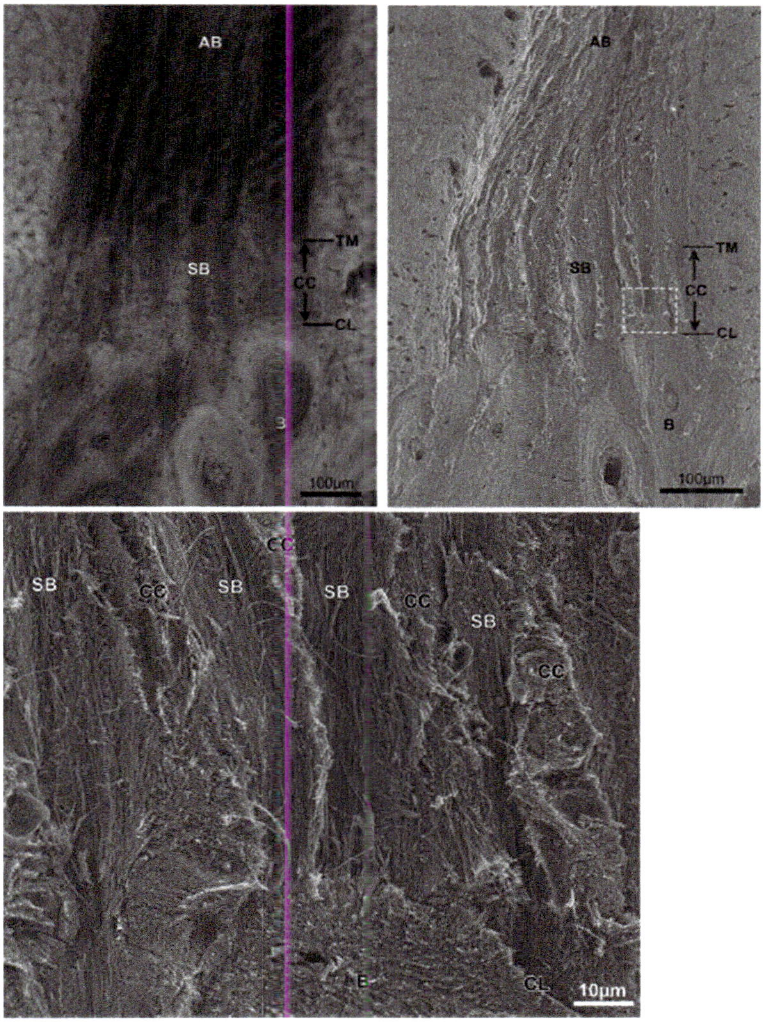

FIG. 3

(top left) Differential interference contrast in-plane optical image showing annular bundle (AB) entering calcified cartilage (CC) and splitting into annular sub-bundles (SB) that extend down to the cement line (CL). *(top right)* Matching low-magnification SEM image of the top left image. *(bottom)* Enlarged view of boxed region in the top right image showing sub-bundles in cartilaginous endplate separated by narrow seams of calcified cartilage. Note also the abrupt juxtapositioning of the fibrillar structures in the sub-bundle and bone. *TM*, tidemark; *B*, bone.

structural feature, and Fig. 4 illustrates it schematically. It has been proposed that this splitting into sub-bundles greatly increases the interface surface area between the annular fibers and the cartilaginous endplate, thus improving the transfer of shear stress across the interface. In fact, on average, the lamellae split into 22 leaves or sub-bundles upon their insertion into the endplate (ranging from 8 to 40). This will obviously have a corresponding increase in the surface area available to transmit the load on the lamellae into the endplate, thus reducing the stress acting upon this interface and increasing the force required to pull the lamella from its attachment point in the endplate. It was also proposed that the matrix of fibers surrounding this branching will support the branching bundles near the calcified junction, thus rendering the angle at which they bend less acute and reducing the stress concentration within these fibers.

A subsequent study that employed novel techniques of loading and decalcification of abstracted annulus-endplate samples showed the importance of the calcified cartilage in strengthening this junction, but also that there is still sufficient structural integrity if this calcification is removed to result in vertebral bone fragments being pulled out with annular fiber bundles (Rodrigues et al., 2015). These factors along with electron microscope images showing a degree of penetration of sub-bundle fibrils into the underlying vertebral bone indicate the importance of both components in the load-bearing ability of the annulus-endplate junction. It was argued that, although the interdigitation mechanism

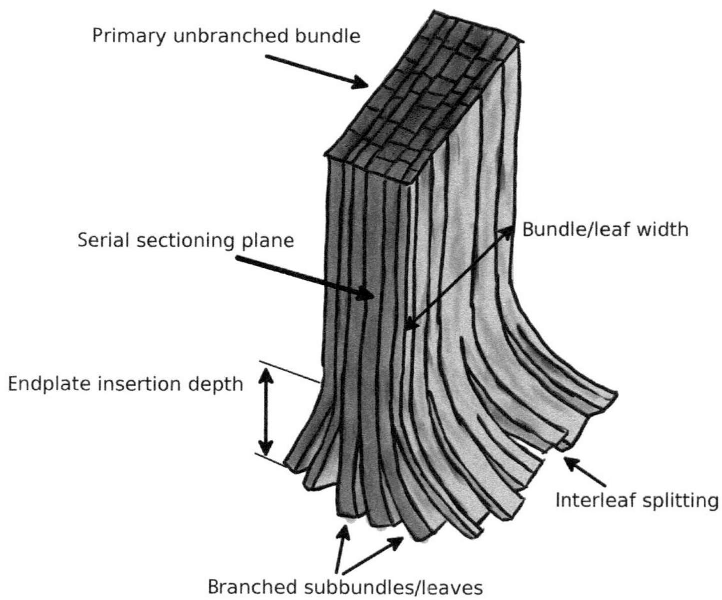

FIG. 4

Structural model illustrating (exaggerated) branching of the primary annular bundle into multiple sub-bundles in the endplate, which creates the "leaf" morphology, thus greatly increasing the interface area between annular lamellae and the endplate matrix.

Adapted from Rodrigues, S. A., Wade, K. R., Thambyah, A., Broom, N. D., 2012. Micromechanics of annulus-end plate integration in the intervertebral disc. Spine J. 12, 143–150.

mentioned previously is important in providing resistance to shear at the cement line, the mechanism is not likely to resist the tensile forces applied in the experiment, with this resistance being much more likely to result from the interweaving of the sub-bundle fibrils with those of the vertebral bone. This lack of resistance to tension has also been shown experimentally in the mid and inner annulus (Lama et al., 2014), suggesting that this fibril-level interweaving is mainly apparent in the outer region, as suggested in Fig. 1. The findings summarized here clearly illustrate the intricate and elegant structure of the annulus-endplate junction. Far from being a simple interface with a planar boundary between the annulus lamellae and the endplates, this junction is highly integrated and is structured so that the stresses from the soft tissue of the annular lamellae are distributed into the hard vertebral bone as smoothly as possible, with the cartilaginous endplate anchoring the fibrils of the lamellae and smoothing their transition. Despite this structural elegance, as mentioned in the introduction, it is vulnerable to failure. The process by which this failure occurs is described subsequently in this chapter, following consideration of nucleus-endplate integration.

NUCLEUS-ENDPLATE INTEGRATION

The predominant concept of the disc presented in most anatomy textbooks and many general papers portrays the disc as a binary structure with the nucleus being an amorphous, gel-like substance possessing little to no connectivity with its surroundings. However, this concept is an oversimplification and has recently been shown to be inaccurate (Wade et al., 2011). As will be seen, the true nature of the nucleus is quite different, and as described in the following section on endplate failure mechanisms, the nature of this structure and how it is integrated with the endplate influences the failure modes of the disc.

The consensus among early researchers such as Schmorl, Weber, von Luschka, and Carlier was that the nucleus is a gelatinous, mucoid structure (Carlier, 1890). By the 1940s, the concept of the nucleus as a loose network of fibrocartilage surrounded by a gelatinous matrix was well established (Keyes and Compere, 1932; Saunders and Inman, 1940; Coventry et al., 1945). It is interesting to note that Keyes and Compere state that the fibers of the nucleus pulposus and annulus fibrosus continue into the cartilaginous endplate (Keyes and Compere, 1932). However, little additional detail was provided in the literature to confirm exactly what these two authors meant by that claim. With the advent of electron microscopy in the late 1940s, it was possible to investigate the structure of the disc in finer detail than was previously possible. In 1951 Sylven stated that the nucleus should be considered as "a three-dimensional lattice containing a loose network of poorly differentiated collagenous fibrils surrounded by an amorphous interfibrillar substance" (Sylven et al., 1951). Subsequent studies further established the basic concept of the disc as a closed pack structure with the nucleus enclosed by the cartilaginous endplates and annulus (Happey, 1970; Hashizume, 1980; Inoue, 1981; Inoue and Takeda, 1975). However, Inoue stated that there is no (fibrillar) connection between the nucleus and the endplates (Inoue and Takeda, 1975). In the opinion of the author, this could be due to the nature of scanning electron microscopy, as any continuity needs to be located on the surface of the sample to be imaged clearly. The structural studies previously mentioned can be considered as the basis of current thought on the structure of the nucleus.

Subsequently, Roberts et al. restated the view that the fibers of the nucleus continue into the endplates (Roberts et al., 1989). Furthermore, Yu et al. identified a network of elastic fibers within the disc, which in the nucleus region appear to run from endplate to endplate as well as radially (Yu, 2002;

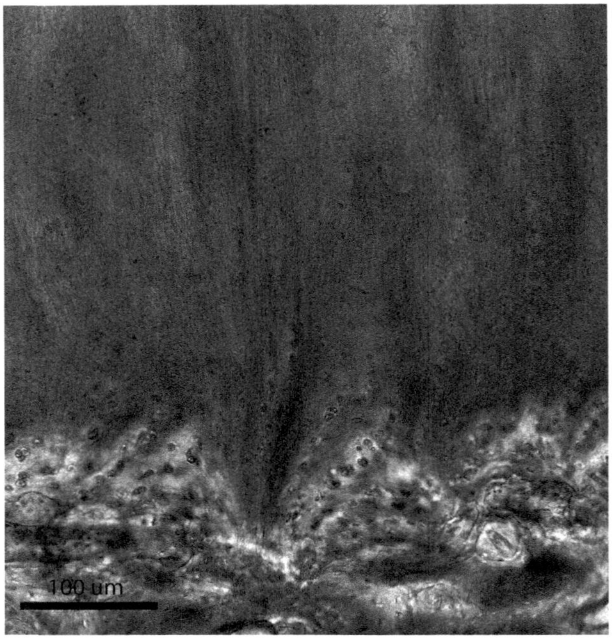

FIG. 5

Differential interference contrast image of the nucleus/endplate region. Note how the nucleus fibers are anchored within the cartilaginous endplate at the large discrete attachment node visible in the center of this image.

Yu et al., 2002; Yu et al., 2001). This extended the earlier findings of Buckwalter who originally identified elastic fibers in the nucleus (Buckwalter et al., 1976). It was suggested that these elastic fibers play some role in the anchorage and mechanical function of the nucleus.

A recent study by the author on ovine discs that employed novel techniques to isolate the nucleus from the annulus showed that it was possible to transmit a substantial load from endplate to endplate through this isolated nucleus (Wade et al., 2011). This finding implied that structural integration existed between the nucleus and endplates, and microstructural investigation showed that this process involved the fibers of the nucleus being gathered into characteristic nodal insertion points within the endplate. This mechanism of integration was first shown at the fiber level via light microscopy (see Fig. 5) and was subsequently confirmed at the ultrastructural level (Wade et al., 2012). The novel "ring severing" technique that enabled the nucleus to be isolated from its surroundings and fixed under load was key to enabling observation of this structure.

FAILURE OF THE ENDPLATES

Because the endplates are a junction between the relatively soft tissues of the disc and the hard bone of the vertebral body, they are subject to high concentrations of stress and are thus a common site of disc

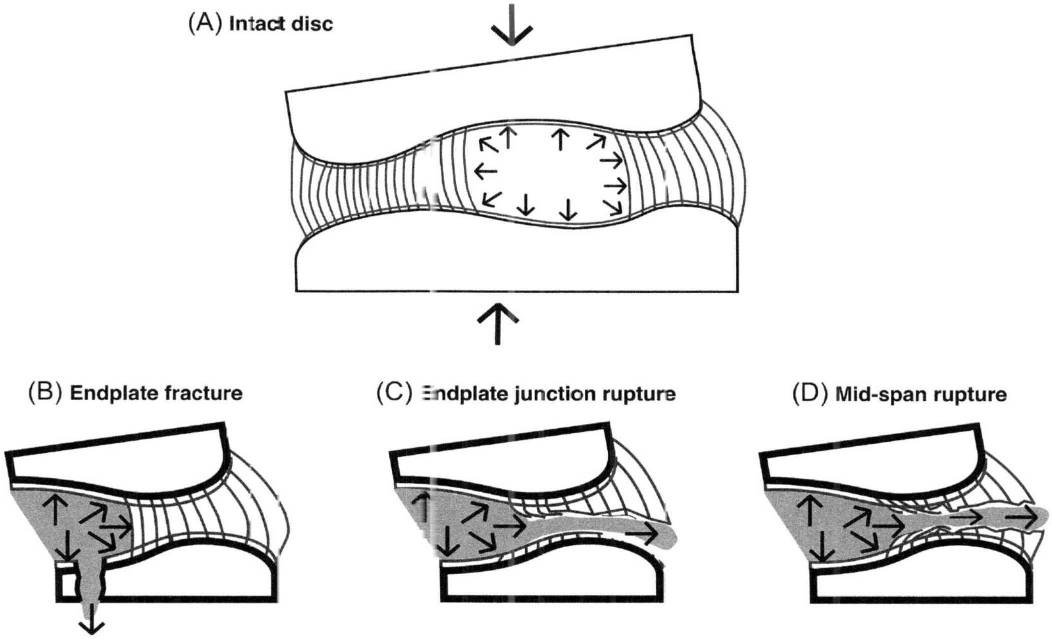

FIG. 6

(A) Shows the effects of compressive loading of a flexed healthy disc in which the posterior annulus is placed in tension and pressure is generated in the nucleus. If overloaded, the disc fails via one (or more) of the mechanisms illustrated below. (B) Endplate fracture in which the vertebral endplate itself suffers failure, allowing disc material to extrude into the vertebral body. (C) Endplate junction rupture in which the annulus separates from its attachment with the vertebra. (D) Mid-span rupture in which the annulus itself is torn, thus allowing migration of inner disc material.

failure. This failure can occur via several mechanisms that are governed by the posture and loading conditions the disc is subjected to. Clinically, disc herniation is known to occur via annulus-endplate junction failure or mid-substance rupture of the annular wall (Rajasekaran et al., 2013). Endplate fracture is also implicated in the development of Schmorl's nodes (Hansson and Roos, 1983; Hilton et al., 1976). Fig. 6 summarizes these modes of failure schematically. The following section outlines the factors that influence the development and progression of these failure mechanisms as identified by in vitro testing and clinical studies.

VERTEBRAL FRACTURE

The vertebral endplates have been described as the "weak link" of the spine in compression as they appear to be the most common initial point of failure of the motion segment in direct compression. Failure of the endplates causes decompression of the nucleus, and thus the compressive load on the

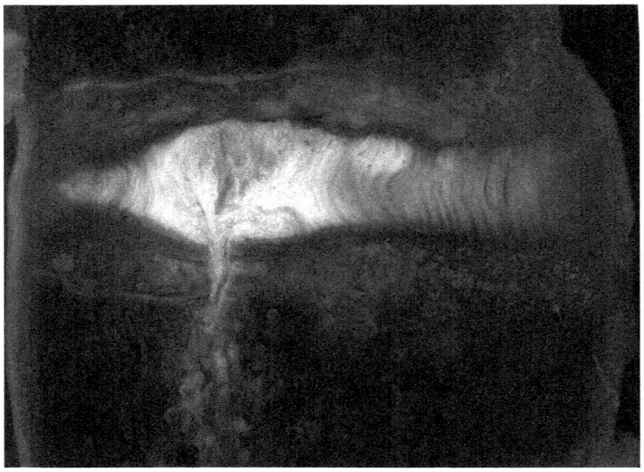

FIG. 7

Macro level view of a typical endplate fracture. The inferior vertebral endplate can be seen to have fractured subjacent to the mid-nucleus allowing the nucleus to be extruded into the vertebral body.

disc is transferred into the annulus, which then tends to bulge inward (Adams et al., 2000; Holm et al., 2004). Often nuclear material is extruded through the damaged endplate into the vertebral body, where calcification occurs and a Schmorl's node is formed (Hansson and Roos, 1983, Hilton et al., 1976). Fig. 7 shows an example. During in vitro testing, when discs are tested in a neutral, that is, an unflexed posture, they almost invariably suffer vertebral fracture (Brown et al., 1957; Lin et al., 1978; Lundin et al., 1998; Roaf, 1960; Rolander and Blair, 1975; Yoganandan et al., 1988). Also, if discs are tested at a very low loading rate, they almost invariably suffer this type of vertebral endplate fracture. This phenomenon is thought to be a consequence of the low loading rate allowing viscoelastic creep or the annulus to redistribute the load that it is subjected to, thereby reinforcing the idea that the vertebral endplate is the weak link of the disc. For herniation to occur, the disc must be subjected to both a non-neutral posture and sufficiently rapid loading for either the annulus or the annulus-endplate junction to suffer failure (Wade et al., 2014; Wade et al., 2015; Adams and Hutton, 1982; Adams and Hutton, 1985; McNally et al., 1993; Veres et al., 2009); the latter mode of failure will be described subsequently. Note that the effects of very rapid (that is, impact) loading on the disc are yet to be determined at the microstructural level in a physiologically relevant scenario. However, experiments using drop towers to subject discs to very rapid loading do indicate that the vertebral endplate and growth plate are vulnerable under such conditions (Oxland et al., 1991; Panjabi et al., 1989), in addition to burst fractures of the vertebral bodies (Yoganandan et al., 1986). In other words, at extreme loading rates, the vertebral endplates may suffer failure, whereas under the conditions in between that is, at moderate loading rates that the soft tissue is more vulnerable.

Another common mode of failure involving vertebral fracture is the rim fracture or rim lesion, which, as the name suggests, involves the failure of the edge of the vertebral body adjacent to the annulus, at the ring apophysis in humans (Osti et al., 1990; Vernon-Roberts and Pirie, 1977). In the opinion of the author, this is a form of annulus-endplate junction failure and is an area where in vitro results from animal models should be interpreted with caution because the structure of the human vertebral body is considerably different from the ovine or porcine models commonly used. This caution is not intended to discredit the use of these animal models, which are valuable in the investigation of overall phenomena, but rather to point out that the less dense human vertebra may be more vulnerable to suffering subchondral fracture when its annulus-endplate junction suffers failure as compared to the denser quadruped vertebral body. However, both adolescent human and immature animal motion segments exhibit similar vulnerabilities in vitro (Lundin et al., 2000).

ANNULUS-ENDPLATE JUNCTION FAILURE

As previously mentioned, the endplate junction is a common location for the initiation of disc herniation because of the complex forces the junction is subject to and the fact that it is a hard-soft junction, which, despite its intricate architecture, is inherently vulnerable to failure. In vitro studies have shown that both loading rate and posture influence the likelihood of herniation, with a posture that does not allow the disc to redistribute load in combination with a rapid enough loading rate to prohibit viscoelastic creep of the disc (Race et al., 2000). However, the analysis of disc failure in these investigations was limited to a macro-level visual examination of the damaged discs, which could not reveal the subtleties of disc failure.

More recent in vitro investigations that explored the effects of posture and loading rate on ovine discs combined with microstructural investigation revealed that loading rate further influences the incidence of endplate junction failure. Relatively low loading rates invariably produced endplate fracture irrespective of posture, likely because viscoelastic effects allowed the disc to creep and redistribute the load it was subjected to. Loading one order of magnitude faster caused discs to suffer herniation; however, the majority of the discs subjected to this rate suffered mid-span rupture of the annulus (Wade et al., 2014). When discs were loaded an additional order of magnitude faster, a significant majority suffered annulus-endplate junction failure (Wade et al., 2015), as shown in Fig. 8. This result was attributed to the rapid loading causing a compliance imbalance across the hard-soft interface of the annulus-endplate junction. A previous study by the same research group, using a novel fluid inflation technique, suggested that rapid pressurization produces a shock wave traveling outward from the nucleus which causes disc failure (Veres et al., 2010b).

Complex loading conditions can also affect the mode of failure of the disc. These complex postures are thought to create unequal loading within the annulus, which leads to one population of fibers being more highly stressed than the other fibers. Therefore this more highly stressed population of fibers suffers failure first. When this failure involves the annulus-endplate junction, it is often possible to observe the failure of the alternate (loaded) lamellae (Veres et al., 2010a). The previously mentioned rate-dependent effect still appears to operate; that is, more rapid loading tends to lead to annulus endplate junction failure. It has also been suggested that these complex postures tend to cause a higher proportion of endplate junction failures (Berger-Roscher et al., 2017). In addition, although the results of discs subjected to compression or simple flexion could lead to the conclusion that failure will occur either in the annular wall or the vertebral endplate, both modes of failure have been observed in discs subjected to complex loading in vitro. This result may arise from the less loaded population of annulus fibers still being able to contain enough pressure in the nucleus to cause the endplate fracture following the failure of the loaded annulus fibers.

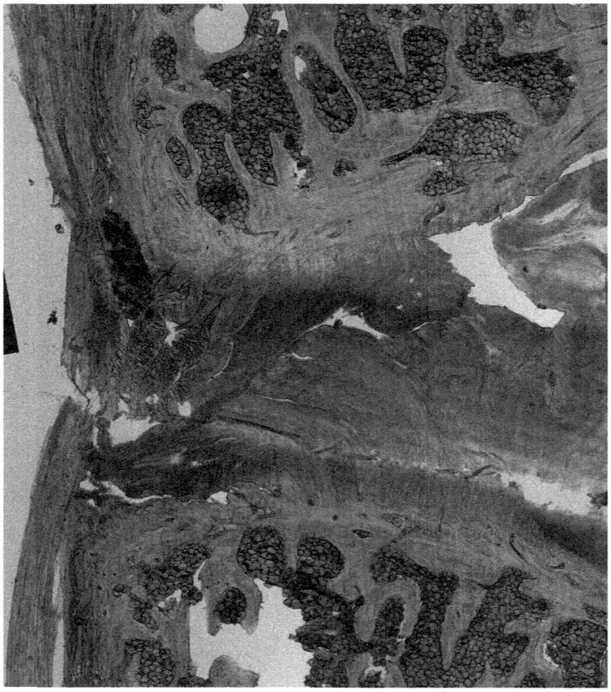

FIG. 8

A sagittal section through the posterior-mediolateral region of a disc that failed via subligamentous herniation. Failure initiated at the annulus-endplate junction and damaged the cartilaginous endplate at both the tidemark and cement line.

NUCLEUS-ENDPLATE JUNCTION

The nucleus-endplate junction can also be damaged during disc herniation; however, the mechanism by which this occurs is quite different from the annulus-endplate junction failure previously described. More specifically, in vitro testing indicates that annulus-endplate junction failure is usually the initiating event during disc herniation, whereas nucleus endplate failure is a consequence of disc herniation. Only extensively compressed discs with substantial herniations were observed to have suffered nucleus-endplate failure, suggesting that extensive displacement of nucleus material causes the nucleus fibers to pull away from the endplates at their attachment nodes. This condition usually occurs at their interface with the tidemark, that is, the calcified portion of the cartilaginous endplate, as was observed recently in vitro. This finding provides a further possible explanation for the clinical observation of fragments of endplate in herniated material (Rajasekaran et al., 2013; Lama et al., 2014).

In this chapter, the endplate was shown to be an elegantly structured region of the spine, which effectively connects the disc to the vertebrae. The mechanisms of integration vary between the outer and inner regions of the disc and are related to the forces acting on these regions: the endplate must both anchor the annulus and contain the hydrostatic pressure generated within the nucleus. Because of the complex forces acting across this hard-soft tissue junction and despite its ingenious and elegant structure, the endplate is a very common site of disc failure. The mechanism by which this occurs depends on the loading conditions the disc is subjected to, and the nature of the disc structure governs how it reacts under these conditions.

Acknowledgments

The author would like to acknowledge that the material in this chapter relating to disc structure has been adapted from his doctoral thesis "Reconsidering the disc nucleus and its surroundings" published in 2012 at The University of Auckland.

REFERENCES

Accadbled, F., Ambard, D., De Gauzy, J.S., Swider, P., 2008. A measurement technique to evaluate the macroscopic permeability of the vertebral end-plate. Med. Eng. Phys. 30, 116–122.

Adams, M.A., Hutton, W.C., 1982. Prolapsed intervertebral disc: a hyperflexion injury. Spine 7, 184–191.

Adams, M.A., Hutton, W.C., 1985. Gradual disc prolapse. Spine 10, 524–531.

Adams, M.A., Freeman, B.J.C., Morrison, H.P., Nelson, I.W., Dolan, P., 2000. Mechanical initiation of intervertebral disc degeneration. Spine 25, 1625–1636.

Adams, M.A., Bogduk, N., Burton, K., Dolan, P., 2002. The Biomechanics of Back Pain. Elsevier Science Limited, London.

Benneker, L.M., Heini, P.F, Alini, M., et al., 2005. 2004 Young investigator award winner: Vertebral endplate marrow contact channel occlusions and intervertebral disc degeneration. Spine 30, 167–73.

Berg-Johansen, B., Fields, A.J., Liebenberg, E.C., et al., 2017. Structure-function relationships at the human spinal disc-vertebra interface. Journal of Orthopaedic Research.

Berger-Roscher, N., Casaroli, G., Rasche, V., Villa, T., Galbusera, F., Wilke, H.J., 2017. Influence of complex loading conditions on intervertebral disc failure. Spine 42, E78–E85.

Brown, T., Hansen, R.J., Yorra, A.J., 1957. Some mechanical tests on the lumbosacral spine with particular reference to the intervertebral discs; a preliminary report. J. Bone Joint Surg. Am. 39A, 1135–1164.

Brown, S., Rodrigues, S., Sharp, C., Wade, K., Broom, N., McCall, I.W., Roberts, S., 2016. Staying connected: structural integration at the intervertebral disc-vertebra interface of human lumbar spines. Eur. Spine J. 26, 248–258.

Buckwalter, J.A., Cooper, R.R., Maynard, J.A., 1976. Elastic fibers in human intervertebral discs. J. Bone Joint Surg. Am. 58, 73–76.

Carlier, E.W., 1890. The fate of the notochord and development of the intervertebral disc in the sheep, with observations on the structure of the adult disc in these animals. J. Anat. Physiol. 24, 573–584.

Coventry, M.B., Ghormley. R.K., Kernohan, J.W., 1945. The intervertebral disc: its microscopic anatomy and pathology. Part I. Anatomy, development and physiology. J. Bone Joint Surg. 105A, 105–112. XXVII.

Crock, H.V., Yoshizawa, H., 1976. The blood supply of the lumbar vertebral column. Clin. Orthop. Relat. Res. 115, 6–21.

Fields, A.J., Sahli, F., Rodriguez, A.G., et al., 2012. Seeing double: A comparison of microstructure, biomechanical function, and adjacent disc health between double- and single-layer vertebral endplates. Spine 37, E1310–E1317.

Francois, R.J., 1975. Ligament insertions into the human lumbar vertebral body. Acta Anat. 467–480.

Francois, R.J., 1982. Three-dimensional architecture of lumbar intervertebral discs. Spine 7, 522–523.
Francois, R.J., Dhem, A., 1974. Microradiographic study of the normal human vertebral body. Acta Anat. 251–265.
Hansson, T., Roos, B., 1983. The amount of bone mineral and Schmorl's nodes in lumbar vertebrae. Spine 8, 266–271.
Happey, F., 1970. Studies of the human intervertebral disc. Eur. Polym. J. 6.
Hashizume, H., 1980. Three-dimensional architecture and development of lumber intervertebral discs. Acta Med. Okayama 34, 301–314.
Hilton, R.C., Ball, J., Benn, R.T., 1976. Vertebral end plate lesions (Schmorl's nodes) in the dorsolumbar spine. Ann. Rheum. Dis. 35, 127–132.
Hirsch, C., Schajowicz, F., 1952. Studies on structural changes in the lumbar annulus Fibrosus. Acta Orthop. 22, 184–231.
Holm, S., Holm, A.K., Ekström, L., Karladani, A., Hansson, T., 2004. Experimental disc degeneration due to endplate injury. J. Spinal Disord. Tech. 17, 64–71.
Hutton, W.C., Murakami, H., Li, J., Elmer, W.A., Yoon, S.T., Minamide, A., Akamaru, T., Tomita, K., 2004. The effect of blocking a nutritional pathway to the intervertebral disc in the dog model. J. Spinal Disord. Tech. 17, 53–63.
Inoue, H., 1981. Three-dimensional architecture of lumbar intervertebral discs. Spine 6, 139–146.
Inoue, H., Takeda, T., 1975. Three dimensional observation of collagen framework of lumbar intervertebral discs. Acta Orthop. Scand. 46, 949–956.
Johnson, E.F., Chetty, K., Moore, I.M., 1982. The distribution and arrangement of elastic fibers in the intervertebral disc of the adult human. J. Anat. 135, 301–309.
Johnson, E.F., Caldwell, R.W., Berryman, H.E., 1984. Elastic fibers in the annulus fibrosus of the dog intervertebral disc. Acta Anat. 118, 238–242.
Johnstone, B., Markopoulos, M., Neame, P., Caterson, B., 1993. Identification and characterization of glycanated and non-glycanated of biglycan and decorin in the human intervertebral disc. Biochem. J. 292, 661–666.
Keyes, D.C., Compere, E.L., 1932. The normal and pathological physiology of the nucleus pulposus of the intervertebral disc: an anatomical, clinical and experimental study. J. Bone Joint Surg. 14, 897–938.
Laffosse, J.M., Kinkpe, C., Gomez-Brouchet, A., Accadbled, F., Viguier, E., Sales De Gauzy, J., Swider, P., 2009. Micro-computed tomography study of the subchondral bone of the vertebral endplates in a porcine model: correlations with histomorphometric parameters. Surg. Radiol. Anat. 1–7.
Lama, P., Zehra, U., Balkovec, C., Claireaux, H.A., Flower, L., Harding, I.J., Dolan, P., Adams, M.A., 2014. Significance of cartilage endplate within herniated disc tissue. Eur. Spine J. 23, 1869–1877.
Lin, H.S., Liu, Y.K., Adams, K.H., 1978. Mechanical response of the lumbar intervertebral joint under physiological (complex) loading. J. Bone Joint Surg. Am 60 (A), 41–55.
Lotz, J.C., Fields, A.J., Liebenberg, E.C., 2013. The Role of the Vertebral End Plate in Low Back Pain. Global Spine Journal 3, 153–64.
Lundin, O., Ekström, L., Hellström, M., Holm, S., Swärd, L., 1998. Injuries in the adolescent porcine spine exposed to mechanical compression. Spine 23, 2574–2579.
Lundin, O., Ekström, L., Hellström, M., Holm, S., Swärd, L., 2000. Exposure of the porcine spine to mechanical compression: differences in injury pattern between adolescents and adults. Eur. Spine J. 9, 466–471.
McNally, D.S., Adams, M.A., 1992. Internal intervertebral disc mechanics as revealed by stress profilometry. Spine 17, 66–73.
McNally, D.S., Adams, M.A., Goodship, A.E., 1993. Can intervertebral disc prolapse be predicted by disc mechanics? Spine 18, 1525–1530.
Moore, R.J., 2006. The vertebral endplate: disc degeneration, disc regeneration. Eur. Spine J. 15, .
Nachemson, A., 1963. The influence of spinal movements on the lumbar intradiscal pressure. Acta Orthop. Scand. 33, 183–207.
Nachemson, A.L., 1981. Disc pressure measurements. Spine 6, 93–97.

Osti, O.L., Vernon-Roberts, B., Fraser, R.D., 1990. 1990 Volvo award in experimental studies: anulus tears and intervertebral disc degeneration: an experimental study using an animal model. Spine 15, 762–767.
Oxland, T.R., Panjabi, M.M., Southern, E.P., Duranceau, J.S., 1991. An anatomic basis for spinal instability: a porcine trauma model. J. Orthop. Res. 9, 452–462.
Panjabi, M.M., Duranceau, J.S., Oxland, T.R., Bowen, C.E., 1989. Multidirectional instabilities of traumatic cervical spine injuries in a porcine model. Spine 14, 1111–1115.
Race, A., Broom, N.D., Robertson, P., 2000. Effect of loading rate and hydration on the mechanical properties of the disc. Spine 25, 662–669.
Rajasekaran, S., Bajaj, N., Tubaki, V., Kanna, R.M., Shetty, A.P., 2013. ISSLS prize winner: the anatomy of failure in lumbar disc herniation: an in-vivo, multi-modal, prospective study of 181 subjects. Spine 38, 1491–1500.
Roaf, R., 1960. A study of the mechanics of spinal injuries. J. Bone Joint Surg. (Br.) 42B, 810–823.
Roberts, S., Menage, J., Urban, J.P.G., 1989. Biochemical and structural properties of the cartilage end-plate and its relation to the intervertebral disc. Spine 14, 166–174.
Roberts, S., Menage, J., Duance, V., Wotton, S., Ayad, S., 1991a. 1991 Volvo award in basic sciences: collagen types around the cells of the intervertebral disc and cartilage end plate: an immunolocalization study. Spine 16, 1030–1038.
Roberts, S., Menage, J., Duance, V., Wotton, S.F., 1991b. Type III collagen in the intervertebral disc. Histochem. J. 23, 503–508.
Roberts, S., Menage, J., Eisenstein, S.M., 1993. The cartilage end-plate and intervertebral disc in scoliosis: calcification and other sequelae. J. Orthop. Res. 11, 747–757.
Roberts, S., Caterson, B., Evans, H., Eisenstein, S.M., 1994. Proteoglycan components of the intervertebral disc and cartilage endplate: an immunolocalization study of animal and human tissues. Histochem. J. 26, 402–411.
Roberts, S., Urban, J.P.G., Evans, H., Eisenstein, S.M., 1996. Transport properties of the human cartilage endplate in relation to its composition and calcification. Spine 21, 415–420.
Roberts, S., McCall, I.W., Menage, J., Haddaway, M.J., Eisenstein, S.M., 1997. Does the thickness of the vertebral subchondral bone reflect the composition of the intervertebral disc? Eur. Spine J. 6, 385–389.
Roberts, S., Bains, M.A., Kwan, A., Menage, J., Eisenstein, S.M., 1998. Type X collagen in the human invertebral disc: an indication of repair or remodelling? Histochem. J. 30, 89–95.
Rodrigues, S.A., Wade, K.R., Thambyah, A., Broom, N.D., 2012. Micromechanics of annulus-end plate integration in the intervertebral disc. Spine J. 12, 143–150.
Rodrigues, S.A., Thambyah, A., Broom, N.D., 2015. A multiscale structural investigation of the annulus-endplate anchorage system and its mechanisms of failure. Spine J. 15, 405–416.
Rodrigues, S.A., Thambyah, A., Broom, N.D., 2017. How maturity influences annulus-endplate integration in the ovine intervertebral disc: a micro- and ultra-structural study. J. Anat. 230, 152–164.
Rolander, S.D., Blair, W.E., 1975. Deformation and fracture of the lumbar vertebral end plate. Orthop. Clin. N. Am. 6, 75–81.
Sato, K., Kikuchi, S., Yonezawa, T., 1999. In vivo intradiscal pressure measurement in healthy individuals and in patients with ongoing back problems. Spine 24, 2468–2474.
Saunders, J.B.D.C.M., Inman, V.T., 1940. Pathology of the intervertebral disk. Arch. Surg. 40, 389–416.
Sylven, B., Paulson, S., Hirsch, C., Snellman, O., 1951. Biophysical and physiological investigations on cartilage and other mesenchymal tissues: II. The ultrastructure of bovine and human nuclei pulposi. J. Bone Joint Surg. Am. 33, 333–340.
Urban, J.P.G., Holm, S., Maroudas, A., Nachemson, A., 1977. Nutrition of the intervertebral disk. An in vivo study of solute transport. Clin. Orthop. Relat. Res. 129, 101–114.
Urban, J.P.G., Holm, S., Maroudas, A., 1978. Diffusion of small solutes into the intervertebral disc: an in vivo study. Biorheology 15, 203–223.
Urban, J.P.G., Smith, S., Fairbank, J.C.T., 2004. Nutrition of the intervertebral disc. Spine 29, 2700–2709.

Veres, S.P., Robertson, P.A., Broom, N.D., 2009. The morphology of acute disc herniation: a clinically relevant model defining the role of flexion. Spine (Phila Pa 1976) 34, 2288–2296.

Veres, S.P., Robertson, P.A., Broom, N.D., 2010a. The influence of torsion on disc herniation when combined with flexion. Eur. Spine J. 19, 1468–1478.

Veres, S.P., Robertson, P.A., Broom, N.D., 2010b. ISSLS prize winner: how loading rate influences disc failure mechanics: a microstructural assessment of internal disruption. Spine 35, 1897–1908.

Vernon-Roberts, B., Pirie, C.J., 1977. Degenerative changes in the intervertebral discs of the lumbar spine and their sequelae. Rheumatol. Rehabil. 16, 13–21.

Wade, K.R., Robertson, P.A., Broom, N.D., 2011. A fresh look at the nucleus-endplate region: new evidence for significant structural integration. Eur. Spine J. 20, 1225–1232.

Wade, K.R., Robertson, P.A., Broom, N.D., 2012. On how nucleus-endplate integration is achieved at the fibrillar level in the ovine lumbar disc. J. Anat. 221, 39–46.

Wade, K.R., Robertson, P.A., Thambyah, A., Broom, N.D., 2014. How healthy discs herniate: a biomechanical and microstructural study investigating the combined effects of compression rate and flexion. Spine 39, 1018–1028.

Wade, K.R., Robertson, P.A., Thambyah, A., Broom, N.D., 2015. 'Surprise' loading in flexion increases the risk of disc herniation due to annulus-endplate junction failure: a mechanical and microstructural investigation. Spine 40, 891–901.

Wilke, H.J., Neef, P., Caimi, M., Hoogland, T., Claes, L.E., 1999. New in vivo measurements of pressures in the intervertebral disc in daily life. Spine 24, 755–762.

Yoganandan, N., Sances Jr, A., Maiman, D.J., Myklebust, J.B., Pech, P., Larson, S.J., 1986. Experimental spinal injuries with vertical impact. Spine 11, 855–860.

Yoganandan, N., Maiman, D.J., Pintar, F., Ray, G., Myklebust, J.B., Sances Jr, A., Larson, S.J., 1988. Microtrauma in the lumbar spine: a cause of low back pain. Neurosurgery 23, 162–168.

Yu, J., 2002. Elastic tissues of the intervertebral disc. Biochem. Soc. Trans. 30, 848–852.

Yu, J., Winlove, C.P., Roberts, S., Urban, J.P.G., 2001. Elastic fibres in bovine intervertebral disc. Int. J. Exp. Pathol. 82, A24–A25.

Yu, J., Peter, C., Roberts, S., Urban, J.P.G., 2002. Elastic fibre organization in the intervertebral discs of the bovine tail. J. Anat. 201, 465–475.

CHAPTER

SPINAL MUSCLES

9

Themis Toumanidou
Universitat Pompeu Fabra, Barcelona, Spain

INTRODUCTION

Muscles play a vital role in the skeletal joint system principally as stabilizers for the upright skeleton and as the basis for a variety of movements. Moreover, they function as shock absorbers that disperse the loads applied and protect the internal organs from external forces. They act individually or in coordination as prime movers, synergists, or antagonists in function of their morphology and the movement performed. Coordination often occurs between muscles that have direct vertebral insertions, like the multifidus, and others that have an indirect effect when they contract, like the abdominal muscles. Knowledge of both morphological and physiological muscle characteristics allows for better interpretations of their role and is essential for the development of more truthful computational models to complement the experimental studies on spine biomechanics.

First, the structure and physical properties of skeletal muscles are summarized and the cross-bridge theory, that is, the principal mechanism that explains muscle force generation, is described briefly to familiarize the reader with the muscle physiology. Then the spinal muscles are presented in groups as distinguished by their location around the vertebral column and the region they attach, that is, cervical, thoracic, or lumbar. The discussion is based on clinical anatomical descriptions of the insertions, innervation patterns, and actions of the superficial and deep muscles of the back, the iliac, and the abdominal muscles. Finally, a review of the reported efforts so far to model the geometry and biomechanics of spinal muscles using numerical methods is included.

PHYSIOLOGY

STRUCTURE

Spinal muscles belong to the skeletal type, that is, muscles attached to the skeleton via tendons. They are able to create voluntary motion of the skeletal system and skin, as well as maintain the trunk posture and distribute the external loads with the purpose of protecting the spine. Skeletal muscles have a hierarchical structure that is highly organized and that can be studied at both a macroscopic and microscopic level (Fig. 1). They are separated by the *fascia*, a band of connective tissue that surrounds and stabilizes the muscles and other internal organs forming an outer wrapping. Interiorly, each muscle is covered by the *epimysium*, a fibrous tissue that in its turn encloses the *perimysium*. The latter is a collagenous sheath that surrounds and protects the fascicles. Each *fascicle* (or fasciculus) is a bundle

Biomechanics of the Spine. https://doi.org/10.1016/B978-0-12-812851-0.00009-4

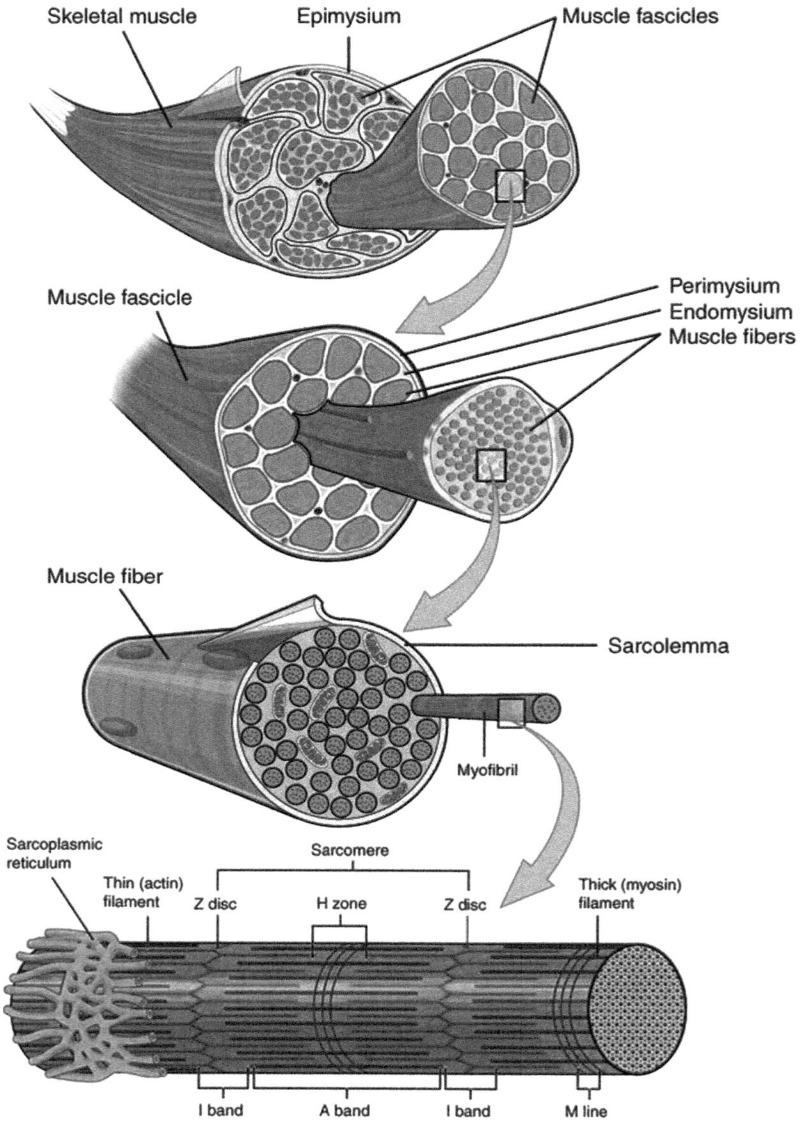

FIG. 1

Structural hierarchy of skeletal muscle.

Adapted from Anatomy and Physiology, Rice University, OpenStax College.

of muscle fibers, also called *myocytes*, bound together via the *endomysium* tissue that provides pathways for the passage of blood vessels and nerves. At a macroscopic level, three general types of fiber architectures exist depending on the angle formed between the muscle longitudinal direction and the force-generating axis, i.e., unipennate (one angle), multipennate (multiple angles), and parallel (zero angle) muscles. Examples of multipennate muscles are the deltoids, and of parallel architecture are the rectus abdominis and psoas major muscles. This physical arrangement of fibers influences the muscle mechanical function. For instance, force transmission to the tendon in parallel muscles is 100% efficient, whereas in the pennate muscles, the force is distributed also in the fiber direction. However, a pennate muscle is stronger because the maximum isometric force σ_0 it can exert depends on its physiological cross-sectional area (PCSA), that is, the area of the cross-section of the muscle perpendicular to its fibers, which is greater in pennate muscles.

At a microscopic level, muscle fibers are the smallest contractile units of the skeletal muscles and develop from a fusion of precursor cells known as myoblasts in the process of myogenesis. They are long, cylindrical, and multinucleated cells that run parallel to each other within a muscle. Like other cells, skeletal muscle fibers have a cytoplasm, the *sarcoplasm*, filled with significant amounts of glycogen, myoglobin, and mitochondria between parallel bundles of *myofibrils*. These rod-like units attach to the *sarcolemma*, that is, the plasma membrane that is the site of action potential conduction, and they are surrounded by the *sarcoplasmic reticulum* that serves to store the calcium ions needed to cause a muscle contraction. Myofibrils consist of long proteins such as actin, myosin, and titin, organized in thick (myosin) and thin (actin) *myofilaments* (Fig. 1). Such filaments repeat along the length of the myofibril forming the *sarcomeres*, the units that attribute the distinctive striated appearance to skeletal muscles. The sliding between such myofilaments to form cross-bridges is the essential mechanism for skeletal muscle contraction and was introduced in the 1950s as the "sliding filament theory."

CROSS-BRIDGE THEORY

The pioneering studies of Huxley and Hanson (1954) and Huxley and Niedergerke (1954) introduced the idea that muscle shortening during contraction of striated muscles is probably caused by sliding between the actin and myosin. The hypothesis proposed that the sliding is caused by the energy produced by the hydrolysis of adenosine triphosphate (ATP) to adenosine diphosphate (ADP) molecules and inorganic phosphate (P_i). A chemical process of this kind causes a sequential formation of cross-bridges, that is, independent force generators such that the whole muscle fiber shortens and the muscle contracts. The mechanism was described as the "cross-bridge theory" and was published in 1957 (Huxley, 1957), and the dynamic nature of this interaction was confirmed in a later study (Huxley, 1963). Fig. 2 shows the stages of the cross-bridge cycle.

PHYSICAL PROPERTIES

Force generation is largely controlled by the intrinsic properties of the muscle and varies with its length and velocity. The force-length (f-l) relation is defined as the relation between the maximum isometric forces generated while the muscle is held at a series of discrete lengths different from its rest length (Fig. 3A). While such concept of dependence on the muscle length was described over a century ago, it was only until 1966 that active force generation was elucidated experimentally as a function of the magnitude of overlap between the actin and myosin filaments (Gordon et al., 1966). In addition to this active force component, a passive force is exerted when a muscle is stretched over its rest length without stimulation and contributes as well to the total f-l curve of the muscle. Muscles function at peak performance at intermediate lengths, that is, when they are neither overstretched nor overcontracted from their rest state.

During muscle contraction, the dynamic properties of the cross-bridges change. The force-velocity (f-v) relation represents the rate of muscle length change to the maximum force a muscle can generate under constant load. Like the f-l relationship, the f-v curve is not continuous but is obtained through the interpolation of force and velocity data measured in distinct experiments for maximal activation

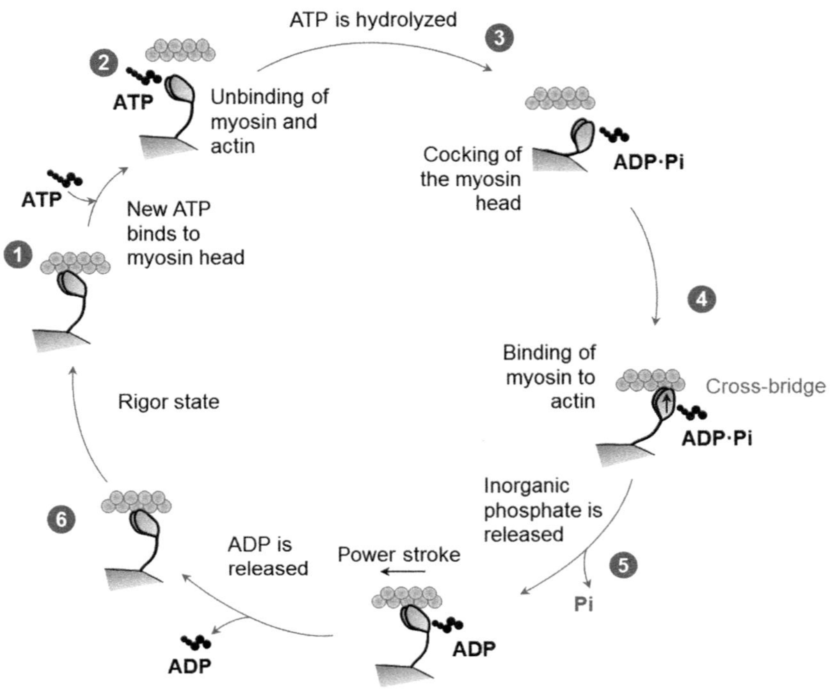

FIG. 2

The myosin and actin configurations of the cross-bridge cycle.

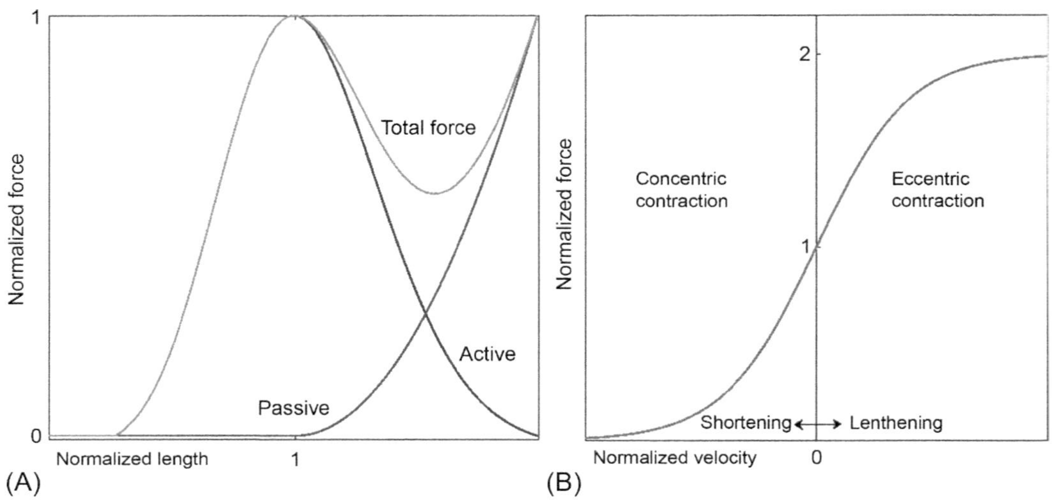

FIG. 3

Normalized (A) force-length, and (B) force-velocity curves of skeletal muscle.

(Fig. 3B). The physiological basis of such a relationship is that the force generated by a muscle depends on the total number of cross-bridges attached. For example, as the muscle shortens (that is, the filaments slide past one another) with increasing velocity, the force produced decreases because of the lower number of cross-bridges attached. Fenn and Marsh were the first to describe such a relationship (Fenn and Marsh, 1935), and the classical papers of Hill (1938) and Katz (1939) determined the effect of load on the speed of shortening through their experiments on frog muscles.

ANATOMY

SUPERFICIAL MUSCLES

The trapezius and latissimus dorsi are the largest of the superficial back muscles (Fig. 4). They cover the superior and latero-inferior region of the back, respectively, and extend longitudinally. These muscles

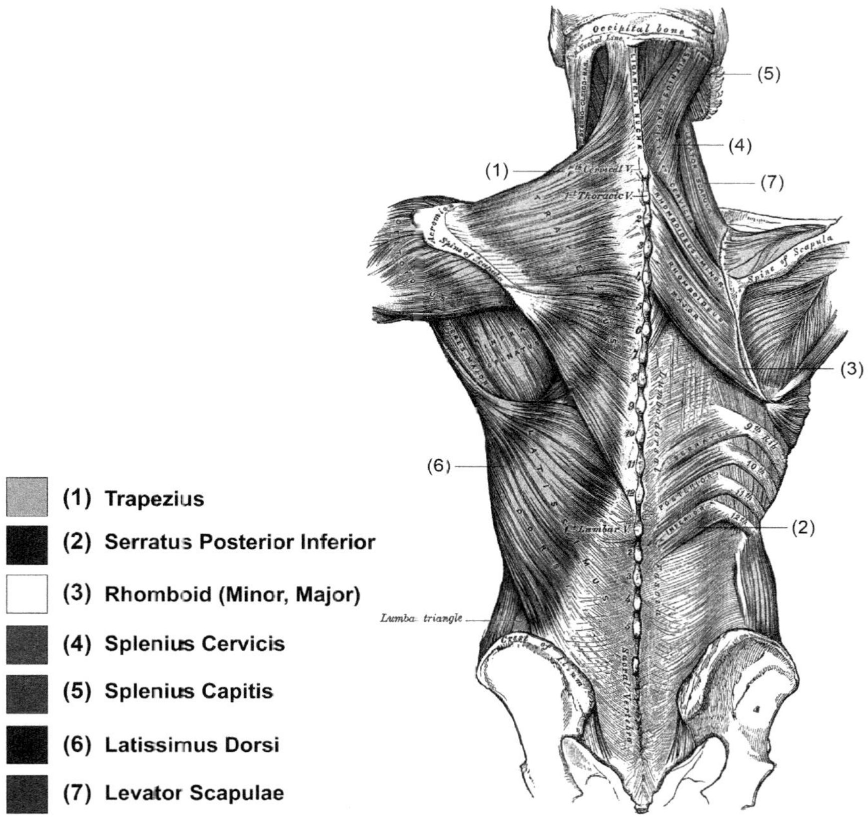

FIG. 4

Superficial muscles of the back.

(Source: Gray's Anatomy/Wikimedia Commons/Public Domain)

connect the upper limb to the vertebral column such as the rhomboid and the levator scapulae muscles that lie deep in the trapezius but are still located superficially to deep long muscles such as the erector spinae.

Trapezius

The trapezius has multiple attachment sites emanating from the occipital bone down to the lower thoracic vertebrae and laterally to the shoulder blade (also called scapula), that is, the flat bone that connects the humerus with the clavicle bone. The motor function of the trapezius is supplied by the accessory nerve (cranial nerve XI). The muscle additionally receives some proprioceptive fibers from the ventral rami of the third (C3) and fourth (C4) cervical nerves.

Because of its size, the trapezius has several actions including scapular or spinal movements depending on whether the spinal origins or the scapulae are stable, respectively. Its function is also determined by the region of the muscle that is contracting. For instance, when the head and neck are stable, the middle fibers retract the scapula while the lower portion depresses and rotates it downward. When the scapula is stabilized, the trapezius muscle acts as an important extensor of the head and neck, and its upper portion is also determined by whether one (unilateral) or both (bilateral) sides are contracted (Bakkum and Cramer, 2014). Dissection studies concluded that by balancing moments, the trapezius muscle relieves the cervical spine of compressive loads, and the downward loads applied to the upper arm are instead sustained by the sternoclavicular joint (Johnson et al., 1994).

Latissimus Dorsi

This muscle's name is based on its wide shape (*latissimus,* meaning broadest in Latin) and its position (*dorsum,* meaning the back in Latin). The most comprehensive architectural study was performed in 1998 by Bogduk and colleagues (Bogduk et al., 1998). According to the latter study, the latissimus dorsi is "a large fan-shaped muscle" that covers the postero-lateral aspect of the thorax with attachments points to the spinous processes and supraspinous ligament of T7-L5 vertebrae, the upper two to three sacral segments, the iliac crest, and the lower three to four ribs via the thoracolumbar fascia (TFL). Rostrally, the latissimus dorsi twists under the teres major muscle and inserts via a thick, flat tendon into the floor of the intertubercular groove of the humerus bone lateral to the attachments of the pectoralis major muscle. In fact, because of such fiber-wrapping around the posterior chest wall, the latissimus dorsi is categorized as a spiral muscle, and the orientation of its fibers is governed largely by the position of the arm.

The latissimus dorsi muscle is innervated by the sixth (C6) through eighth (C8) cervical nerves via the thoracodorsal branch of the brachial plexus. When contracted, it is responsible for adduction, extension, and medial rotation of the arm developing high maximum total forces. Besides its obvious contribution to move the upper limb or raise the entire trunk in brachiating, its attachments to the lumbar spine implied an indirect biomechanical effect on the motion of the lumbar spine as well. However, previous studies concluded that the possible extensor moments exerted on the lumbar region by the latissimus dorsi during lifting were not sizable (~ 4%–5% of the total moment) (Bogduk et al., 1998; McGill and Norman, 1988). Accordingly, it was suggested that the latissimus dorsi muscle belonged to the shoulder and was not designed as an extensor of the back.

Rhomboid Major and Minor

The two rhomboid muscles are diamond-shaped (rhomboid) with parallel muscle fibers that run in an infero-lateral direction and attach caudally onto the scapula's medial border level (rhomboid minor) to or inferior (rhomboid major) to the root of the scapular spine. The rhomboid major is the larger of the two and arises from the thoracic spinous process (T2 through T5) and the supraspinous ligament via tendinous fibers. It is located directly inferior to the rhomboid minor (Fig. 4), which stems from the inferior border of the ligamentum nuchae and the spinous processes of the C7 and T1 vertebrae. Both muscles lie deep in the trapezius muscle but still superficial to the long spinal muscles, such as the erector spinae. The nerve supply for both rhomboid muscles comes from the anterior primary ramus through the dorsal scapular nerve (C5).

The parallel arrangement of their fibers causes the two rhomboid muscles to have similar action in elevating and pulling the scapula toward the spine (scapular adduction). During activation of the shoulder muscles, the rhomboid major and minor act as stabilizers of the scapula by keeping it attached to the rib cage. They also work together with the levator scapulae muscle to rotate the scapula such that the lateral angle moves inferiorly, thereby acting as antagonists to the engaged trapezius.

Levator Scapulae

The levator scapulae muscle is situated at the back superiorly to the rhomboid minor muscle and on the side of the neck, as illustrated in Fig. 4. It originates by tendinous slips from the cervical transverse processes from the C1 through the C4 vertebrae and inserts on the scapula (superior angle) above the root of scapular spine where the rhomboid minor attaches. The levator scapulae is innervated by the ventral rami of the third (C3) and fourth (C4) cervical nerves and by a branch of the dorsal scapular nerve (C5). As its name suggests (levator), its main function is to help elevate the scapula and, in tandem with the rhomboid muscles, rotate the point of the shoulder down if the cervical spine is fixed. When the scapula is stable, unilateral contraction of the levator scapulae rotates and flexes the cervical spine laterally, whereas contraction of both sides may produce extension of the cervical spine.

Serratus Posterior Superior and Posterior Inferior

The serratus posterior superior is a thin, quadrangle-shaped muscle located at the upper back part of the thorax deep in the rhomboid muscles. Specifically, it arises from the spinous processes of the last cervical (C7) and first three thoracic (T1–T3) vertebrae and the supraspinous ligament. Its fibers descend infero-laterally and eventually attach via four fleshy insertions onto the superior borders of the second through fifth ribs. Its primary function to elevate such ribs is supplied by the ventral rami of the T2–T5 nerves (intercostal nerves) and may help with inspiration.

With a similar morphology, the serratus posterior inferior muscle lies at the junction of the thoracic and lumbar regions. Its fibers run supero-laterally from the supraspinal ligament and spinous processes of the T11 to L2 vertebrae to insert via fleshy digitations into the lower border of the 9th to 12th ribs (Fig. 4). The muscle attaches to the thoracic levels and is innervated by the intercostal nerves and the subcostal nerve that enables acting to depress the lower ribs.

The serratus posterior muscles are generally considered clinically insignificant; however, their actions have generated controversy among anatomists and physiologists. More specifically, on the basis of their attachments to the ribs, a non-negligible contribution in respiration might be supported, such as in forced expiration when the serratus posterior inferior muscles contract. Nonetheless, electromyographic (EMG) and other studies have not ascribed them with any respiratory role and have suggested that the serratus posterior muscles may function primarily in proprioception (Vilensky et al., 2001; Loukas et al., 2008).

DEEP MUSCLES

The intrinsic group of back muscles includes the splenius, erector spinae, transversospinalis, and short intersegmental muscles, that is, interspinales and intertransversarii, all located behind the plane of the transverse processes. The largest among them is the erector spinae (or sacrospinalis) that consists of a series of longitudinal muscles spanning the entire length of the spine. The erector spinae muscles are the primary extensors of the spine and head. Acting bilaterally, they straighten the back returning it to the upright position, and participate in controlling the flexion of the trunk by contracting and relaxing in a coordinated fashion. When contracted unilaterally, they act as lateral flexors of the vertebral column. According to Gray, erector spinae splits in the upper lumbar region into three vertical muscle columns from lateral to medial: *iliocostalis, longissimus,* and *spinalis* (Drake et al., 2004). Each of these muscles can be further subdivided regionally depending on where they attach (capitis, cervicis, thoracis, and lumborum) (Table 1). Macintosh and Bogduk (Macintosh and Bogduk, 1987) performed a series of elegant dissections and found that the disposition of both the iliocostalis lumborum and longissimus thoracis included thoracic and lumbar fibers. They called each part *"pars thoracis"* or *"pars lumborum,"* respectively. At that time, these observations contradicted descriptions that did not recognize the presence of lumbar insertions, more recent evidence from the Visible Human Project corroborated such classifications (Daggfeldt et al., 2000). The descriptions presented in this chapter follow the organization suggested for the morphology of the erector spinae as listed in Table 1.

Among the muscles that comprise the transversospinalis group and are presented in this section, that is, the semispinalis, multifidus, and rotatores, detailed anatomical and biomechanical descriptions are given for the multifidus by virtue of its multi-segmental morphology and its high clinical relevance in spinal disorders.

Splenius Capitis and Splenius Cervicis

The splenius muscles are located at the most superficial layer of deep back muscles and link the skull with the vertebral column. Their fibers are divided into two parallel parts that ascend in the supero-lateral direction. The splenius cervicis begins at the thoracic spinous process of T3 through T6 and inserts into the transverse processes of the upper three or four cervical vertebrae deep in the origins of the levator scapulae. The splenius capitis arises from the lower half of the nuchal ligament and the spinous processes of the C7 through T3 or T4 vertebrae. Its fibers attach to the mastoid process of the temporal bone and the occiput. Both splenius muscles are innervated by the dorsal rami via the mid-cervical (capitis) and cervical (cervicis) spinal nerves. When contracted unilaterally, the function of the splenius muscles is lateral flexion and rotation of the cervical spine (neck) toward the side of contraction. In co-contraction of both sides, the splenius muscles extend the head (capitis) and neck (cervical).

Table 1 The Muscles of the Erector Spinae (Sacrospinalis) Group

Iliocostalis	Cervicis	–
	Thoracis	–
	Lumborum	pars lumborum
		pars thoracis
Longissimus	Capitis	–
	Cervicis	–
	Thoracis	pars lumborum
		pars thoracis
Spinalis	Capitis	–
	Cervicis	–
	Thoracis	–

Iliocostalis Cervicis

The iliocostalis cervicis is the smallest of the three parts of the iliocostalis (Fig. 5). It originates from the superior aspect of the angle of the third, fourth, fifth, and sixth ribs and courses superiorly to attach onto the posterior tubercles of the transverse process of the fourth through sixth cervical vertebrae (C4 to C6). The muscle is innervated by the dorsal rami of the spinal nerves and acts

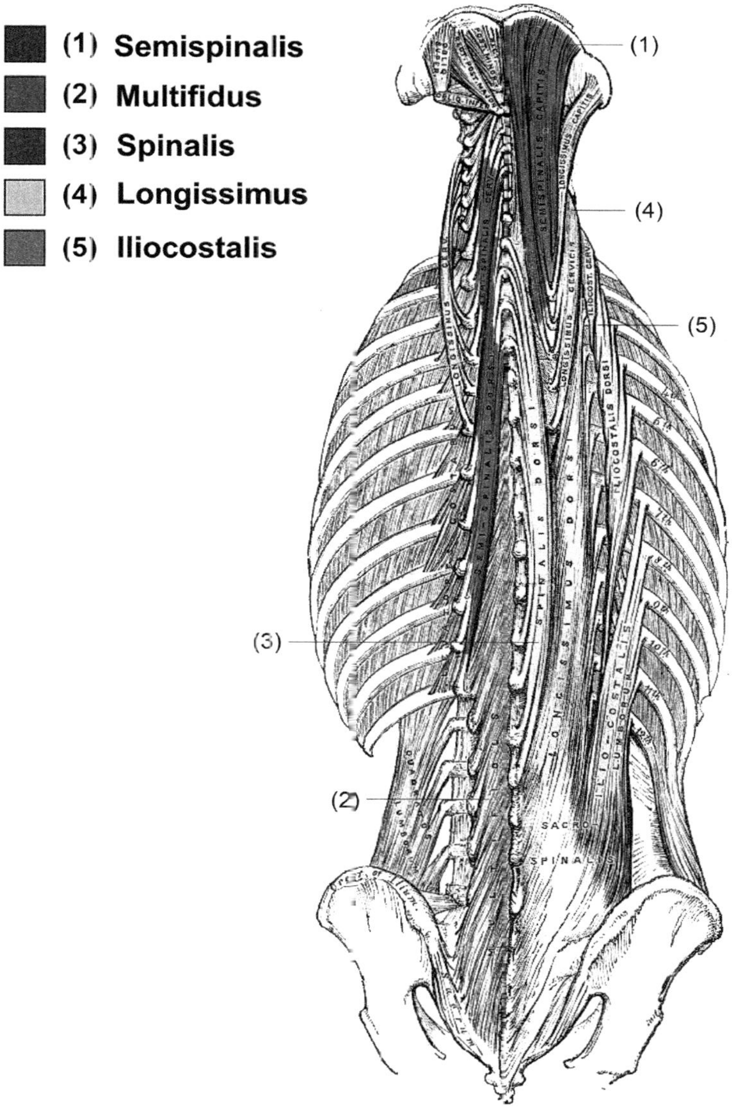

FIG. 5

Deep spinal muscles.

(Source: Gray's Anatomy/Wikimedia Commons/Public Domain)

synergistically to the splenius cervicis when contracting bilaterally to laterally flex and extend the lower cervical spine.

Iliocostalis Thoracis

Also called *iliocostalis dorsi,* this muscle arises from the superior aspect of the angles of the seventh to twelfth ribs and ascends to attach onto the angles of the first six ribs and the transverse process of the last cervical vertebra (C7). The iliocostalis thoracis acts on the thoracic region receiving supplies from the lateral branches of the dorsal rami.

Iliocostalis Lumborum

The iliocostalis lumborum is the most inferior and lateral portion among the erector spinae muscles innervated by the dorsal rami of lumbar and lower thoracic nerves (Fig. 5).

Based on the descriptions made by Bogduk, the iliocostalis lumborum's lumbar fibers (iliocostalis lumborum pars lumborum) have rostral attachments between the tips of the transverse process of the L1 through L4 vertebrae and an area extending on the middle layer of the TFL (Bogduk, 2005). The literature does not report any attachment onto the L5 vertebra, and it has been suggested that this "lacking" part is represented by the iliolumbar ligament. Caudally, the muscle inserts directly into the iliac crest and the posterior surface of the posterior superior iliac spine. The most lateral of the iliac insertions lies medial to the attachment of the lateral raphe of the TFL. When contracted bilaterally, the lumbar fibers of the iliocostalis lumborum act principally as extensors and axial rotators especially of the lower lumbar levels. In unilateral contraction, they produce lateral flexion and axial rotation and extension simultaneously using their attachments onto the transverse process as lever arms.

The thoracic fibers of iliocostalis lumborum derive via flattened (ribbon-like) tendons from the angles of the lower seven or eight ribs, possess long muscle bellies (8–10 cm (Bogduk, 2005)) and, before attaching caudally to the ilium and sacrum, they form tendons that contribute to the erector spinae aponeurosis. Bogduk described this aponeurosis as "a broad sheet of tendinous fibers that is attached to the ilium, the sacrum, and the lumbar and sacral spinous processes." Because they have no direct attachment to the lumbar spine, when contracted bilaterally, they exert an indirect "bowstring effect" that accentuates the lumbar lordosis (Middleditch and Oliver, 2005). Acting unilaterally, the thoracic and lumbar fibers have similar function in laterally bending the lumbar region of the spine and can also have a small axial rotator role.

Longissimus Capitis

The longissimus capitis muscle arises by tendons from the transverse processes of the T1 to T5 and the articular processes of C4 through C7 vertebrae and is inserted into the mastoid process of the temporal bone (Fig. 5). It is innervated by the dorsal rami of the upper thoracic and cervical spinal nerves, and its primary function is as a posterior sagittal rotator of the head. If contracted unilaterally, the longissimus capitis produces lateral flexion and rotation to the side of contraction.

Longissimus Cervicis

The longissimus cervicis is located medially to the iliocostalis cervicis and derives from the upper thoracic vertebrae (transverse process of T1 to T5) via long tendons. It runs superiorly to end on the posterior tubercle of the transverse and articular processes of the cervical vertebrae (C2 through C6). Because of its cervical attachments, the longissimus cervicis acts to extend (bilateral contraction), or extend and laterally flex (unilateral contraction) the neck. The nerve supply comes from the lateral branches of the upper thoracic and lower cervical divisions of the dorsal rami.

Longissimus Thoracis

The longissimus thoracis (dorsi) illustrated in Fig. 5 is the largest among the erector spinae muscles and the longest muscle of the back. It is innervated by lateral branches of the thoracic and lumbar dorsal rami.

Macintosh and Bogduk noted that the longissimus thoracis pars lumborum, that is, the lumbar fibers of the muscle, arises from the accessory process and the adjacent medial end of the transverse process of the L1 through L5 vertebrae (Macintosh and Bogduk, 1987). These fibers lie progressively more dorsally (i.e., the ones arising from the L2 cover those from L3 and below) and insert into the ilium (L5 fibers) and the posterior superior iliac spine (L1–L4 fibers) via a muscular and a tendinous insertion (lumbar intermuscular aponeurosis), respectively. Contrary to the implications of previous anatomical studies (Bustami, 1986), Bogduk found no attachment of the lumbar fibers to the erector spinae aponeurosis and suggested that the longissimus thoracis pars lumborum acted independently from the rest of the erector spinae. The longissimus thoracis expresses a function of combined actions, that is, posterior sagittal rotation and horizontal translation to draw the lumbar vertebrae backward simultaneously (Bogduk, 2005). EMG measurements in flexion confirmed high activity of the muscle that probably reflects its function to restore the anterior translation of the lumbar vertebrae (Arjmand and Shirazi-Adl, 2006). Its action as an extensor increases rostrally in line with its longer fascicles, whereas its capacity for posterior translation when contracted bilaterally is greater at lower lumbar levels.

The thoracic fibers of the longissimus thoracis (longissimus thoracis pars thoracis) attach proximally to the transverse process and ribs of 11 or 12 thoracic levels through tendons that might merge medially from T5 down to T12. Their fascicle muscle bellies are small, and the ones from the upper levels overlap those from the lower levels. Eventually, longissimus thoracis pars thoracis attaches via tendon insertions that form the medial half of the erector spinae aponeurosis into the lumbar and sacral spinous process, the sacrum, and the ilium (Bogduk, 2005). In fact, such attachments together with the tendinous insertions of thoracic fibers of the iliocostalis lumborum form almost exclusively the erector spinae aponeurosis, with a small additional contribution from the upper fibers of the multifidus muscle. The longissimus thoracis pars thoracis is designed to act principally on the thoracic vertebrae and ribs. Moreover, its portion that originates from the lower six thoracic levels spans the entire lumbar region and in bilateral contraction can indirectly increase the lumbar lordosis through the erector spinae aponeurosis. When contracted unilaterally, the longissimus thoracis pars thoracis acts as a lateral flexor principally of the thoracic and secondarily the lumbar region of the vertebral column (Middleditch and Oliver, 2005).

Spinalis Capitis

The spinalis capitis, also called *biventer cervicis*, is innervated by upper thoracic and lower cervical nerves of the dorsal rami. It is usually difficult to differentiate from the semispinalis capitis situated laterally. It is considered to blend with the latter and attach through it onto the occiput between the superior and inferior nuchal lines. In contraction of both sides, the spinalis capitis acts as an extensor of the head, whereas when it is contracted unilaterally, its function is to laterally flex and rotate the head as well as laterally flex the cervical spine to the side of contraction.

Spinalis Cervicis

The spinalis cervicis *(spinalis colli)* is a small and frequently absent muscle (Bakkum and Cramer, 2014). When present, it is derived from the spinous process of the upper T1 through T6 vertebrae and ends on the C2 to C4 cervical spinous process. The action of spinalis cervicis is to extend the cervical spine.

Spinalis Thoracis

Also called *spinalis dorsi*, this portion is the most highly developed of the spinalis muscle group. It is innervated by the thoracic nerves of the dorsal rami and is situated at the medial side of the longissimus dorsi with which it blends (Fig. 5). The spinalis thoracis fibers begin in the lower thoracic and upper lumbar vertebrae (transverse process of T11 through L2) and ascend to insert into the spinous process of the upper thoracic vertebrae. This small muscle is designed to act as an extensor of the thoracic spine.

Semispinalis Capitis

The semispinalis capitis is located medial to the longissimus cervicis and capitis (Fig. 5), and lies deep to the splenius. It is a thick muscle and represents the most developed portion of the semispinalis muscle group (Kamibayashi and Richmond, 1998). It arises from the transverse process of the last cervical vertebrae (C7) and the upper thoracic (T1–T6) and the articular process of C4 to C6, and ascends to insert between the occipital superior and inferior nuchal lines. Its innervation comes from the dorsal rami similar to the spinalis capitis with which it blends. When both sides of the muscle are contracted, they act to extend the head, whereas in unilateral contraction, a slight rotation of the head to the opposite side occurs (contralateral axial rotation).

Semispinalis Cervicis

The semispinalis cervicis begins from the transverse process of T1 to T5 or T6 vertebrae and mainly terminates on the cervical spinous processes of the axis through C5. The muscle is supplied by the C6 to C8 spinal nerves of the dorsal rami and acts as a neck extensor (bilateral contraction), lateral flexor, and contralateral rotator of the cervical spine (unilateral contraction).

Semispinalis Thoracis

This is the thinnest muscle of the semispinalis group. The semispinalis thoracis (dorsi) consists of thin, fleshy fascicles with long tendons. Inferiorly, it arises from the T1–T6 transverse process through small tendons and inserts superiorly to the spinous processes of C6 through T4. It acts similarly to the semispinalis cervicis but on the thoracic region of the spine receiving innervation from the medial branches of the T1–T6 spinal nerves of the dorsal rami.

Multifidus

The multifidus muscle group lies deep in the semispinalis group, fills up the groove on either side between the transverse and spinous processes of the vertebrae, and spans the lumbosacral junction (Fig. 5). It comprises multiple fleshy and tendinous fascicles that begin from the dorsal part of the sacrum, posterior superior iliac spine, and lumbar mammillary processes (*lumbar multifidus)*, the transverse processes of all thoracic vertebrae (*thoracic multifidus)*, and the articular processes of lower cervical vertebrae (*cervical multifidus)*. Fascicles ascend two to four vertebrae levels until they insert into the spinous processes of all vertebrae from L5 up to C2.

Multifidus is best developed in the lumbar region, and the extensive literature available on its morphology has reported different descriptions of the number (Valenica and Munro, 1984; Shindo, 1995; Lonnemann et al., 2008b; Macintosh and Bogduk, 1986), attachment site (Drake et al., 2004; Lewin et al., 1962), and orientation (De Foa et al., 1989; Biedermann et al., 1991; Hollinshead and Rosse, 1997) of its fascicles. However, there is a consensus on the multi-segmental arrangement of the lumbar multifidus in five distinct bands (one per lumbar level) overlapping one another (Macintosh and Bogduk, 1986). At each lumbar level, the fibers are arranged in three layers (superficial, intermediate, and deep) except for the L5/S1 level, where no intermediate fascicles have been found (Rosatelli et al., 2008). The deep (laminar) fascicles are the shortest in length connecting adjacent vertebrae, whereas the remaining two layers form the bulk of the lumbar multifidus with long fascicles that have a postero-anterior orientation and radiate caudally until they attach into lower processes, the iliac crest and the sacrum. The most superficial fibers from the upper lumbar levels contribute a small number of fibers to the deep surface of the erector spinae aponeurosis.

Such fascicle disposition along with the segmental innervation (medial branch of the dorsal rami of that segment's spinal nerve) implies that the muscle is designed to act principally on a particular vertebral segment. Indeed, the long superficial fascicles lying behind the lordotic curve may create a "bowstring effect" between the segments they attach. Thus, indirectly they would tend to accentuate the lumbar lordosis, resulting in the development of compressive and tensile loads on any interposed vertebrae and intervertebral discs (Bogduk, 2005). In the lumbar region, the multifidus produces primarily extension of the lumbar spine (Macintosh and Bogduk, 1986). Rotation happens secondarily and simultaneously with extension when, for instance, the oblique abdominal muscles contract to rotate the trunk and indirectly involve some flexion of the spine. As an opposing action to such flexion, the multifidus contracts to maintain a pure axial rotation (Ng et al., 2002) acting as a stabilizer that might also minimize the shear and compression loads on the facets (Lonnemann et al., 2008a).

Rotatores

The rotatores group includes small and quadrilateral-shaped muscles that are located beneath the multifidus and run from sacrum to axis the entire length of the vertebral column. The rotatores are most prominent in the thoracic region and are only poorly developed in the cervical and lumbar region, where it is difficult to differentiate them from the multifidus muscles. They consist of a *brevis* group that arises from the transverse processes and attaches superiorly to the root of the spinous process immediately above, and a more superficial *longus* portion with longer fibers that attach two segments above. The rotatores muscles are segmentally innervated by the medial branches of the dorsal rami, and their main function is to stabilize the spine. When they act unilaterally, they contribute to the rotation of the vertebral column to the opposite side, whereas during bilateral contraction they help extend the spine.

Short Intersegmental

The interspinales are short paired muscles connecting the spinous processes of adjacent vertebrae. They lie on either side of the interspinous ligament and are present in all spinal regions. The interspinales are innervated by the medial branches of the dorsal rami and make only a small contribution to extending the spine (Bogduk, 2005).

The intertransversarii are likewise small and paired muscles extending between adjacent transverse processes. They are not always considered as true muscles, and they act as stabilizers of the adjoining vertebrae during large spinal movements. In the cervical region where they are most prominent, they usually begin at C1 and continue to T1 involving fascicles that pass between the adjacent anterior (*anterior intertransversarii)* and posterior *(posterior intertransversarii)* tubercles. In the thoracic region, they are found only in the lower thoracic levels and are supplied by the spinal nerves of the dorsal rami. The lumbar intertransversarii muscles are present between all vertebrae and are innervated by the lumbar nerves of the dorsal rami. Because of their small size and their very close proximity to the axes of rotation, any biomechanical action attributed would be speculative. Nevertheless, the high density of muscle spindles (especially in the cervical region) may give the intertransversarii a proprioceptive function as monitors of the movements of the spine, allowing more effective action of the surrounding muscles (Bogduk, 2005).

ILIAC MUSCLES

Another group with a prominent role in spine biomechanics consists of the iliac muscles: the psoas major, psoas minor, and quadratus lumborum (Fig. 6). Their fibers connect the spine to the lower limbs through insertions onto the pelvis or the femur. Because of their antero-lateral attachments, the psoas and quadratus lumborum muscles are classified as posterior abdominal wall muscles. Nonetheless,

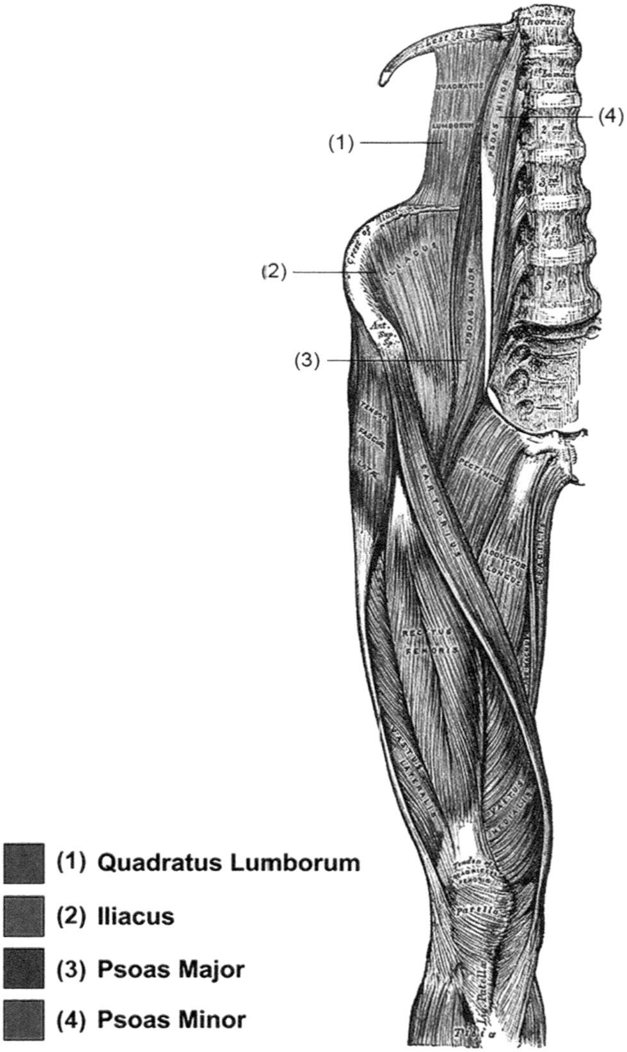

FIG. 6

Iliac muscles.

(Source: Gray's Anatomy/Wikimedia Commons/Public Domain)

because they have been suggested to act on the lower region of the spine, they are often included in the group of back muscles.

Psoas Major

The psoas major is a long muscle situated on the antero-lateral aspect of the vertebral column (Fig. 6). Although it is essentially considered as a muscle of the thigh, in absence of hip movement, the muscle can act on the lumbar spine (Bogduk et al., 1992). Its fascicles have approximately the same length and begin from the anterior aspect of the transverse processes, the antero-lateral surfaces of the L1 and L5 vertebrae and the rim of the intervertebral disc with the adjacent vertebrae from T12/L1 to L4/L5 level. The psoas major fascicles descend along the pelvic brim, deep to the inguinal ligament and insert through a tendon onto the lesser trochanter of the femur. Such femoral insertion is common with the iliacus muscle (Fig. 6), and together they form the *iliopsoas* (Sajko and Stuber, 2009). The medial fascia that encloses the psoas forms a tendinous arch over the lateral aspects of the vertebrae allowing passage between this arch and the spine for nerves and vessels. As it surrounds the muscle, the fascia passes under the medial arcuate ligament superiorly and continues to the diaphragm. It becomes thicker caudally and is continued by the iliac fascia under the inguinal ligament where both form the iliopectineal arch. The iliopsoas muscle is supplied by the ventral rami of the L2-L3 spinal nerves.

Although in vivo measurements would allow exploring other functions of the psoas major apart from its role as a flexor of the hip joint, the deep location of the muscle in the body makes it difficult to capture its activity in vivo. Some preliminary evidence suggested its role as a stabilizer of the lumbar spine in upright positions (Nachemson, 1966; Nachemson, 1968; Santaguida and McGill, 1995; Penning, 2000; Andersson et al., 1995; Jemmett et al., 2004). To date, textbooks report a wide inconsistency involving lateral flexion of the lumbar spine (Woodburne and Burkel, 1988; Romanes, 1981), control of the lumbar lordosis (Bogduk et al., 1992; Hadjipavlou et al., 1996), and the pelvic tilt (Hadjipavlou et al., 1996).

Psoas Minor

The psoas minor is a thin, long muscle absent in about 40% of the population. When present, it is situated on the anterior surface of the psoas major and derives from the T12 and L1 vertebrae and the interposing intervertebral disc (Fig. 6). Via a long tendon, it runs caudally and inserts on the pectin pubis and iliopectineal eminence. It is innervated by the L1 spinal nerves of the ventral rami, and its function as a trunk flexor is not appreciable.

Quadratus Lumborum

The quadratus lumborum is a wide, irregularly shaped muscle with a quadrilateral form that lies along the tips of the lumbar transverse processes. It consists of various oblique and longitudinally running fibers that connect inferiorly the ilium to the twelfth rib and the tips of the transverse processes of L1 to L4 (Fig. 6). The function of this muscle remains obscure mainly because of its inconstant structure, that is, the absence of some fascicles in some subjects (Phillips et al., 2008). It depresses the lowermost ribs and stabilizes the origin of the diaphragm during respiration and contributes to lateral flexion of the lumbar spine to a small extent (Bogduk, 2005). Its position behind the axis of sagittal rotation of the lumbar spine contributes to the quadratus lumborum's role of trunk extensor when both sides are contracted. Nonetheless, the magnitude of forces produced by this muscle are minimal compared to the principal back extensors, that is, the erector spinae and multifidus (Phillips et al., 2008).

THORACOLUMBAR FASCIA

Given its clinical significance, the TLF is discussed here briefly. Also referred to as *lumbodorsal fascia*, it is an investing membrane that extends from the thoracic region down to the sacrum. The TLF envelops the erector spinae and quadratus lumborum muscles and anteriorly merges with the aponeurosis of the transverse and internal oblique abdominal muscles. It consists of three separate layers: anterior, middle, and posterior, with the two latter ones reported to influence the biomechanics of the lumbar region. The anterior layer is quite thin, is derived from the fascia of the quadratus lumborum, and inserts onto the transverse processes of the spine. With similar spinal insertions, the middle layer lies behind the quadratus lumborum and was reported as having a stabilizing influence toward the neutral zone and in lateral bending (Tesh et al., 1987; Barker et al., 2006). The posterior layer is primarily attached to the spinous processes of the thoracic and lumbar levels and covers posteriorly the erector spinae before blending laterally with the other two layers to form a dense, lateral union called lateral raphe. The posterior is the thickest of the three layers and has its own two laminae: the superficial lamina formed by the aponeurosis of the latissimus dorsi, and the deep one made up of collagen fiber bands passing caudo-laterally from the midline (Bogduk, 2005). The two laminae fuse in the lateral raphe not only with the middle layer but also with the transverse abdominal muscle, thereby providing an indirect attachment for the latter to the lumbar spinous processes.

Several authors have debated the biomechanics of the TLF and its ability to extend the spine by producing forces that might be comparable to those exerted by other lumbar muscles (Bogduk and Macintosh, 1984; Tesh et al., 1987; McGill and Norman, 1988; Gatton et al., 2010). Another anatomic study concluded that the pelvic and lower limb muscles interact with the spinal muscles via the TLF, thereby enabling effective load transfer between the trunk, pelvis, legs, and arms (Vleeming et al., 1995).

ABDOMINAL MUSCLES

Even though the muscles of the antero-lateral abdominal wall have no direct attachments to the spine, they are important in several movements of the trunk, such as flexion and axial rotation. These muscles lie in front of the lumbar spine and connect the thorax to the pelvis (Fig. 7). From superficial to deep, there are the external abdominal oblique, internal abdominal oblique, and transversus abdominis, whose aponeurosis *(rectus sheath)* encloses the rectus abdominis muscles. The abdominal muscles principally act to retain the abdominal viscera and ribs in place via compression creating intra-abdominal pressure.

External Abdominal Oblique

The external abdominal oblique is located on the latero-anterior aspect of the abdomen and is the largest and most superficial of the group (Fig. 7). It consists of eight fleshy slips that emanate from the inferior borders of the fifth to twelfth ribs and course inferiorly and anteriorly in an oblique arrangement to attach close to the cartilages of the ribs. Its slips originating from the lowest three ribs blend with the latissimus dorsi muscle and descend almost vertically to attach onto the anterior half of the outer lip of the iliac crest. Those from the upper and middle levels interdigitate with the serratus anterior muscles, and because they pass the line that connects the umbilicus and the anterior superior iliac spine, they become aponeurotic. The external oblique aponeurosis is a thin, strong sheet of connective tissue that covers the whole abdomen and whose fibers are directed infero-medially. It is formed by both external oblique muscles that blend in the midline with the linea alba. In its inferior

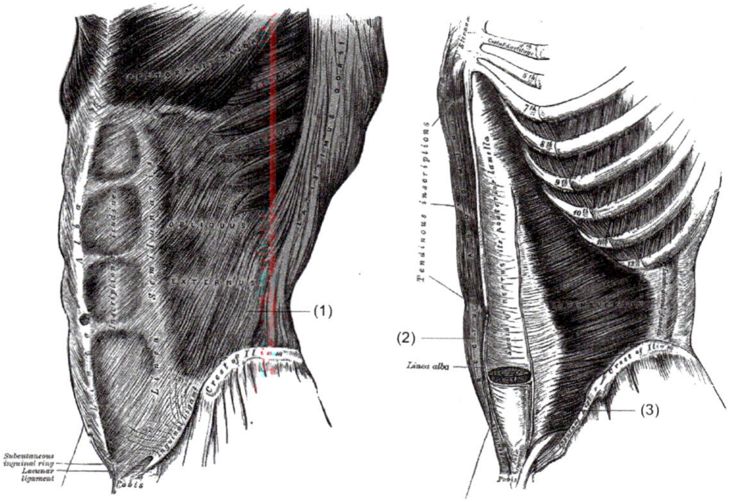

(1) External Abdominal Oblique

(2) Rectus Abdominis

(3) Transversus Abdominis

FIG. 7

Abdominal muscles.

(Source: Gray's Anatomy/Wikimedia Commons/Public Domain)

portion, the external oblique aponeurosis forms the inguinal ligament, its reflected portion, and the lacunar ligament. The external oblique muscle is innervated by the ventral branches of the T7 to T12 spinal nerves. This muscle depresses the lower ribs, aids expiration, and increases the intra-abdominal pressure that is important among other things for diagnostic purposes (e.g., Valsalva maneuver). With unilateral contraction, the external abdominal oblique can help in lateral flexion and rotation of the trunk to the opposite side.

Internal Abdominal Oblique

The internal oblique muscle lies immediately beneath the previously mentioned external oblique with fibers coursing perpendicular to the latter. They begin from the TFL, the anterior two-thirds of the iliac crest, and the lateral part of the inguinal ligament and ascend supero-medially to insert into the lower ends of the ninth or tenth to twelfth ribs and the linea alba through the internal oblique aponeurosis. The lowest fibers have tendinous ends that attach to the pubic crest and blend with the transversus abdominis aponeurosis to form the conjoint tendon. The nerve supplies of the internal oblique come from the T7 to L1 spinal nerves of the ventral rami. Like the external oblique muscle, it performs as an accessory muscle of respiration acting as antagonist to the diaphragm to help compress the abdominal cavity during exhalation when the thorax and pelvis are fixed. Secondarily, in unilateral contraction, it acts with the external oblique of the opposite side to rotate and laterally flex the trunk.

Transversus Abdominis

The transversus abdominis muscle is the innermost muscle among the antero-lateral abdominal muscles and is located deep in the internal oblique (Fig. 7). It consists of muscular fibers that arise from the lateral third of the inguinal ligament, the anterior two-thirds of the outer lip of the iliac crest, the TFL, and the inner surfaces of the lower six costal (rib) cartilages interdigitating with the diaphragm (Bakkum and Cramer, 2014). Such fibers extend almost horizontally forming the transversus abdominis aponeurosis that attaches medially to the linea alba and inferiorly to the pubic crest via the conjoint tendon. Apart from the lower one-fourth of the muscle, the rest of its fibers lie behind the rectus abdominis muscle.

The transversus abdominis is innervated by the ventral rami of the T7–T11 or L1 spinal nerves and helps to provide thoracic and pelvic stability (Hodges, 1999) opposing the effects of gravity. The transversus abdominis's role in respiratory movement was also reviewed previously (De Troyer et al., 1990). Given the consistent attachments of the middle fibers of the transversus abdominis to the TFL (Willard et al., 2012), it has the greatest potential to influence the lumbar segmental control (Barker et al., 2007). Nevertheless, the extensor moment they could produce on the lumbar spine upon maximum contraction was seriously questioned on theoretical grounds (Macintosh et al., 1987). A recent study found that, although similar voluntary activation was seen between normal and symptomatic groups, patients with chronic back pain exhibited atrophy of the transverse abdominis muscles (Kim et al., 2013).

Rectus Abdominis

The rectus abdominis muscle is a long, paired muscle extending vertically the entire length of the anterior abdominal wall and is enclosed by the aponeurosis of the other abdominal muscles (rectal sheath). The two sides of the rectus are separated in the midline by the linea alba and are laterally confined by the linea semilunaris (Fig. 7). Its fibers attach superiorly to the costal cartilages of the fifth to seventh ribs and the xiphoid process (sternum), and insert inferiorly onto the pubic crest and the symphysis. The rectus abdominis is traversed by three tendinous intersections, that is, horizontal bands of connective tissue that separate the muscle into distinct muscle bellies at the levels of the umbilicus, the end of the sternum, and between these points. It is supplied by the ventral rami of the T6 through T12 spinal nerves and primarily assists in respiration and protection of the internal organs like the other muscles of the group. Moreover, the rectus abdominis acts as a flexor of the trunk and the lumbar spine moving the pelvis toward the thorax when the latter is fixed, and vice versa.

Table 2 lists the deep and abdominal wall muscles according to their action on the thoracolumbar spine in bilateral or unilateral contraction.

MODELING

The importance of neck muscles as stabilizers of the cervical spine during loading is widely known and has been documented through experimental measurements (Magnusson et al., 1999; Siegmund et al., 2003; Pinsault and Vuillerme, 2010) and numerical studies (Brolin et al., 2005; Ejima et al., 2005; Kimpara et al., 2006; Hedenstierna et al., 2009). The most common technique used to evaluate muscle activity was via EMG (Eckner et al., 2014; Lo Martire et al., 2017). However, because of the anatomical complexity of the cervical musculature and the restrictions in the impact levels to test them, numerical models were suggested as an effective alternative.

Table 2 Deep and Abdominal Wall Muscles Acting on the Thoracolumbar Spine			
Bilateral contraction		**Unilateral contraction**	
Flexion	**Extension**	**Lateral Flexion**	**Axial Rotation**
Psoas major External abd. Oblique Internal abd. Oblique Rectus abdominis	Iliocostalis thoracis Iliocostalis lumborum Longissimus thoracis Spinalis thoracis Semispinalis thoracis Multifidus Rotatores Quadratus lumborum	Iliocostalis thoracis Iliocostalis lumborum Longissimus thoracis Semispinalis thoracis Multifidus Psoas major Quadratus lumborum External abd. Oblique Internal abd. Oblique Rectus abdominis	*Same side:* Iliocostalis thoracis Iliocostalis lumborum Internal abd. Oblique *Opposite side:* Semispinalis thoracis Multifidus Rotatores External abd. Oblique

The finite element (FE) method allows simulating several cervical muscles simultaneously and under any acceleration level providing also predictions of strain and insights on the interaction between the musculature and the ligamentous spine. In the majority of the FE models, neck muscles were represented using deformable discrete elements (Brolin et al., 2005; Östh et al., 2017) and the Hill-type model for both active and passive forces, or only passive forces "to simulate an unaware condition prior to impact" (Kimpara et al., 2006). Hill's phenomenological model (Hill, 1938) consists of a passive, elastic spring connected in parallel to a contractile element (Fig. 8A) where the active force is generated as a result of the cross-bridges formed and according to the f-l and f-v relations discussed previously. This modeling approach remains the most reported one in skeletal muscle modeling mainly because of its computational simplicity and its ability to calculate adequate results, such as nodal forces and displacements in the axial direction, taking into account the muscle's physiological properties.

Although some discrete models used sliding contact interaction (Stemper et al., 2004) or multisegment muscles (Brolin et al., 2005; Panzer, 2006; Östh et al., 2017) to avoid non-anatomical loading lines of action in hyperflexion/hyperextension, the non-uniform deformation between these segment springs might lead to instabilities. Solid element muscle models (Ejima et al., 2005; Hedenstierna et al., 2008) addressed this point by modeling the wrapping of neck muscles that cover multiple joints and follow the geometry of the spine. In the latter study, 24 muscle groups were modeled, including the cervicis and capitis portions of the deep muscles described in the previous section. Continuum mechanical material descriptions were used to account for properties such as tissue inertia, shear stiffness, and friction between muscles that could affect the kinematics of the cervical spine. The authors in (Hedenstierna and Halldin, 2008) proposed a combination of solid, nonlinear elements with hyperviscoelastic passive response and truss contractile elements with Hill's model properties that was able to model active and passive forces at constant rate elongations, isometric and concentric contractions (Fig. 8B). In the context of an FE neck model, comparisons between a continuum and a discrete muscle model in simulated impacts reported similar behavior in spine kinematics but stiffer response and improved strain predictions of the continuum models in rear-end impact (Hedenstierna and Halldin,

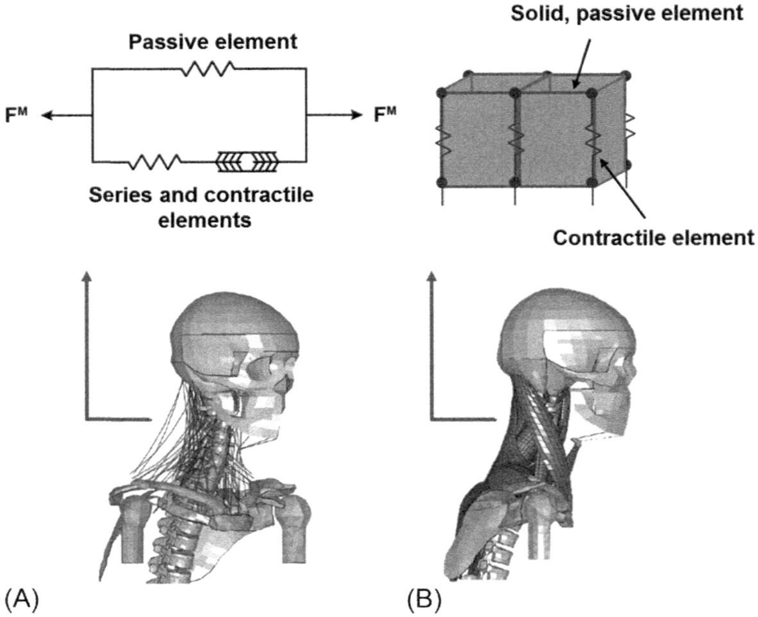

FIG. 8

(A) Hill's model used in discrete neck muscle model; (B) Combination of solid elements and Hill contractile springs used in continuum muscle models.

Adapted from Hedenstierna, S., Halldin, P., 2008. How does a three-dimensional continuum muscle model affect the kinematics and muscle strains of a finite element neck model compared to a discrete muscle model in rear-end, frontal, and lateral impacts. Spine 33(8), E236–E245. Available at: http://www.ncbi.nlm.nih.gov/pubmed/18404033.

2008). Importantly, these models can be computationally time-consuming, so their use may be preferred rather than discrete models when the benefits of continuum mechanical properties or refined meshes justify such a choice for practical use, for example, for localized strain predictions.

In the thoracolumbar spine, the majority of the numerical models reported for decades did not model the musculature. Instead, they were represented by models that accounted for the stabilizing action of the muscles via segmental compressive forces (follower loads) applied as punctual forces (Rohlmann et al., 2009), via a path of connector elements along the axial axis (Schmidt et al., 2010), or a surface pressure on the uppermost vertebral endplate (Noailly et al., 2011). The accuracy of such approximations was questioned because it depended on the location of the follower load path (Dreischarf et al., 2010), and it was unclear whether they could represent the balance of body weight and muscle effects.

Efforts to simulate the thoracolumbar muscles through explicit models were reported with increasing complexity and realism incorporating fascicles of deep iliac and abdominal oblique muscles (Zander et al., 2001; Arjmand and Shirazi-Adl, 2006; Christophy et al., 2012; Han et al., 2013; Dao et al., 2014; Toumanidou and Noailly, 2015). Chronologically, the loads transferred by the muscles to the thoracolumbar spine were mainly calculated using EMG, kinematical data, and optimization methods until the first published attempt in 2012 to consider muscle dynamics (Christophy et al., 2012) (see the Chapter 15). The thoracolumbar muscle model made of 238 fascicles (Fig. 9A) developed in OpenSim (Delp et al., 2007) employed Thelen's model for the musculo-tendon actuators, that is, a simplified Hill's model that allows for age-based adjustments in muscle mechanics through four parameters (Thelen, 2003) (Fig. 9B). The authors in the cited study did not report on force estimations and did not consider continuum mechanical material properties. However, it remains the most anatomically detailed thoracolumbar muscle model in the literature.

Recently, researchers proposed a three-dimensional formulation to represent passive and active muscle forces for a lumbar spine FE model by combining continuum theories and the Hill's model (Toumanidou and Noailly, 2015). Inspired by Hedenstierna et al. (2008), the authors in the cited work used the proposed formulation to study the continuum passive response of the muscles by decoupling its deviatoric and dilatational parts to represent a hyperelastic solid reinforced with fibers, that is, springs defined as Hill-type elements able to generate active forces (Fig. 10). The activation was controlled via one parameter, and preliminary results based on the use of a discrete muscle model showed the ability to achieve realistic force predictions in simulated flexion and standing postures. The approach proposed did not consider viscous dampers and can be coupled to both discrete and solid muscle models.

To date, no continuum models of the thoracolumbar muscles have been published to the best of the author's knowledge. The increased anatomical complexity of lumbar muscles and their deep location around the vertebral column remain significant obstacles for the quantification of geometrical characteristics and shape via imaging modalities. Muscle modeling in the thoracolumbar region of the spine remains underexplored even though muscle dysfunction and altered characteristics have been identified experimentally in patients with back pain and were reported to lead to spinal instability (Panjabi, 2006; Peter Reeves et al., 2007; D'hooge et al., 2013).

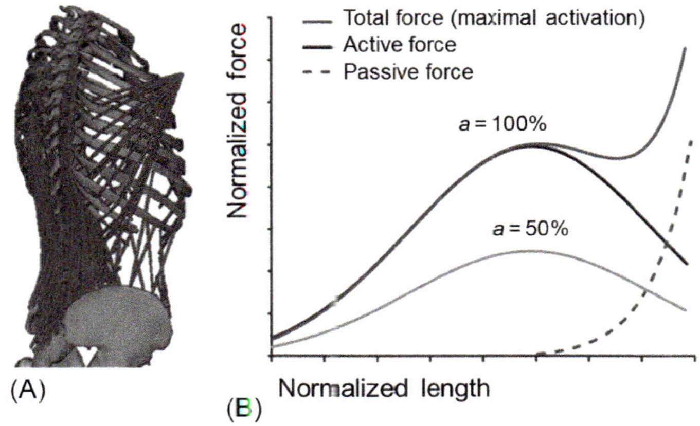

FIG. 9

(A) Thoracolumbar muscle model with discrete elements; (B) Thelen's muscle model.

Adapted from Christophy, M., et al., 2012. A musculoskeletal model for the lumbar spine. Biomech. Model. Mechanobiol. 11(1–2), 19–34.

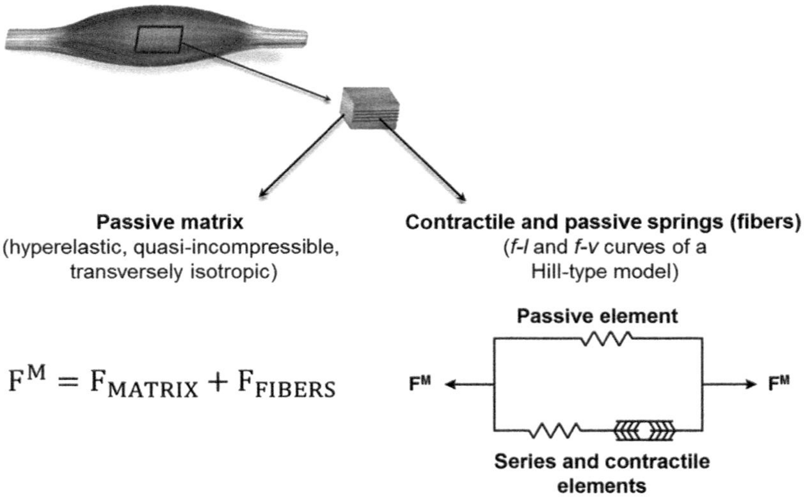

FIG. 10

Continuum muscle material model proposed by Toumanidou and Noailly (2015). The total force is a sum of the passive matrix and the active and passive fiber contributions.

CONCLUSION

Human morphology involves numerous muscle tissues, and the total number influencing the spine is especially high because of the ones that function in coordination to stabilize the spine. In this chapter, the most important geometrical and functional characteristics of the spinal muscles were illustrated through anatomical descriptions and a brief overview of the cross-bridge theory and the Hill's muscle model. The objective was to cover all large groups that have been captured experimentally to act as stabilizers during occupational activities, such as trunk flexion (erector spinae-abdominals), or that can influence injury severity, for instance in whiplash.

A review unearthed more extensive literature on existing computational approaches for modeling of neck muscles than for modeling of lumbar spine muscles. From discrete muscle elements with straight lines of actions and passive forces to continuum muscle geometries curving around the cervical spine, the available neck muscle models were assessed within the context of cervical spine assemblies. Reliable insights were provided about the interactions between the muscles and the surrounding tissues, and possible modes of validation were suggested, for example, through sled tests. Such explorations can offer useful information for establishing injury criteria or for designing rehabilitation programs for patients with chronic neck pain.

Analysis of the thoracolumbar muscles' functional interactions with the spine needs further research because most muscles' deep location and multi-segmental organization, such as in the multifidus, do not favor three-dimensional reconstruction from images. Still, the discrete muscle models coupled with Hill-type material laws provide valuable attempts to complement EMG data and enhance the understanding of lumbar muscle function, such as if and when the muscles are activated, and how their response is altered based on conditions like muscle fatigue, or age-related deterioration. Nevertheless, defining the parameter number involved in such muscle models must be done with caution in order for calibration of the parameter values to be possible and to deal with the difficulty of performing direct force measurements.

REFERENCES

Andersson, E.A., et al., 1995. The role of the psoas and iliacus muscles for stability and movement of the lumbar spine, pelvis and hip. Scand. J. Med. Sci. Sports 5 (1), 10–16.

Arjmand, N., Shirazi-Adl, A., 2006. Model and in vivo studies on human trunk load partitioning and stability in isometric forward flexions. J. Biomech. 39 (3), 510–521. Available at http://www.ncbi.nlm.nih.gov/pubmed/16389091Accessed July 26, 2011.

Bakkum, B.W., Cramer, G.D., 2014. Muscles that influence the spine. In: Clinical Anatomy of the Spine, Spinal Cord, and Ans. third ed. . Elsevier Inc, St. Louis, MO, pp. 98–134.

Barker, P.J., et al., 2006. Effects of tensioning the lumbar fasciae on segmental stiffness during flexion and extension. Spine 31 (4), 397–405.

Barker, P.J., et al., 2007. The middle layer of lumbar fascia and attachments to lumbar transverse processes: Implications for segmental control and fracture. Eur. Spine J. 16 (12), 2232–2237.

Biedermann, H.J., De Foa, J.L., Forrest, W.J., 1991. Muscle fibre directions of iliocostalis and multifidus: male-female differences. J. Anat. 179, 163–167.

Bogduk, N., 2005. Clinical Anatomy of the Lumbar Spine and Sacrum, fourth ed. Elsevier, Edinburgh.

Bogduk, N., Macintosh, J.E., 1984. The applied anatomy of the thoracolumbar fascia. Spine 9, 164–170.

Bogduk, N., Pearcy, M.J., Hadfield, G., 1992. Anatomy and biomechanics of psoas major. Clin. Biomech. 7, 109–119.

Bogduk, N., Johnson, G., Spalding, D., 1998. The morphology and biomechanics of latissimus dorsi. Clin. Biomech. (Bristol, Avon) 13 (6), 377–385.

Brolin, K., Halldin, P., Leijonhufvud, I., 2005. The effect of muscle activation on neck response. Traffic Inj. Prev. 6 (1), 67–76.

Bustami, F., 1986. A new description of the lumbar erector spinae in man. J Anatom. 81–91.

Christophy, M., et al., 2012. A musculoskeletal model for the lumbar spine. Biomech. Model. Mechanobiol. 11 (1–2), 19–34.

D'hooge, R., et al., 2013. Lumbar muscle dysfunction during remission of unilateral recurrent nonspecific low-back pain: evaluation with muscle functional MRI. Clin. J. Pain 29 (3), 187–194. http://www.ncbi.nlm.nih.gov/pubmed/23369927.

Daggfeldt, K., Huang, Q.M., Thorstensson, A., 2000. The visible human anatomy of the lumbar erector spinae. Spine 25 (21), 2719–2725. Available at http://www.ncbi.nlm.nih.gov/pubmed/11064515.

Dao, T.T., et al., 2014. Estimation of patient specific lumbar spine muscle forces using multi-physical musculoskeletal model and dynamic MRI. Knowledge Systems Eng. 96–97, 411–422. Available at: https://doi.org/10.1016/j.datak.2015.04.001.

De Foa, J.L., Forrest, W.J., Biedermann, H.J., 1989. Muscle fibre direction of longissimus, iliocostalis and multifidus: landmark-derived reference lines. J. Anat. 163, 243–247. Available at: http://www.pubmedcentral.nih.gov/articlerender.fcgi?artid=1256533&tool=pmcentrez&rendertype=abstract.

De Troyer, A., et al., 1990. Transversus abdominis muscle function in humans. J. Appl. Physiol. 68 (3), 1010–1016. Available at: http://www.ncbi.nlm.nih.gov/entrez/query.fcgi?cmd=Retrieve&db=PubMed&dopt=Citation&list_uids=2140344.

Delp, S.L., et al., 2007. OpenSim: open-source software to create and analyze dynamic simulations of movement. IEEE Trans. Biomed. Eng. 54 (11), 1940–1950. Available at http://www.ncbi.nlm.nih.gov/pubmed/18018689.

Drake, R.L., Vogl, A.W., Mitchell, A.W.M., 2004. Gray's Anatomy for students, second ed. Churchill Livingstone, England.

Dreischarf, M., et al., 2010. A non-optimized follower load path may cause considerable intervertebral rotations. J. Biomech. 43 (13), 2625–2628.

Eckner, J.T., et al., 2014. Effect of neck muscle strength and anticipatory cervical muscle activation on the kinematic response of the head to impulsive loads. Am. J. Sports Med. 42 (3), 566–576. Available at: http://www.scopus.com/inward/record.url?eid=2-s2.0-84897612717&partnerID=tZOtx3y1.

Ejima, S., et al., 2005. In: Development and validation of the human neck muscle model under impact loading. IRCOBI Conference, Prague (Czech Republic).

Fenn, W.O., Marsh, B.S., 1935. Muscular force at different speeds of shortening. J. Physiol. 85 (3), 277–297.

Gatton, M.L., et al., 2010. A three-dimensional mathematical model of the thoracolumbar fascia and an estimate of its biomechanical effect. J. Biomech. 43 (14), 2792–2797.

Gordon, A.M., Huxley, A.F., Julian, F.J., 1966. The variation in isometric tension with sarcomere length in vertebrate muscle fibres. J. Physiol. 184 (1), 170–192. Available at: http://www.pubmedcentral.nih.gov/articlerender.fcgi?artid=1357553&tool=pmcentrez&rendertype=abstract.

Hadjipavlou, A.G., Farfan, H.F., Simmons, J.W., 1996. The functioning spine. In: Farfan, H.F., Simmons, J.W., Hadjipavlou, A.G. (Eds.), The Sciatic Syndrome. Slack, Thorofare,NJ, pp. 41–73.

Han, K.-S., et al., 2013. Lumbar spinal loads vary with body height and weight. Med. Eng. Phys. 35 (7), 969–977. Available at http://www.ncbi.nlm.nih.gov/pubmed/23040051.

Hedenstierna, S., Halldin, P., 2008. How does a three-dimensional continuum muscle model affect the kinematics and muscle strains of a finite element neck model compared to a discrete muscle model in rear-end, frontal, and lateral impacts. Spine 33 (8), E236–45. Available at http://www.ncbi.nlm.nih.gov/pubmed/18404093.

Hedenstierna, S., Halldin, P., Brolin, K., 2008. Evaluation of a combination of continuum and truss finite elements in a model of passive and active muscle tissue. Comput. Methods Biomech. Biomed. Engin. 11 (6), 627–639. Available at http://www.ncbi.nlm.nih.gov/pubmed/18642161.

Hedenstierna, S., Halldin, P., Siegmund, G.P., 2009. Neck muscle load distribution in lateral, frontal, and rear-end impacts: a three-dimensional finite element analysis. Spine 34 (24), 2626–2633. Available at http://www.ncbi.nlm.nih.gov/pubmed/19910765.

Hill, A.V., 1938. The heat of shortening and the dynamic constants of muscle. Proc. R. Soc. B Biol. Sci. 126 (843), 136–195. Available at: http://rspb.royalsocietypublishing.org/cgi/doi/10.1098/rspb.1938.0050.

Hodges, P.W., 1999. Is there a role for transversus abdominis in lumbo-pelvic stability? Man. Ther. 4 (2), 74–86. Available at http://www.researchgate.net/publication/12791526_Is_there_a_role_for_transversus_abdominis_in_lumbo-pelvic_stability.

Hollinshead, W., Rosse, C., 1997. Textbook of Anatomy, fifth ed. Lippincott Williams & Wilkins, Baltimore.

Huxley, H.E., 1957. The double array of filaments in cross-striated muscle. J. Biophys. Biochem. Cytol. 3 (5), 631–673.

Huxley, H.E., 1963. Electron microscope studies on the structure of natural and synthetic protein filaments from striated muscle. J. Mol. Biol. 7, 281–308.

Huxley, H.E., Hanson, J., 1954. Changes in the cross-striations of muscle during contraction and stretch and their structural interpretation. Nature 173 (4412), 973–976.

Huxley, A.F., Niedergerke, R., 1954. Structural changes in muscle during contraction; interference microscoopy of living muscle fibres. Nature 173 (4412), 971–973.

Jemmett, R.S., Macdonald, D.A., Agur, A.M.R., 2004. Anatomical relationships between selected segmental muscles of the lumbar spine in the context of multi-planar segmental motion: a preliminary investigation. Man. Ther. 9 (4), 203–210. Available at http://www.ncbi.nlm.nih.gov/pubmed/15522645.

Johnson, G., et al., 1994. Anatomy and actions of the trapezius muscle. Clin. Biomech. 9 (1), 44–50.

Kamibayashi, L.K., Richmond, F.J., 1998. Morphometry of human neck muscles. Spine 23 (12), 1314–1323.

Katz, B.Y.B., 1939. The relation between force and speed in muscular contraction. J. Physiol. 45–64.

Kim, K.H., et al., 2013. Differences in Transversus Abdominis muscle function between chronic low back pain patients and healthy subjects at maximum expiration: measurement with real-time ultrasonography. J. Phys. Ther. Sci. 25 (7), 861–863.

Kimpara, H., et al., 2006. Investigation of Anteroposterior head-neck responses during severe frontal impacts using a brain-spinal cord complex FE model. Stapp Car Crash J. 50, 509–544.

Lewin, T., Moffet, B., Viidik, A., 1962. The morphology of the lumbar synovial intervertebral joints. Acta Morphologica Scandinavica 4, 299–319.

Lo Martire, R., et al., 2017. Neck muscle EMG-force relationship and its reliability during isometric contractions. Sports Med. Open 3 (1), 16. Available at: http://sportsmedicine-open.springeropen.com/articles/10.1186/s40798-017-0083-2.

Lonnemann, M.E., Paris, S.V., Gorniak, G.C., 2008a. A morphological comparison of the human lumbar multifidus by chemical dissection. J. Man. Manip. Ther. 16 (4), E84–E92.

Lonnemann, M.E., Paris, S.V., Gorniak, G.C., 2008b. Multifidus by chemical dissection. J. Man. Manip. Ther. 16 (4), 84–92.

Loukas, M., et al., 2008. An anatomic investigation of the serratus posterior superior and serratus posterior inferior muscles. Surg. Radiol. Anat. 30 (2), 119–123.

Macintosh, J.E., Bogduk, N., 1986. The biomechanics of the lumbar multifidus. Clin. Biomech. 1 (4), 205–213.

Macintosh, J.E., Bogduk, N., 1987. 1987 Volvo award in basic science. The morphology of the lumbar erector spinae. Spine 12 (7), 658–668.

Macintosh, J.E., Bogduk, N., Gracovetsky, S., 1987. The biomechanics of the thoracolumbar fascia. Clin. Biomech. 2 (2), 78–83.

Magnusson, M.L., et al., 1999. Cervical electromyographic activity during low-speed rear impact. Eur. Spine J. 8 (2), 118–125.

McGill, S.M., Norman, R.W., 1988. Potential of lumbodorsal fascia forces to generate back extension moments during squat lifts. J. Biomed. Eng. 10 (4), 312–318.

Middleditch, A., Oliver, J., 2005. Structure of the vertebral column. In: Functional Anatomy of the Spine. Elsevier, Edinburgh, pp. 47–51.

Nachemson, A.L.F., 1966. Electromyographic studies on the vertebral portion of the psoas muscle. Acta Orthop. Scand. 37, 177–190.

Nachemson, A.L.F., 1968. The possible importance of the psoas muscle for stabilization of the lumbar spine. Acta Orthop. Scand. 39, 47–57.

Ng, J.K.F., et al., 2002. EMG activity of trunk muscles and torque output during isometric axial rotation exertion: a comparison between back pain patients and matched controls. J. Orthop. Res. 20 (1), 112–121.

Noailly, J., et al., 2011. Silico evaluation of a new composite disc substitute with a L3-L5 lumbar spine finite element model. Eur. Spine J.. Available at http://www.ncbi.nlm.nih.gov/pubmed/21380572. (Accessed March 15, 2011).

Östh, J., et al., 2017. A female head–neck model for rear impact simulations. J. Biomech. 51, 49–56. Available at http://linkinghub.elsevier.com/retrieve/pii/S0021929016312568.

Panjabi, M.M., 2006. A hypothesis of chronic back pain: ligament subfailure injuries lead to muscle control dysfunction. Eur. Spine J. 15 (5), 668–676.

Panzer, M.B., 2006. Numerical Modelling of the Human Cervical Spine in Frontal Impact. University of Waterloo, Waterloo, Ontario, Canada. Available at: http://books.google.ie/books/about/Numerical_Modelling_of_the_Human_Cervica.html?id=Ni7MH8IonDAC&pgis=1.

Penning, L., 2000. Psoas muscle and lumbar spine stability: a concept uniting existing controversies. Critical review and hypothesis. Eur. Spine J. 9 (6), 577–585.

Peter Reeves, N., Narendra, K.S., Cholewicki, J., 2007. Spine stability: the six blind men and the elephant. Clin. Biomech. 22 (3), 266–274.

Phillips, S., Mercer, S., Bogduk, N., 2008. Anatomy and biomechanics of quadratus lumborum. Proc. Inst. Mech. Eng. H J. Eng. Med. 222 (2), 151–159. Available at: http://journals.pepublishing.com/openurl.asp?genre=article&id=doi:10.1243/09544119JEIM266.

Pinsault, N., Vuillerme, N., 2010. Degradation of cervical joint position sense following muscular fatigue in humans. Spine 35 (3), 294–297.

Rohlmann, A., et al., 2009. Applying a follower load delivers realistic results for simulating standing. J. Biomech. 42 (10), 1520–1526.

Romanes, G.J., 1981. Cunningham's Textbook of Anatomy, twelfth ed. Oxford University Press, Toronto.

Rosatelli, A.L., Ravichandiran, K., Agur, A.M., 2008. Three-dimensional study of the musculotendinous architecture of lumbar multifidus and its functional implications. Clin. anatom. (New York, N.Y.) 21 (6), 539–546. Available at http://www.ncbi.nlm.nih.gov/pubmed/18627104.

Sajko, S., Stuber, K., 2009. Psoas major: a case report and review of its anatomy, biomechanics, and clinical implications. J. Can. Chiropr. Assoc. 53 (4), 311–318. Available at: http://www.pubmedcentral.nih.gov/articlerender.fcgi?artid=2796950&tool=pmcentrez&rendertype=abstract.

Santaguida, P.L., McGill, S.M., 1995. The psoas major muscle: a three-dimensional geometric study. J. Biomech. 28 (3), 339–345.

Schmidt, H., et al., 2010. Response analysis of the lumbar spine during regular daily activities—a finite element analysis. J. Biomech. 43 (10), 1849–1856. Available at http://www.ncbi.nlm.nih.gov/pubmed/20394933.

Shindo, H., 1995. Anatomical study of the lumbar multifidus muscle and its innervation in human adults and fetuses. Nihon Ika Daigaku Zasshi. 62 (5), 439–446.

Siegmund, G.P., et al., 2003. Rapid neck muscle adaptation alters the head kinematics of aware and unaware subjects undergoing multiple whiplash-like perturbations. J. Biomech. 36 (4), 473–482.

Stemper, B.D., et al., 2004. In: Effects of reflexive cervical muscle contraction on whiplash kinematics and injury. IRCOBI Conference, Graz (Austria). pp. 269–282.

Tesh, K., Dunn, J., Evans, J.H., 1987. The abdominal muscles and vertebral stability. Spine 12 (5), 501–508.

Thelen, D.G., 2003. Adjustment of muscle mechanics model parameters to simulate dynamic contractions in older adults. J. Biomech. Eng. 125 (1), 70–77.

Toumanidou, T., Noailly, J., 2015. Musculoskeletal modeling of the lumbar spine to explore functional interactions between back muscle loads and intervertebral disk multiphysics. Front. Bioeng. Biotechnol. 3, 1–13.

Valenica, F., Munro, R., 1984. Morphology of the lumbar multifidus in man. J. Anat. 139, 196.

Vilensky, J.A., et al., 2001. Serratus posterior muscles: anatomy, clinical relevance, and function. Clin. Anat. 14 (4), 237–241.

Vleeming, A., et al., 1995. The posterior layer of the thoracolumbar fascia. Spine 20 (7), 753–758.

Willard, F.H., et al., 2012. The thoracolumbar fascia: anatomy, function and clinical considerations. J. Anat. 221 (6), 507–536.

Woodburne, R.T., Burkel, W.E., 1988. Essentials of Human Anatomy. Oxford University Press, New York.

Zander, T., et al., 2001. Estimation of muscle forces in the lumbar spine during upper-body inclination. Clin. Biomech. (Bristol, Avon) 16 (Suppl. 1), S73–80. Available at http://www.ncbi.nlm.nih.gov/pubmed/11275345.

SECTION

3

METHODS

CHAPTER

IN VIVO STUDIES: SPINAL IMAGING

10

Tito Bassani*, Luca M. Sconfienza*,†, Fabio Galbusera*

*IRCCS Galeazzi Orthopedic Institute, Milan, Italy** *University of Milano, Milan, Italy*†

IMAGING IN THE DIAGNOSIS AND TREATMENT OF SPINAL DISORDERS

Since the discovery of X-ray imaging and its introduction into clinical practice, imaging techniques have represented a keystone for the diagnosis of spinal disorders, as well as for follow-up with patients during conservative treatment or after surgery. In comparison with other commonly investigated musculoskeletal districts, such as the hip and the knee, spinal imaging is more challenging because of the complexity of the vertebral anatomy. This situation is characterized by a complex morphology of the vertebrae and a complicated ligamentous apparatus and is moreover significantly variable depending on the spinal region being considered (i.e., cervical, thoracic, or lumbar spine). Nevertheless, the currently available imaging modalities offer effective methods for investigating the shape and morphology of the spine and the composition of the biological tissues, thus providing fundamental information for diagnosis and biomechanical assessment.

IMAGING MODALITIES—BASICS AND HISTORICAL PERSPECTIVE

The introduction of radiography was a milestone in medicine. Spine X-ray images were created shortly after the X-ray's discovery by Roentgen in 1895 (Dewing, 1962), but they were not documented in scientific publications. One of the first known images is reported by Cushing (Cushing, 1898), who showed coronal and sagittal projections of a bullet in the cervical spine. Radiography is based on the transmission of X-rays, a high-energy form of electromagnetic radiation able to penetrate living tissues, through the body of the patient. Biological tissues absorb the radiation depending on their density; therefore the amount of X-rays transmitted is high in correspondence to tissues with lower densities, such as the lungs and soft tissues, and is low when crossing more dense tissues, such as bone. An X-ray detector (originally photographic film, today most commonly digital solutions such as image plates of flat panel detectors) then measures the transmitted radiation and transforms it into a planar image. X-ray images clearly highlight the morphology of the skeletal system and thus have been widely used for the detection of fractures (Kendler et al., 2016), assessment of deformities (Cobb, 1948), as well as degenerative changes to bone tissue (Wilke et al., 2006). However, planar X-ray imaging projects onto two-dimensional plane anatomic structures that are actually in three dimensions, and does not therefore allow for a full volumetric assessment of the anatomy. Furthermore, conventional radiography does not provide sufficient contrast for diagnostic imaging of soft tissues, such as muscles, intervertebral discs,

Biomechanics of the Spine. https://doi.org/10.1016/B978-0-12-812851-0.00010-0

and ligaments. Being highly energetic electromagnetic waves, X-rays have the effect of ionizing the living matter. This implies well-documented direct and indirect adverse effects on the DNA of patients exposed to X-rays (Hall and Brenner, 2008), potentially leading to carcinogenesis. It is known that, at diagnostic levels, the occurrence of DNA damage is a stochastic event; that is, the higher the administered radiation dose, the higher the probability of DNA damage. This issue should be carefully considered particularly in patients needing repeated examinations and in younger patients in whom the high cellular turnover and long life expectancy potentially increase the risk (Pearce et al., 2012).

The potential of X-rays to obtain tomograms, that is, images of a specific body section rather than planar projections, was foreseen in the early days of radiography. Conventional tomography exploited a rotational motion of the X-ray source around a fulcrum during exposure in order to blur the anatomical structures out of the point of X-ray focalization (Hesselink, 1988). Conventional tomography has been generally superseded today by computed tomography (CT), except for panoramic radiographs for dental imaging and tomosynthesis, which is mainly used for breast imaging. However, other motion blur-based techniques are still in use in spinal imaging, for example, to remove the shadows of ribs in planar X-rays by exploiting their physiological motion during respiration.

A tremendous advance in tomographic imaging was made by the introduction of CT (CAT) scanners in 1973 (Hounsfield, 1973; Ambrose, 1973). In the original implementation of the CT scanner, the creation of the image of an anatomical section was based on the acquisition of several linear projections obtained by rotating the X-ray source and a detector around the investigated object, each time by a given small angle. The set of the acquired projections for all scanning angles, known as a sinogram, can be mathematically calculated based on the original tissue density by using the Radon transform (Radon, 1986), which was published in 1917 long before the development of CT scanners. The determination of the local tissue densities from the sinogram can therefore be performed by applying an inverse Radon transform. Several analytical solutions have been presented to solve this problem (Kak and Slaney, 2001), the best-known one being the filtered back-projection algorithm, which was introduced by Johann Radon himself and is theoretically exact if an infinite number of projections is available. The original CT scanner by Hounsfield and Cormack, however, did not use an analytical method to invert the sinogram, but relied on a less efficient iterative method, the algebraic reconstruction technique (Gordon et al., 1970), which is more suited to obtain a reconstructed image from a small number of projections. Technological and computational advances then enabled CT scanners to simultaneously acquire several slices, decreasing the thickness of each slice while increasing the in-plane resolution, and thus the image quality in general. The speed of acquisition and reconstruction also improved greatly, and real-time scanning as well as imaging of moving structures such as the heart are now in widespread use. It is worth noting that the radiation dose administered with a CT examination is higher than that with planar X-ray imaging, thus its use should be limited to very specific cases.

A revolution in the diagnosis of spinal disorders was brought about by the introduction of magnetic resonance imaging (MRI). The phenomenon of MR was discovered in 1946 (Bloch et al., 1946; Purcell et al., 1946) and has been used since to obtain quantitative information about the composition of substances. In the early 1970s, scientists (Lauterbur, 1989; Mansfield and Maudsley, 1977; Damadian, 1971) presented ideas on obtaining body images based on the MR signal, which were then implemented in a physical prototype scanner by Damadian et al. (1976). Initially tested in only the brain, clinical applications of MRI for spinal imaging soon followed in the 1980s (Modic et al., 1988, 1984; Edelman et al., 1985).

MRI does not involve ionizing radiations and is therefore inherently safer for both the patient and the radiographer. The technique is based on the interaction of atomic nuclei (typically hydrogen in medical applications) with an external magnetic field and on their capability to absorb and emit electromagnetic waves in the radio frequency (RF) range (Bernstein et al., 2004). Atomic nuclei possess a spin, that is, an angular momentum, which is normally randomly oriented and aligns with an external magnetic field, if present, either parallel to or opposite the direction of the field. The parallel state has a lower energy, and therefore the number of parallel spins is slightly higher than that of the antiparallel spins, resulting in a macroscopic magnetic moment in the direction of the external field proportional to the intensity of the latter. To acquire an MRI scan an RF pulse is applied, which changes the direction of the macroscopic magnetization vector by decreasing its longitudinal component and increasing its transverse one. The RF pulse is then removed, and the magnetization vector returns to its original alignment parallel to the external field and dismissing energy, which induces a signal in a receiving antenna in the form of a coil placed around the patient's region of interest. The rate of regrowth of the longitudinal magnetization depends on the tissue properties and is characterized by a time constant, called T1, whereas the transverse magnetization exponentially decays with a time constant (T2*), because of both the magnetic properties of the tissue and the inhomogeneity of the magnetic field. Another time constant (T2) relates only to the decay of the transverse magnetization because of the tissue properties, without the field inhomogeneity effects, and is therefore more relevant for clinical evaluations. To obtain an image from the signal acquired with the coil, it is necessary to correlate it with the spatial location of the sources. This is achieved mainly by adding a longitudinal gradient during the RF pulse to select the slice of interest. followed by additional gradients in the other spatial directions to modulate the signal.

As previously mentioned, T1 and T2 values depend on tissue composition; therefore the creation of an image in which the gray level is proportional to these time constants achieves good contrast between tissues such as bone, cartilage, ligaments, fat, and liquid. In general, compared to CT scans, an MRI allows for better discrimination between tissues with similar density but different composition, such as the intervertebral disc and the spinal cord. Several pulse sequences have been developed to obtain different types of images, such as to highlight either the T1 or the T2 of the tissue; to suppress the signal arising from specific tissues, such as fat, blood, or water; and to optimize specific parameters, such as reducing the acquisition time or the noise generated by the gradients to improve a patient's comfort.

The magnetic field strength of commercially available scanners range from 0.18 to 3T, with 1.5T being the strength for most commonplace systems. It has been demonstrated that the higher the field strength, the higher the image quality. However, for most clinical applications, the superiority of 3.0T systems over 1.5T systems has not been demonstrated.

SPINAL MORPHOLOGY

The measurement of size, shape, and orientation of vertebrae and intervertebral discs from radiological images has a crucial clinical and biomechanical relevance. As a matter of fact, disc degeneration is associated with loss of the intervertebral height and of the lordotic angle and flattening of the endplates, and the evaluation of these conditions plays a fundamental role in grading the degenerative disease (Boos et al., 2002). Similarly, spinal deformities such as scoliosis and hyperkyphosis involve morphological alterations, such as wedging of vertebrae and intervertebral discs, as well as changes in the vertebral orientations (Schlosser et al., 2014), which are commonly evaluated for the diagnosis and staging of the pathologies. Typically, these measurements are conducted on planar X-rays. These images are usually taken in the upright position, which better mimics the average physiological loads acting on the spine and can highlight disorders such as scoliotic curves, spondylolisthesis, and instability (Boos and Aebi, 2008). Whereas the morphology of vertebrae is independent of the posture because of their high stiffness, measurements about intervertebral discs and vertebral orientations are strongly influenced by body weight and muscle forces. To these clinical aims, planar X-ray imaging is therefore superior to conventional CT and MRI scans, in which the orthostatic position is typically not achievable.

Measurements on planar X-rays are usually conducted manually with the help of computer programs, which enable the evaluation of angles and relative lengths (Akbar et al., 2013). Typically, the user may draw points on the vertebral corners and/or lines along the endplates and the anterior and posterior margins of the vertebral bodies. These types of measurements involve an unavoidable degree of imprecision, which can be quantified on average by 3–5 degrees (Ricart et al., 2011; Shea et al., 1998; Wills et al., 2007). Indeed, vertebral contours appearing on X-ray projections are dependent on the vertebral orientation relative to the plane of acquisition, and even small changes in a patient's positioning may have a marked effect on the appearance of the image (Fig. 1). To reduce the impact of such inaccuracies in lateral X-rays, Frobin and colleagues introduced a set of auxiliary markers and tested the reliability and reproducibility of the improved measurements with good success (Frobin et al., 1997). However, that method did not have a major impact on the clinical practice; in fact, surgeons and radiologists tend to simply draw lines along the endplates or on the posterior vertebral wall. For some applications, such as measuring the Cobb angle in patients with scoliosis, such an approach is believed to be sufficiently accurate for diagnosing the pathology or monitoring its progression (Kim et al., 2010).

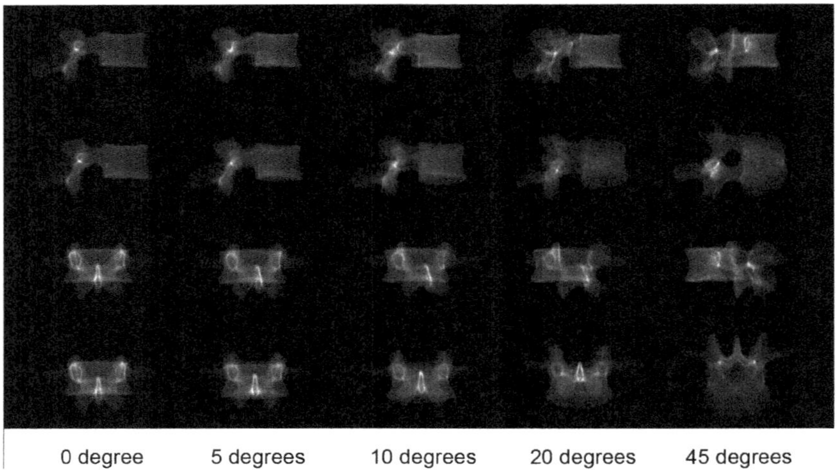

FIG. 1

Radiographic views of the same L4 vertebra from different points of view, depicting the differences in the visible contours and highlighting the consequent difficulties in locating anatomical landmarks. First row: sagittal view progressively rotated by 5, 10, 20, and 45 degrees around the longitudinal axis. Second row: sagittal view progressively around the anteroposterior axis. Third row: coronal view rotated around the longitudinal axis. Fourth row: coronal view rotated around the latero-lateral axis.

Obtaining realistic linear measurements on planar X-rays is challenging because of the effects of magnification and distortion. Because the source is typically small and X-rays travel in straight lines, the objects and anatomical structures that are more distant from the detector are magnified more than the closer ones are (Fig. 2). To minimize these effects, the structures for which accurate measurements are required should be placed as close as possible to the detector. Also, the radiation source can be located at a distance sufficient enough from the object to minimize the angle of the X-rays (teleradiology) (van der Plaats, 2012). In clinical practice, a minimal distance of about 2 m between the source and the detector is usually considered sufficient enough to ensure practical parallelism of the X-rays, and therefore minimal geometric distortions. Note that a larger tube-detector distance requires longer exposures and may involve motion artifacts. If a sufficiently powerful X-ray tube is available, this problem can be avoided by using higher tensions for the exposure (van der Plaats, 2012). Because the effects of magnification prevent a direct correspondence between the length measurements on an X-ray's projection and the actual measurements, applying radio-opaque markers of a known size to a patient's skin is commonly used as reference.

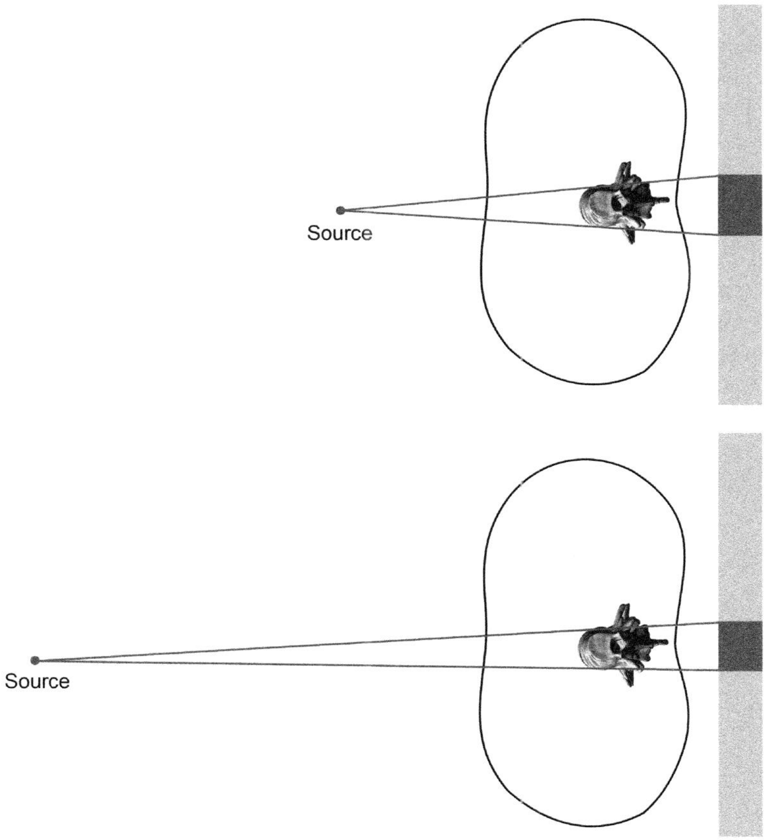

FIG. 2

Magnification and distortion in planar X-ray projections. If the X-ray's source is located at a sufficient distance from the object (bottom), the rays are almost parallel and the magnification effect is minimized.

SPINAL MOTION

The in vivo measurement of the segmental range of motion, typically in flexion-extension movements, has gained clinical relevance in the assessment of degenerative disorders of the lumbar spine, which may be characterized by an instability phase in which an abnormal motion of the degenerated segment can be detected (Kirkaldy-Willis and Farfan, 1982; Panjabi, 2003). Furthermore, after decompression surgery to treat spinal stenosis, iatrogenic destabilization due to tissue removal may occur, which is commonly evaluated by measuring the segmental range of motion (Guha et al., 2015). Another way to apply the in vivo range of motion is to assess spinal mobility after implanting motion preservation devices, such as flexible rods and artificial discs, in both the lumbar spine and the cervical spine (Berg et al., 2011).

The range of motion is typically measured on two dynamic X-rays while a patient performs flexion-extension movements; that is, radiographs are taken with a patient in the standing posture while trying to bend forward or extend backward as far as possible (Pieper et al., 2014). The two X-rays are then superimposed so that one vertebra, considered as the reference, is perfectly overlapping in the two images. The rotation of the upper vertebra is then measured by tracing lines along its lower endplate (Fig. 3) (White and Panjabi, 1990). Another common method to assess instability of dynamic X-rays is to measure the anteroposterior relative displacement of two vertebrae, assessed at the posteroinferior edge of the upper vertebral body (Fig. 3). In the lumbar spine, instability is then diagnosed if the rotation exceeds 10 degrees or the anteroposterior displacement is greater than 3 mm (White and Panjabi, 1990; Dupuis et al., 1985; Muto et al., 2016).

When manually performed, these types of measurements suffer from the previously mentioned inaccuracies, which can be much more significant in magnitude than the ranges of motion under investigation. If the aim of the measurement is to diagnose instability, even a qualitative assessment may be sufficient; on the contrary, evaluation of motion-preserving implants generally requires a precise measurement, especially if conducted for regulatory purposes. In this respect, computer-aided tools able to automatically identify vertebral contours, landmarks, or patterns, and thus able to improve measurement repeatability and accuracy, are currently considered as state-of-the-art tools. The proprietary image analysis software QMA (Medical Metrics, Inc., Houston, Texas), which is able to locate landmarks

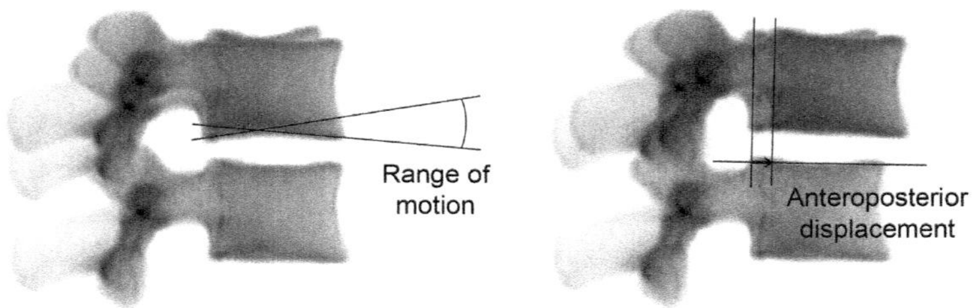

FIG. 3

Measurement of the range of motion (left) and of the anteroposterior displacement (right) on sagittal X-rays. First, the flexion and the extension image are superimposed so that the lower vertebra is perfectly overlapping. The rotation of the upper vertebra is measured by tracing lines along its lower endplate (left). The anteroposterior relative displacement is assessed at the posteroinferior edge of the upper vertebral body.

in dynamic X-rays as well as in fluoroscopic videos by means of contour and pattern analysis (Reitman et al., 2004), has been thoroughly validated by comparison to motion capture systems (Zhao et al., 2005) and is widely employed today. The QMA program is based on the generalized Hough transform, a computational method based on template matching that is widely employed to locate an object in an image independently of its scale and orientation (Ballard, 1991). After locating the landmarks, the QMA automatically calculates the transformations of the vertebrae and outputs the translations, rotations, and location of the center of rotation during the motion. FXA (ACES GmbH, Stuttgart, Germany) is also proprietary software that provides similar results and that has been validated against synthetic images as well as motion capture data (Schulze et al., 2011); however, the numerical methods exploited by this computer program are not documented. In a recent scientific study, Volkheimer et al. (2017) used a novel, fully automatic method based on the SURF algorithm to match interest points in flexion and extension X-rays to compute the vertebral transformations during in vitro testing of cadaveric specimens (Fig. 4), with excellent accuracy and repeatability.

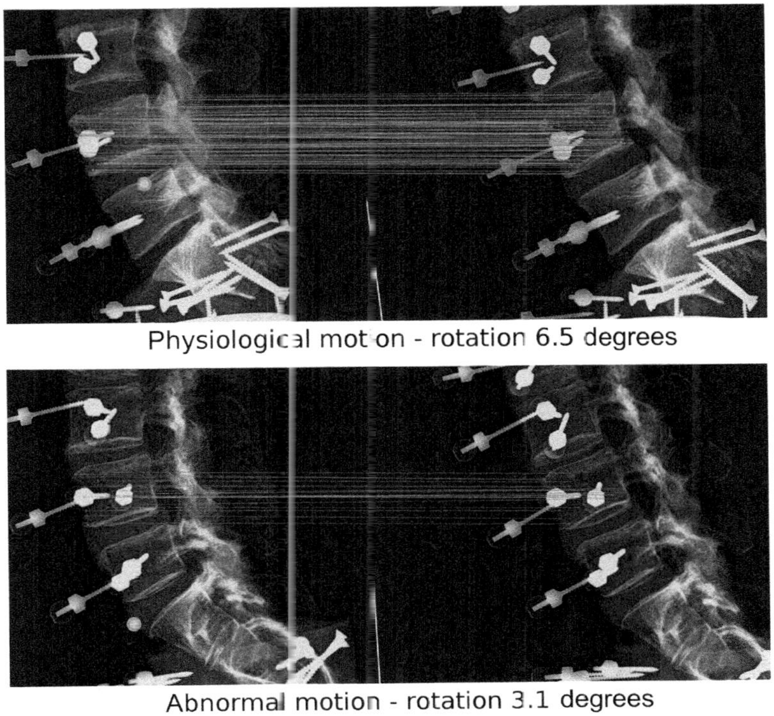

FIG. 4

Vertebral motion analyzed by means of automated software, which identifies and matches interest points *(small circles)*. The program then computes the rigid transformation and locates the center of rotation *(large circle)*. Top: a physiological flexion motion with the center of rotation located in the middle of the lower vertebral endplate; bottom: an abnormal motion depicting instability of the motion segment.

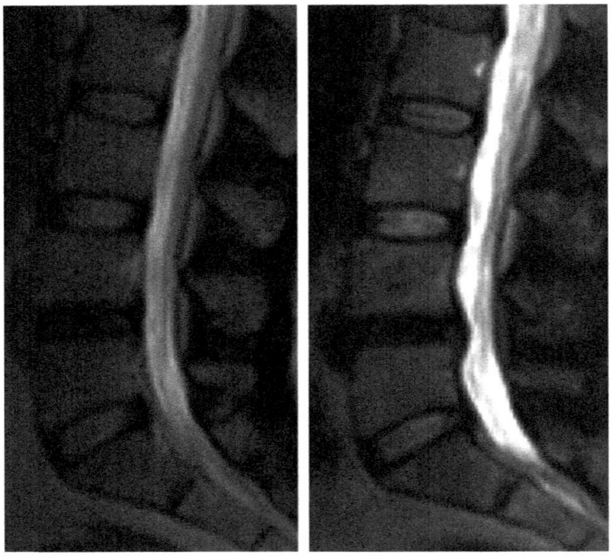

FIG. 5

T2-weighted MRI images of the lumbar spine of the same patient in the supine position (left) and in standing (right), which highlights a disc protrusion at L4-L5 not visible without axial loading.

Reproduced with permission from Muto, M., Giurazza, F., Guarnieri, G., Izzo, R., Diano, A., 2016. Neuroimaging of spinal instability. Magn. Reson. Imaging Clin. N. Am. 24(3), 485–494.

A different approach to assess spinal instability is based on MRI, which has the advantage of avoiding ionizing radiation. Conventional MRI is performed in the supine or prone position because of the design of commercially available magnets, which do not allow for upright imaging. Since shortly after their commercial introduction, low-field, open magnets have been used for spinal imaging in the seated and standing positions (Wildermuth et al., 1998) with interesting results that, however, did not prove the clinical superiority of upright versus conventional MR examinations (Leone et al., 2007). Later technologies improved image quality and performed true dynamic studies in flexion-extension positions, as well as during other movements (Alyas et al., 2008). Despite being inherently limited by the low magnetic field intensities achieved by the current open magnet designs, dynamic MRI can support the diagnosis of disorders invisible in supine scans (McGregor et al., 2002; Jinkins et al., 2005) (Fig. 5).

SPINAL DEFORMITIES

As previously mentioned, the radiological assessment of spinal deformities, such as adolescent and adult scoliosis as well as disorders of the sagittal alignment, is typically conducted on planar X-rays. In scoliotic patients, a coronal projection is required to grade the deformity and to monitor curve progression over time. In many cases (e.g., before and after surgical correction), a sagittal image is also acquired to measure radiological parameters such as thoracic kyphosis, lumbar lordosis, and pelvic incidence. Similarly, in elderly patients with sagittal disorders, a coronal projection may be acquired to assess a possible degenerative scoliosis (Lafage et al., 2015). Traditionally, the two views, which are acquired independently, are taken in slightly different postures and have different magnification factors and cannot therefore be exploited in combination to extract three-dimensional information about the deformity. Furthermore, because of the limited field of view of most hardware, several images need to be captured and stitched together to create a single image of the whole spine, thus requiring a substantial dose of radiation (van der Plaats, 2012).

The EOS Imaging System was designed to overcome these limitations and is capable of simultaneously acquiring a coronal and a sagittal whole-body view of a patient in an upright position. These images are perfectly orthogonal and in true 1:1 scale, thus allowing for linear measurements without the need for radio-opaque markers (Wybier and Bossard, 2013) and for three-dimensional reconstruction by means of proprietary software (Carreau et al., 2014) (Fig. 6). The imaging system exploits an innovative technology to reduce the radiation dose to a fraction of that necessary for a standard X-ray

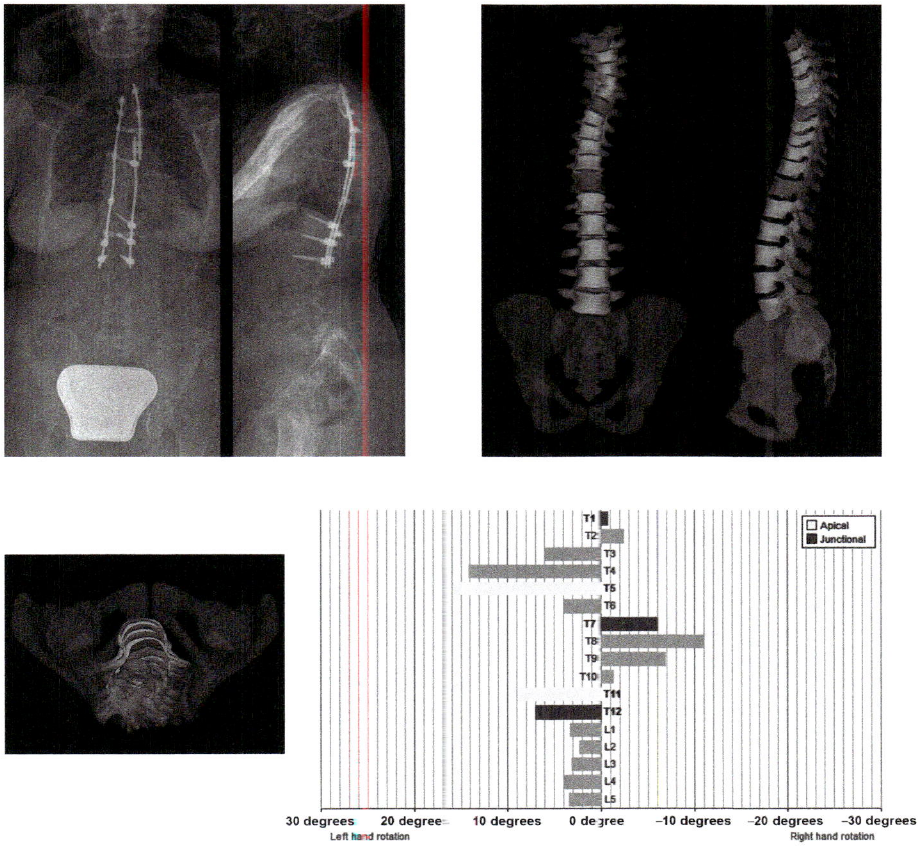

FIG. 6

Biplanar radiography by means of the EOS Imaging System of a patient suffering from adolescent idiopathic scoliosis subjected to surgical correction. Top: coronal and sagittal X-rays (left) and three-dimensional reconstruction in the same views (right). Bottom: axial view of the reconstructed spine (left) and diagram of the calculated vertebral rotations in the axial plane.

projection (Yvert et al., 2015). Furthermore, a micro-dose protocol specifically designed for the follow-up of adolescent scoliotic patients was recently presented (Ilharreborde et al., 2016).

Planar X-ray views are also employed to assess the stiffness of spinal deformities and the distinction between flexible and nonflexible curves, which are important concerns when planning surgical correction of the deformities. To this end, the most consolidated approach is to acquire side-bending views, that is, images of the patient in the supine or standing position during maximal passive side bending in both directions (Boos and Aebi, 2008; Ni et al., 2011) (Fig. 7). This type of imaging highlights the major curves, which rotate rigidly during bending, in contrast with the flexible curves which undergo deformation. A similar technique exploits a rigid fulcrum positioned under the scoliotic curve, with the patient lying on the side (Luk et al., 2008; Hay et al., 2008). Traction applied by means of loading frames, under general anesthesia (Davis et al., 2004) or not (Polly and Sturmk, 1998), has also been used for the same purpose. A recent paper proposed a suspension test performed inside the EOS Imaging System, with results comparable to a traction test and reduced radiation exposure (Hirsch et al., 2015).

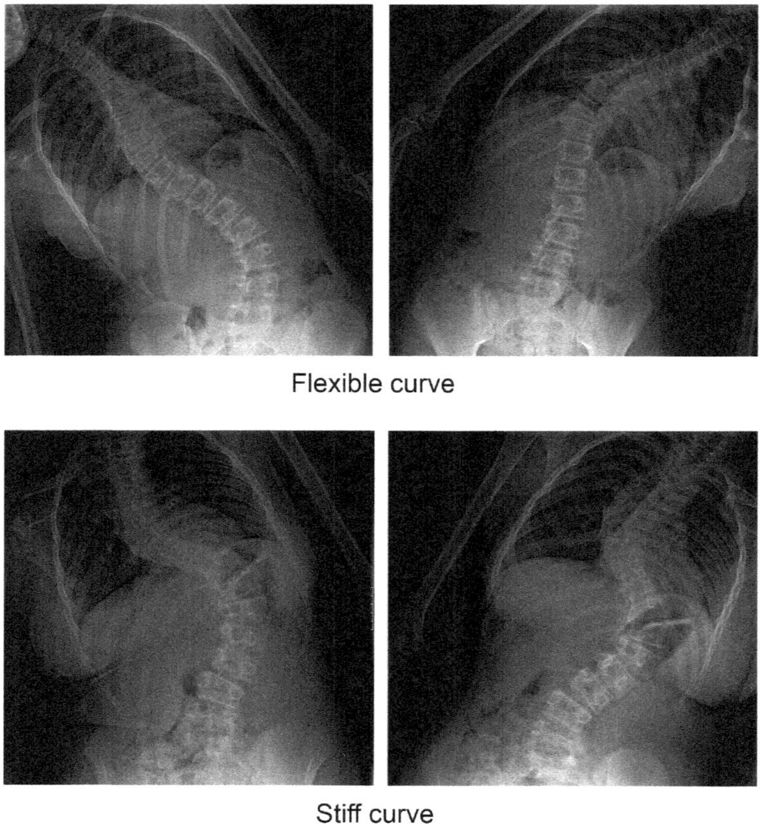

FIG. 7

Side bending X-rays of adolescent scoliotic patients depicting a flexible scoliotic curve (top) as well as a stiff one, in which the deformity is negligibly altered during bending (bottom).

Noninvasive imaging techniques, that is, those not based on ionizing radiation, are also used for the diagnosis and follow-up of spinal deformities. Surface topography techniques enable the creation of a detailed height map of the back surface and for the precise localization of anatomical landmarks based on the surface curvature. These techniques are based on various principles, such as laser (Hill et al., 2002), structured light scanning (Betsch et al., 2015; Hackenberg et al., 2003), and moirè patterns (Willner, 1979; Porto et al., 2010) (Fig. 8). These methods have been used for decades, especially with pediatric and adolescent patients for whom the risk of radiation-induced damage is more relevant.

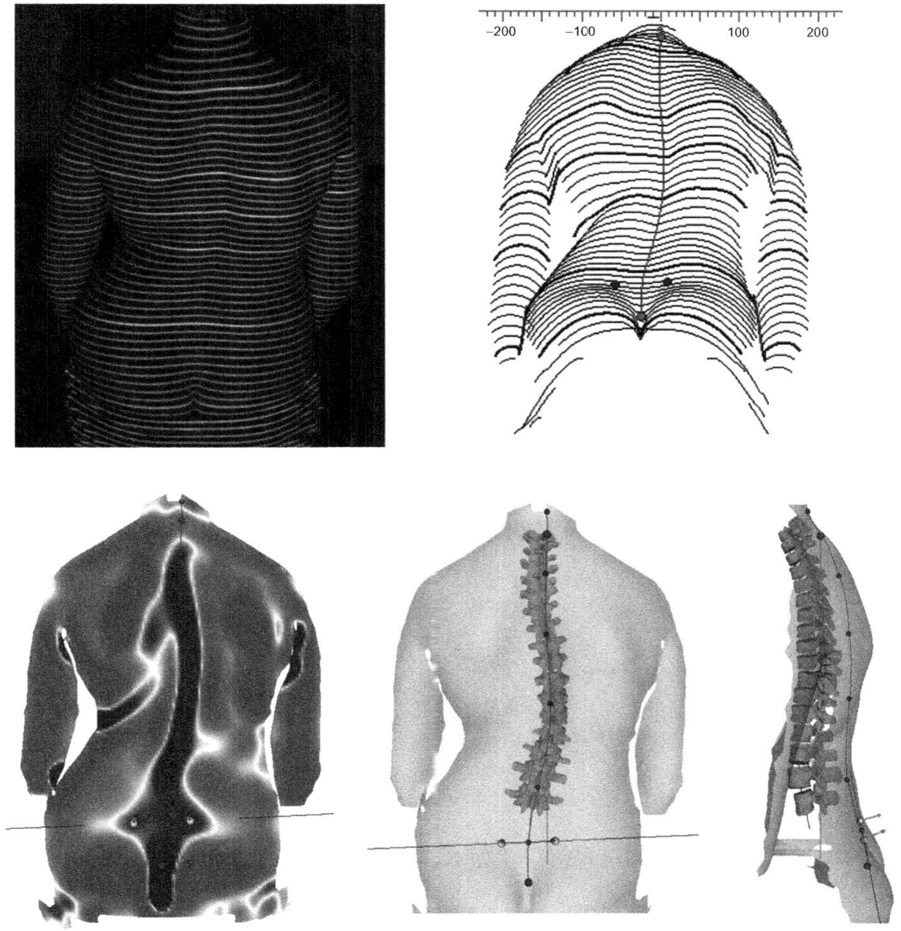

FIG. 8

Surface topography of an adolescent scoliotic patient obtained by means of the Formetric 4D (DIERS International GmBH, Schlagenbach, Germany). Top: structured light scan (left) and reconstruction of the back surface (right). Bottom: analysis of the curvature of the surface (left) and estimated shape of the thoracolumbar spine (right).

Surface topography achieves accurate and repeatable characterization of the back, but it has not shown a strong correlation with radiological parameters such as the Cobb angle. Moreover, because of errors within a 6- to 10-degree range, surface topography techniques should not be considered as a valid alternative to radiological assessment (Knott et al., 2016; Frerich et al., 2012). However, positive reports have been published (Mohokum et al., 2015), and in order to reduce cumulative exposure to radiation, surface topography methods are expected to find a role in monitoring the progression of scoliosis (Knott et al., 2016; Komeili et al., 2015).

The use of ultrasound imaging in scoliotic patients has been proposed (Letts et al., 1988) and refined for the measurement of the Cobb angle in the coronal plane and three-dimensionally (Purnama et al., 2010; Nguyen et al., 2015). The technique has inherent advantages, such as being totally safe for patients and its relatively low cost. Recent studies showed nonnegligible errors in a 5- to 10-degree range if compared to planar X-rays, but good reliability and reproducibility of the measurements in both the prone (Wang et al., 2015) and the standing postures (Zheng et al., 2016), thus confirming the technique's potential for clinical applications.

IMAGING OF THE INTERVERTEBRAL DISC

Because disc degenerative changes one of the most frequent causes of back and neck pain, the intervertebral disc is frequently examined via imaging methods. Because its density is rather low and homogeneous, methods based on X-rays are not well suited for disc imaging. Nevertheless, Wilke and colleagues introduced and validated a quantitative method to grade intervertebral disc degeneration based on sagittal X-ray projections, in both the cervical spine and the lumbar spine (Wilke et al., 2006; Kettler et al., 2006). Being radiographic disc imaging not achievable, the grading systems are based on bone-related parameters that can be assessed on radiographic images, such as the height of the intervertebral space, the presence of osteophytes, and diffuse endplate sclerosis. The methods were widely used to grade the quality of cadaveric specimens before they were used for in vitro testing. Recently, Volkheimer et al. showed that, unlike MRI imaging, X-ray grading of disc degeneration is not capable of capturing subtle degenerative changes, such as those in the early degeneration stage (Volkheimer et al., 2017) (Fig. 9).

An imaging-related technique employing X-rays that was widely used in the past and is still employed in selected patients is provocative discography, which involves injecting a radiopaque agent under pressure into the disc under fluoroscopic or CT guidance (Carragee and Alamin, 2001). If the disc is a source of pain, the injection will induce a pain response in the patient. Subsequently, a CT scan identifies tears and lesions in the intervertebral disc. This technique has been criticized because of its high false positive rate (Carragee et al., 1999), especially in subjects with somatization disorders, chronic pain, or ongoing compensation litigation (Carragee et al., 2006), and the evidence of adverse effects in the long term (Carragee et al., 2009). Indeed, it has been observed that disc puncture involves a faster progression of disc degeneration and a higher risk of disc herniations. Furthermore, discography is an invasive procedure, potentially prone to complications. Despite these limitations, because of a lack of valid alternatives for diagnosing discogenic back pain, provocative discography may still be used in patients with severe low back pain, especially as a selection tool for fusion surgery.

The current gold standard for intervertebral disc imaging is MRI. In clinical practice, T2-weighted images are commonly used to grade disc degeneration by means of the validated Pfirrmann score (Pfirrmann et al., 2001) (Fig. 9). To classify discs into five categories, this grading system considers three

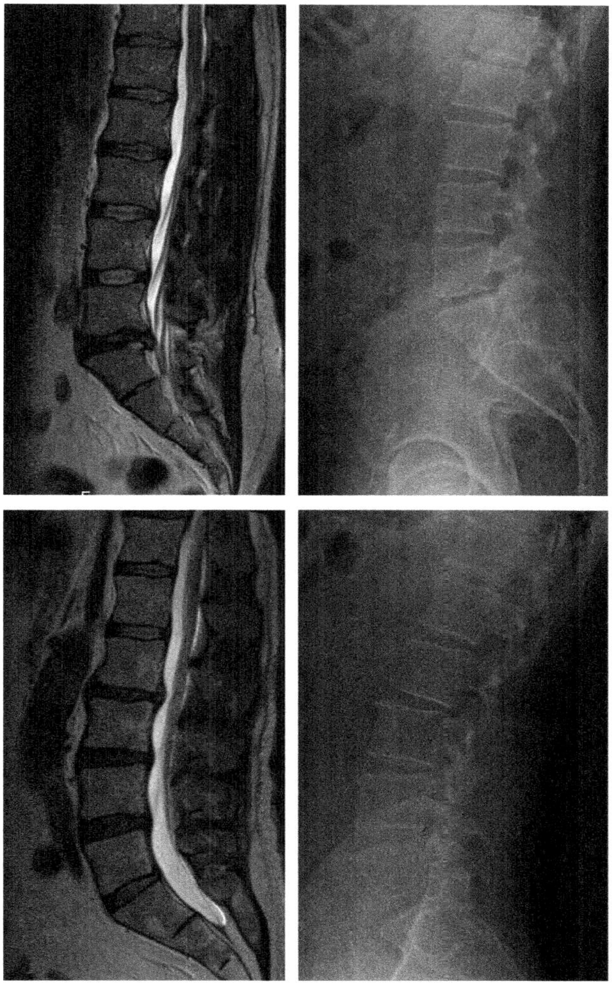

FIG. 9

Comparison of sagittal X-rays (left) and T2-weighted MRI scans (right) in two adult patients suffering from low back pain. The MRI of the first patient (top) shows a disc herniation and reduced signal level at L5-S1 and degenerative signs at the upper levels, whereas no major degenerative features are visible in the X-rays. Similarly, the second patient (bottom) has a protrusion at L2-L3 and reduced signal intensity at the lower levels, whereas the X-ray's image shows only anterior osteophytosis at L2-L3. The difference in the lumbar lordosis between MRI (supine) and X-rays (standing) is evident.

parameters (signal intensity in comparison with cerebrospinal fluid, disc height, and distinction between nucleus pulposus and annulus fibrosus). Intra- and interobserver reliability of the grading system proved to be very good to excellent (Pfirrmann et al., 2001). However, a certain degree of subjectivity, also related to the experience of the reader, cannot be avoided. To overcome this limitation, modified versions (Griffith et al., 2007) and quantitative indices based on T2-weighted (Videman et al., 2017) images have been introduced. In addition, the T2-mapping technique, in which the T2 relaxation time is quantitatively assessed, has been applied to disc imaging, especially for the evaluation of the water content of the nucleus pulposus, which is a major parameter involved in disc degeneration (Marinelli et al., 2010; Takashima et al., 2012; Trattnig et al., 2010).

T2 relaxation times are not strongly correlated with the proteoglycan content (Marinelli et al., 2009), which is another major variable characterizing the stage of disc degeneration, especially to identify its early stages in which the water content is less affected (Sivan et al., 2014). To investigate the loss of proteoglycans, T1ρ MRI imaging has been introduced. T1ρ, a time constant obtained by a specific MRI sequence named spin-lock, has been shown to be linearly correlated with proteoglycan content (Johannessen et al., 2006) and is currently under investigation for the quantitative grading of disc degeneration (Wang et al., 2013; Zobel et al., 2012). Recent studies indicate that T1ρ might be better suited for imaging the annulus fibrosus with respect to T2-mapping, whereas the two techniques provide similar results in the nucleus pulposus (Wang et al., 2013).

Other MRI techniques are under investigation for a detailed characterization of the composition of the intervertebral disc. Nuclear magnetic resonance spectroscopy, which was introduced before MRI imaging (Bloch et al., 1946; Purcell et al., 1946), has been applied to assess metabolic markers of disc degeneration in disc samples with dedicated spectrometers (Keshari et al., 2005) as well as in vivo by means of clinical scanners (Zuo et al., 2009). Diffusion tensor imaging (DTI) is an MR technique based on the diffusivity of water molecules in tissues. Mainly used for cancer imaging, in which diffusivity of water molecules is remarkably reduced compared to normal tissues, DTI has been widely used for fiber tracking in the brain. In intervertebral discs, promising results have been shown, demonstrating that annular fibers of normal and pathologic discs have different arrangements (Zhang et al., 2012). Finally, sodium MRI, in which sodium nuclei are used instead of protons to generate an image, is also being tested to evaluate the concentration of proteoglycans (Noebauer-Huhmann et al., 2012; Wang et al., 2010; Insko et al., 2002), but using it in clinical applications is not yet feasible.

REFERENCES

Akbar, M., Terran, J., Ames, C.P., Lafage, V., Schwab, F., 2013. Use of Surgimap spine in sagittal plane analysis, osteotomy planning, and correction calculation. Neurosurg. Clin. N. Am. 24 (2), 163–172.

Alyas, F., Connell, D., Saifuddin, A., 2008. Upright positional MRI of the lumbar spine. Clin. Radiol. 63 (9), 1035–1048.

Ambrose, J., 1973. Computerized transverse axial scanning (tomography). 2. Clinical application. Br. J. Radiol. 46 (552), 1023–1047.

Ballard, D.H., 1991. Generalizing the Hough transform to detect arbitrary shapes. Pattern Recogn. 13 (2), 183–194.

Berg, S., Tropp, H.T., Leivseth, G., 2011. Disc height and motion patterns in the lumbar spine in patients operated with total disc replacement or fusion for discogenic back pain. Results from a randomized controlled trial. Spine J. 11 (11), 991–998.

Bernstein, M.A., King, K.F., Zhou, X.J., 2004. Handbook of MRI Pulse Sequences. Academic Press, Burlington.

Betsch, M., Wild, M., Rath, B., Tingart, M., Schulze, A., Quack, V., 2015. Radiation-free diagnosis of scoliosis: an overview of the surface and spine topography. Der Orthopade 44 (11), 845–851.

Bloch, F., Hanse, W.W., Packard, M., 1946. The nuclear induction experiment. Phys. Rev. 70, 474–485.

Boos, N., Aebi, M., 2008. Spinal Disorders—Fundamentals of Diagnosis and Treatment. Springer, Heidelberg.

Boos, N., Weissbach, S., Rohrbach, H., Weiler, C., Spratt, K.F., Nerlich, A.G., 2002. Classification of age-related changes in lumbar intervertebral discs: 2002 Volvo award in basic science. Spine 27 (23), 2631–2644.

Carragee, E.J., Alamin, T.F., 2001. Discography. A review. Spine J. 1 (5), 364–372.

Carragee, E.J., Alamin, T.F., Carragee, J.M., 2006. Low-pressure positive discography in subjects asymptomatic of significant low back pain illness. Spine 31 (5), 505–509.

Carragee, E.J., Don, A.S., Hurwitz, E.L., Cuellar, J.M., Carrino, J.A., Herzog, R., 2009. 2009 ISSLS prize winner: does discography cause accelerated progression of degeneration changes in the lumbar disc: a ten-year matched cohort study. Spine 34 (21), 2338–2345.

Carragee, E.J., Tanner, C.M., Yang, B., Brito, J.L., Truong, T., 1999. False-positive findings on lumbar discography. Reliability of subjective concordance assessment during provocative disc injection. Spine 24 (23), 2542–2547.

Carreau, J.H., Bastrom, T., Petcharaporn, M., Schulte, C., Marks, M., Illes, T., Somoskeoy, S., Newton, P.O., 2014. Computer-generated, three-dimensional spine model from Biplanar radiographs: a validity study in idiopathic scoliosis curves greater than 50 degrees. Spine Deform. 2 (2), 81–88.

Cobb, J.C., 1948. Outline for the study of scoliosis. J. Am. Acad. Orthop. Surg. 7, 261–275.

Cushing, H.W., 1898. Haematomyelia from gunshot wounds of the spine. Am. J. Med. Sci. 115, 654–683.

Damadian, R., 1971. Tumor detection by nuclear magnetic resonance. Science 171 (3976), 1151–1153.

Damadian, R., Minkoff, L., Goldsmith, M., Stanford, M., Koutcher, J., 1976. Field focusing nuclear magnetic resonance (FONAR): visualization of a tumor in a live animal. Science 194 (4272), 1430–1432.

Davis, B.J., Gadgil, A., Trivedi, J., Ahmed, e.B., 2004. Traction radiography performed under general anesthetic: a new technique for assessing idiopathic scoliosis curves. Spine 29 (21), 2466–2470.

Dewing, S.B., 1962. Modern Radiology in Historical Perspective. Charles C Thomas, Springfield.

Dupuis, P.R., Yong-Hing, K., Cassidy, J.D., Kirkaldy-Willis, W.H., 1985. Radiologic diagnosis of degenerative lumbar spinal instability. Spine 10 (3), 262–276.

Edelman, R.R., Shoukimas, G.M., Stark, D.D., Davis, K.R., New, P.F., Saini, S., Rosenthal, D.I., Wismer, G.L., Brady, T.J., 1985. High-resolution surface-coil imaging of lumbar disk disease. AJR Am. J. Roentgenol. 144 (6), 1123–1129.

Frerich, J.M., Hertzler, K., Knott, P., Mardjetko, S., 2012. Comparison of radiographic and surface topography measurements in adolescents with idiopathic scoliosis. Open Orthopaed. J. 6, 261–265.

Frobin, W., Brinckmann, P., Biggemann, M., Tillotson, M., Burton, K., 1997. precision measurement of disc height, vertebral height and sagittal plane displacement from lateral radiographic views of the lumbar spine. Clin. Biomech. (Bristol, Avon) 12 (Suppl 1), S1–S63.

Gordon, R., Bender, R., Herman, G.T., 1970. Algebraic reconstruction techniques (ART) for three-dimensional electron microscopy and x-ray photography. J. Theor. Biol. 29 (3), 471–481.

Griffith, J.F., Wang, Y.X., Antonio, G.E., Choi, K.C., Yu, A., Ahuja, A.T., Leung, P.C., 2007. Modified Pfirrmann grading system for lumbar intervertebral disc degeneration. Spine 32 (24), E708–12.

Guha, D., Heary, R.F., Shamji, M.F., 2015. Iatrogenic spondylolisthesis following laminectomy for degenerative lumbar stenosis: systematic review and current concepts. Neurosurg. Focus 39 (4), E9.

Hackenberg, L., Hierholzer, E., Potzl, W., Gotze, C., Liljenqvist, U., 2003. Rasterstereographic back shape analysis in idiopathic scoliosis after anterior correction and fusion. Clin. Biomech. (Bristol, Avon) 18 (1), 1–8.

Hall, E.J., Brenner, D.J., 2008. Cancer risks from diagnostic radiology. Br. J. Radiol. 81 (965), 362–378.

Hay, D., Izatt, M.T., Adam, C.J., Labrom, R.D., Askin, G.N., 2008. The use of fulcrum bending radiographs in anterior thoracic scoliosis correction: a consecutive series of 90 patients. Spine 33 (9), 999–1005.

Hesselink, J.R., 1988. Spine imaging: history, achievements, remaining frontiers. AJR Am. J. Roentgenol. 150 (6), 1223–1229.

Hill, D.L., Berg, D.C., Raso, V.J., Lou, E., Durdle, N.G., Mahood, J.K., Moreau, M.J., 2002. Evaluation of a laser scanner for surface topography. Stud. Health Technol. Inform. 88, 90–94.

Hirsch, C., Ilharreborde, B., Mazda, K., 2015. EOS suspension test for the assessment of spinal flexibility in adolescent idiopathic scoliosis. Eur. Spine J. 24 (7) 1408–1414.

Hounsfield, G.N., 1973. Computerized transverse axial scanning (tomography). 1. Description of system. Br. J. Radiol. 46 (552), 1016–1022.

Ilharreborde, B., Ferrero, E., Alison, M., Mazda, K., 2016. EOS microdose protocol for the radiological follow-up of adolescent idiopathic scoliosis. Eur. Spine J. 25 (2), 526–531.

Insko, E.K., Clayton, D.B., Elliott, M.A., 2002. In vivo sodium MR imaging of the intervertebral disk at 4 T. Acad. Radiol. 9 (7), 800–804.

Jinkins, J.R., Dworkin, J.S., Damadian, R.V., 2005. Upright, weight-bearing, dynamic-kinetic MRI of the spine: initial results. Eur. Radiol. 15 (9), 1815–1825.

Johannessen, W., Auerbach, J.D., Wheaton, A.J., Kurji, A., Borthakur, A., Reddy, R., Elliott, D.M., 2006. Assessment of human disc degeneration and proteoglycan content using T1rho-weighted magnetic resonance imaging. Spine 31 (11), 1253–1257.

Kak, A.C., Slaney, M., 2001. Principles of Computerized Tomographic Imaging. Society for Industrial and Applied Mathematics, Philadelphia, PA.

Kendler, D.L., Bauer, D.C., Davison, K.S., Dian, L., Hanley, D.A., Harris, S.T., McClung, M.R., Miller, P.D., Schousboe, J.T., Yuen, C.K., Lewiecki, E.M., 2016. Vertebral fractures: clinical importance and management. Am. J. Med. 129 (2), 221.e1–221.e10.

Keshari, K.R., Zektzer, A.S., Swanson, M.G., Majumdar, S., Lotz, J.C., Kurhanewicz, J., 2005. Characterization of intervertebral disc degeneration by high-resolution magic angle spinning (HR-MAS) spectroscopy. Magn. Reson. Med. 53 (3), 519–527.

Kettler, A., Rohlmann, F., Neidlinger-Wilke, C., Werner, K., Claes, L., Wilke, H.J., 2006. Validity and interobserver agreement of a new radiographic grading system for intervertebral disc degeneration: part II cervical spine. Eur. Spine J. 15 (6), 732–741.

Kim, H., Kim, H.S., Moon, E.S., Yoon, C.S., Chung, T.S., Song, H.T., Suh, J.S., Lee, Y.H., Kim, S., 2010. Scoliosis imaging: what radiologists should know. Radiographics 30 (7), 1823–1842.

Kirkaldy-Willis, W.H., Farfan, H.F., 1982. Instability of the lumbar spine. Clin. Orthop. Relat. Res. 165 (165), 110–123.

Knott, P., Sturm, P., Lonner, B., Cahill, P., Betsch, M., McCarthy, R., Kelly, M., Lenke, L., Betz, R., 2016. Multicenter comparison of 3D spinal measurements using surface topography with those from conventional radiography. Spine Deform. 4 (2), 98–103.

Komeili, A., Westover, L., Parent, E.C., El-Rich, M., Adeeb, S., 2015. Monitoring for idiopathic scoliosis curve progression using surface topography asymmetry analysis of the torso in adolescents. Spine J. 15 (4), 743–751.

Lafage, V., Diebo, B.G., Schwab, F., 2015. Sagittal Spino-Pelvic Alignment: From Theory to Clinical Application. Editorial Medica Panamericana, Madrid.

Lauterbur, P.C., 1989. Image formation by induced local interactions. Examples employing nuclear magnetic resonance 1973. Clin. Orthop. Relat. Res. 244 (244), 3–6.

Leone, A., Guglielmi, G., Cassar-Pullicino, V.N., Bonomo, L., 2007. Lumbar intervertebral instability: a review. Radiology 245 (1), 62–77.

Letts, M., Quanbury, A., Gouw, G., Kolsun, W., Letts, E., 1988. Computerized ultrasonic digitization in the measurement of spinal curvature. Spine 13 (10), 1106–1110.

Luk, K.D., Don, A.S., Chong, C.S., Wong, Y.W., Cheung, K.M., 2008. Selection of fusion levels in adolescent idiopathic scoliosis using fulcrum bending prediction: a prospective study. Spine 33 (20), 2192–2198.

Mansfield, P., Maudsley, A.A., 1977. Medical imaging by NMR. Br. J. Radiol. 50 (591), 188–194.

Marinelli, N.L., Haughton, V.M., Anderson, P.A., 2010. T2 relaxation times correlated with stage of lumbar intervertebral disk degeneration and patient age. AJNR Am. J. Neuroradiol. 31 (7), 1278–1282.

Marinelli, N.L., Haughton, V.M., Munoz, A., Anderson, P.A., 2009. T2 relaxation times of intervertebral disc tissue correlated with water content and proteoglycan content. Spine 34 (5), 520–524.

McGregor, A.H., Anderton, L., Gedroyc, W.M., Johnson, J., Hughes, S.P., 2002. The use of interventional open MRI to assess the kinematics of the lumbar spine in patients with spondylolisthesis. Spine 27 (14), 1582–1586.

Modic, M.T., Masaryk, T.J., Ross, J.S., Carter, J.R., 1988. Imaging of degenerative disk disease. Radiology 168 (1), 177–186.

Modic, M.T., Pavlicek, W., Weinstein, M.A., Boumphrey, F., Ngo, F., Hardy, R., Duchesneau, P.M., 1984. Magnetic resonance imaging of intervertebral disk disease. Clinical and pulse sequence considerations. Radiology 152 (1), 103–111.

Mohokum, M., Schulein, S., Skwara, A., 2015. The validity of rasterstereography: a systematic review. Orthod. Rev. 7 (3), 5899.

Muto, M., Giurazza, F., Guarnieri, G., Izzo, R., Diano, A., 2016. Neuroimaging of spinal instability. Magn. Reson. Imaging Clin. N. Am. 24 (3), 485–494.

Nguyen, D.V., Vo, Q.N., Le, L.H., Lou, E.H., 2015. Validation of 3D surface reconstruction of vertebrae and spinal column using 3D ultrasound data—a pilot study. Med. Eng. Phys. 37 (2), 239–244.

Ni, H.J., Su, J.C., Lu, Y.H., Zhu, X.D., He, S.S., Wu, D.J., Xu, J., Yang, C.W., Wang, C.F., Zhao, Y.C., Li, M., 2011. Using side-bending radiographs to determine the distal fusion level in patients with single thoracic idiopathic scoliosis undergoing posterior correction with pedicle screws. J. Spinal Disord. Tech. 24 (7), 437–443.

Noebauer-Huhmann, I., Juras, V., Pfirrmann, C.W., Szomolanyi, P., Zbyn, S., Messner, A., Wimmer, J., Weber, M., Friedrich, K.M., Stelzeneder, D., 2012. Sodium MR imaging of the lumbar intervertebral disk at 7 T: correlation with T2 mapping and modified Pfirrmann score at 3 T—preliminary results. Radiology 265 (2), 555–564.

Panjabi, M.M., 2003. Clinical spinal instability and low back pain. J. Electromyogr. Kinesiol. 13 (4), 371–379.

Pearce, M.S., Salotti, J.A., Little, M.P., McHugh, K., Lee, C., Kim, K.P., Howe, N.L., Ronckers, C.M., Rajaraman, P., Sir Craft, A.W., Parker, L., Berrington de Gonzalez, A., 2012. Radiation exposure from CT scans in childhood and subsequent risk of leukaemia and brain tumours: a retrospective cohort study. Lancet 380 (9840), 499–505.

Pfirrmann, C.W., Metzdorf, A., Zanetti, M., Hodler, J., Boos, N., 2001. Magnetic resonance classification of lumbar intervertebral disc degeneration. Spine 26 (17), 1873–1878.

Pieper, C.C., Groetz, S.F., Nadal, J., Schild, H.H., Niggemann, P.D., 2014. Radiographic evaluation of ventral instability in lumbar spondylolisthesis: do we need extension radiographs in routine exams? Eur. Spine J. 23 (1), 96–101.

Polly Jr., D.W., Sturm, P.F., 1998. Traction versus supine side bending. Which technique best determines curve flexibility? Spine 23 (7), 804–808.

Porto, F., Gurgel, J.L., Russomano, T., Farinatti Pde, T., 2010. Moire topography: characteristics and clinical application. Gait Posture 32 (3), 422–424.

Purcell, E.M., Torrey, H.C., Pound, R.V., 1946. Resonance absorption by nuclear moments in a solid. Phys. Rev. 69, 37–38.

Purnama, K.E., Wilkinson, M.H., Veldhuizen, A.G., van Ooijen, P.M., Lubbers, J., Burgerhof, J.G., Sardjono, T.A., Verkerke, G.J., 2010. A framework for human spine imaging using a freehand 3D ultrasound system. Technol. Health Care 18 (1), 1–17.

Radon, J., 1986. On the determination of functions from their integral values along certain manifolds. IEEE Trans. Med. Imaging 5 (4), 170–176.

Reitman, C.A., Mauro, K.M., Nguyen, L., Ziegler, J.M., Hipp, J.A., 2004. Intervertebral motion between flexion and extension in asymptomatic individuals. Spine 29 (24), 2832–2843.

Ricart, P.A., Andres, T.M., Apazidis, A., Errico, T.J., Trobisch, P.D., 2011. Validity of Cobb angle measurements using digitally photographed radiographs. Spine J. 11 (10), 942–946.

Schlosser, T.P., van Stralen, M., Brink, R.C., Chu, W.C., Lam, T.P., Vincken, K.L., Castelein, R.M., Cheng, J.C., 2014. Three-dimensional characterization of torsion and asymmetry of the intervertebral discs versus vertebral bodies in adolescent idiopathic scoliosis. Spine 39 (19), E1159–66.

Schulze, M., Trautwein, F., Vordemvenne, T., Raschke, M., Heuer, F., 2011. A method to perform spinal motion analysis from functional X-ray images. J. Biomech. 44 (9), 1740–1746.

Shea, K.G., Stevens, P.M., Nelson, M., Smith, J.T., Masters, K.S., Yandow, S., 1998. A comparison of manual versus computer-assisted radiographic measurement. Intraobserver measurement variability for Cobb angles. Spine 23 (5), 551–555.

Sivan, S.S., Hayes, A.J., Wachtel, E., Caterson, B., Merkher, Y., Maroudas, A., Brown, S., Roberts, S., 2014. Biochemical composition and turnover of the extracellular matrix of the normal and degenerate intervertebral disc. Eur. Spine J. 23 (3), 344–353.

Takashima, H., Takebayashi, T., Yoshimoto, M., Terashima, Y., Tsuda, H., Ida, K., Yamashita, T., 2012. Correlation between T2 relaxation time and intervertebral disk degeneration. Skeletal Radiol. 41 (2), 163–167.

Trattnig, S., Stelzeneder, D., Goed, S., Reissegger, M., Mamisch, T.C., Paternostro-Sluga, T., Weber, M., Szomolanyi, P., Welsch, G.H., 2010. Lumbar intervertebral disc abnormalities: comparison of quantitative T2 mapping with conventional MR at 3.0 T. Eur. Radiol. 20 (11), 2715–2722.

van der Plaats, G.J., 2012. Medical X-ray Techniques in Diagnostic Radiology: A Textbook for Radiographers and Radiological Technicians. Springer Science & Business Media, Houten, Netherlands.

Videman, T., Battie, M.C., Gibbons, L.E., Gill, K., 2017. A new quantitative measure of disc degeneration. Spine J. 17 (5), 746–753.

Volkheimer, D., Galbusera, F., Liebsch, C., Schlegel, S., Rohlmann, F., Kleiner, S., Wilke, H.J., 2017. Is intervertebral disc degeneration related to segmental instability? An evaluation with two different grading systems based on clinical imaging. Acta Radiol. https://doi.org/10.1177/0284185117715284.

Wang, Y.J., Zhao, F., Griffith, J.F., Mok, G.S., Leung, J.C., Ahuja, A.T., Yuan, J., 2013. T1rho and T2 relaxation times for lumbar disc degeneration: an in vivo comparative study at 3.0-Tesla MRI. Eur. Radiol. 23 (1), 228–234.

Wang, C., McArdle, E., Fenty, M., Witschey, W., Elliott, M., Sochor, M., Reddy, R., Borthakur, A., 2010. Validation of sodium magnetic resonance imaging of intervertebral disc. Spine 35 (5), 505–510.

Wang, Q., Li, M., Lou, E.H., Wong, M.S., 2015. Reliability and validity study of clinical ultrasound imaging on lateral curvature of adolescent idiopathic scoliosis. PLoS One 10 (8), e0135264.

White, A.A., Panjabi, M.M., 1990. Clinical Biomechanics of the Spine, second ed. Lippincott Williams & Wilkins, Philadelphia, PA.

Wildermuth, S., Zanetti, M., Duewell, S., Schmid, M.R., Romanowski, B., Benini, A., Boni, T., Hodler, J., 1998. Lumbar spine: quantitative and qualitative assessment of positional (upright flexion and extension) MR imaging and myelography. Radiology 207 (2), 391–398.

Wilke, H.J., Rohlmann, F., Neidlinger-Wilke, C., Werner, K., Claes, L., Kettler, A., 2006. Validity and interobserver agreement of a new radiographic grading system for intervertebral disc degeneration: part I Lumbar spine. Eur. Spine J. 15 (6), 720–730.

Willner, S., 1979. Moire topography—a method for school screening of scoliosis. Arch. Orthop. Trauma Surg. 95 (3), 181–185.

Wills, B.P., Auerbach, J.D., Zhu, X., Caird, M.S., Horn, B.D., Flynn, J.M., Drummond, D.S., Dormans, J.P., Ecker, M.L., 2007. Comparison of cobb angle measurement of scoliosis radiographs with preselected end vertebrae: traditional versus digital acquisition. Spine 32 (1), 98–105.

Wybier, M., Bossard, P., 2013. Musculoskeletal imaging in progress: the EOS imaging system. Joint Bone Spine 80 (3), 238–243.

Yvert, M., Diallo, A., Bessou, P., Rehel, J.L., Lhomme, E., Chateil, J.F., 2015. Radiography of scoliosis: comparative dose levels and image quality between a dynamic flat-panel detector and a slot-scanning device (EOS system). Diagn. Interv. Imaging 96 (11), 1177–1188.

Zhao, K., Yang, C., Zhao, C., An, K.N., 2005. Assessment of non-invasive intervertebral motion measurements in the lumbar spine. J. Biomech. 38 (9), 1943–1946.

Zhang, Z., Chan, Q., Anthony, M.P., Samartzis, D., Cheung, K.M., Khong, P.L., Kim, M., 2012. Age-related diffusion patterns in human lumbar intervertebral discs: a pilot study in asymptomatic subjects. Magn. Reson. Imaging 30 (22), 181–188.

Zheng, Y.P., Lee, T.T., Lai, K.K., Yip, B.H., Zhou, G.Q., Jiang, W.W., Cheung, J.C., Wong, M.S., Ng, B.K., Cheng, J.C., Lam, T.P., 2016. A reliability and validity study for Scolioscan: a radiation-free scoliosis assessment system using 3D ultrasound imaging. Scoliosis Spinal Disord. 11, 13. https://doi.org/10.1186/s13013-016-0074-y (eCollection 2016).

Zobel, B.B., Vadala, G., Del Vescovo, R., Battisti, S., Martina, F.M., Stellato, L., Leoncini, E., Borthakur, A., Denaro, V., 2012. T1rho magnetic resonance imaging quantification of early lumbar intervertebral disc degeneration in healthy young adults. Spine 37 (14), 1224–1230.

Zuo, J., Saadat, E., Romero, A., Loo, K., Li, X., Link, T.M., Kurhanewicz, J., Majumdar, S., 2009. Assessment of intervertebral disc degeneration with magnetic resonance single-voxel spectroscopy. Magn. Reson. Med. 62 (5), 1140–1146.

CHAPTER

IN VIVO MEASUREMENTS: MOTION ANALYSIS

11

Elena Stucovitz, Jacopo Vitale, Fabio Galbusera
IRCCS Galeazzi Orthopedic Institute, Milan, Italy

INTRODUCTION

Human motion analysis is aimed at providing quantitative information related to the mechanics of the musculoskeletal system while various motor tasks are performed, with a specific focus on the relative movement between adjacent bones or joint kinematics. Kinematics is the measurement of motion, that is, of displacements, velocities, and accelerations during a specific movement. Human motion analysis became one of the most used techniques for understanding musculoskeletal system physiopathology, and its use continues to grow. Among several technological options for investigating human motion, two categories are particularly noteworthy here: optoelectronic systems and wearable devices. Their main differences lie in cost, suitability, accuracy, and type of data provided. All systems need to be able to capture the body kinematics, by recording the position and orientation of the body segments, the angles of the joints, and the corresponding linear and angular velocities and accelerations (Whittle, 2007). The selection of the best system to use for a specific application should be based on the specific research question (Perry, 1992), but the selection is often dictated by the direct availability of certain hardware and by the expertise of the staff. In the following discussion, the categories are described and their main clinical application for spine disorders is presented.

OPTOELECTRONIC SYSTEMS

Optoelectronic systems currently represent the most popular technological solution for the quantitative assessment of human movement. Most of these systems consist of a set of infrared cameras (at least two) that capture the position of several passive reflective spherical markers on the subject's skin that correspond to the anatomical sites of interest (Benedetti and Cappozzo, 1994). In the most common implementations, each camera has an array of lights emitting diodes mounted around the lens and producing an infrared stroboscopic illumination (Chiari et al., 2004).

Optoelectronic systems use the two-dimensional (2D) centroid coordinates of the marker on the image plane of each camera and the data from each camera's position to reconstruct the position of a marker through a process called stereophotogrammetry. Given a marker moving in the acquisition space, the system reconstructs the marker's position instant by instant by providing a characteristic spatial triad (x, y, z), according to the laboratory's frame of reference. To do so, it is necessary that at least two cameras observe the same marker simultaneously. Pattern recognition software then

Biomechanics of the Spine. https://doi.org/10.1016/B978-0-12-812851-0.00011-2

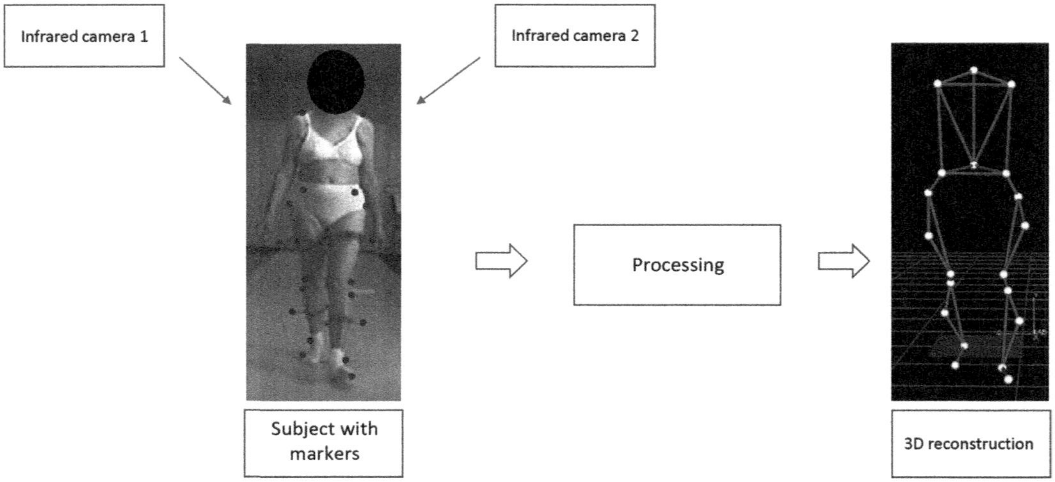

FIG. 1

Acquisition of the body motion of a subject by means of an optoelectronic system. Markers are placed on the subject, and their positions are acquired by at least two infrared cameras. The stereophotogrammetry reconstruction allows for the calculation of the 3D coordinates of each marker during the motion.

enables a noninvasive estimation of the instantaneous position of each marker in a three-dimensional (3D) measurement volume, that is, the kinematics in time and space of the anatomical site of interest (Chiari et al., 2004) (Fig. 1).

LIMITATIONS AND ERRORS IN STEREOPHOTOGRAMMETRY

Motion capture using 3D optoelectronic systems has several clinical applications, but is constrained by technical issues and errors. The accuracy and precision of a stereophotogrammetric system are affected by three types of errors: instrumental errors, marker placement errors, and soft tissue artifacts, which are the primary factors restricting the widespread use of motion analysis in musculoskeletal research (Andriacchi and Alexander, 2000).

- *Instrumental errors*: Arm swinging, walking aids, or subjects' movements reduce marker visibility during acquisitions (Chiari et al., 2004). The quality of the acquisition can be improved by creating a good laboratory setup, increasing the number of infrared cameras and optimizing their placement, and using a proper acquisition volume. Moreover, electronic noise, marker flickering, and software processing may contribute to amplifying and enlarging instrumental errors. These errors may arise at different levels of the measurement chain, and several compensation techniques are used routinely to limit them. First, an accurate infrared camera calibration procedure minimizes optical distortion and systematic errors. Second, smoothing and filtering procedures reduce random errors, and missing markers can be reinserted through dedicated software (Chiari et al., 2004).

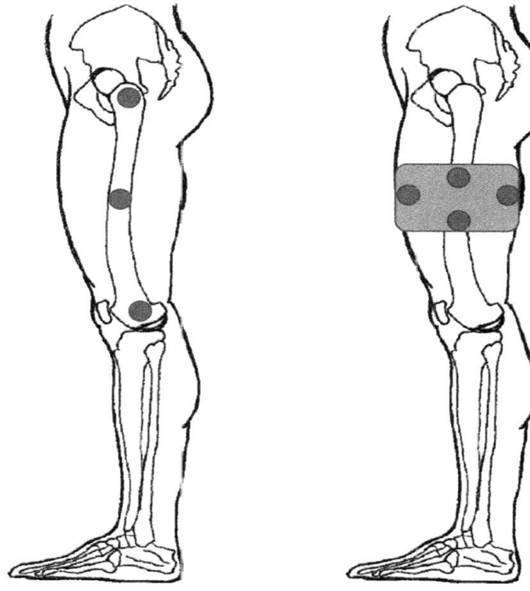

FIG. 2

Marker placement based on the anatomical frame (left) and the technical frame (right).

- *Marker placement*: Marker positioning directly affects the accuracy of the measurements. Marker placement employs two distinct approaches: use of an anatomical frame of reference or use of a technical frame of reference (Fig. 2). The anatomical frame of reference system relates to each body segment as defined by the anatomic landmarks upon which the markers are placed, whereas the technical frame of reference system relates to each body segment as defined by at least three markers that are not aligned and mounted on a support (Della Croce et al., 2005).

The markers' position on the technical frame fulfills certain technical requirements, such as achieving visibility to at least two infrared cameras and diminishing relative movement between the markers themselves and the underlying bone of interest. However, the position of the markers has no direct correlation with the segment of interest's anatomical landmarks and is therefore somewhat arbitrary, thus suffering from low repeatability. Because of this limitation, in order to extract the motion of the body segment, a further calibration is required to define the relationship between the technical and the anatomical frames (Cappozzo et al., 2005). The anatomical frame, on the other hand, is specifically defined to offer greater repeatability, both intra- and intersubject. Anatomical markers are positioned on the subject's skin in correspondence with external, palpable anatomical landmarks associated with specific bone locations (Cappozzo et al., 2005). Even though in clinical practice the most used protocols refer to anatomical frames rather than to technical frames, the determination of marker locations still suffers from inaccuracies and imprecisions. Determinants of the errors are that the irregular surface of the bones makes palpable recognition of the anatomical landmarks of interest difficult, that the thick soft tissues tend to cover such landmarks, and that the low standardization of the palpation procedures used to locate the anatomical landmarks leads to a subjective, operator-dependent approach to properly arranging the markers (Della Croce et al., 2005).

- *Soft tissue artifacts*: One of the most common problems with soft tissue relates to the movement caused by muscular contractions, skin deformation, and displacement between markers and the bone of interest. The relative movement between markers and bones creates an artifact (i.e., a soft tissue artifact, or STA) able to influence the kinematic assessment and is considered to be the most critical cause of error in stereophotogrammetric systems (Leardini et al., 2005). Because of its nature, an STA has a frequency content comparable to that of bone motion, and canceling it through a filtering technique is impossible (Leardini et al., 2005). As a matter of fact, the amount of STAs is difficult to quantify, especially regarding the spine. It is worth noting that spinal movements differ from other skeletal joint movements, such as in the hip or knee, because the spine consists of several segments which allow movement relative to one another, with a different range of motion (ROM) in several planes (Zemp et al., 2014). In their work, Leardini et al. (2005) suggested identifying and calibrating STA models, and then applying such models to the specific subject under analysis to compensate for the artifact. Today several techniques, such as skin surface models, joint motion constraints, and extended Kalman filters, have been used to evaluate, minimize, and compensate for STAs (Cerveri et al., 2005); however, further research is required to make such techniques appropriate for use on the spine.

Previous studies on trunk movements have reported instrumental errors ranging between 0.1 mm and 5.3 mm (Ehara et al., 1995; Haggard and Wing, 1990; Klein and DeHaven, 1995; Thornton et al. 1998). The number of STAs related to spinal markers ranged from 9.2 to 10.7 mm during flexion-extension (Zemp et al., 2014), whereas average errors for the pelvis ranged from 14.7 to 24.8 mm (Della Croce et al., 1999). However, those errors were highly dependent on the protocol used, and might be reduced by using optimization algorithms and protocols specifically designed for each research question (Rast et al., 2016).

SPINE PROTOCOLS

Trunk movements play a key role for most human activities and contribute to the movement of the whole body (Maaswinkel et al., 2016). In general, the trunk stabilizes the body and allows the limbs to operate properly (Needham et al., 2016). Differences in the trunk kinematics of healthy and subjects and subjects suffering from pathological spinal conditions have been observed (Ferrarin et al., 2004). Also, the trunk's role during locomotion is especially emphasized in diseases where the trunk is deformed and where it is responsible for several compensatory mechanisms (Nadeau et al., 2003). Nonetheless, the trunk has rarely been studied as an independent research subject. Several optoelectronic systems and protocols are used in different ways to monitor trunk kinematics, with different aims and results; and different models are used to track and analyze trunk motions. The use of 3D rotations between rigid body regions and 2D projection angles has been reported (Frigo et al., 2003). Single vertebrae are difficult to track in 3D (Konz et al., 2006); consequently, only the trunk's overall motion was reported in most papers. Indeed, most of the models for gait analysis considered the trunk as a single rigid body (Davis, 2005; Gage et al., 1996). Nevertheless, protocols aimed at detecting the segmental motions have been presented (Frigo et al., 2003) (Fig. 3). Most protocols used single skin markers located on anatomical landmarks, but rigid clusters also have been utilized (Konz et al., 2006).

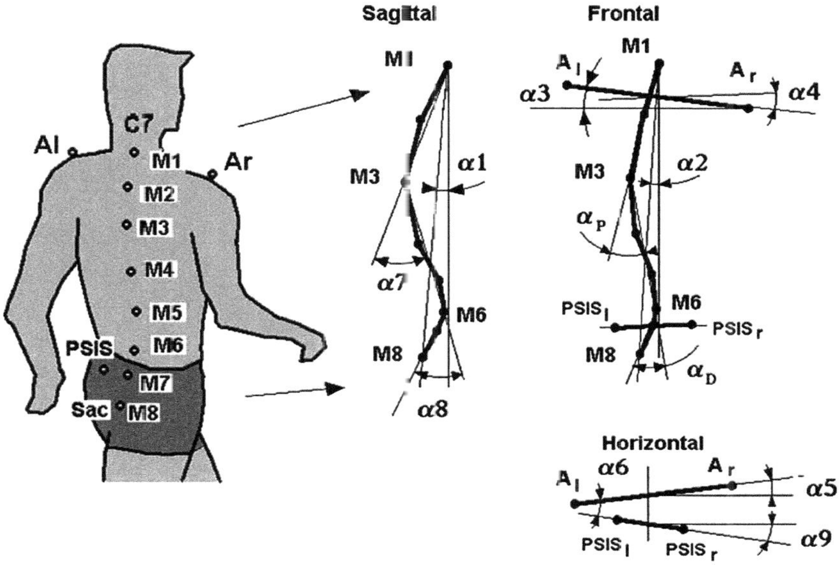

FIG. 3

Protocol presented by Frigo et al. (2003) to determine the motion of the trunk during walking. Markers are positioned on the spinous process of C7 (M1), at the apex of kyphosis (M3), at the apex of kyphosis (M6), and at the lowest edge of the sacrum (M8). Markers M2, M4, M5, and M7 are positioned in between, with uniform spacing. Based on the coordinate of the markers, the angles $\alpha 1$–$\alpha 9$, α_P, and α_D describing the spinal curvature in the sagittal and in the coronal planes are calculated as shown.

Reproduced with permission from Frigo, C., Carabalona, R., Dalla Mura, M., Negrini, S., 2003. The upper body segmental movements during walking by young females. Clin. Biomech. 18, 419–425.

Furthermore, the variety of anatomical coordinate systems and joint conventions creates confusion in terminology. In a recent study, eight different protocols for human trunk kinematics during locomotion were quantitatively compared (Leardini et al., 2009). The results showed very different patterns and ranges of motion among the protocols (Fig. 4), highlighting the necessity to carefully determine the number of markers involved and the reference frame definitions before interpreting the results in clinical decision-making.

The available protocols can be classified based on the anatomical regions covered: the whole trunk (Davis, 2005), the spine, and the shoulder line (Frigo et al., 2003) as well as upper trunk, lower trunk, and lumbar spine (Crosbie et al., 1997). In general, models based on thorax and spine markers are more suitable for describing only postural static and dynamic abnormalities in the sagittal plane (Wu et al., 2005), whereas marker clusters used in a large area in the frontal plane can provide information on coronal and rotational deformities and motion disorders (Fantozzi et al., 2003). Mixing these two approaches can give more comprehensive information about the 3D motion of different segments of this complex body region. A study by Leardini et al. (2011) described the multisegmental trunk motion during locomotion in healthy subjects by analysis of a larger number of segments within the spine, thorax, and shoulders, which were assumed to be rigid. The aim of the study was to establish how these trunk

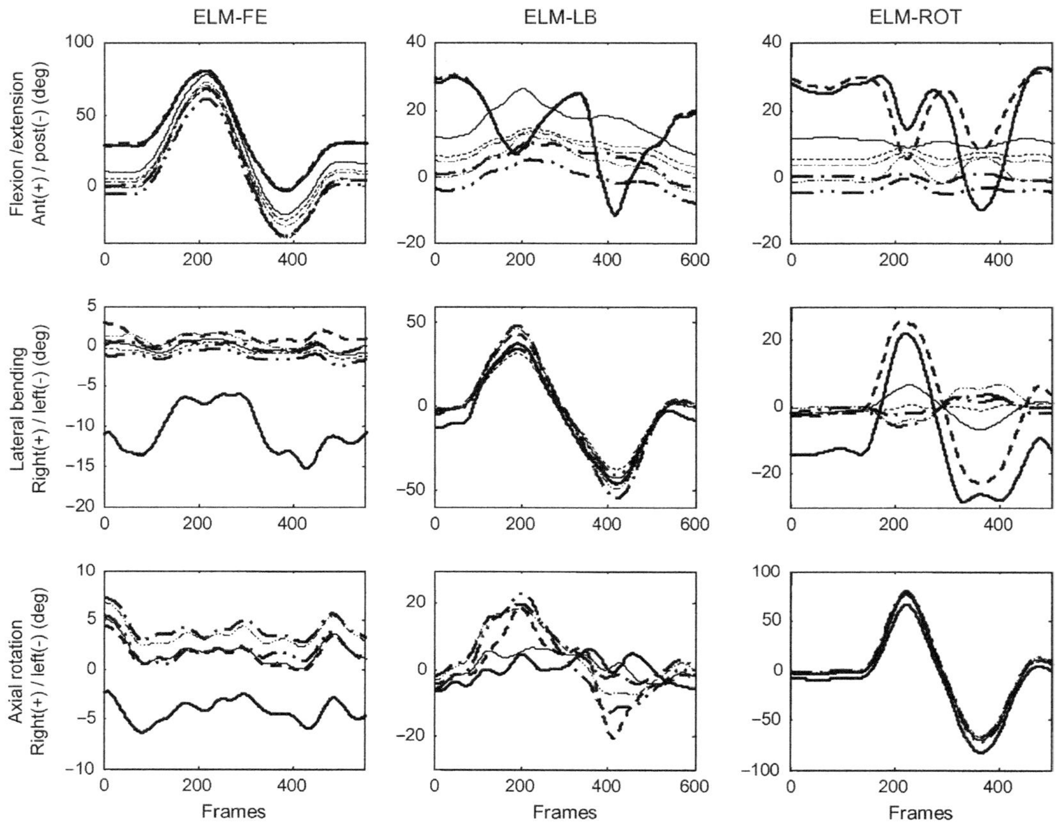

FIG. 4

Motion of the trunk determined with eight different protocols while the subject performs three exercises: flexion-extension of the trunk (ELM-FE), lateral bending (ELM-LB) and axial rotation (ELM-ROT). The reconstructed rotations of the trunk around the three main axes (FLEXION/EXTENSION, LATERAL BENDING, and AXIAL ROTATION) highlight the marked differences among the eight protocols.

Reproduced with permission from Leardini, A., Biagi, F., Belvedere, C., Benedetti, M.G., 2009. Quantitative comparison of current models for trunk motion in human movement analysis. Clin. Biomech. 24, 542–550.

segments contribute to the necessary oscillations of the upper body in the three anatomical planes. The marker-set included 14 markers: the right and left anterior superior iliac spines and posterior superior iliac spines, the incisura jugularis, the xiphoid process, the midpoint between the two scapulae, the right and left acromions, the seventh cervical vertebra, the second thoracic vertebra, and the first, third, and fifth lumbar vertebrae (Fig. 5). The thorax was tracked with four thoracic markers, a separate shoulder line motion was reported with respect to the thorax, and a five-marker model determined the spine motion. The novel model showed a high intrasubject repeatability and intersubject variability with simpler models.

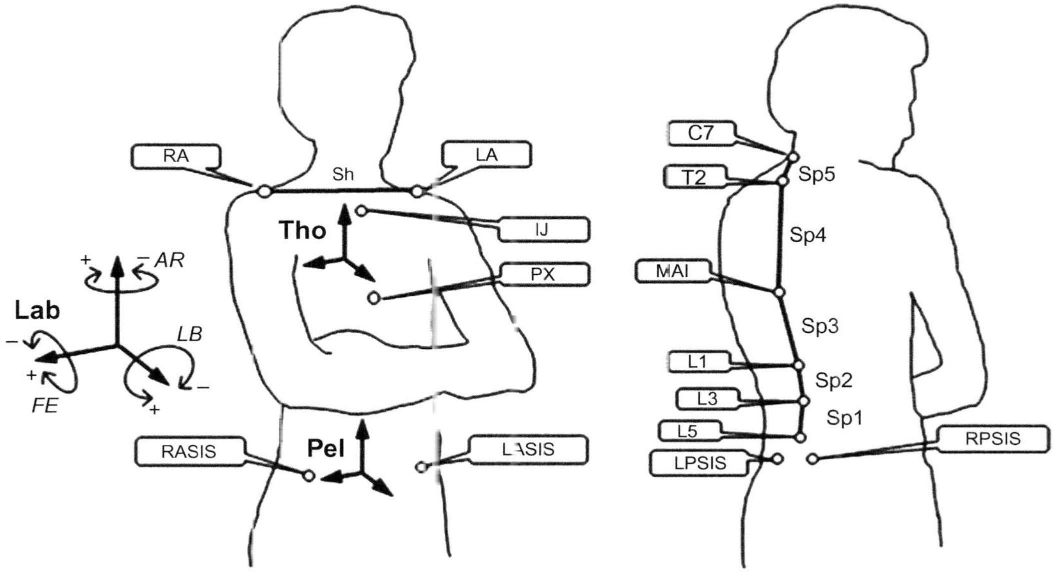

FIG. 5

Location of the markers in the protocol proposed by Leardini et al. (2011). RASIS-LASIS: right and left anterior superior iliac spine; RPSIS-LPSIS: right and left posterior superior iliac spines; IJ: incisura jugularis; PX: xiphoid process; RA-LA: right and left acromions; MAI: midpoint between the inferior angles of most caudal points of the two scapulae; C7-T2-L1-L3-L5: spinous process of C7, T2, L1, L3, and L5, respectively

Reproduced with permission from Leardini, A., Biagi, F., Belvedere, C., Benedetti, M.G., 2009. Quantitative comparison of current models for trunk motion in human movement analysis. Clin. Biomech. 24, 542–550.

Trunk motion was also studied during movements not related to gait (Negrini et al., 2016). A few studies focused on daily activities, such as standing up movements, trunk flexion-extension, circumduction, and weight lifting (Bourigua et al., 2014; Cheng et al., 2013).

CLINICAL APPLICATIONS OF MOTION ANALYSIS

Motion analysis is a fundamental instrument for the quantification of human motion and function during daily activities and is an essential tool in making clinical decisions (Andriacchi and Alexander, 2000) and in conducting clinical research (McGinley et al., 2009). Even though motion analysis is not used for medical diagnosis, in several cases, it can provide a useful, objective measurement and quantification of the mobility state of patients suffering from various disorders, and thus help in the selection of a proper and specific treatment and in the measurement of its outcomes.

Currently, locomotion is the task most commonly evaluated through motion analysis systems, as it represents the simplest and most basic daily activity performed at any age. Clinical gait analysis is typically aimed at pointing out differences between normal and abnormal gaits and at assessing changes during walking within a certain period of time (Baker, 2006). Repeated gait measurements are used to estimate response to therapeutic interventions such as surgery, physiotherapy, medications, and orthotics. Variability between "before" and "after" measurements may be due to treatment effects, or measurement variations, or a combination of the two (Schwartz et al., 2004).

To date, gait analysis is one of the most popular investigative techniques used to test patients with central nervous system disorders associated with spasticity. However, research has demonstrated a strong potential for the use of gait analysis to obtain valuable clinical information related to several other disorders and medical conditions, such as those related to rheumatology, orthopedics, and endocrinology (Simon, 2004).

Sutherland (1964) and Perry (1992) were two pioneers in the use of gait analysis techniques to assist in the treatment of cerebral palsy. The orthopedic field employed gait analysis to investigate gait recovery after certain interventions, particularly after surgical replacement of a hip or knee. In the past several decades, assessment of trunk posture and motion has also gained importance in the clinical practice (Don et al., 2012).

As previously mentioned, trunk movement assists the whole body mechanics in achieving optimal locomotion. Muscles of the back and abdomen play an important, active role in gait balance by controlling anteroposterior trunk sways and shoulder girdle rotations (Saunders et al., 2005). Therefore pathological conditions such as Parkinson's disease, spinal cord injuries, obesity, those requiring prostheses, and particularly spinal deformities can significantly affect natural walking patterns (Simon et al., 2015). The following paragraphs describe the use of motion analysis to investigate two spinal disorders that have been reported in the literature, that is, idiopathic scoliosis and low back pain.

IDIOPATHIC SCOLIOSIS

Idiopathic scoliosis is a three-dimensional deformity of the spine and rib cage that alters trunk symmetry and can lead to changes in the coordination between body regions and to a pathological gait pattern (Kramers-de Quervain et al., 2004). It has been claimed that gait alterations are related to the severity of the scoliosis curve: the higher the spinal deformity, the higher the asymmetry of the gait (Syczewska et al., 2012). Moreover, scoliosis modifies the center of mass and the distribution of weight; thus patients with this pathology show an extremely asymmetric trunk motion in the transverse plane and less stability than an age-matched normal population shows (Kramers-de Quervain et al., 2004). Trunk rotation is essential to maintain the horizontal lines of the pelvis and head progression, which are symmetrically oriented in normal subjects. In patients with scoliosis, trunk rotation was shown to be significantly different than that of a control group, in both adolescent (Mahaudens et al., 2009) and adult patients (Engsberg et al., 2001).

LOW BACK PAIN

Low back pain (LBP) represents the fifth most common reason for medical treatment in the United States (Waterman et al., 2012). Past studies have demonstrated a high reliability in using gait analysis to detect physical impairments in patients with LBP (Arendt-Nielsen et al., 1996; Toosizadeh et al., 2015). Several differences in gait parameters between patients with LBP and healthy controls were observed. LBP patients exhibited a significantly slower walking speed, which was more pronounced for those subjects with higher self-reported low back pain (Al-Obaidi et al., 2003). Furthermore, LBP patients could not reach the natural and physiological trunk rotation and showed more variability and fewer tight movements in the frontal plane (Lamoth et al., 2006; Vogt et al., 2001). These patterns of movements were deemed to be the result of a compensatory mechanism to avoid pain: because rotation is smoothened, lateral bending motion increases to maintain the role of the trunk as a gait stabilizer. Nevertheless, the differences between normal and LBP patients were not observed in all situations (Sánchez-Zuriaga et al., 2015).

WEARABLE SENSORS

Even though optoelectronic systems are the instruments most commonly used for motion analysis, they have several limitations, mainly related to their cost and bulk, to the long time needed for postprocessing, and to the difficulties in using such instruments outdoors. In the past two decades, nonoptical wearable sensors have shown great potential for motion analysis, overcoming the restrictions of optoelectronic systems (Tao et al., 2012). Wearable sensors are inexpensive and are portable devices applied or attached to patients' skin at the area of interest, and they enable nonstop recording of the movements during task performances or daily activities.

Electrogoniometers are devices that provide angular measurements between body segments based on the potentiometers' resistance changes in consequence to motion. Their main limits are related to skin motion artifacts and the need for individual calibration. Such instruments were used to study lumbar spine (Mieritz et al., 2014) and cervical spine (Petersen et al., 2008) motion. Good reliability outcomes were reported for the lumbar measurements, whereas less satisfactory results were obtained for the cervical spine.

Ultrasound transmitter devices are tracking systems based on the application of rigid plates with ultrasound microphone triads on top; they are applied at the anatomical points of interest and are able to measure the relative motion between the local coordinate system that each triad determines (Vogt et al., 2003). These instruments have been used for monitoring rotational pelvic motion, even though their main use is for the detection of cervical or lumbar motion during easy and basic neck actions (Don et al., 2012).

Williams et al. (2010) used fiber-optic sensors to investigate the curvature of the lumbar spine. In those devices, the light flows through a specific optical wire, and the light intensity is modulated by the spine bending. Therefore changes in the spine curvature can be analyzed by interpreting the light intensity modifications. Because the fiber optic wire cannot flex appropriately, this system is not considered to be a suitable solution to measure lateral bending and axial rotation.

Inertial sensors are portable and cheap wearable sensors consisting of a very small box containing a variable organization of a 3D sensing unit: accelerometers, magnetometers, and gyroscopes. Briefly, the accelerometer measures acceleration along its sensitive axes, the gyroscope measures the angular velocity, and the magnetometer estimates body segment orientation changes with respect to the magnetic North (Tao et al., 2012). Thanks to specific algorithms, the 3D real-time orientation of every sensor in the box with respect to a global system of reference can be obtained (Don et al., 2012). Inertial sensors have been widely used to measure the motion of the spine. Theobald et al. (2011) as well as Duc et al. (2014) used them for measuring the cervical motion, finding a reliable and objective tool for spine shape evaluation (Fig. 6). Goodvin et al. (2006) used them for 3D real-time measurement of spinal motion, estimating a tracking error of the segment orientation of about less than 3.1 degrees.

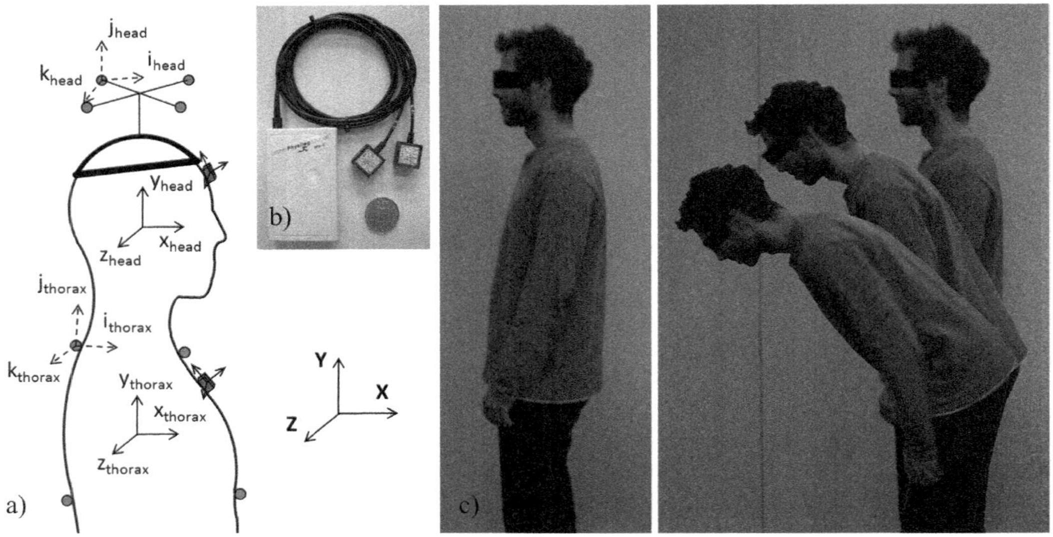

FIG. 6

Wearable sensors used to assess the cervical motion by Duc et al. (2014). Two inertial sensors are mounted on the head and on the chest, allowing determining the orientation of the two body segments with respect to the laboratory system of reference (A). The sensors are linked to a data logger worn at the waist (B). The system was used to determine the alignment of the neck in standing and its motion during trunk flexion (C).

The main limitation of inertial sensors is related both to the motion between the skin and the bone of interest and to the slippage because of straps used for the sensors' positioning.

Wearable sensors greatly changed the way motion analysis is investigated, mainly because of the small size of the devices, the possibility of using them anywhere without the need of an infrastructure, and their wide range of application. Ongoing research and development are aimed at improving the accuracy and stability of the signals, the reliability of analytical algorithms to reconstruct the motion, and the reduction of the sensitivity to STAs.

REFERENCES

Al-Obaidi, S.M., Al-Zoabi, B., Al-Shuwaie, N., Al-Zaabie, N., Nelson, R.M., 2003. The influence of pain and pain-related fear and disability beliefs on walking velocity in chronic low back pain. Int. J. Rehabil. Res. 26 (2), 101–108.

Andriacchi, T.P., Alexander, E.J., 2000. Studies of human locomotion: past, present and future. J. Biomech. 33, 1217–1224.

Arendt-Nielsen, L., Graven-Nielsen, T., Svarrer, H., Svensson, P., 1996. The influence of low back pain on muscle activity and coordination during gait: a clinical and experimental study. Pain 64 (2), 231–240.

Baker, R., 2006. Gait analysis methods in rehabilitation. J. Neuro Eng. Rehabil. 3, 4.

Benedetti, M.G., Cappozzo, A., 1994. Anatomical Landmark Definition and Identification in Computer Aided Movement Analysis in a Rehabilitation Context II. Università degli studi La Sapienza, Internal Report, Rome, Italy, pp.1–31.

Bourigua, I., Simoneau, E.M., Leteneur, S., Gillet, C., Ido, G., Barbier, F., 2014. Chronic low back pain sufferers exhibit freezing-like behaviors when asked to move their trunk as fast as possible. Spine J. 14, 1291–1299.

Cappozzo, A., Della Croce, U., Leardini, A., Chiari, L., 2005. Human movement analysis using stereophotogrammetry part 1: theoretical background. Gait Posture 21, 186–196.

Cerveri, P., Pedotti, A., Ferrigno, G., 2005. Kinematics models to reduce the effect of skin artifacts on marker based human motion estimation. J. Biomech. 38, 2228–2236.

Cheng, S.C., Hsu, C.H., Ting, Y.T., Kuo, L.C., Lin, R.M., Su, F.C., 2013. Developing functional workspace for the movement of trunk circumduction in healthy young subjects: a reliability study. Biomed. Eng. Online 12, 4.

Chiari, L., Della Croce, U., Leardini, A., Cappozzo, A., 2004. Human movement analysis using stereophotogrammetry part 2: instrumental errors. Gait Posture 21, 197–211.

Crosbie, J., Vachalathiti, R., Smith, R., 1997. Patterns of spinal motion during walking. Gait Posture 5, 6–12.

Davis III, R.B., 2005. Newington Trunk Model. (Personal Communication).

Della Croce, U., Cappozzo, A., Kerrigan, D.C., 1999. Pelvis and lower limb anatomical landmark calibration precision and its propagation to bone geometry and joint angles. Med. Biol. Eng. Comput. 37, 155–161.

Della Croce, U., Leardini, A., Chiari, L., Cappozzo, A., 2005. Human movement analysis using stereophotogrammetry part 4: assessment of anatomical landmark misplacement and its effects on joint kinematics. Gait Posture 21, 226–237.

Don, R., Capodaglio, P., Cimolin, V., Benedetti, M.G. D'Osualdo, F., Frigo, C., Vismara, L., Negrini, S., 2012. Instrumental measures of spinal function: is it worth? A state of the art from a clinical perspective. Eur. J. Rehabil. Med. 48, 255–273.

Duc, C., Salvia, P., Lubansu, A., Feipel, V., Aminian, K., 2014. A wearable inertial system to assess the cervical spine mobility: comparison with an optoelectronic-based motion capture evaluation. Med. Eng. Phys. 36 (1), 49–56.

Ehara, Y., Fujimoto, H., Miyazaki, S., Tanaka, S., Yamamoto, S., 1995. Comparison of the performance of 3D camera systems. Gait Posture 3, 166–169.

Engsberg, J.R., Bridwell, K.H., Reitenbach, A.K., Uhrich, M.L., Baldus, C., Blanke, K., et al., 2001. Preoperative gait comparisons between adults undergoing long spinal deformity fusion surgery (thoracic to L4, L5, or sacrum) and controls. Spine 26, 2020–2028.

Fantozzi, S., Benedetti, M.G., Leardini, A., Banks, S.A., Cappello, A., Assirelli, D., et al., 2003. Fluoroscopic and gait analysis of the functional performance in stair ascent of two total knee replacement designs. Gait Posture 17, 225–234.

Ferrarin, M., Rizzone, M., Lopiano, L., Recalcati, M., Pedotti, A., 2004. Effects of subthalamic nucleus stimulation and L-dopa in trunk kinematics of patients with Parkinson's disease. Gait Posture 19, 164–171.

Frigo, C., Carabalona, R., Dalla Mura, M., Negrini, S., 2003. The upper body segmental movements during walking by young females. Clin. Biomech. 18, 419–425.

Gage, J.R., DeLuca, P.A., Renshaw, T.S., 1996. Gait analysis: principle and applications with emphasis on its use in cerebral palsy. Instr. Course Lect. 45, 491–507.

Goodvin, C., Park, E.J., Huang, K., Sakaki, K., 2006. Development of a real-time three-dimensional spinal motion measurement system for clinical practice. Med. Biol. Eng. Comput. 44, 1061–1075.

Haggard, P., Wing, A.M., 1990. Assessing and reporting the accuracy of position measurements make with optical tracking systems. J. Mot. Behav. 22, 315–321.

Klein, P.J., DeHaven, J.J., 1995. Accuracy of three-dimensional linear and angular estimates obtained with the ariel performance analysis system. Arch. Phys. Med. Rehabil. 76, 183–189.

Konz, R., Fatone, S., Stine, R.L., Ganju, A., Gard, S., Ondra, S., 2006. A kinematic model to assess spinal motion during walking. Spine 31, E898–E906.

Kramers-de Quervain, I.A., Müller, R., Stacoff, A., Grob, D., Stüssi, E., 2004. Gait analysis in patients with idiopathic scoliosis. Eur. Spine J. 13 (5), 449–456.

Lamoth, C.J., Meijer, O.G., Daffertshofer, A., Wuisman, P.I., Beek, P.J., 2006. Effects of chronic low back pain on trunk coordination and back muscle activity during walking: changes in motor control. Eur. Spine J. 15, 23–40.

Leardini, A., Biagi, F., Belvedere, C., Benedetti, M.G., 2009. Quantitative comparison of current models for trunk motion in human movement analysis. Clin. Biomech. 24, 542–550.

Leardini, A., Biagi, F., Merlo, A., Belvedere, C., Benedetti, M.G., 2011. Multi-segment trunk kinematics during locomotion and elementary exercises. Clin. Biomech. 26, 562–571.

Leardini, A., Chiari, L., Della Croce, U., Cappozzo, A., 2005. Human movement analysis using stereophotogrammetry. Part 3. Soft tissue artifact assessment and compensation. Gait Posture 21, 212–225.

Maaswinkel, E., Griffioen, M., Perez, R.S.G.M., van Dieën, J.H., 2016. Methods for assessment of trunk stabilization, a systematic review. J. Electromyogr. Kinesiol. 26, 18–35.

Mahaudens, P., Banse, X., Mousny, M., Detrembleur, C., 2009. Gait in adolescent idiopathic scoliosis: kinematics and electromyographic analysis. Eur. Spine J. 18, 512–521.

McGinley, J.L., Baker, R., Wolf, R., Morris, M.E., 2009. The reliability of three-dimensional kinematic gait measurements: a systematic review. Gait Posture 29, 360–369.

Mieritz, R.M., Bronfort, G., Jakobsen, M.D., Aagaard, P., Hartvigsen, J., 2014. Reliability and measurement error of sagittal spinal motion parameters in 220 chronic low back pain patients using a 3D measurement device. Spine J. 14, 1835–1843.

Nadeau, S., Amblard, B., Mesure, S., Bourbonnais, D., 2003. Head and trunk stabilization strategies during forward and backward walking in healthy adults. Gait Posture 18, 134–142.

Needham, R., Stebbins, J., Chockalingam, N., 2016. Three-dimensional kinematics of the lumbar spine during gait using marker-based systems: a systematic review. J. Med. Eng. Technol. 40, 172–185.

Negrini, S., Piovanelli, B., Amici, C., Cappellini, V., Bovi, G., Ferrarin, M., Zaina, F., Borboni, A., 2016. Trunk motion analysis: a systematic review from a clinical and methodological perspective. Eur. J. Phys. Rehabil. Med. 52 (4), 583–592.

Perry, J., 1992. Gait Analysis: Normal and Pathological Function. SLACK Incorporated, Thorofare, NJ.

Petersen, C.M., Schuit, D., Johnson, R.D., Knecht, H., Levine, P., 2008. Agreement of measures obtained radiographically and by the OSI CA-6000 spine motion analyzer for cervical spinal motion. Man. Ther. 13, 200–205.

Rast, F.M., Graf, E.S., Meichtry, A., Kool, J., Bauer, C., 2016. Between-day reliability of three dimensional motion analysis of the trunk: a comparison of marker based protocols. J. Biomech. 49, 807–811.

Sánchez-Zuriaga, D., López-Pascual, J., Garrido-Jaén, D., García-Mas, M.A., 2015. A comparison of lumbo-pelvic motion patterns and erector spinae behavior between asymptomatic subjects and patients with recurrent low back pain during pain-free periods. J. Manipulative Physiol. Ther. 38, 130–137.

Saunders, S.W., Schache, A., Rath, D., Hodges, P.W., 2005. Changes in three-dimensional lumbo-pelvic kinematics and trunk muscle activity with speed and mode of locomotion. Clin. Biomech. 20, 784–793.

Schwartz, M.H., Trost, J.P., Wervey, R.A., 2004. Measurement and management of errors in quantitative gait data. Gait Posture 20, 196–203.

Simon, A.L., Ilharreborde, B., Souchet, P., Kaufman, K.R., 2015. Dynamic balance assessment during gait in spinal pathologies—a literature review. Orthop. Traumatol. Surg. Res. 101 (2), 235–246.

Simon, S.R., 2004. Quantification of human motion: gait analysis-benefits and limitations to its application to clinical problems. J. Biomech. 37, 1869–1880.

Sutherland, D.H., 1964. Gait Disorders in Childhood and Adolescence. Lippincott Williams & Wilkins, Baltimore, MD.

Syczewska, M., Graff, K., Kalinowska, M., Szczerbik, E., Domaniecki, J., 2012. Influence of the structural deformity of the spine on the gait pathology in scoliotic patients. Gait Posture 35 (2), 209–213.

Tao, W., Liu, T., Zheng, R., Feng, H., 2012. Gait analysis using wearable sensors. Sensors 12, 2255–2283.

Theobald, P.S., Jones, M.D., Williams, J.M., 2011. Do inertial sensors represent a viable method to reliably measure cervical spine range of motion? Man. Ther. 17, 92–96.

Thornton, M.J., Morrisey, M.C., Coutts, F.J., 1998. Some effects of camera placement on the accuracy of the Kinemetrix three-dimensional motion analysis system. Clin. Biomech. 13, 452–454.

Toosizadeh, N., Yen, T.C., Howe, C., Dohmd, M., Mohler, J., Najafi, B., 2015. Gait behaviors as an objective surgical outcome in low back disorders: a systematic review. Clin. Biomech. 30 (6), 528–536.

Vogt, L., Pfeifer, K., Portscher, M., Banzer, W., 2001. Influences of nonspecific low back pain on three-dimensional lumbar spine kinematics in locomotion. Spine 26, 1910–1919.

Vogt, L., Portscher, M., Brettmann, K., Pfeifer, K., Banzer, W., 2003. Cross validation of marker configurations to measure pelvic kinematics in gait. Gait Posture 18, 178–184.

Waterman, B.R., Belmont Jr., P.J., Schoenfeld, A.J., 2012. Low back pain in the United States: incidence and risk factors for presentation in the emergency setting. Spine J. 12 (1), 63–70.

Whittle, M., 2007. Gait Analysis: An Introduction. Butterworth-Heinemann, Oxford.

Williams, J.M., Haq, I., Lee, R.Y., 2010. Dynamic measurement of lumbar curvature using fiber-optic sensors. Med. Eng. Phys. 32, 1043–1049.

Wu, G., van der Helm, F.C., Veeger, H.E., Makhsous, M., Van Roy, P., Anglin, C., et al., 2005. ISB recommendation on definitions of joint coordinate systems of various joints for the reporting of human joint motion—part II: shoulder, elbow, wrist and hand. J. Biomech. 38, 981–992.

Zemp, R., List, R., Gulay, T., Elsig, J.P., Naxera, J., et al., 2014. Soft tissue artefacts of the human back: comparison of the sagittal curvature of the spine measured using skin markers and an open upright MRI. PLoS One 9, e95426. https://doi.org/10.1371/journal.pone.0095426.

FURTHER READING

Cappozzo, A., Catani, F., Leardini, A., Benedetti, M.G., Della Croce, U., 1996. Position and orientation in space of bones during movement: experimental artefacts. Clin. Biomech. 11 (2), 90–100.

Peters, A., Galna, B., Sangeux, M., Morris, M., Baker, R., 2010. Quantification of soft tissue artifacts in lower limb human motion analysis: a systematic review. Gait Posture 31, 1–8.

Vander Linden, D.W., Carlson, S.J., Hubbard, R.L., 1992. Reproducibility and accuracy of angle measurement obtained under static conditions with the motion analysis video system. Phys. Ther. 72, 300–305.

CHAPTER

IN VITRO TESTING OF CADAVERIC SPECIMENS

12

Fabio Galbusera*, David Volkheimer[†], Hans-Joachim Wilke[†]

IRCCS Galeazzi Orthopedic Institute, Milan, Italy Institute of Orthopaedic Research and Biomechanics, Ulm University, Ulm, Germany[†]*

INTRODUCTION

The scientific study of spinal anatomy and biomechanics on cadavers goes back to Leonardo da Vinci in the 15th century and then on to Andreas Vesalius and Giovanni Alfonso Borelli in the following two centuries. However, mechanical tests on spine specimens were not reported until the 20th century (Naderi et al., 2007) when investigations were done concerning vertebral fractures in pilots ejecting from military aircrafts. After the end of World War II several laboratories around the world built on the experience gained during the war and started to systematically use cadaveric specimens to investigate clinically relevant research questions, such as injury mechanisms and spinal kinematics. By the 1950s, in vitro spine research began to make an effective contribution to the diagnosis and treatment of spinal disorders (Evans and Lissner, 1959; Hirsch, 1955; Hirsch and Nachemson, 1954; Lysell, 1969; Oxland, 2016).

Today in vitro testing of cadaveric specimens constitutes one of the pillars of biomechanical research. The technological evolution of actuators as well as control and measuring systems allowed the development of equipment capable of applying any combination of loading and boundary conditions and of measuring the resulting motion and reaction forces with outstanding precision. In combination with other methodologies such as biomedical imaging, motion analysis, and numerical simulations, in vitro testing is currently used to investigate various aspects of the spine's biomechanical responses and their individual components, and for the preclinical evaluation of novel implants and surgical techniques.

SPECIMENS

The gold standard for in vitro testing is to use fresh or fresh a frozen human specimens (Fig. 1) (Wilke et al., 1998). Studies have shown that multiple freezing and storage, even for prolonged periods, do not significantly alter the mechanical properties of spine specimens and should therefore be considered a safe practice (Panjabi et al., 1985). Naturally, specimens harvested from elderly donors, which are commonly available for research, usually show degenerative signs and pathologies such as osteoporosis and bone metastases. Such aspects should be considered in the planning phase of the experiment, and precise inclusion and exclusion criteria for donors and specimens matching the aim of the study should be defined. Collecting X-rays, CT, and MRI scans of the specimens before testing and analyzing their bone mineral density are recommended whenever possible to evaluate their suitability for experiments.

Biomechanics of the Spine. https://doi.org/10.1016/B978-0-12-812851-0.00012-4

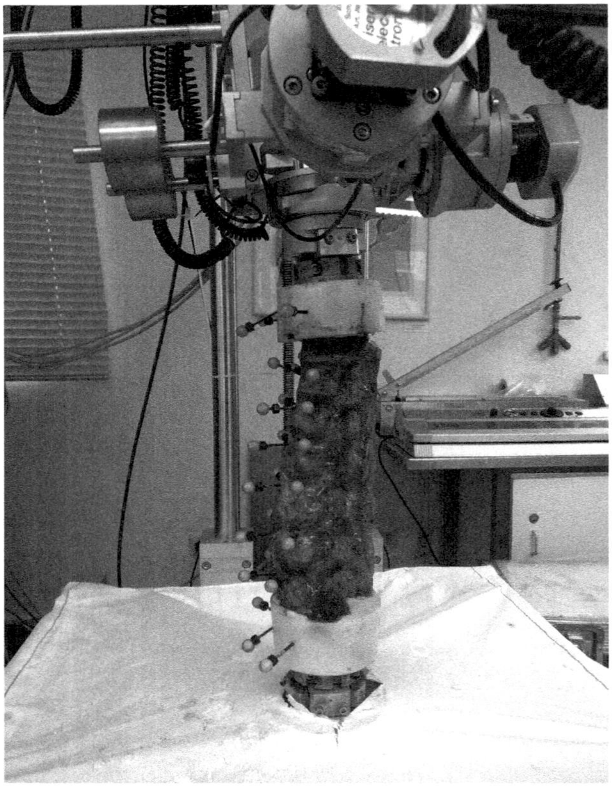

FIG. 1

Fresh human lumbar specimen mounted in a spine tester. The extremities are potted in a polymethylmethacrylate embedding medium to facilitate clamping to the loading device. The specimen is fixed at the caudal end, whereas loads are applied to the cranial end. Passive reflective markers are screwed to each vertebra in order to capture motion via an optoelectronic system.

Although valuable for education purposes and in anatomical studies, the use of specimens fixed in formalin should be avoided because of the soft tissues' increased stiffness, loss of nonlinearity, and viscoelastic properties following the fixation procedure (Wilke et al., 1996). Specimens fixed according to Thiel should not be used as well, even if their properties have a closer similarity to those of fresh spines (Wilke et al., 2011).

Large animal specimens such as ovine, caprine, bovine, and porcine ones have been frequently employed in biomechanical research (see the Chapter 16). In general, their use is very limited and should be discouraged because of their marked differences in anatomy, loading environments, and mechanical properties, which are well documented in the literature (Kettler et al., 2007). However, a few exceptions to this general rule exist. Animal specimens might be considered when a high number of very similar specimens is needed for a test, such as when the number of groups to be compared is large. Indeed, results from specimens harvested from animals of similar age and body weight provide better repeatability than do human samples, in which the age at the time of death, the body size, and possible pathologies cannot be easily controlled and only achieved by selection. Animal specimens are frequently used for cyclic testing because their bone strength is generally higher than bone in human samples (Kettler et al., 2007), which enables them to sustain long protocols without damage to the bony components. In addition, animal specimens are valuable for preliminary testing such as during the development of the experimental setup, as well as for screening experiments prior to tests conducted on human specimens. In general, animal models should not be used for the in vitro preclinical validation of novel spinal implants; in any case, their suitability must be critically analyzed based on the specific research question (Wilke et al., 1998).

LOADING PROTOCOLS

A biomechanical investigation conducted on a spinal specimen, either for basic research or to test an implant or a surgical treatment, should ideally replicate the complex loading and constraint conditions acting in vivo, which are determined by the combination of gravity, external loads, and the action of the trunk muscles (Wilke et al., 1998). However, for several reasons, a simplified loading environment is preferred in most cases. First, knowledge about the muscle loads in different postures and during various activities is relatively limited because they cannot be measured directly in vivo, and they only marginally cover the intersubject variability, the effect of spinal disorders, as well as possible iatrogenic changes if a surgical procedure is the subject of investigation. Second, the use of a realistic loading setup customized to a specific specimen would hinder the repeatability and comparability of the results among different specimens. Furthermore, the realization of a realistic loading setup involves a high degree of technical complexity, and thus costs and time for the development, which might not be affordable for some research institutions.

The two key loading modes commonly simulated in an in vitro test are compression related to the action of the body weight and the counteracting muscle forces required to keep equilibrium and stability as well as bending in the three principal anatomical planes, that is, in flexion-extension, lateral bending, and axial rotation (Rohlmann et al., 2009b). In some studies, the combined effect of compression and bending was achieved by imposing an eccentric compressive load on the specimen (Anasetti et al., 2010; Umehara et al., 2000; Akamaru et al., 2003) or an inclination of the specimen with respect to the loading axis (Adams et al., 1996a,b). These loading setups, however, generate a bending moment that is not constant along the specimen because of the variable lever arm of the force with respect to the axis of rotation of each motion segment, resulting in bending loads that are dependent on the length and curvature of the specimen (Crawford et al., 1995).

A different approach involves the application of pure bending moments to one end of the specimens, usually the uppermost vertebra, whereas the other vertebra is rigidly constrained (Wilke et al., 1998, 2001; Panjabi et al., 1994; Yamamoto et al., 1989). By using this method, the bending moment is constant along the spinal specimens and constant throughout the whole test and in all specimens included in the study, thus allowing for optimal comparability of the results. Indeed, there is consensus that pure moments are the preferred loading mode for testing surgical treatments and spinal implants. Typical values for the applied moments are 7.5 Nm for lumbar specimens, 5 Nm for thoracic specimens, and 2.5 Nm for specimens harvested from the lower cervical spine (C3-C7) (Wilke et al., 1998).

Despite these advantages, no physiological compression of the spine is simulated when applying pure moments. To improve this aspect, axial preloads superimposed to the pure moments have been used (Cripton et al., 2000; Cheng et al., 2007; Lindsey et al., 2003; Swanson et al., 2003). In short specimens undergoing small rotations, this approach may be feasible because the axial preload has a negligible lever arm with respect to the center of rotation of the specimen when adjusted properly; however, in longer specimens or when large rotations take place, the axial preload may generate bending moments and thus rotational artifacts (Cripton et al., 2001). As a matter of fact, a compressive force of 80 to 100N, thus markedly lower than the physiological spinal loads, was shown to be sufficient to generate buckling of lumbar specimens (Patwardhan et al., 1999).

To overcome this issue, the follower load concept was introduced by Patwardhan et al. (1999) and currently constitutes the most widely used approach to simulate compressive loads in spine specimens (Miura et al., 2002; Goel et al., 2006; Zirbel et al., 2013). This experimental technique was developed with the aim of minimizing internal shear forces and bending moments, thus simulating a load path that follows the curvature of the specimen and ideally resulting in pure compression. Values ranging from 400 to 1175N have been used to simulate lumbar loads in various postures (Goel et al., 2006; Rohlmann et al., 2001, 2009a; Wilke et al., 2003). The use of this technique allows applying loads up to 1200N without causing buckling of the specimen (Patwardhan et al., 1999), and it can be coupled to the application of pure moments in flexion-extension, whereas its use in lateral bending or axial rotation would result in large artifacts and is therefore avoided. Artifacts have been described also in the case of a nonoptimized load path, which would result in considerable intervertebral rotations (Dreischarf et al., 2010). Also, it should be noted that the optimal loading path would need to be adjusted during motion to follow the displacements of the centers of rotation, but this issue, which would potentially generate other rotational artifacts, was ignored in all previous studies.

In vitro testing has been frequently used to assess the possible effect of spinal instrumentation on the adjacent segments. Although it is still a matter of debate, several papers showed that early degeneration of the adjacent motion segments might be induced by rigid stabilization (Lee et al., 2014; Park et al., 2004; Miyakoshi et al., 2000; Wiltse et al., 1999; Hambly et al., 1998; Schulte et al., 2007) and that implants able to preserve a certain mobility of the operated segments might prevent such a phenomenon (Buckland et al., 2016; Korovessis et al., 2002). To test this hypothesis, flexibility testing with pure moments has been questionably used in several studies (Pfeiffer et al., 1997; Bastian et al., 2001; Dahl et al., 2013; Delank et al., 2010). As a matter of fact, because all levels are subjected to the same bending moment independently of the surgery, loading with pure moments inherently impedes any effect of the implant on the other segments (Volkheimer et al., 2015). To solve this issue, an alternative testing method named hybrid protocol has been developed (Panjabi, 2007). With this loading protocol, a multisegmental specimen is first tested with pure moments in its intact state, that is, before the implantation of the device, and the resulting range of motion is recorded. Then the device is implanted and the specimen retested by applying a pure moment until the global range of motion matches the one of the specimen in the intact state. With this protocol, an implant that stiffens the motion segment will induce a larger motion of the other segments, and vice versa. The hybrid protocol has been widely used (Panjabi et al., 2007a,b; Tan et al., 2008), but also has generated several criticisms (Volkheimer et al., 2015). As a matter of fact, its foundation hypothesis, which purposed that the global motion of the spine would be unaffected by the implantation of the device, was never clearly confirmed in clinical studies, especially regarding the lumbar spine (Malakoutian et al., 2015). Furthermore, the resulting hypermobility of the adjacent segments would be dependent on the length of the specimens, with an evident impact on the clinical relevance of the test results.

Testing setups aimed at replicating the action of the muscles in a more realistic way the action of the muscles, and thus even with the potential to predict the effect of an implant on the adjacent segments, have also been published. Following the steps of prior works (Panjabi et al., 1989), Wilke et al. (1994) connected the cranial extremity of the specimen to a plate that was linked to cables pulled to mimic the action of the muscles. Analogous methods have also been presented for cervical specimens (Panjabi et al., 2001). Such complex testing protocols, however, are not widely used and still suffer from a lack of standardization and limited validation. Therefore their use should be recommended only for specific research purposes, and more consolidated protocols such as pure moments with or without follower load should be recommended for most cases of implant testing.

SPINE TESTERS

The scientific literature describing in vitro testing of spine specimens is very extensive, and dozens of loading apparatuses have been presented, ranging from simple frames based on dead weights to robotic systems. In general, all setups share several components: actuators (such as motors or dead weights) to apply the loads; load cells to measure the loads that are actually applied; a mechanical system to fix the specimen, typically at its caudal end; and a cranial fixture that allows leaving degrees of freedom open while constraining others, depending on the aim of the test. Furthermore, systems able to measure the intersegmental motion of the specimens are commonly employing optical or ultrasonic waves.

In general, because of the coupling behavior of the spine (i.e., motion in one plane is accompanied by motion in the other motion planes), unconstrained loading devices that leave the remaining five degrees of freedom of the movable end of spinal specimens free (i.e., those along or around which the motion is not induced) are preferred (Wilke et al., 1998). However, because of their simplicity, constrained setups have also been employed, but have lost popularity in recent years. In fact, commercially available materials testing systems can be easily adapted to impose simple loading scenarios such as eccentric loads (Anasetti et al., 2010) or can be combined with external hardware to simulate more refined loading and boundary conditions (Crawford et al., 1995; Holsgrove et al., 2014). Indeed, some manufacturers of materials testing systems provide commercial add-ons to their loading frames that are able to reproduce loading scenarios that can be used to test spine specimens and implants in an unconstrained manner (MTS Systems Corporation, 2017; Moghadas et al., 2015).

A simple unconstrained loading system able to apply pure moments, which has been used in several research studies, is the one available at the ENSAM (Paris, France) (Lavaste et al., 1990; Templier et al., 1999); it consists of a jig able to apply loads to the cranial end of the specimen by means of bars pressing on the pedicles and on the vertebral body without constraining its motion, whereas the caudal end is fixed. Loads are applied by means of dead weights connected to cables and then to the loading bars via pulleys. To measure the motion of the specimen, linear and angular displacement sensors have been used. It should be noted that the loading device and protocol obtained an ISO quality certification. Other devices based on similar principles were also produced and were used to investigate clinically relevant questions (Lund et al., 1998; Goel et al., 1985; Dickey and Kerr, 2003).

Following preliminary in vitro work conducted by means of dead weights (Wilke et al., 1992, 1994), built a universal spine tester in Ulm, Germany, that was able to apply any combination of quasistatic loads to spinal specimens, either mono-segmental or multisegmental, in a purely unconstrained manner or with restriction of specific degrees of freedom (Fig. 2). Furthermore, the spine tester allows

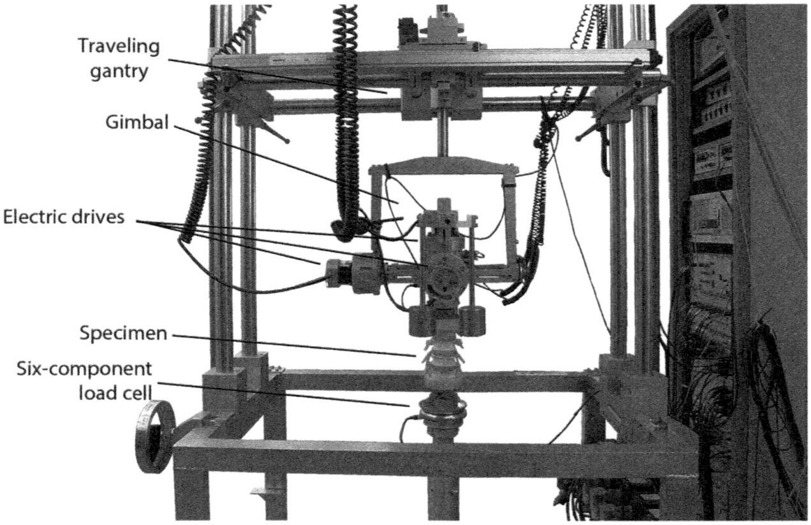

FIG. 2

The Ulm spine tester, able to apply any combination of quasistatic loads to spinal specimens in a purely unconstrained manner or with restriction of specific degrees of freedom.

modifying the loading and constraint conditions, such as the direction of the load without the need of removing the specimen from the frame and thus minimizing the risk of errors and required testing time. The loading device consists of a frame solidly connected to an adjustable mounting table, which is used to fix the caudal end of the specimen, and a gimbal to be linked to its cranial end. The gimbal allows free rotation around the three axes and translation in the vertical direction. To permit translations in the axial plane, the gimbal is connected to a traveling gantry, which also allows translations in the transversal plane. The gimbal is used to apply the bending moments through stepper motors. In its original implementation, displacement and rotational transducers were used to capture the motion of the specimen; in recent studies, an optoelectronic system was used for this purpose, which allows capturing the intersegmental motion rather than only the total motion. Passive reflective markers are screwed to each vertebra of the specimen and then localized in the three-dimensional space by means of infrared cameras (Fig. 3). Several spine testers, which were subsequently developed, share some of their construction principles with the Ulm device (Lysack et al., 2000; Schmoelz et al., 2009).

Goertzen et al. (2004) developed a simpler mechanical system able to apply a pure moment while leaving the other five degrees of freedom unconstrained (Fig. 4). The equipment consists of a motor, an arm including universal joints to transmit the moment to the specimen, a torque load cell connected to the arm, and a fixture for the specimen, including a six-component load cell. After one axis is tested, this implementation requires moving the motor and the articulating arm to proceed to the next loading direction.

A different approach to the implementation of an unconstrained pure moments protocol is achieved by using robots. Woo et al. (1993) and Hurschler et al. (2003) employed a general-purpose industrial robot equipped with a six-component load cell. Appropriate programming of the robot achieved an accurate control of the load in the desired direction as well as negligible reaction moments and forces in

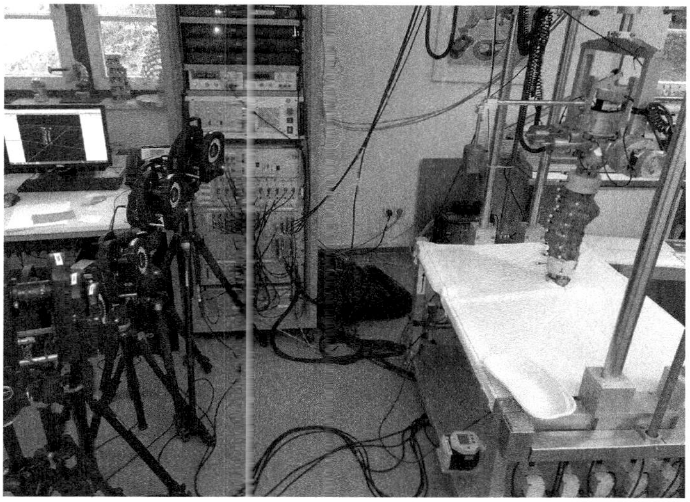

FIG. 3

Infrared cameras (left) used to capture the motion of a specimen equipped with passive reflective markers, mounted in the Ulm spine tester (right).

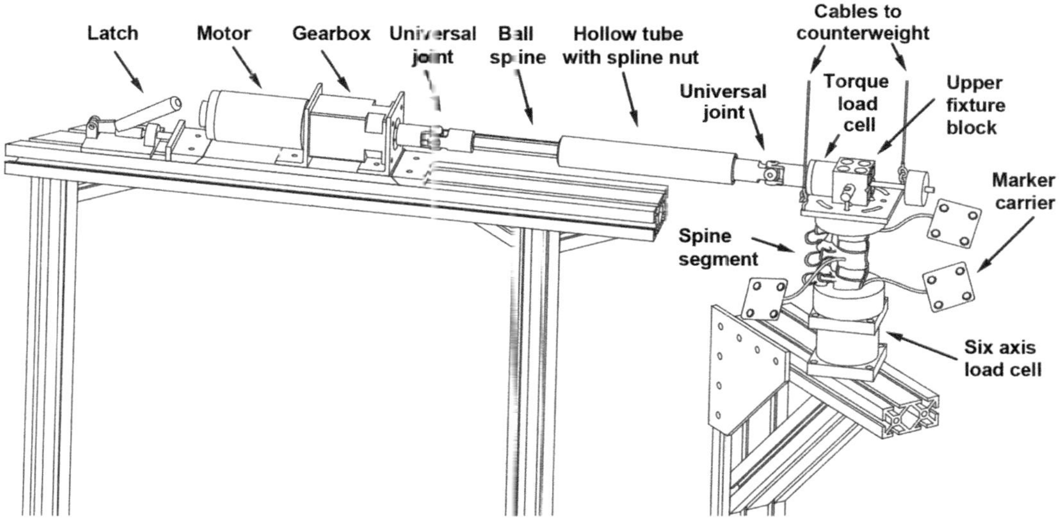

FIG. 4

The spine tester developed by Goertzen and colleagues, able to apply a pure moment while leaving the other five degrees of freedom unconstrained.

Reproduced with permission from Goertzen, D.J., Lane, C., Oxland, T.R., 2004. Neutral zone and range of motion in the spine are greater with stepwise loading than with a continuous loading protocol. An in vitro porcine investigation. J. Biomech. 37(2) 257–261.

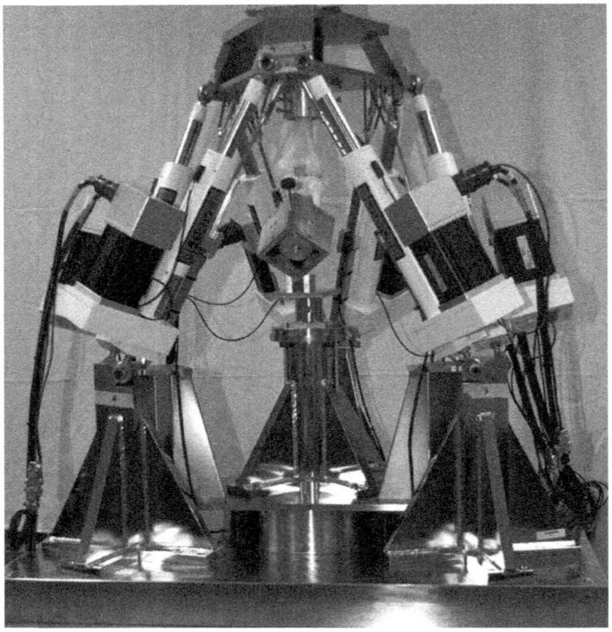

FIG. 5

The hexapod developed by the research group in Adelaide.

Reproduced with permission from Lawless, I., Ding, B., Cazzolato, B., Costi, J., 2014. Adaptive velocity-based six degree of freedom load control for real-time unconstrained biomechanical testing. J. Biomech. 47(12) 3241–3247.

the other directions. This approach was further developed by Stokes et al. (2002) in Burlington (United States) as well as by Ding et al. (2011) in Adelaide (Australia), who developed robotic loading frames based on the Stewart platform with the specific aim of testing spines and other specimens as well as implants of orthopedic interest (Fig. 5). The hexapods employ six linear actuators, which precisely position a mobile upper plate linked to the cranial end of a spine specimen relative to a fixed lower plate, which is connected to the caudal end. A six-component load cell is used to measure reaction forces and moments. Similar to the case of industrial robots, a sophisticated control software is required to ensure an accurate loading and respect of the unconstrained degrees of freedom. Such innovative platforms required the development of novel control schemes capable of coping with physiological loading rates in real time (Walker and Dickey, 2007; Lawless et al., 2014).

MEASUREMENT OF THE INTRADISCAL PRESSURE AND STRESS

Pressure transducers were first used in human volunteers at the end of the 1950s to compare the stresses in the spine in different postures and while performing different activities (Nachemson, 1959). Several papers reported values of the intradiscal pressure measured in vivo in the cervical (Pospiech et al.,

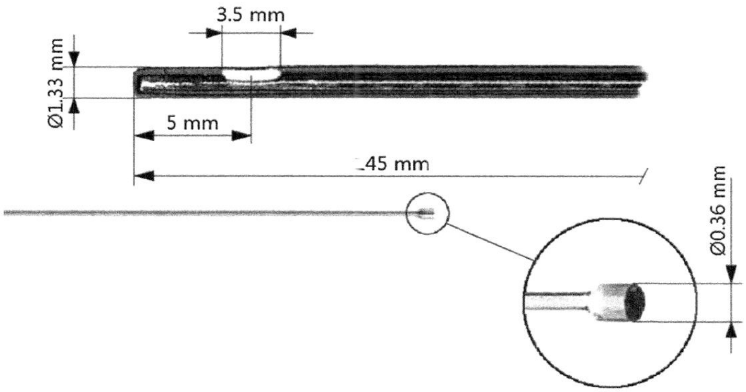

FIG. 6

Two examples of industrial pressure transducers of different sizes that can be used to measure the intradiscal pressure in live human subjects as well as in cadaveric specimens.

Reproduced with permission from Bashkuev, M., Vergroesen, P.A., Dreischarf, M., Schilling, C., van der Veen, Albert J, Schmidt, H., Kingma, I., 2016. Intradiscal pressure measurements: a challenge or a routine? J. Biomech. 49(6) 864–868.

1999), thoracic (Polga et al., 2004), and lumbar spines (Wilke et al., 1999), typically in the center of the nucleus pulposus either in healthy subjects or in patients suffering from intervertebral disc degeneration (Sato et al., 1999). This body of literature provides valuable set of references for the validation of the loads, which need to be applied by the loading apparatus, as well as for the evaluation of the effect of spinal implants on the load distribution within the motion segment by assuming a linear relationship between intradiscal pressure and compressive loads acting on the spine (Brinckmann and Grootenboer, 1991; Nachemson, 1960). Indeed, intradiscal pressure measurements were used to compare posterior instrumentation with different degrees of stiffness (Schmoelz et al., 2003) and to quantify the effect of spinal implants on the adjacent levels (Weinhoffer et al., 1995; Eck et al., 2002; Dmitriev et al., 2005; Cunningham et al., 1997). Industrial pressure transducers that can be used for in vitro testing of spine specimens are available (Bashkuev et al., 2016) and can be used easily in combination with purposely developed spine testers (Fig. 6).

Stress profilometry is a technique in which the pressure inside the intervertebral disc is measured by pulling a sensor along the disc or by using a transducer that has a sensor along the entire length of a needle rather than at a specific point, which is typically done along the mid-sagittal diameter of the disc (McNally and Adams, 1992; Adams et al., 1996a, b). By using this technique, researchers could depict the differences in the stresses acting in the nucleus pulposus and the annulus fibrosus. Furthermore, the hydrostatic nature of the intradiscal pressure could be evaluated by comparing the readings of the transducer in the sagittal and transverse planes, simply by rotating the needle 90 degrees. Results showed that in healthy discs, the pressure is hydrostatic, uniform in the nucleus pulposus and in the inner annulus, and decreases in the outer annulus. In contrast, degenerated discs revealed a lower and nonuniform pressure in the nucleus, as well as local stress peaks especially in the posterior annulus, which have been associated with discogenic pain (Adams et al., 1996a; Adams and Roughley, 2006; Adams et al., 2009).

CYCLIC LOADING

Several clinically relevant research questions, such as the investigation of pedicle screw fatigue or loosening, cage subsidence, and intervertebral disc herniation, are not adequately answered by quasistatic noncyclic testing, but require the simulation of repetitive loading. Implant loosening and subsidence do not typically result from the application of a single load with high magnitude, but are the effect of accumulating damage of the biological tissues under cyclic loading. Such effects are especially relevant in elderly subjects exhibiting a low bone mineral density (Galbusera et al., 2015). The number of loading cycles to which the spine is subjected during daily living is extremely high; Ashman et al. estimated that, in an average individual, up to 1 million loading cycles can be reached in a period of 4 months (Ashman et al., 1988). Despite such high numbers of loading cycles can be simulated when spinal implants are tested alone, for example under fatigue testing (see Chapter 13), the use of biological specimens that are subjected to their natural degradation when kept at room temperature forces researchers to limit the maximum duration of an experiment within a range of some hours. As a matter of fact, Wilke suggested that the whole duration of an in vitro experiment should not exceed 20h, including specimen preparation (Wilke et al., 1998). Because such a duration is not compatible with the simulation of a physiological number of cycles at reasonable frequencies, most experiments featuring repetitive loading available in the literature do not exceed 10,000–20,000 loading cycles and are therefore inherently limited in their clinical relevance.

The loads that act on the spine during daily activities may contribute to the accumulation of damage over time. These conditions are complex and variable and therefore are not easily replicable in an in vitro experiment; therefore simplifications and assumptions need to be considered. A typical approach is to observe the mechanisms responsible for the failure in vivo and to replicate them in vitro. This approach led, for example, to the development of quasistatic pull-out tests to assess the anchoring stability of pedicle screws, which are known to be "pulled out" of the pedicle when loosened (Fig. 7) (Halvorson et al., 1994; Zdeblick et al., 1993; Costa et al., 2013). Today, although pull-out tests are still common, more physiological experimental protocols involving cyclic loading are routinely performed. The toggle test, in which the head of the screw is cyclically loaded in the cranio-caudal direction, showing a loosening pattern expanding at the screw tip similarly to clinical observations is emerging as an accepted protocol to compare different screw designs or fixation options, for example, techniques for screw augmentation (Fig. 7) (Kueny et al., 2014; Reichle et al., 2002; Dawson et al., 2003). Note that the toggle test, similarly to the pull-out test, does not require a purposely developed loading apparatus, whereas a standard uniaxial materials testing system equipped with specific fixtures is sufficient.

Another example of a simple setup that can be implemented by adding specific equipment to a uniaxial testing system is the one presented by Kettler et al. (2001) to simulate subsidence of intervertebral cages. The authors applied a cyclic eccentric compressive load to a specimen, which slowly rotated around the cranio-caudal axis so that the direction of the resulting bending moment changed continuously over time (Fig. 8). By using such a method, the authors hypothesized that the test would effectively simulate the complex motions and loads arising from various daily activities.

Failure of the biological tissues themselves without any spinal implant, such as a herniation of the intervertebral disc, was also tested in vitro. Early equipment used for this purpose was limited to uniaxial testing (Farfan et al., 1970; Lin et al., 1978). Subsequently, special fixtures were used in combination with uniaxial testing systems to investigate the failure mechanisms under loading conditions in which compression was associated with either flexion, lateral bending, or axial rotation (Liu et al.,

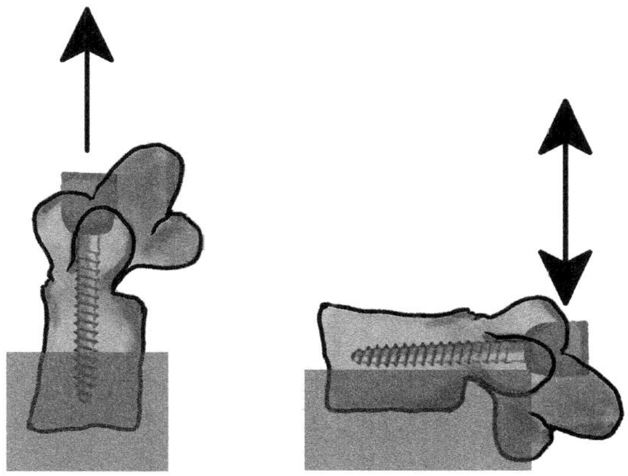

FIG. 7

Schematic representation of the axial pull-out test (left) and of the toggle test (right), which are commonly used to evaluate the stability of pedicle screws.

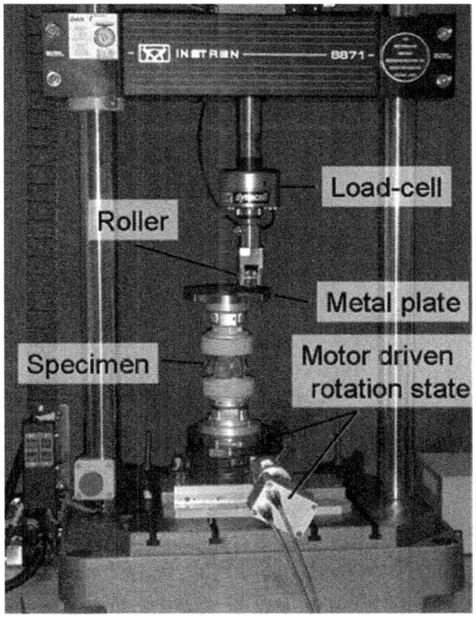

FIG. 8

The loading setup developed in Ulm to test subsidence of cages, consisting of a standard uniaxial materials testing system additionally equipped with special fixtures and a rotating plate.

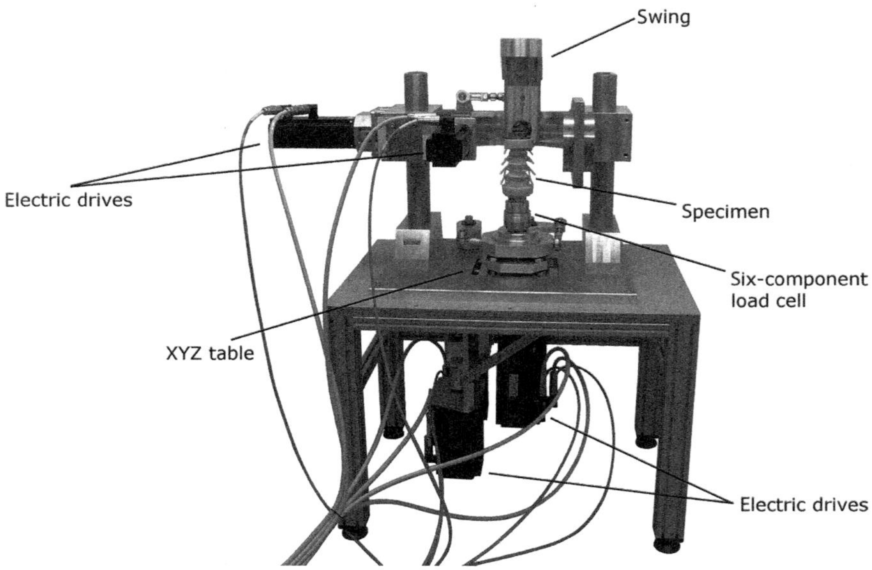

FIG. 9

The six-axis dynamic testing machine developed in Ulm with the aim of reaching both high load magnitudes and movement velocities.

1983, Adams and Hutton, 1983). Later, proper multiaxial testing allowing for the study of complex loading scenarios was introduced (Callaghan and McGill, 2001). In principle, all spinal testers described in the preceding paragraph might be used for this purpose, but the technical characteristics of some of these testers prevent their use at high loading rates, such as those required for cyclic destructive testing. In general, robotic systems are designed to be used at high speeds and have been employed to perform repetitive loading of spine specimens (Woo et al., 1993; Hurschler et al., 2003; Ding et al., 2011), although with limitations. In fact, the industrial robots were capable of reaching high movement speeds, but with a limited maximal load; on the contrary, the hexapod system was able to apply extremely high loads (up to 21 kN and 2.5 KNm) with precision, which is currently unmatched by any other dynamic system, but with a rotational speed limited to a maximum of 135 degrees/s. Recently, Wilke et al. presented a six-axis dynamic testing machine purposely developed to reach high load magnitudes (although lower than the hexapod, 2 kN and 40 Nm) and high movement speeds (400 degrees/s) (Fig. 9) (Wilke et al., 2016). The system is powered by electric motors and was used to create disc herniations and endplate junction failures in sheep discs. Together with the hexapod, this machine represents the current state-of-the-art machines for dynamic testing of spine specimens.

FLUID FLOW AND TIME-DEPENDENT RESPONSE

Recently, several papers presenting in vitro testing of spine specimens focused on the exchange of fluid between the intervertebral disc and the surrounding structures, which may play a critical role in the degenerative process (Urban et al., 1982; Holm et al., 1981; Ferguson et al., 2004). The intervertebral disc is the largest avascular structure in the human body, and the nutrition of its cells is dependent on solute diffusion as well as on convection, that is, pumping of fluids through the vertebral endplates due to cyclic compressive loading. In addition to this pumping effect, because of its osmotic potential, the intervertebral disc imbibes fluid during bedrest and exudates fluid during the day when spine loading increases the intradiscal pressure (Wilke et al., 1999; Urban et al., 2004). Reports showed that the fluid exchange during daily activities, such as walking, is scarce and that the net loss during the day may reach 3% to 10% of the total fluid content of the intervertebral disc, with potentially significant consequences for both the cell nutrition and the mechanical properties of the motion segment (Holm et al., 1981).

The research group led by Adams conducted several studies in which human lumbar motion segments were subjected to sustained compressive loads and cyclic loading. Among several noteworthy results are those from researchers who observed that the fluid loss was not homogeneously distributed inside the disc (McMillan et al., 1996), that it may raise potentially painful stress concentrations (Adams et al., 1996a, b), and that it determines compressive stress gradients that are dependent on the degree of disc degeneration and may be responsible for progressive delamination and collapse of the annulus (Stefanakis et al., 2014). Another research group active in the field is located in Amsterdam; this group published several relevant studies about the role of the endplates and the different types of mechanical loading in determining fluid exchange and pressure inside the disc (Vergroesen et al., 2014, 2016; Van der Veen et al., 2007).

From a technical point of view, the loading and boundary conditions used in these studies are rather simple, and most of them could be performed with standard uniaxial materials testing systems without the need for additional hardware. However, their capability to represent the conditions that act in vivo are a matter of heated debate (Schmidt et al., 2016). Studies conducted on live human volunteers and patients showed that the intervertebral disc is fully capable of recovering its height and fluid content after a night of bedrest (Wilke et al., 1999; Tyrrell et al., 1985), whereas experimental investigations reported that full recovery can be achieved only after longer times (O'Connell et al., 2011). Several researchers believe that such discrepancies are due to clogging of the blood vessels immediately after death, which prevents a physiological level of fluid flow through the endplates. However, no definitive evidence to prove this hypothesis has been provided. An alternative suggestion, or rather a possible concomitant cause of the differences between in vivo and in vitro results, was formulated by Schmidt and colleagues, who suggested that an excessive fluid content in cadaveric specimens might hamper the recovery of disc height and fluid content when the compressive load is removed (Schmidt et al., 2016). The authors therefore recommended accurate preconditioning of the specimens prior to testing to ensure that their hydration does not exceed physiological levels.

REFERENCES

Adams, M., Hutton, W., 1983. The mechanics of prolapsed intervertebral disc. Int. Orthop. 6 (4), 249–253.

Adams, M.A., Dolan, P., McNally, D.S., 2009. The internal mechanical functioning of intervertebral discs and articular cartilage, and its relevance to matrix biology. Matrix Biol. 28 (7), 384–389.

Adams, M.A., McMillan, D.W., Green, T.P., Dolan, P., 1996a. Sustained loading generates stress concentrations in lumbar intervertebral discs. Spine 21 (4), 434–438.

Adams, M.A., McNally, D.S., Dolan, P., 1996b. Stress' distributions inside intervertebral discs. The effects of age and degeneration. J. Bone Joint Surg 78 (6), 965–972.

Adams, M.A., Roughley, P.J., 2006. What is intervertebral disc degeneration, and what causes it? Spine 31 (18), 2151–2161.

Akamaru, T., Kawahara, N., Tim Yoon, S., Minamide, A., Su Kim, K., Tomita, K., Hutton, W.C., 2003. Adjacent segment motion after a simulated lumbar fusion in different sagittal alignments: a biomechanical analysis. Spine 28 (14), 1560–1566.

Anasetti, F., Galbusera, F., Aziz, H.N., Bellini, C.M., Addis, A., Villa, T., Teli, M., Lovi, A., Brayda-Bruno, M., 2010. Spine stability after implantation of an interspinous device: an in vitro and finite element biomechanical study. J. Neurosurg. Spine 13 (5), 568–575.

Ashman, R.B., Birch, J.G., Bone, L.B., Corin, J.D., Herring, J.A., CHARLES E JOHNSTON, I., Ritterbush, J.F., Roach, J.W., 1988. Mechanical testing of spinal instrumentation. Clin. Orthop. Relat. Res. (227)113-125.

Bashkuev, M., Vergroesen, P.A., Dreischarf, M., Schilling, C., van der Veen, Albert, J., Schmidt, H., Kingma, I., 2016. Intradiscal pressure measurements: a challenge or a routine? J. Biomech. 49 (6), 864–868.

Bastian, L., Lange, U., Knop, C., Tusch, G., Blauth, M., 2001. Evaluation of the mobility of adjacent segments after posterior thoracolumbar fixation: a biomechanical study. Eur. Spine J. 10 (4), 295–300.

Brinckmann, P., Grootenboer, H., 1991. Change of disc height, radial disc bulge, and Intradiscal pressure from discectomy an in vitro investigation on human lumbar discs. Spine 16 (6), 641–646.

Buckland, A.J., Baker, J.F., Roach, R.P., Spivak, J.M., 2016. Cervical disc replacement—emerging equivalency to anterior cervical discectomy and fusion. Int. Orthop. 40 (6), 1329–1334.

Callaghan, J.P., McGill, S.M., 2001. Intervertebral disc herniation: studies on a porcine model exposed to highly repetitive flexion/extension motion with compressive force. Clin. Biomech. 16 (1), 28–37.

Cheng, B.C., Gordon, J., Cheng, J., Welch, W.C., 2007. Immediate biomechanical effects of lumbar posterior dynamic stabilization above a circumferential fusion. Spine 32 (23), 2551–2557.

Costa, F., Villa, T., Anasetti, F., Tomei, M., Ortolina, A., Cardia, A., La Barbera, L., Fornari, M., Galbusera, F., 2013. Primary stability of pedicle screws depends on the screw positioning and alignment. Spine J. 13 (12), 1934–1939.

Crawford, N.R., Brantley, A.G., Dickman, C.A., Koeneman, E.J., 1995. An apparatus for applying pure nonconstraining moments to spine segments in vitro. Spine 20 (19), 2097–2100.

Cripton, P.A., Bruehlmann, S.B., Orr, T.E., Oxland, T.R., Nolte, L., 2000. In vitro axial preload application during spine flexibility testing: towards reduced apparatus-related artefacts. J. Biomech. 33 (12), 1559–1568.

Cripton, P.A., Dumas, G.A., Nolte, L.P., 2001. A minimally disruptive technique for measuring intervertebral disc pressure in vitro: application to the cervical spine. J. Biomech. 34 (4), 545–549.

Cunningham, B.W., Kotani, Y., McNulty, P.S., Cappuccino, A., McAfee, P.C., 1997. The effect of spinal destabilization and instrumentation on lumbar intradiscal pressure: an in vitro biomechanical analysis. Spine 22 (22), 2655–2663.

Dahl, M.C., Ellingson, A.M., Mehta, H.P., Huelman, J.H., Nuckley, D.J., 2013. The biomechanics of a multilevel lumbar spine hybrid using nucleus replacement in conjunction with fusion. Spine J. 13 (2), 175–183.

Dawson, J.M., Boschert, P., Macenski, M.M., Rand, N., 2003. Clinical relevance of pull-out strength testing of pedicle screws. In: Spinal Implants: Are We Evaluating Them Appropriately? ASTM International, West Conshohocken, PA.

Delank, K., Gercek, E., Kuhn, S., Hartmann, F., Hely, H., Röllinghoff, M., Rothschild, M., Stützer, H., Sobottke, R., Eysel, P., 2010. How does spinal canal decompression and dorsal stabilization affect segmental mobility? A biomechanical study. Arch. Orthop. Trauma Surg. 130 (2), 285–292.

Dickey, J., Kerr, D., 2003. Effect of specimen length: are the mechanics of individual motion segments comparable in functional spinal units and multisegment specimens? Med. Eng. Phys. 25 (3), 221–227.

Ding, B., Stanley, R.M., Cazzolato, B.S., Costi, J.J., 2011. In: Real-time FPGA control of a hexapod robot for 6-DOF biomechanical testing. IECON 2011-37th Annual Conference on IEEE Industrial Electronics Society. pp. 252.

Dmitriev, A.E., Cunningham, B.W., Hu, N., Sell, G., Vigna, F., McAfee, P.C., 2005. Adjacent level intradiscal pressure and segmental kinematics following a cervical total disc arthroplasty: an in vitro human cadaveric model. Spine 30 (10), 1165–1172.

Dreischarf, M., Zander, T., Bergmann, G., Rohlmann, A., 2010. A non-optimized follower load path may cause considerable intervertebral rotations. J. Biomech. 43 (13), 2625–2628.

Eck, J.C., Humphreys, S.C., Lim, T.H., Jeong, S.T., Kim, J.G., Hodges, S.D., An, H.S., 2002. Biomechanical study on the effect of cervical spine fusion on adjacent-level intradiscal pressure and segmental motion. Spine 27 (22), 2431–2434.

Evans, F.G., Lissner, H., 1959. Biomechanical studies on the lumbar spine and pelvis. JBJS 41 (2), 278–290.

Farfan, H., Cossette, J., Robertson, G., Wells, R., Kraus, H., 1970. The effects of torsion on the lumbar intervertebral joints: the role of torsion in the production of disc degeneration. JBJS 52 (3), 468–497.

Ferguson, S.J., Ito, K., Nolte, L.P., 2004. Fluid flow and convective transport of solutes within the intervertebral disc. J. Biomech. 37 (2), 213–221.

Galbusera, F., Volkheimer, D., Reitmaier, S., Berger-Roscher, N., Kienle, A., Wilke, H., 2015. Pedicle screw loosening: a clinically relevant complication? Eur. Spine J. 24 (5), 1005–1016.

Goel, V.K., Goyal, S., Clark, C., Nishiyama, K., Nye, T., 1985. Kinematics of the whole lumbar spine. Effect of discectomy. Spine 10 (6), 543–554.

Goel, V.K., Panjabi, M.M., Patwardhan, A.G., Dooris, A.P., Serhan, H., American Society for Testing and Materials, 2006. Test protocols for evaluation of spinal implants. J. Bone Joint Surg 88 (Suppl 2), 103–109.

Goertzen, D.J., Lane, C., Oxland, T.R., 2004. Neutral zone and range of motion in the spine are greater with stepwise loading than with a continuous loading protocol. An in vitro porcine investigation. J. Biomech. 37 (2), 257–261.

Halvorson, T.L., Kelley, L.A., Thomas, K.A., Whitecloud 3rd, T.S., Cook, S.D., 1994. Effects of bone mineral density on pedicle screw fixation. Spine 19 (21), 2415–2420.

Hambly, M.F., Wiltse, L.L., Raghavan, N., Schneiderman, G., Koenig, C., 1998. The transition zone above a lumbosacral fusion. Spine 23 (16), 1785–1792.

Hirsch, C., 1955. The reaction of intervertebral discs to compression forces. JBJS 37 (6), 1188–1196.

Hirsch, C., Nachemson, A., 1954. New observations on the mechanical behavior of lumbar discs. Acta Orthop. Scand. 23 (4), 254–283.

Holm, S., Maroudas, A., Urban, J.P., Selstam, G., Nachemson, A., 1981. Nutrition of the intervertebral disc: solute transport and metabolism. Connect. Tissue Res. 8 (2), 101–119.

Holsgrove, T.P., Gheduzzi, S., Gill, H.S., Miles, A.W., 2014. The development of a dynamic, six-axis spine simulator. Spine J. 14 (7), 1308–1317.

Hurschler, C., Pott, L., Gosse, F., 2003. Validierung eines roboterassistierten Prüfverfahrens für In-vitro-Wirbelsäulenprüfung. Z. Orthop. Ihre Grenzgeb. 141 (S1). O31_1.

Kettler, A., Liakos, L., Haegele, B., Wilke, H.J., 2007. Are the spines of calf, pig and sheep suitable models for pre-clinical implant tests? Eur. Spine J. 16 (12), 2186–2192.

Kettler, A., Wilke, H., Claes, L., 2001. Effects of neck movements on stability and subsidence in cervical interbody fusion: an in vitro study. J. Neurosurg. Spine 94 (1), 97–107.

Korovessis, P., Papazisis, Z., Lambiris, E., 2002. The role of rigid vs dynamic instrumentation for stabilization of the degenerative lumbosacral spine. Stud. Health Technol. Inform. 91, 457–461.

Kueny, R.A., Kolb, J.P., Lehmann, W., Püschel, K., Morlock, M.M., Huber, G., 2014. Influence of the screw augmentation technique and a diameter increase on pedicle screw fixation in the osteoporotic spine: pullout versus fatigue testing. Eur. Spine J. 23 (10), 2196–2202.

Lavaste, F., Asselineau, A., Diop, A., Grandjean, J., Laurain, J., Skalli, W., Roy-Camille, R., 1990. Protocole expérimental pour la caractérisation mécanique de segments rachidiens et de matériels d'ostéosynthèse dorso-lombaires. Rachis 2 (6), 435–446.

Lawless, I., Ding, B., Cazzolato, B., Costi, J., 2014. Adaptive velocity-based six degree of freedom load control for real-time unconstrained biomechanical testing. J. Biomech. 47 (12), 3241–3247.

Lee, J.C., Kim, Y., Soh, J.W., Shin, B.J., 2014. Risk factors of adjacent segment disease requiring surgery after lumbar spinal fusion: comparison of posterior lumbar interbody fusion and posterolateral fusion. Spine 39 (5), E339–45.

Lin, H., Liu, Y., Adams, K., 1978. Mechanical response of the lumbar intervertebral joint under physiological (complex) loading. JBJS 60 (1), 41–55.

Lindsey, D.P., Swanson, K.E., Fuchs, P., Hsu, K.Y., Zucherman, J.F., Yerby, S.A., 2003. The effects of an interspinous implant on the kinematics of the instrumented and adjacent levels in the lumbar spine. Spine 28 (19), 2192–2197.

Liu, Y.K., Njus, G., Buckwalter, J., Wakano, K., 1983. Fatigue response of lumbar intervertebral joints under axial cyclic loading. Spine 8 (8), 857–865.

Lund, T., Oxland, T.R., Jost, B., Cripton, P., Grassmann, S., Etter, C., Nolte, L.P., 1998. Interbody cage stabilisation in the lumbar spine: biomechanical evaluation of cage design, posterior instrumentation and bone density. J. Bone Joint Surg 80 (2), 351–359.

Lysack, J.T., Dickey, J.P., Dumas, G.A., Yen, D., 2000. A continuous pure moment loading apparatus for biomechanical testing of multi-segment spine specimens. J. Biomech. 33 (6), 765–770.

Lysell, E., 1969. Motion in the cervical spine: an experimental study on autopsy specimens. Acta Orthop. Scand. 40 (sup123), 1–61.

Malakoutian, M., Volkheimer, D., Street, J., Dvorak, M.F., Wilke, H.J., Oxland, T.R., 2015. Do in vivo kinematic studies provide insight into adjacent segment degeneration? A qualitative systematic literature review. Eur. Spine J. 24 (9), 1865–1881.

McMillan, D.W., Garbutt, G., Adams, M.A., 1996. Effect of sustained loading on the water content of intervertebral discs: implications for disc metabolism. Ann. Rheum. Dis. 55 (12), 880–887.

McNally, D.S., Adams, M.A., 1992. Internal intervertebral disc mechanics as revealed by stress profilometry. Spine 17 (1), 66–73.

Miura, T., Panjabi, M.M., Cripton, P.A., 2002. A method to simulate in vivo cervical spine kinematics using in vitro compressive preload. Spine 27 (1), 43–48.

Miyakoshi, N., Abe, E., Shimada, Y., Okuyama, K., Suzuki, T., Sato, K., 2000. Outcome of one-level posterior lumbar interbody fusion for spondylolisthesis and postoperative intervertebral disc degeneration adjacent to the fusion. Spine 25 (14), 1837–1842.

Moghadas, P., Mahomed, A., Shepherd, D.E., Hukins, D.W., 2015. Wear of the Charité® lumbar intervertebral disc replacement investigated using an electro-mechanical spine simulator. Proc. Inst. Mech. Eng. J. Eng. Med. 229 (3), 264–268.

MTS Systems Corporation, 2017. Available https://www.mts.com/cs/groups/public/documents/library/dev_004724.pdf.

Nachemson, A., 1960. Lumbar intradiscal pressure: experimental studies on post-mortem material. Acta Orthop. Scand. 31 (sup 43), 1–104.

Nachemson, A., 1959. Measurement of intradiscal pressure. Acta Orthop. Scand. 28, 269–289.

Naderi, S., Andalkar, N., Benzel, E.C., 2007. History of spine biomechanics: part II—from the renaissance to the 20th century. Neurosurgery 60 (2), 392–404.

O'Connell, G.D., Jacobs, N.T., Sen, S., Vresilovic, E.J., Elliott, D.M., 2011. Axial creep loading and unloaded recovery of the human intervertebral disc and the effect of degeneration. J. Mech. Behav. Biomed. Mater. 4 (7), 933–942.

Oxland, T.R., 2016. Fundamental biomechanics of the spine—what we have learned in the past 25 years and future directions. J. Biomech. 49 (6), 817–832.

Panjabi, M.M., 2007. Hybrid multidirectional test method to evaluate spinal adjacent-level effects. Clin. Biomech. 22 (3), 257–265.

Panjabi, M.M., Krag, M., Summers, D., Videman, T. 1985. Biomechanical time-tolerance of fresh cadaveric human spine specimens. J. Orthop. Res. 3 (3), 292–300.

Panjabi, M.M., Miura, T., Cripton, P.A., Wang, J., Nain, A.S., DuBois, C., 2001. Development of a system for in vitro neck muscle force replication in whole cervical spine experiments. Spine 26 (20), 2214–2219.

Panjabi, M.M., Oxland, T., Yamamoto, I., Crisco, J. 1994. Mechanical behavior of the human lumbar and lumbosacral spine as shown by three-dimensional load-displacement curves. JBJS 76 (3), 413–424.

Panjabi, M., Abumi, K., Duranceau, J., Oxland, T., 1989. Spinal stability and intersegmental muscle forces: a biomechanical model. Spine 14 (2), 194–200.

Panjabi, M., Henderson, G., Abjornson, C., Yue, J., 2007a. Multidirectional testing of one- and two-level ProDisc-L versus simulated fusions. Spine 32 (12), 1311–1319.

Panjabi, M., Malcolmson, G., Teng, E., Tominaga, Y., Henderson, G., Serhan, H., 2007b. Hybrid testing of lumbar CHARITE discs versus fusions. Spine 32 (9), 959–966. (discussion 967).

Park, P., Garton, H.J., Gala, V.C., Hoff, J.T., McGillicuddy, J.E., 2004. Adjacent segment disease after lumbar or lumbosacral fusion: review of the literature. Spine 29 (17), 1938–1944.

Patwardhan, A.G., Havey, R.M., Meade, K.P., Lee, B., Dunlap, B., 1999. A follower load increases the load-carrying capacity of the lumbar spine in compression. Spine 24 (10), 1003–1009.

Pfeiffer, M., Hoffman, H., Goel, V., Weinstein, J., Griss, P., 1997. In vitro testing of a new transpedicular stabilization technique. Eur. Spine J. 6 (4), 249–255.

Polga, D.J., Beaubien, B.P., Kallemeier, P.M., Schellhas, K.P., Lew, W.D., Buttermann, G.R., Wood, K.B., 2004. Measurement of in vivo intradiscal pressure in healthy thoracic intervertebral discs. Spine 29 (12), 1320–1324.

Pospiech, J., Stolke, D., Wilke, H.J., Claes, L.E., 1999. Intradiscal pressure recordings in the cervical spine. Neurosurgery 44 (2), 379–384. discussion 384–5.

Reichle, E., Morlock, M., Sellenschloh, K., Eggers, C., 2002. Zur Definition der PedikelfehllagePrimärstabilität und Lockerungsverhalten von Pedikelschrauben in Abhängigkeit von ihrer Lage: spongiöse Verankerung, kortikale Verankerung, Perforation und Fehllage. Orthopade 31 (4), 402–405.

Rohlmann, A., Neller, S., Claes, L., Bergmann, G., Wilke, H., 2001. Influence of a follower load on intradiscal pressure and intersegmental rotation of the lumbar spine. Spine 26 (24), E557–E561.

Rohlmann, A., Zander, T., Rao, M., Bergmann, G., 2009a. Applying a follower load delivers realistic results for simulating standing. J. Biomech. 42 (10), 1520–1526.

Rohlmann, A., Zander, T., Rao, M., Bergmann, G., 2009b. Realistic loading conditions for upper body bending. J. Biomech. 42 (7), 884–890.

Sato, K., Kikuchi, S., Yonezawa, T., 1999. In vivo intradiscal pressure measurement in healthy individuals and in patients with ongoing back problems. Spine 24 (23), 2468–2474.

Schmidt, H., Reitmaier, S., Graichen, F., Shirazi-Adl, A., 2016. Review of the fluid flow within intervertebral discs-how could in vitro measurements replicate in vivo? J. Biomech. 49 (14), 3133–3146.

Schmoelz, W., Onder, U., Martin, A., Von Strempel, A., 2009. Non-fusion instrumentation of the lumbar spine with a hinged pedicle screw rod system: an in vitro experiment. Eur. Spine J. 18 (10), 1478–1485.

Schmoelz, W., Huber, J.F., Nydegger, T., Dipl-Ing, C., L. & Wilke, H.J., 2003. Dynamic stabilization of the lumbar spine and its effects on adjacent segments: an in vitro experiment. J. Spinal Disord. Tech. 16 (4), 418–423.

Schulte, T.L., Leistra, F., Bullmann, V., Osada, N., Vieth, V., Marquardt, B., Lerner, T., Liljenqvist, U., Hackenberg, L., 2007. Disc height reduction in adjacent segments and clinical outcome 10 years after lumbar 360 fusion. Eur. Spine J. 16 (12), 2152–2158.

Stefanakis, M., Luo, J., Pollintine, P., Dolan, P., Adams, M.A., 2014. ISSLS prize winner: mechanical influences in progressive intervertebral disc degeneration. Spine 39 (17), 1365–1372.

Stokes, I.A., Gardner-Morse, M., Churchill, D., Laible, J.P., 2002. Measurement of a spinal motion segment stiffness matrix. J. Biomech. 35 (4), 517–521.

Swanson, K.E., Lindsey, D.P., Hsu, K.Y., Zucherman, J.F., Yerby, S.A., 2003. The effects of an interspinous implant on intervertebral disc pressures. Spine 28 (1), 26–32.

Tan, J.S., Singh, S., Zhu, Q.A., Dvorak, M.F., Fisher, C.G., Oxland, T.R., 2008. The effect of cement augmentation and extension of posterior instrumentation on stabilization and adjacent level effects in the elderly spine. Spine 33 (25), 2728–2740.

Templier, A., Skalli, W., Lemaire, J., Diop, A., Lavaste, F., 1999. Three-dimensional finite-element modelling and improvement of a bispherical intervertebral disc prosthesis. Eur. J. Orthop. Surg. Traumatol. 9 (2), 51–58.

Tyrrell, A.R., Reilly, T., Troup, J.D., 1985. Circadian variation in stature and the effects of spinal loading. Spine 10 (2), 161–164.

Umehara, S., Zindrick, M.R., Patwardhan, A.G., Havey, R.M., Vrbos, L.A., Knight, G.W., Miyano, S., Kirincic, M., Kaneda, K., Lorenz, M.A., 2000. The biomechanical effect of postoperative hypolordosis in instrumented lumbar fusion on instrumented and adjacent spinal segments. Spine 25 (13), 1617–1624.

Urban, J.P., Holm, S., Maroudas, A., Nachemson, A., 1982. Nutrition of the intervertebral disc: effect of fluid flow on solute transport. Clin. Orthop. Relat. Res. 170 (170), 296–302.

Urban, J.P., Smith, S., Fairbank, J.C., 2004. Nutrition of the intervertebral disc. Spine 29 (23), 2700–2709.

Van der Veen, A., Van Dieën, J., Nadort, A., Stam, B., Smit, T., 2007. Intervertebral disc recovery after dynamic or static loading in vitro: is there a role for the endplate? J. Biomech. 40 (10), 2230–2235.

Vergroesen, P.A., van der Veen, Albert, J., Emanuel, K.S., van Dieën, J.H., Smit, T.H., 2016. The poro-elastic behaviour of the intervertebral disc: a new perspective on diurnal fluid flow. J. Biomech. 49 (6), 857–863.

Vergroesen, P.A., van der Veen, Albert, J., van Royen, B.J., Kingma, I., Smit, T.H., 2014. Intradiscal pressure depends on recent loading and correlates with disc height and compressive stiffness. Eur. Spine J. 23 (11), 2359–2368.

Volkheimer, D., Malakoutian, M., Oxland, T.R., Wilke, H.J., 2015. Limitations of current in vitro test protocols for investigation of instrumented adjacent segment biomechanics: critical analysis of the literature. Eur. Spine J. 24 (9), 1882–1892.

Walker, M.R., Dickey, J.P., 2007. New methodology for multi-dimensional spinal joint testing with a parallel robot. Med. Biol. Eng. Comput. 45 (3), 297–304.

Weinhoffer, S.L., Guyer, R.D., Herbert, M., Griffith, S.L., 1995. Intradiscal pressure measurements above an instrumented fusion: a cadaveric study. Spine 20 (5), 526–531.

Wilke, H., Kienle, A., Maile, S., Rasche, V., Berger-Roscher, N., 2016. A new dynamic six degrees of freedom disc-loading simulator allows to provoke disc damage and herniation. Eur. Spine J. 25 (5), 1363–1372.

Wilke, H., Krischak, S., Claes, L.E., 1996. Formalin fixation strongly influences biomechanical properties of the spine. J. Biomech. 29 (12), 1629–1631.

Wilke, H., Rohlmann, A., Neller, S., Schulthei, M., Bergmann, G., Graichen, F., Claes, L.E., 2001. Is it possible to simulate physiologic loading conditions by applying pure moments? A comparison of in vivo and in vitro load components in an internal fixator. Spine 26 (6), 636–642.

Wilke, H., Werner, K., Häussler, K., Reinehr, M., Böckers, T.M., 2011. Thiel-fixation preserves the non-linear load–deformation characteristic of spinal motion segments, but increases their flexibility. J. Mech. Behav. Biomed. Mater. 4 (8), 2133–2137.

Wilke, H., Fischer, K., Kugler, A., Magerl, F., Claes, L., Wörsdörfer, O., 1992. In vitro investigations of internal fixation systems of the upper cervical spine. Eur. Spine J. 1 (3), 185–190.

Wilke, H., Rohlmann, A., Neller, S., Graichen, F., Claes, L., Bergmann, G., 2003. ISSLS prize winner: a novel approach to determine trunk muscle forces during flexion and extension: a comparison of data from an in vitro experiment and in vivo measurements. Spine 28 (23), 2585–2593.

Wilke, H.J., Claes, L., Schmitt, H., Wolf, S., 1994. A universal spine tester for in vitro experiments with muscle force simulation. Eur. Spine J. 3 (2), 91–97.

Wilke, H.J., Neef, P., Caimi, M., Hoogland, T., Claes, L.E., 1999. New in vivo measurements of pressures in the intervertebral disc in daily life. Spine 24 (8), 755–762.

Wilke, H.J., Wenger, K., Claes, L., 1998. Testing criteria for spinal implants: recommendations for the standardization of in vitro stability testing of spinal implants. Eur. Spine J. 7 (2), 148–154.

Wiltse, L.L., Radecki, S.E., Biel, H.M., DiMartino, P.P., Oas, R.A., Farjalla, G., Ravessoud, F.A., Wohletz, C., 1999. Comparative study of the incidence and severity of degenerative change in the transition zones after instrumented versus noninstrumented fusions of the lumbar spine. Clin. Spine Surg. 12 (1), 27–33.

Woo, S.L., Livesay, G.A., Arai, S., Tsukamoto, Y., 1993. The use of robotics technology to study human joint kinematics: a new methodology. J. Biomech. Eng. 115, 211.

Yamamoto, I., Panjabi, M.M., Crisco, T., Oxland, T., 1989. Three-dimensional movements of the whole lumbar spine and lumbosacral joint. Spine 14 (11), 1256–1260.

Zdeblick, T.A., Kunz, D.N., Cooke, M.E., McCabe, R., 1993. Pedicle screw pullout strength: correlation with Insertional torque. Spine 18 (12), 1673–1676.

Zirbel, S.A., Stolworthy, D.K., Howell, L.L., Bowden, A.E., 2013. Intervertebral disc degeneration alters lumbar spine segmental stiffness in all modes of loading under a compressive follower load. Spine J..

CHAPTER

STANDARD TESTING

13

Annette Kienle*, Nicolas Graf*, Tomaso Villa[†,‡], Luigi La Barbera[†,‡]
SpineServ GmbH & Co. KG, Ulm, Germany Politecnico di Milano, Milan, Italy[†] IRCCS Galeazzi Orthopedic Institute, Milan, Italy[‡]*

INTRODUCTION: ADVANTAGES AND DISADVANTAGES OF STANDARD TESTING

Implantable medical devices must meet specific testing requirements before they can be put on the market. These requirements are defined by respected regulatory bodies such as the United States Food and Drug Administration (FDA) and the European Notified Bodies. To improve the comparability between testing laboratories, testing is carried out according to national or international standards wherever possible. The main aim of standard testing is to investigate the safety and effectiveness of medical devices. The test results are part of the submission process used in seeking approval from the appropriate regulatory body.

In Europe and North America, ASTM (American Society for Testing and Materials) and ISO (International Organization for Standardization) standards need to be followed for the most part. The process of standard development and revision plays an important role in better understanding the advantages and disadvantages of standard testing.

In the case of the American Society for Testing and Materials, task groups develop the basics for the draft standards. These drafts pass through different hierarchical levels (subcommittee, main committee, society) for review and voting. This whole process may take more than a year. Everyone is free to join ASTM and take part in standard development. ISO standards are developed by independent technical experts who elaborate a draft version. Then a voting process begins and continues until a consensus is reached. ISO standard development takes about three years from the first draft to the published version. Thus, in case of a new type of implant, it may take years until a suitable standard test method is available. To overcome this problem, it may be appropriate to test implants according to custom testing procedures. These procedures should take into consideration the loads that are expected to act on the device in vivo.

Loading a spinal implant in vivo is very complex and is rarely heard of. In an attempt to improve repeatability and increase reproducibility, standard test methods greatly simplify this complexity. This simplification is one of the main disadvantages of standard testing because the results do not directly reflect the situation in vivo. Therefore comparative data are almost always necessary, which means that the mechanical safety and effectiveness of a new medical device must be compared to a well-known and successfully used implant. In the United States, this process is called a 510(*k*) submission or Premarket

Biomechanics of the Spine. https://doi.org/10.1016/B978-0-12-812851-0.00013-6

Table 1 Advantages and Disadvantages of Preclinical Implant Testing According to Standard Test Methods

Advantages	Disadvantages
– Standard testing procedures are internationally known and widely accepted.	– No standards are available for new and innovative implants. Because drafting and approval of a new standard can take years, standard development always lags behind implant development. Innovative implants therefore often have to be tested according to custom, implant-specific testing procedures. Such testing can be of high quality but usually requires more efforts to convince the regulatory bodies.
– Standard testing can easily be conducted.	– Easy test conduction is mostly associated with strong simplifications of the real loading in vivo. Therefore interpretation of the test results is difficult. For that reason, commonly comparative data is required from a predicate device.
– The interlaboratory comparability is increased.	– Some standards leave flexibility for the user, which again reduces the interlaboratory comparability.

Notification. But such comparative data are also often required for CE Marking. This comparison must demonstrate that the device to be marketed is at least as safe and effective (i.e., substantially equivalent to) as the legally marketed device (predicate device).

On the other hand, the biggest advantage of the standard testing methods is their international recognition (Table 1).

OVERVIEW OF CURRENT AVAILABLE STANDARDS FOR TESTING SPINAL DEVICES

The standards for preclinical testing of implantable devices for spinal surgery are numerous. Table 2 shows the latest revisions of all the current available documents with a short description of the main peculiarities of each standard.

ISO and ASTM are responsible for designing and issuing the standards, with slightly different approaches, and some general comments can be drawn from the analysis of the standards, as follows:

- ISO standards tend to give a complete description of the testing condition that must be applied, including the geometry of the setup, the entity of the loads to be applied, and in the case of dynamic tests, the run-out cycles; however, performance levels for the tested devices are not always given (e.g., the maximum amount of debris that can be accepted during wear tests is not defined).
- ASTM standards generally denote methods for running tests for comparison purposes; this is a main requirement of the FDA for the introduction of new devices on the US market. In this light, some details (e.g., regarding the definition of the load to be applied in fatigue testing) are missing, and it's up to the manufacturer to justify the test protocol and to demonstrate the equivalence of the new device with the predicates.
- In the latest revisions of the standards, some cross-references between ISO and ASTM standards designed for testing similar devices are reported (e.g., in ISO 12189), often to give evidence of

the differences between the two approaches and to explain under which conditions one should use one standard rather than the other.

- In the latest revisions of the ASTM standards, some reference tables containing results of static tests on predicate devices were introduced to enhance the importance of a statistical approach; however, these data can also be used to compare the performances of the new device with predicate device performances.
- Useful description of the rationale underlying the choice of the prescribed setup and bibliographic references are always reported in both ASTM and ISO standards.

STUDY CASES

INTERPRETATION OF CURRENT STANDARDS WITH FEM METHODOLOGY (APPLICATION TO ASTM F 1717/ISO 12189)

The assessment of the fatigue performances of a screw-rod system in the preclinical phase is mainly based on the use of two standards, ASTM F 1717 and ISO 12189. Both standards simulate a 2-FSUs application, but whereas the former uses a vertebrectomy model without any anterior support, the latter uses calibrated springs interposed between each couple of vertebra to simulate an intervertebral disc (Fig. 1).

As shown in Table 2, the ASTM standards do not specify the load to be applied, and the maximum load to reach the run-out has to be determined by the user. This engineering-based approach generally leads to finding and applying a load so low that it is hardly comparable to clinical use (Villa et al., 2014a, b). On the other hand, the ISO standards impose a maximum load of 2 kN until the same run-out is reached.

Comparison of the results from the two different procedures may present difficulties, perhaps some that are impossible to overcome. Differences in the applied load and the typology of the anterior support (absent or of different stiffness) will produce different internal states of stress in the implant that, in turn, will strongly affect their fatigue behavior. Moreover, as much as possible, the test's results should predict the device's behavior after it is implanted in the patient. This is generally accomplished if the geometry of the setup, the loads that are applied, and the number of cycles are representative of the ones the device will be subjected to in vivo. In this light, we wondered, on the one hand, whether the ASTM experimental configuration is correctly representative of the physiological anatomy of the thoracolumbar spine and, on the other hand, whether the internal loads arising in the rods using the two procedures are representative of those that arise when everyday life activities are carried out.

Analysis of the Anatomical Parameters Influencing ASTM F1717 Results

In regard to the suitability of the ASTM standard to propose a methodology that accurately represents the anatomy of the spine, an investigation was conducted to (i) compare the set of geometrical parameters imposed by the standard with their physiological value in the thoracolumbar tract; and (ii) determine a worst-case scenario for the stress state in the instrumentation deriving from the superposition of all parameters (La Barbera et al., 2014).

The values of the anatomical parameters of the thoracolumbar tract were extrapolated from both literature data and direct measurements on patients subjected to stereoradiographic X-ray imaging

Table 2 Current ASTM and ISO Standards for Preclinical Evaluation of Spinal Devices With Details of the Test Protocol

Standard—Year of Release	Year of Reapproval	Title	Tested Implant	Mechanical Evaluation
ISO 18192-17, part 3	Draft	Wear of total intervertebral spinal disc prostheses. Impingement-wear testing and corresponding environmental conditions for test of lumbar prostheses under adverse kinematic conditions.	IVD prosthesis	Wear
ISO 18192-10, part 2	–	Wear of total intervertebral spinal disc prostheses. Nucleus replacements.	IVD prosthesis—nucleus replacements	Fatigue, wear
ISO 18192-08, part 1	2011	Wear of total intervertebral spinal disc prostheses. Loading and displacement parameters for wear testing and corresponding environmental conditions for test.	IVD prosthesis	Wear
ISO 12189-08	2011	Fatigue test method for spinal implant assemblies using an anterior support.	Posterior stabilization devices	Fatigue
ASTM F2624-12	2016	Standard test method for static, dynamic, and wear assessment of extra-discal single level spinal constructs.	Extra-discal constructs (motion preserving, rigid fixation devices)	Static, fatigue,wear
ASTM F2789-10	2015	Standard guide for mechanical and functional characterization of nucleus devices.	Nucleus devices	Static, fatigue
ASTM F2790-10	2014	Standard practice for static and dynamic characterization of motion preserving lumbar total facet prostheses.	Total facets prostheses	Static, fatigue
ASTM F2706-08	2017	Standard test methods for occipital-cervical and occipital-cervical-thoracic spinal implant constructs in a vertebrectomy model.	Occipital-cervical, occipital-cervical-thoracic constructs	Static, fatigue
ASTM F2694-07	2016	Standard practice for functional and wear evaluation of motion-preserving lumbar total facet prostheses	Components for total facets prostheses	Wear
ASTM F2423-05	2016	Standard guide for functional, kinematic, and wear assessment of total disc prostheses.	Total disc prosthesis	Wear, fatigue
ASTM F2346-05	2011	Standard test methods for static and dynamic characterization of spinal artificial discs.	IVD devices	Static, fatigue
ASTM F2267-03	2011	Standard test method for measuring load induced subsidence of intervertebral body fusion device under static axial compression.	Intervertebral bony fusion devices	Static
ASTM F2193-02	2014	Standard specifications and test methods for components used in the surgical fixation of the spinal skeletal system.	Components for spinal fixation (screws, hooks, plates, rods)	Static, fatigue
ASTM F2077-00	2014	Test methods for intervertebral body fusion devices.	Intervertebral body fusion devices	Static, fatigue
ASTM F1798-97	2013	Standard test method for evaluating the static and fatigue properties of interconnection mechanisms and subassemblies used in spinal arthrodesis implants.	Interconnection mechanisms and subassemblies for spinal arthrodesis	Static, fatigue
ASTM F1717-96	2016	Standard test methods for spinal implant constructs in a vertebrectomy model.	Posterior fixation devices	Static, fatigue

[a] *AC, axial compression; AR, axial rotation; AT, axial torsion; B, bending; CB, compression bending; CS, compression shear; FE, flexion-extension;*

Functional Evaluation	Spinal Region	Number of FSUs	Simple Motions Mode ([a])	Simple Loading Mode ([a])	Combined Loading Mode ([a])	Environmental Conditions	Run-Out Cycles (millions)
Impingement	Cervical, lumbar	1	FE, LB, and AR (assigned)	AC (assigned)	F + LB	Yes	1
–	Cervical, lumbar	1	FE, LB, and AR (assigned)	AC (assigned), CS	FE + LB + AR + AC + CS	Yes	10
–	Cervical, lumbar	1	FE, LB, and AR (assigned)	AC (assigned), CS	FE + LB + AR + AC + CS	Yes	10
–	Cervical, thoracolumbar	1, 2 or more	–	CB (assigned)	–	Yes	5
–	Lumbar	1	FE, LB, and AR	S, CB, TB	–	–	10 or 5 (for motion-preserving or fixation devices)
Kinematics, subsidence	Lumbar	1	FE, LB, AR	AC, AT, S, B	–	Yes	10
–	Lumbar	1	F, E, LB, AR	S, CB,TB	–	Yes	10
–	Occipital-cervical, occipital-cervical-thoracic	2	AR	CB, TB	–	Yes	5
Kinematics	Lumbar	1	F, E, LB, AR (indicative ranges)	AC (indicative range)	LB + AR + A, FE + LB + A	Yes	10
Kinematics	Cervical, lumbar	1	FE, LB, AR (indicative ranges)	AC (indicative range)	–	Yes	10
–	Not specified	1	–	AC, CS	–	–	Yes
Subsidence	Not specified	–	–	AC	–	–	–
–	Not specified	–	–	B, AT	–	–	2.5
–	Not specified	1	AR	AC, CS	–	–	5
Loosening	Not specified	–	–	AC, AT, S, B	–	–	2.5
–	Cervical, thoracolumbar	2	AR	CB, TB	–	Yes	5

FSU, functional spinal unit; LB, lateral bending; PE, polyethilene; S, shear; TB, tension bending.

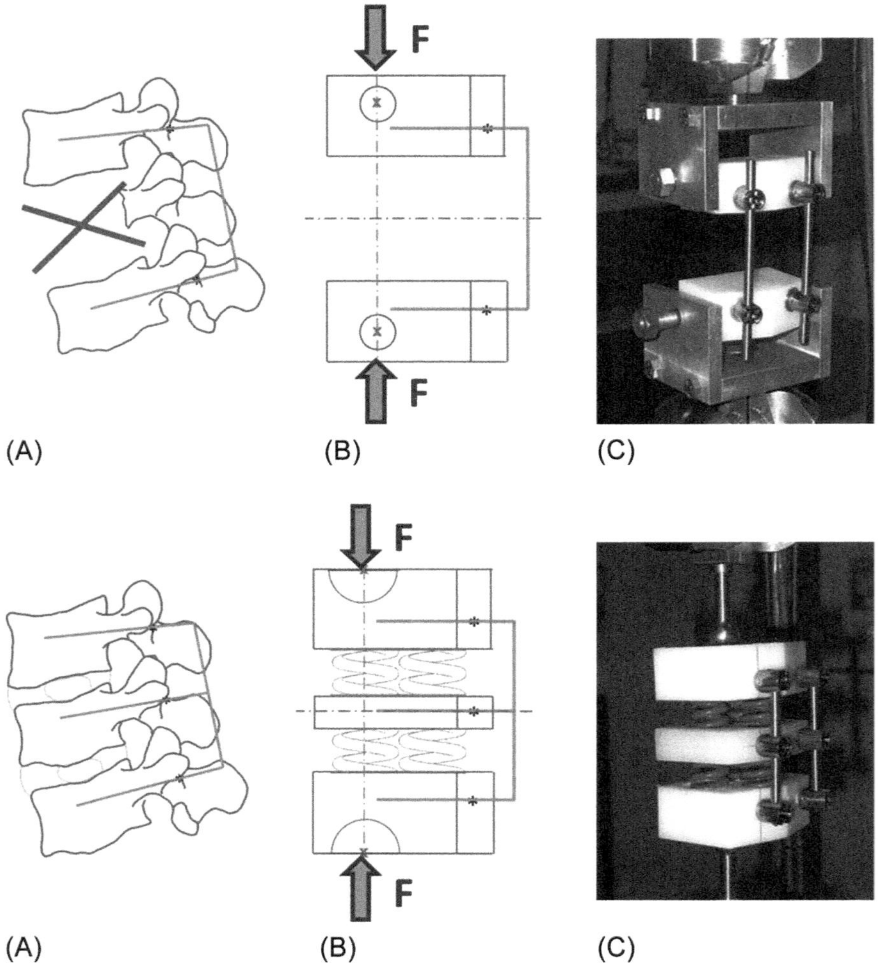

FIG. 1

ASTM F1717 (top) and ISO 12189 (bottom) experimental setups: anatomical sketch (A), sketch taken from the standard (B), specimens during the tests (C).

(Fig. 2); in parallel, a parametric finite element model of the ASTM F1717 setup was built and discretized, assuming its linear elastic material properties (Fig. 3). The simulations were run by imposing a vertical load of 300 N and using a systematic approach so that each anatomical parameter investigated varied between its minimum and maximum values, while all the other parameters were kept fixed. Stresses on the screw head and on the rod in the transverse plane were calculated this way, and results were presented in terms of a von Mises stress increase normalized on those calculated for the reference configuration proposed by ASTM. Fig. 2 shows an example of the results obtained for a particular parameter (the Block Moment Arm, BMA).

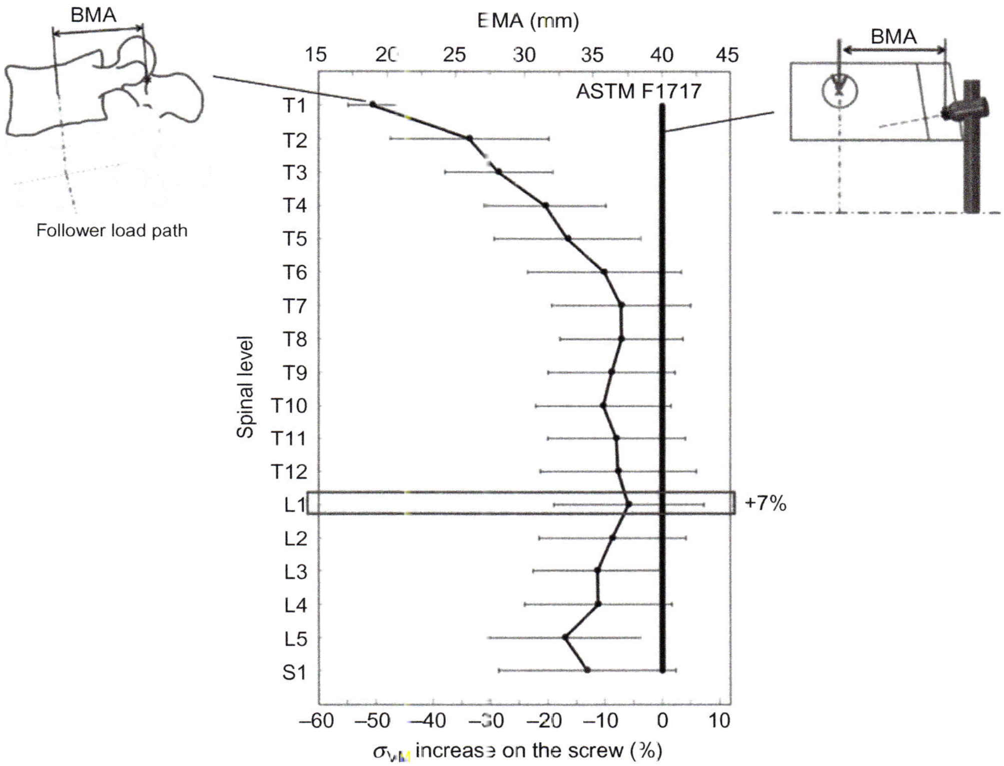

FIG. 2

BMA as a function of the thoracolumbar level: comparison between the value proposed by ASTM F1717 and data measured on healthy subjects. BMA is the lever arm of the eccentric load in ASTM setup (right), interpreted as the lever arm of the follower load in a healthy subject's spine (left). The lower horizontal axis reports the percentage of stress increase on the pedicle screw for each value of BMA compared to ASTM reference configuration.

A further step was taken to identify the worst-case scenario for each spinal level, which meant identifying the combination of the different parameters' value that gave the highest increase of stresses in the instrumentation. Also, in this case, a comparison of the state of stress arising when ASTM proposals are followed is presented (Fig. 4).

Results showed that two anatomical parameters had the strongest influence over the stress arising in the instrumentation: the previously mentioned BMA and the center of fixation to rotation (CoFR). The latter is the vertical distance between the screw insertion point in the PE block and the center of rotation of the cylindrical pin used to apply the load in the ASTM setup, whereas in the anatomical model, it is the distance between the screw insertion point and the instantaneous center of rotation of the FSU. The results demonstrate that the setup proposed by ASTM represents well the stress on the device when it is implanted at the L1 level in an average patient taken from a physiological population (Fig. 4). However,

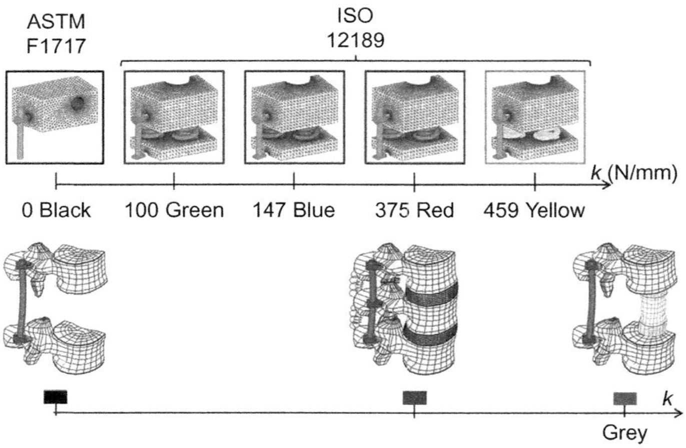

FIG. 3

FE model of ISO 12189 standard (top): calibrated springs with increasing axial stiffness (k) and different colcr codes (green, blue, red, and yellow) from left to right are used to mimic the support provided by the anterior spine, potentially describing a wide range of instability scenarios (bottom) from a full corpectomy to a rigid titanium cage. The ASTM F1717 model can be considered as a limit model without springs. Only one quarte· of each setup was described, assuming its symmetry in the horizontal and sagittal planes.

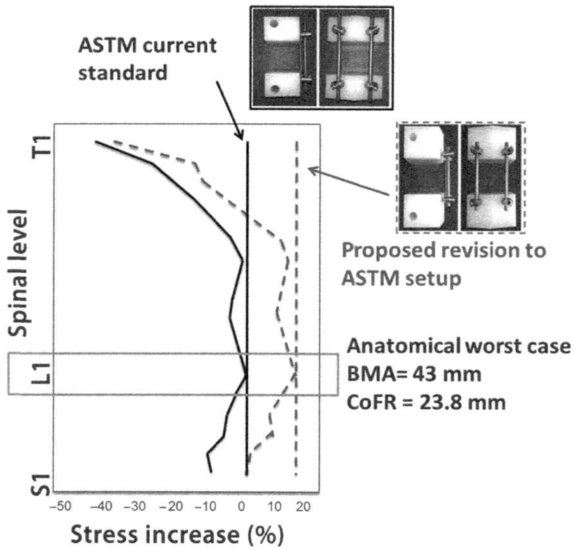

FIG. 4

Results from FE analyses. The solid line indicates the average case (combination of the physiological mean values for each parameter); the dashed line indicates the worst case (combination of physiological worst values for each parameter).

things drastically change when, at the same level, the upper boundary of the anatomical parameters of this population is considered. In such a patient, stress in the instrumentation can increase more than 15% on the screw, suggesting a possible revision of the setup that is able to take into account the actual worst-case scenario (La Barbera et al., 2014). A more recent experimental and numerical study investigated the effect of such a revision on a spinal fixator already cleared for commercial use. It was shown that the revised ASTM standard led to a significant 3.2-million-cycle drop in the fatigue performance compared to that of the same implant tested according to current ASTM recommendations (La Barbera and Villa, 2017).

Analysis of the Anatomical and Mechanical Parameters Influencing ISO 12189 Results

A similar numerical analysis was performed on the ISO 12189 standard to determine the effect of mechanical and anatomical parameters on the loads occurring on the posterior stabilization device (La Barbera et al., 2016a; La Barbera and Villa, 2016). An FE model of the ISO standard setup was built, and a simple vertical load up to 2000 N was applied on the construct (Fig. 3). The stiffness of the calibrated springs could be varied among few standard axial stiffness values, whereas the ASTM model represented a limited case where stiffness is set to zero. A stepwise validation procedure was developed to ensure the reliability of the FE predictions for the experimental data (La Barbera et al., 2015; La Barbera and Villa, 2016).

Simulations demonstrated the primary role played by the anterior support stiffness and its position on the sagittal plane (namely, the anterior support arm, or ASA) over other anatomical parameters. Spring stiffness alone contributed a stress increase up to 360%, whereas BMA reached 10% (La Barbera et al., 2016a). A worst-case combination of anatomical parameters could lead to a maximum stress increase of 22.4% at L1, compared to the current ISO standard.

Analysis of the Suitability of Standards to Represent Daily Activities

In order to study the internal loads arising in the posterior implant during ISO and ASTM standard testing, an FE study was conducted (La Barbera et al., 2016b, 2017). Results were compared to the predictions obtained on an average L2-L4 spine segment that, through corpectomy, had been instrumented with a posterior fixator, a physiological anterior support, and a rigid anterior cage (Fig. 3). Moreover, data were compared to published in vivo measurements of real patients performing everyday activities.

Numerical predictions were represented as a function of the vertical load and were compared to published in vivo measurements taken on patients bearing instrumented spinal fixators as they performed simple daily activities, as shown in Fig. 5 (Wilke et al., 2001; Rohlmann et al., 1994, 1997, 2000).

Results of all the simulations are presented on simple graphs that make it easy to compare the internal actions in the rods calculated with the FE models to the ones measured in the patients. These graphs can be used in many ways. Here are some examples that refer to the calculation of the bending moment arising in the rods:

(i) Case A in Fig. 6—To verify which daily activity produces stresses in the rods closer to the ones arising in the ISO in vitro experiments and when the suggested springs (red) and vertical load (200 N) are used. In this case, we found that the ISO setup reproduces quite well a 2-FSUs stabilization of a real patient, as well as of an average lumbar spine, undergoing flexion of the upper body.

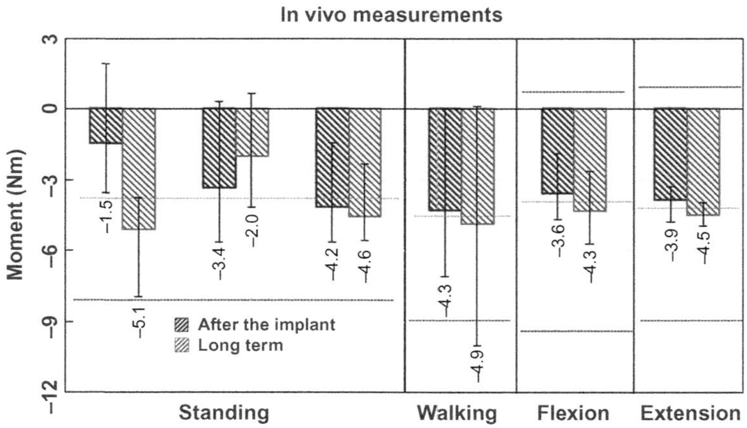

FIG. 5

Results from direct in vivo measurements performed by Wilke et al. (2001) and Rohlmann et al. (1994, 1997, 2000).

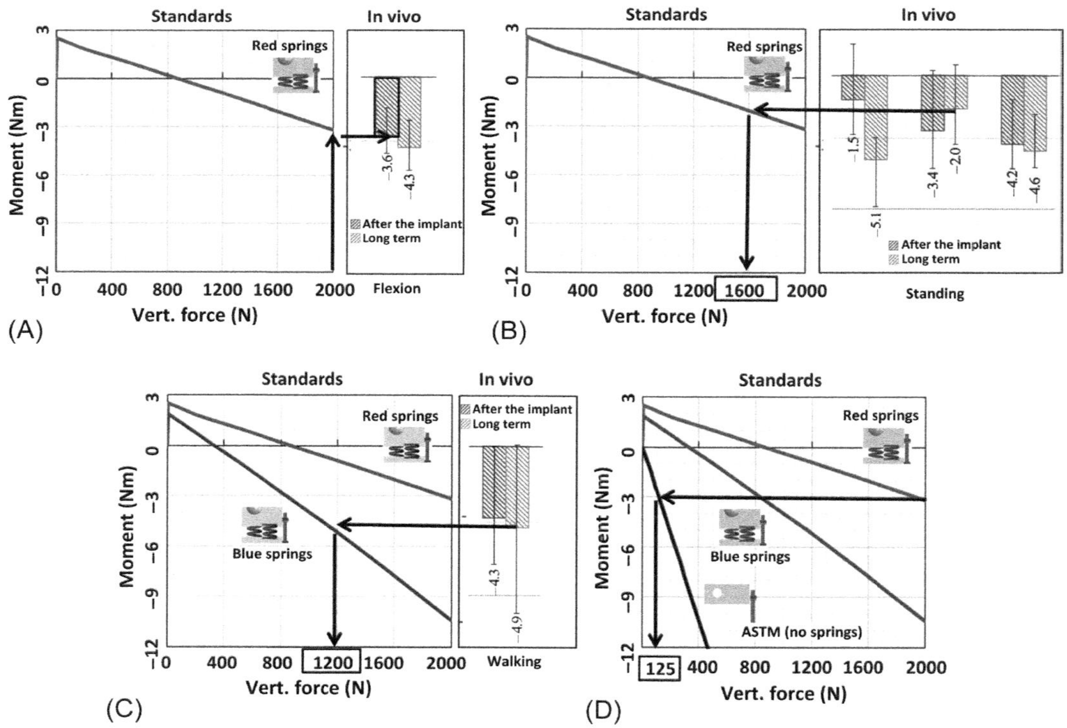

FIG. 6

Results of Case (A), case (B), case (C), and case (D). Please refer to the text for the complete explanation of the four cases.

(ii) Case B in Fig. 6—To identify which vertical load has to be applied, again using the suggested red springs, for example, when standing is the activity to be simulated. In this case, the suggested value for the vertical load is about 1600 N.

(iii) Case C in Fig. 6—To identify which typology of springs must be used when we want to simulate a walking patient. In this case, it was not possible to find a value for the vertical load to be applied to a construct that used red springs, and blue springs can be adopted for a 1200 N vertical load.

(iv) Case D in Fig. 6—To identify which load to apply to an ASTM construct that gives the same stress state as the ISO one when standard proposals are used (red springs, 2000 N vertical load), or to associate the state of stress deriving from the application of a vertical load to an ASTM setup to a particular daily activity. This result is particularly useful for comparison purposes, because results of fatigue testing of predicate devices are often available only according to the ASTM standard. Such a testing method is not representative of a relevant clinical condition, because vertebrectomy is generally associated with the presence of an anterior support. Thus the graph allows matching these available results to real clinical data measured during the daily activities of representative patients.

IMPLEMENTATION OF A NEW STANDARD PROCEDURE: TESTS FOR EVALUATION OF SCREW LOOSENING

Mechanical failure of a posterior spinal fixation device happens from time to time. Cases of broken screws or rods occur, and in other cases, the polyaxial head of the pedicle screws or the set screws loosen. Therefore testing according to several standard procedures is required before a fixation device can enter the marketplace. Whereas testing according to ASTM F1717 includes the whole fixation system, ASTM F1798 focuses on subassemblies, such as one rod plus one screw to which the rod is attached. ASTM F2193 finally describes how to test single components of a fixation device. A single screw, for example, can be tested for its bending characteristics. None of these standards, however, take into account anchorage of the screw onto the vertebral bone. From a clinical point of view, this anchorage plays an important role because screw loosening is a common clinical complication requiring surgical intervention (Galbusera et al., 2015). To overcome this clinical complication, new and innovative pedicle screws have been developed. Some screws are expandable; others are equipped with innovative thread designs or are cannulated for injection of bone cement or other materials. The only standardized possibility to investigate the effectiveness of the anchorage between a screw and a bone is to conduct a pull-out test according to ASTM F543 (Fig. 7).

According to this standard, the screw is drilled into a bone substitute, usually polyurethane foam, and then pulled out. In vivo, however, a screw not only experiences pull-out loads. Actually, a screw that is part of a posterior spinal fixation device also experiences repetitive bending and shear loading. This loading condition is not reflected in the standard pull-out test (Sandén, 2010; Costa et al., 2013; Kueny et al., 2014). Therefore, in the past, some researchers have published more physiological test setups (Zindrick et al., 1986; Law et al., 1993; Tan et al., 2004; Kueny et al., 2014). Some of these setups are called toggle-tests. In this type of test, the screw, which is implanted into a cadaveric vertebra, is loaded cyclically in cranio-caudal direction. This loading condition is already much closer to reality than a static pull-out test. However, standardization is difficult as long as testing is carried out by using cadaveric specimens. Cadaveric specimens are scarce, and the available ones are mostly from elderly

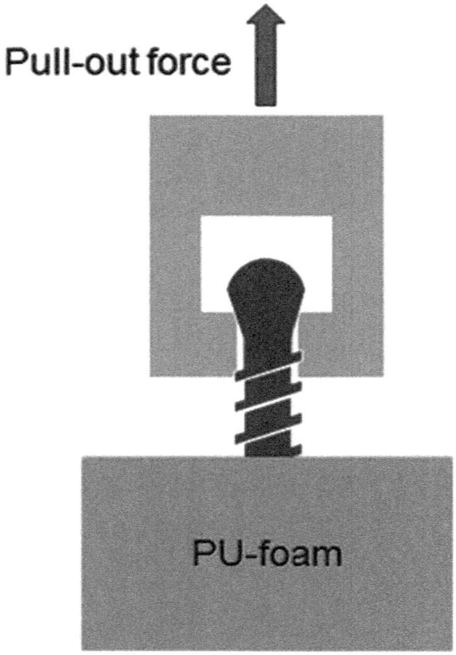

FIG. 7

Sketch of the pull-out setup according to ASTM F543. The screw is drilled into a vertebral body substitute (polyurethane foam) and then pulled out by application of a continuously increasing load.

donors, and they usually do not represent the age of a typical patient. A standard test method, however, needs to be reproducible and conductible under the same conditions. For that reason, a mechanical test setup needed to be developed that does the following:

- Allows testing of various screw designs and anchorage systems.
- Mimics the in vivo loading situation at the screw-bone interface as closely as possible
- Is validated.
- Can be conducted easily.
- Can be conducted on any uniaxial materials testing machine.
- Is independent of the testing laboratory to the greatest possible extent.
- Can be conducted economically.

Based on these requirements, the new setup (Fig. 8) consists of a vertebral body substitute made of polyurethane foam (Open Cell Block 15 PCF, Sawbones, representing healthy cancellous bone). Additionally, a pedicle substitute made of carbon fiber-reinforced epoxy resin was designed. Its inner and outer geometry was adapted to those of the human lumbar pedicles (Zindrick et al., 1987). This pedicle substitute was needed to allow the pedicle to deform elastically as it presumably does in vivo. The vertebral body and pedicle substitutes were rigidly clamped to each other. The screw was drilled through the substitute pedicle into the substitute vertebra.

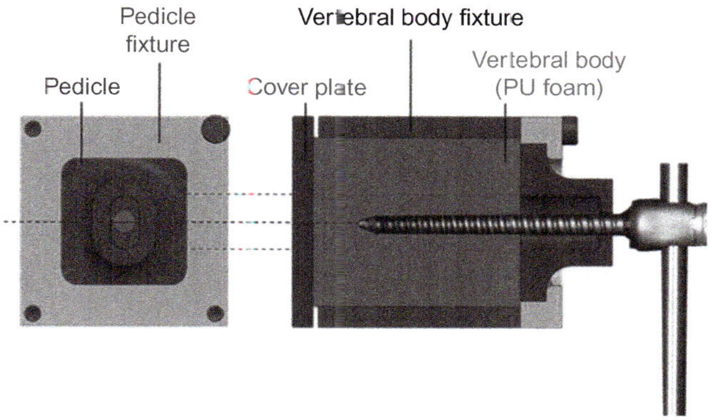

FIG. 8

Sketch of the vertebral body substitute made of polyurethane foam, which is clamped to a pedicle substitute made of a carbon fiber-reinforced epoxy resin.

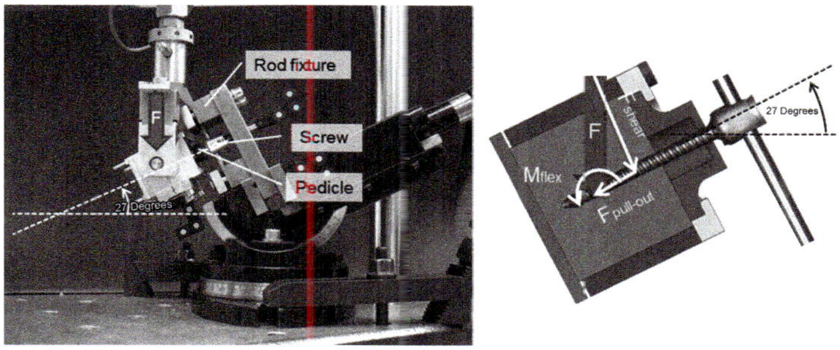

FIG. 9

New test setup, which incorporates shear forces, bending moments, and pull-out forces. The relationship between these three loading components is based on the literature.

Then a rod was fixed to the head of the screw. The rod was rigidly fixed to the bottom plate of the materials testing machine under an inclination of 27 degrees (Fig. 9). This inclination had been used in other standard testing procedures (ASTM F2077) and was selected to generate a physiological relationship between shear load, bending moment, and pull-out force (Rohlmann et al., 1994, 2000).

The load was applied to the vertebral body substitute using hinge pins around which the vertebral body was allowed to rotate during testing.

The validation of the vertebral body and pedicle substitutes was based on a comparative study. Six screws (S4, Aesculap) were tested using the new substitutes in the new test setup as previously described, and another six screws were tested using the new test setup up but with cadaveric vertebral bodies (mean age 58 ± 7 years, mean bone mineral density 75.4 ± 4.0 mg/ccm, 4 × male, 2 × female) instead of the vertebral body and pedicle substitutes. The load was increased stepwise every 50,000 loading cycles until significant loosening was observed. The first load step consisted of a 100 N inferior shear load, a 50 N pull-out load, and a 3.8 Nm flexion moment.

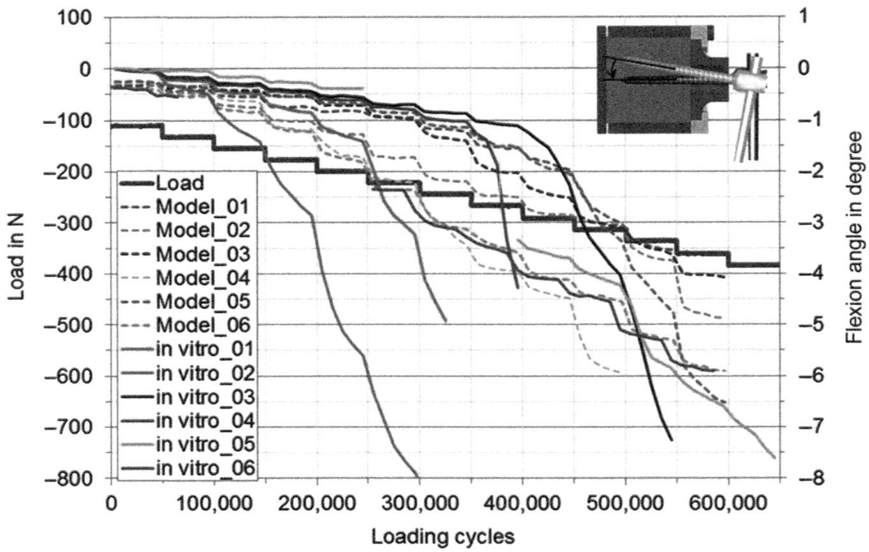

FIG. 10

Flexion angle between screw and vertebra at each load level for each of the six test runs with the vertebral body and pedicle substitutes and with the cadaveric specimens.

The results showed that there was less variability between the displacement curves of six test runs in the group with the vertebral body and pedicle substitutes (Fig. 10).

Also, the displacement in these six test runs was within the range of the displacement of the six test runs with the cadaveric specimens. However, a detailed analysis of the relative motions between the screw and the vertebra revealed that the screw never loosened at its entry point into the pedicle in the substitutes, whereas it sometimes loosened in the cadaveric specimens.

In conclusion, the new vertebral body and pedicle substitutes allow more standardized and reproducible testing. In combination with the new setup, which allows application of shear, bending and pull-out loads at the same time, the substitutes may have the potential to become an easily conductible and reproducible testing procedure for investigation of the anchorage between screw and vertebra. In the future, further comparative tests with the various new screw designs will be needed to prove their general usability.

CONCLUSIONS/FUTURE TRENDS

Notified bodies request standards during experimental preclinical testing of a spinal device.

Despite simplifications of the real in vivo conditions are necessary to improve comparability and repeatability of each standard method, specific aspects deserve a much deeper analysis. Standards should be continuously questioned, and an increasing awareness needs to be developed in order to have clinically relevant testing procedures that are really predictive of in vivo performances.

REFERENCES

Costa, F., Villa, T., Anasetti, F., Tomei, M., Ortolina, A., Cardia, A., La Barbera, L., Fornari, M., Galbusera, F., 2013. Primary stability of pedicle screws depends on the screw positioning and alignment. Spine J. 13 (12), 1934–1939. https://doi.org/10.1016/j.spinee.2013.03.046.

Galbusera, F., Volkheimer, D., Reitmaier, S., et al., 2015. Pedicle screw loosening: a clinically relevant complication? Eur. Spine J. 24 (5), 1005–1016.

Kueny, R.A., Kolb, J.P., Lehmann, W., et al., 2014. Influence of the screw augmentation technique and a diameter increase on pedicle screw fixation in the osteoporotic spine: pullout versus fatigue testing. Eur. Spine J. 23 (10), 2196–2202.

La Barbera, L., Villa, T., 2016. ISO 12189 standard for the preclinical evaluation of posterior spinal stabilization devices—I: assembly procedure and validation. Proc Inst Mech Eng H 230 (2), 122–133. https://doi.org/10.1177/0954411915621587.

La Barbera, L., Villa, T., 2017. Toward the definition of a new worst-case paradigm for the preclinical evaluation of posterior spine stabilization devices. Proc Inst Mech Eng H. 231 (2), 176–185. https://doi.org/10.1177/0954411916684365.

La Barbera, L., Galbusera, F., Villa, T., Wilke, H.-J., 2014. ASTM F1717 standard for the preclinical evaluation of posterior spinal fixators: can we improve it? Proc Inst Mech Eng H 228 (10), 1014–1026. https://doi.org/10.1177/0954411914554244.

La Barbera, L., Ottardi, C., Villa, T., 2015. Comparative analysis of international standards for the fatigue testing of posterior spinal fixation systems: the importance of preload in ISO 12189. Spine J. 15 (10), 2290–2296. https://doi.org/10.1016/j.spinee.2015.07.461.

La Barbera, L., Costa, F., Villa, T., 2016a. ISO 12189 standard for the preclinical evaluation of posterior spinal stabilization devices—II: a parametric comparative study. Proc Inst Mech Eng H 230 (2), 134–144. https://doi.org/10.1177/0954411915621588.

La Barbera, L., Galbusera, F., Wilke, H.-J., Villa, T., 2016b. Preclinical evaluation of posterior spine stabilization devices: can the current standards represent basic everyday life activities? Eur. Spine J. 25 (9), 2909–2918. https://doi.org/10.1007/s00586-016-4622-1.

La Barbera, L., Galbusera, F., Wilke, H.J., Villa, T., 2017. Preclinical evaluation of posterior spine stabilization devices: can we compare in vitro and in vivo loads on the instrumentation? Eur. Spine J. 26 (1), 200–209. https://doi.org/10.1007/s00586-016-4766-z.

Law, M., Tencer, A.F., Anderson, P.A., 1993. Caudo-Cephalad loading of pedicle screws: mechanisms of loosening and methods of augmentaion. Spine 18 (16), 2438–2443.

Rohlmann, A., Bergmann, G., Graichen, F., 1994. A spinal fixation device for in vivo load measurement. J. Biomech. 27 (7), 961–967.

Rohlmann, A., Bergmann, G., Graichen, F., Weber, U., 1997. Comparison of loads on internal spinal fixation devices measured in vitro and in vivo. Med. Eng. Phys. 19 (6), 539–546.

Rohlmann, A., Graichen, F., Weber, U., et al., 2000. Monitoring in vivo implant loads with a telemeterized internal spinal fixation device. Spine 25 (23), 2981–2986.

Sandén, B., Olerud, C., Larsson, S., et al., 2010. Insertion torque is not a good predictor of pedicle screw loosening after spinal instrumentation: a prospective study in 8 patients. Patient Saf. Surg. 4 (1), 14.

Tan, J.S., Kwon, B.K., Dvorak, M.F., et al., 2004. Pedicle screw motion in the osteoporotic spine after augmentation with laminar hooks, sublaminar wires, or calcium phosphate cement: a comparative analysis. Spine 29 (16), 1723–1730.

Villa, T., La Barbera, L., Galbusera, F., 2014a. Comparative analysis of international standards for the fatigue testing of posterior spinal fixation systems. Spine J. 14 (4) 695–704. https://doi.org/10.1016/j.spinee.2013.08.032.

Villa, T., La Barbera, L., Galbusera, F., 2014b. Reply to the letter to the editor entitled: response to "Comparative analysis of international standards for the fatigue testing of posterior spinal fixation systems". Spine J. 14 (12), 3068. https://doi.org/10.1016/j.spinee.2014.08.008

Wilke, H.J., Rohlmann, A., Neller, S., Schultheiss, M., Bergmann, G., Graichen, F., Claes, L., 2001. Is it possible to simulate physiologic loading conditions by applying pure moments? A comparison of in vivo and in vitro load components in an internal fixator. Spine 26 (6), 636–642.
Zindrick, M.R., Wiltse, L.L., Widell, E.H., et al., 1986. A biomechanical study of intrapeduncular screw fixation in the lumbosacral spine. Clin. Orthop. Relat. Res. (203)99–112.
Zindrick, M.R., Wiltse, L.L., Doornik, A., et al., 1987. Analysis of the morphometric characteristics of the thoracic and lumbar pedicles. Spine 12 (2), 160–166.

CHAPTER

MATHEMATICAL AND FINITE ELEMENT MODELING

14

Fabio Galbusera*, Frank Niemeyer†

IRCCS Galeazzi Orthopedic Institute, Milan, Italy Institute of Orthopaedic Research and Biomechanics, Ulm University, Ulm, Germany†*

HISTORICAL PERSPECTIVE

In the early days, biomechanics—like other fields of science—was limited mostly to a qualitative description of foundational concepts that were typically identified intuitively. For example, Galilei noticed the hollow bones of birds and explained his observations with the need for being lightweight and providing sufficient mechanical strength at the same time (Humphrey, 2003; Dumont, 2010). Descartes described the animal body as a machine governed by simple mechanical laws, while such laws were still unknown (Humphrey, 2003). Describing the statics and motion of animals and humans in mathematical terms required tools and methods that became available only after the publication of the laws of motion by Newton in 1687, their generalization by Euler in 1750, and other foundational studies such as Hooke's works about elasticity. A lecture by Young, published in 1809, about the physiology of the heart and arteries is an early, valuable example of this new research methodology of combining purposely designed experiments with a theoretical framework (Young, 1809). Theoretical biomechanical studies went hand in hand with the development of differential calculus and continuum mechanics. Indeed, the theory of partial differential equations emerged as a powerful tool to describe all kinds of physical phenomena in a multidimensional space, including those relevant for biomechanics such as elasticity. Mathematical models of the biomechanics of anatomical structures based on the equations of continuum mechanics started to appear in the 20th century, but the lack of both numerical solution techniques and computational tools such as computers limited the applicability of these models to practical problems of clinical interest. Nevertheless, analytical techniques for solving the equations of continuum mechanics have been applied to biomechanical problems for decades. As a matter of fact, analytical models have the unique advantage of providing exact solutions of the equations in the whole modeling domain, and thus facilitate the investigation of the effect of parameters such as geometric and material properties. For example, in 1980 Broberg and Von Essen published a pioneering analytical model of the intervertebral disc featuring an incompressible nucleus pulposus and 11 fiber layers with realistic inclination (Broberg and von Essen, 1980). This study constitutes one of the foundations for the subsequent developments, including modern numerical models. The use of analytical techniques continued in the 1990s (McNally and Arridge, 1995; Misra and Samanta, 1988) and is still relevant today (Demers et al., 2016).

The development of the finite element method, which revolutionized the engineering world in the second half of the 20th century because of the method's capability of solving physical problems with complex geometries, material properties, and boundary conditions, was evolutionary and can be traced

Biomechanics of the Spine. https://doi.org/10.1016/B978-0-12-812851-0.00014-8

back to the works of scientists such as Ritz and Galerkin at the beginning of the century, who proposed approximate methods for the solution of partial differential equations (Ritz, 1909; Milne and Milne, 1953). The popularity of the method for solving biomechanical problems constantly increased as computers improved and became more accessible to the scientific community. Today personal computers can be used to solve finite element analyses, which was unfeasible even on mainframes a few decades ago, and powerful open-source codes are available to everyone on the Internet. Together with in vitro testing and in vivo studies, finite element modeling has become one the pillars of biomechanical research.

THE ROLE OF THE FINITE ELEMENT METHOD IN BIOMECHANICS

Soon after its first applications in structural, civil, and aerospace engineering, the finite element method's ability to deal with realistic geometries and to simulate complex material constitutive laws as well as arbitrarily complicated loading and boundary conditions made it an obvious choice for the investigation of biomechanical problems, including anatomical structures and implants (Gilbertson et al., 1995). Finite element models provide access to quantities, which are hard or even impossible to be determined experimentally, such as local stresses and strains. Modern finite element programs provide a user-friendly graphical interface able to support both preprocessing and postprocessing of models, and the programs are often integrated with CAD software, facilitating the development and the virtual testing of implants.

Finite element modeling has constantly increased in popularity and is now arguably the most commonly employed investigation technique in biomechanics. In fact, the use of alternative or complementary methods, such as in vitro experimental studies and in vivo studies on animals or patients, have high barriers such as the cost of specimens and animals; the need for specific facilities and testing machinery; issues related to the acquisition, management, and disposal of potentially infected tissues; ethical restrictions; and so on. Although finite element modeling is less affected by such barriers, the resources required for the creation of a valid model are often underestimated (Gilbertson et al., 1995). Licenses for commercial software are expensive, especially when the finite element code is used for for-profit applications rather than for academic research. Nevertheless, valuable free and open-source software is currently available; FEBio, which was specifically designed for biomechanical applications and is free for noncommercial use, is worth mentioning in this respect (Maas et al., 2012). Furthermore, developing, validating, and solving a finite element model are very time-consuming tasks, are often underestimated, and require specific engineering skills and competences.

THE FINITE ELEMENT METHOD

A detailed description of the finite element method for the approximate solution of partial differential equations is beyond the scope of this book and can be found elsewhere (Zienkiewicz et al., 1977; Reddy, 1993). The basic principles of the method and its strengths and limitations when applied to biomechanical investigations are summarized below.

In order to perform numerical simulations using the finite element method, we first need come up with a mathematical description (i.e., a model), abstracting from physical reality. Many physical

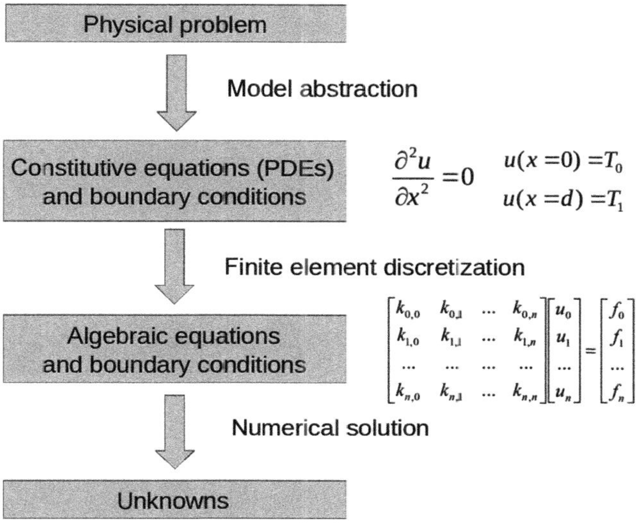

FIG. 1

Schematic representation of the creation of a mathematical model of a physical problem, its discretization with the finite element method, and its solutions.

phenomena, such as elasticity, fluid flow, and heat transfer, can be effectively modeled by a set of governing equations, typically partial differential equations, and boundary conditions (Fig. 1). The first historical example of this process of model abstraction leading to a set of partial differential equations is Fourier's law, a simple partial differential equation first reported by Jean Baptiste Fourier in 1807 (Fourier, 1822), which accurately describes the conduction of heat in solids. The heat equation, derived from Fourier's law and the conservation of energy, links the temperature, which can be regarded as the "solution," to a heat source term, which can be associated with an external load. Fourier based his theory on the postulation that the temperature of a solid in a specific location was dependent only on the conditions in the locations immediately adjacent to it, that is, upstream or downstream, thereby leading straightforwardly to a differential formulation of the problem (Reddy and Gartling, 2010). To deal with the boundaries of the physical domain, Fourier's theoretical framework introduced the concept of "boundary conditions," which are assumed to be known a priori. Interestingly, the whole formulation did not require making any assumption about the nature of heat, which was heavily debated at that time. Initially, Fourier's work was not well received, mostly because of controversies about the solution of the differential equations rather than the governing equation themselves. Over time, however, Fourier's modeling approach gained acceptance and was increasingly employed and expanded to other physical phenomena such as electricity, fluid flow in porous media, diffusion of solutes, and finally continuum mechanics (Reddy and Gartling, 2010). Regarding the latter case, in the most common formulation of the governing equations, the point displacements (and rotations) act as the solution, whereas the mechanical forces (and moments) represent the external loads.

After a set of governing equations and boundary conditions has been established, the first step toward a finite element analysis is to reformulate the equations in a weak form (Zienkiewicz et al., 1977;

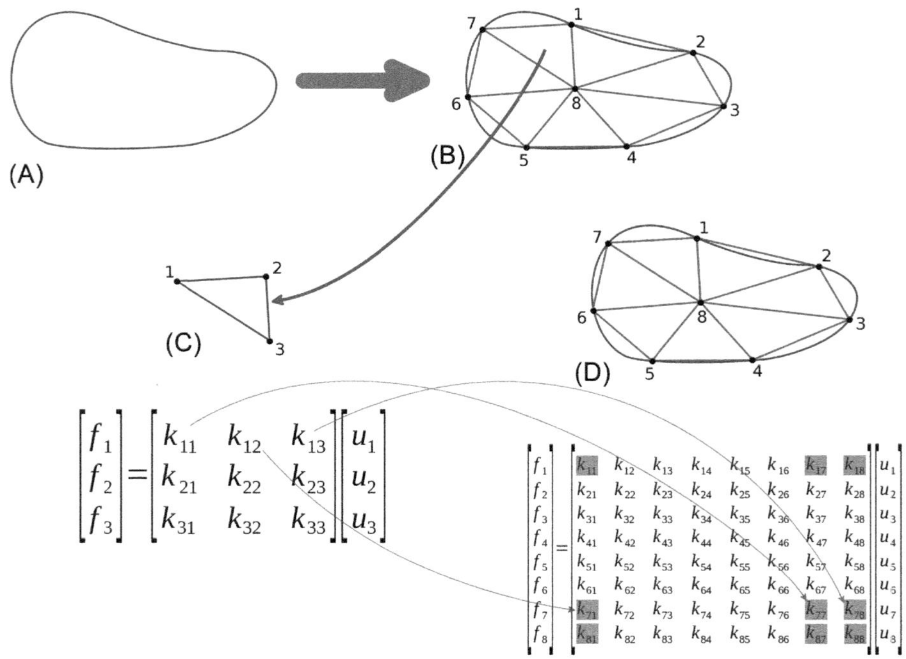

FIG. 2

Schematic representation of the calculation of the stiffness matrix of a structure. The computational domain (A) is discretized in a finite number of elements connected at their nodes (B). For each element, its stiffness matrix relating the loads to the unknowns is built (C). The matrices calculated for the elements are then assembled in the global stiffness matrix of the whole structure, which relates all loads and unknowns of all elements (D). Each item of the element stiffness matrix contributes to the correspondent item in the global matrix based on the node numbering and element connectivity. In the representative example shown in the figure, the local node 1 corresponds to the global node 7; the item k_{11} in the element stiffness matrix therefore contributes to the item k_{77} in the global matrix.

Courant, 1943), which is more suited to the subsequent step of discretization aimed at converting the differential equations to algebraic ones. By subdividing the spatial domain into a finite number of small elements, discretization reduces the dimensionality of the solution from infinite to finite (Fig. 2). In practical applications, elements have a triangular or quadrilateral shape in two-dimensional problems, whereas tetrahedrons, wedges, and hexahedrons are used in three-dimensional applications. Within each element, the exact (continuous) solution of the problem is approximated by a predefined shape function, which interpolates the solution between the vertices of the element, called *nodes*. At any point of the element, the value of the solution (also called *unknown* or *degree of freedom*) is therefore completely defined by the values at the element nodes and by the shape functions. The quality of the discretization thus depends on how well the shape functions are able to approximate the exact solution of the problem. In general, reducing the element size and using a more complex form of the shape functions, for example, quadratic versus linear polynomial, are effective in improving the accuracy of the numerical solution.

For each element, a stiffness matrix describes the relationship between the degrees of freedom and the external loads. This square and symmetrical stiffness matrix is based on the discretized form of the governing equations; its number of rows and columns correspond to the number of degrees of freedom in the element. The values of the stiffness matrix are calculated based on the material properties, that is, the thermal diffusivity for the heat equation or elastic properties (elastic moduli and Poisson's ratios) for Hooke's elasticity. The matrix product between the element stiffness matrix and the vector of the unknowns (i.e., displacements) at each node yields the vector of the nodal external loads (i.e., forces or moments).

Elements are connected to each other at nodes, where the values of the shape functions in the adjacent elements match, forming a finite element mesh. A global stiffness matrix, representing the whole model, is assembled from the stiffness matrices of each element, depending on their interconnectivity (Fig. 2). This global stiffness matrix mathematically links the vector of the nodal unknowns in the whole model to the vector of the external loads; therefore once assembled, the solution of the physical problem (i.e., the calculation of the values of the unknowns based on the values of the external loads) is merely a matter of solving an algebraic system of equations.

Before the algebraic system can be solved, however, the boundary conditions need to be incorporated (Reddy, 1993). Dirichlet boundary conditions force the solution to attain certain a priori prescribed values for a subset of the degrees of freedom in the model. This kind of boundary condition is the most commonly employed one in structural mechanics problems, for example, by assigning a fixed displacement to a set of nodes. Neumann boundary conditions are commonly used to control fluxes in thermal and fluid dynamics simulations and imply fixing the derivative of the unknown at specific nodes. Homogeneous Dirichlet boundary conditions, that is, the prescribed values of the degrees of freedom are zero, can be imposed by removing the corresponding rows and columns from the stiffness matrix. An alternative technique, which works also for nonhomogeneous boundary conditions in which a nonnull value of the unknown is imposed, is the penalty method. In this technique, the values on the main diagonal of the model stiffness matrix, which correspond to the degrees of freedom that need to be constrained, are multiplied by a large number called the *penalty factor*. If a nonhomogeneous boundary condition is imposed, the correspondent entry in the vector of the external loads is modified accordingly by adding the value of the prescribed solution multiplied by the penalty factor.

The resulting stiffness matrix of the whole model is called *sparse* because for each line only the entry on the main diagonal and those entries relevant to the adjacent degrees of freedom are nonnull. Solution methods designed specifically to efficiently deal with sparse matrices, either direct ones such as those based on LU decomposition or iterative ones such as the conjugate gradient method and its derivatives, have been developed over the years and are now commonly used in finite element software (Saad, 2003).

CREATING A FINITE ELEMENT MODEL

Several software tools, which can be used to create finite element models for investigating biomechanical research questions, such as the stabilization of the spine with instrumentation, are currently available both commercially and in the public domain. For some specific applications for which existing software is unsuitable, researchers may decide to develop new computer programs in-house, usually by taking advantage of existing software libraries as building blocks. Both low-level libraries designed to perform linear algebra operations (e.g., BLAS, http://www.netlib.org/blas/; LAPACK, http://www.netlib.org/lapack/; and ATLAS, http://math-atlas.sourceforge.net/) and high-level libraries dedicated to finite element computations (e.g., deal.II, Bangerth et al., 2007; and FreeFem++, Hecht, 2012) are widely used for such purposes. The majority of users, however, employ existing finite element programs, either commercial ones such as ABAQUS (Simulia, Dassault Systèmes, Johnston, Rhode Island) and ANSYS (ANSYS, Inc., Canonsburg, Pennsylvania) or those available under free licenses such as FEBio and Elmer (Lyly et al., 1999).

The first stage of the construction of a structural finite element model to be used in biomechanics is the creation of its geometry. For anatomical structures such as the spine, this step is typically based on the three-dimensional reconstruction of a stack of tomographic images such as CT or MRI scans (Kluess, 2010). Because of their better contrast and resolution, CT data are commonly preferred over MRI data for the reconstruction of bone. Several studies investigated the relationship between the local Hounsfield units available in CT scans and the mechanical properties of the bone tissue (Rho et al., 1995; Schreiber et al., 2011), thus providing a direct way to model local tissue properties in the finite element model. Various preprocessing software tools such as MIMICS (Materialize, Leuven, Belgium) and Bonemat (http://www.bonemat.org/) offer functionality for mapping gray values to local stiffness values. MRI scans provide valuable information about soft tissues such as intervertebral discs and ligaments, which have a relatively low and homogeneous density, resulting in a low contrast in CT imaging. Although CT and MRI scans of the same patients can be combined into a single data set by means of registration (Zitova and Flusser, 2003) in order to exploit the advantages of both techniques, the vast majority of available finite element models of the spine are built based on either CT or MRI scans.

In order to build three-dimensional models of the various anatomical structures, the images need to be segmented, that is, each voxel of the computational domain is assigned (labeled) to one of the considered structures (Pal and Pal, 1993) (Fig. 3). To this aim, various manual and automated tools are available in preprocessing software. *Thresholding* based on the voxel gray level is often the first step used for labeling bone tissue in CT images. Other common segmenting methods are *active contours* and *region growing*. Although research in this field is very active and a large number of sophisticated segmentation methods are currently available, a completely automated segmentation of spine CT or MRI scans is still

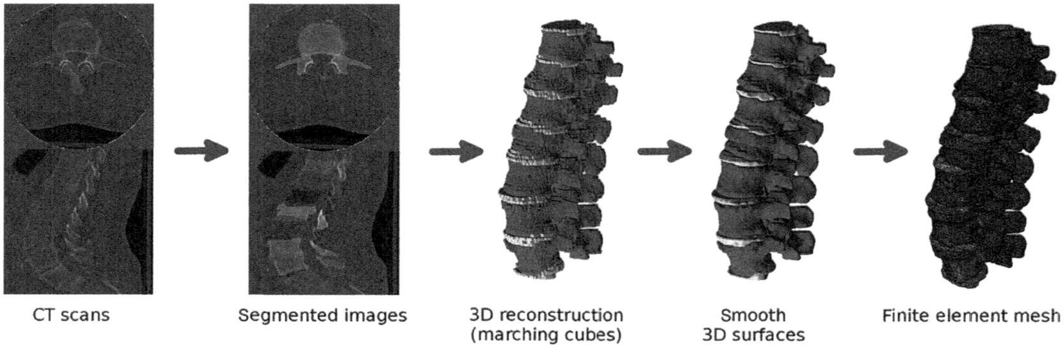

FIG. 3

Procedure to build a finite element model of an anatomical structure based on CT scans.

hardly feasible because of the complex morphology and strong interpatient variability of this anatomical structure; therefore image segmentation always requires some degree of user intervention.

Using computational methods such as the marching cubes algorithm, a segmented stack of images can then be transformed into a three-dimensional model consisting of triangulated surface meshes describing the boundaries of the anatomical structures (Lorensen and Cline, 1987) (Fig. 3). Smoothing filters can be used to remove artificial edges and sharp angles related to the finite voxel size in the CT/MRI scans.

Modeling artificial structures such as implants does not require the segmentation of images because descriptions of their geometries are generally available via CAD models. In contrast with reconstructed CT/MRI images, implant models typically do not consist of a polygonal mesh, but rather of nonuniform rational basis spline (NURBS) surfaces, which are defined analytically and are easier to manipulate and modify (Hughes et al., 2005). It should be noted that polygonal meshes can be converted to NURBS by means of specific software, but this step is not performed often. After proper positioning of the implant, virtual implantation is performed by means of Boolean operations, that is, by subtracting the implant volume from that representing the biological tissue.

Finally, a finite element mesh of the model can be created. In this step, elements and nodes discretizing the computational domain as described in the previous paragraphs are created. Three-dimensional tetrahedral meshes can be generated automatically by means of algorithms derived from Delaunay tetrahedralization, adapted to respect the domain boundaries (Chew, 1989). Tetrahedral meshes obtained by Delaunay methods are unstructured, that is, the connectivity between adjacent elements is irregular. On the other hand, the generation of structured meshes, that is, meshes featuring regular element connectivity, typically requires user intervention in order to achieve high-quality resulting elements, although (semi-)automated methods exist (Anderson and Wendt, 1995). Three-dimensional structured meshes usually consist of hexahedral elements instead of tetrahedral ones and provide computational advantages such as a more efficient memory usage and the possibility of employing algebraic solvers specifically designed to manage such structures. Conversely, unstructured meshes offer higher flexibility in dealing with complex geometries, which is typical for anatomical structures. Hybrid meshes including structured blocks connected to unstructured meshes are also used, but mostly for computational fluid dynamics rather than for structural problems.

Preprocessing software usually offers user-friendly graphical interfaces for assigning boundary and loading conditions to the model. Typically, either external surfaces, edges and vertices, or single nodes and elements can be selected with the mouse or from lists. Dirichlet boundary conditions such as prescribed displacements and external loads, either concentrated or distributed, can be imposed directly on the selected geometrical entities. Similar selection tools are also used to assign constitutive laws and material properties to the various components of the model, as well as to specify interaction properties such as contacts. It should be noted that many advanced users prefer to construct finite element models by directly editing the text files including the model definition whenever possible, rather than using the graphical user interfaces of preprocessing software.

VERIFICATION AND VALIDATION

Viceconti and colleagues efficaciously stated that "numerical analysis is easy to do poorly and very hard to do well" (Viceconti et al., 2005). Indeed, although a skilled user can set up and run a finite element model in a matter of hours, obtaining results that can be trusted for deriving clinically relevant

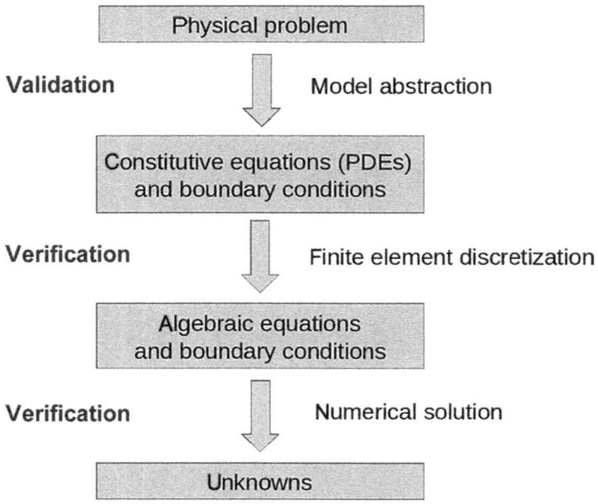

FIG. 4

Schematic representation of verification and validation of a finite element model. Validation ensures that the equations represent well the physical problem; verification aims at assuring that the equations are solved correctly.

conclusions is a very challenging and time-consuming task that may require several months of work. Verification and validation of the model are mandatory before the numerical predictions can be considered valid and therefore safe to interpret in clinical terms.

In synthesis, verification is about solving the equations correctly, whereas validation ensures that the right equations are solved (Viceconti et al., 2005; Roache, 1998) (Fig. 4). Verification is therefore aimed at ensuring that the numerical accuracy in solving the model equations is acceptable (Henninger et al., 2010). Assuming that the finite element code is free of bugs and reliable, in case of the solution of a linear problem verification basically consists of ensuring that the finite element discretization does not significantly affect the predicted result. To this aim, one can perform a mesh sensitivity analysis by running the model with different element sizes and comparing the predictions. A convergence analysis is a slightly more elaborate approach, where the mesh is iteratively refined until the difference between two subsequent simulations falls below some arbitrarily chosen error bound, indicating that the numerical approximation is sufficiently close to the unknown exact solution. In the case of nonlinear problems—that is, models including nonlinear material laws or nonlinear boundary conditions like contacts, or models taking large deformations into account—additional verification steps may be required.

Validation is typically more complex and is aimed at testing whether the mathematical model, independent of its discretization and numerical solution, is a valid representation of the physics of the real biomechanical problem (Viceconti et al., 2005; Henninger et al., 2010). The gold standard for model validation is direct validation (i.e., a quantitative comparison of the predictions to in vitro experiments in a controlled laboratory environment) specifically designed to represent the in vivo case, but with boundary and loading conditions that can be exactly replicated in the finite element model (Henninger et al., 2010). Indirect validation, on the other hand, is based on data, which cannot be controlled by the user, such as data retrieved from the literature or from clinical studies in which quality control, sources of errors, and degree of variability are not completely known. Although direct validation should be preferred whenever possible, high costs, lack of facilities, and the inability to measure the quantities of interest may make indirect validation an acceptable choice in some circumstances.

STRUCTURAL NUMERICAL MODELS OF THE SPINE

Currently, the literature provides information on hundreds of finite element models of the spine that have been used to predict the biomechanical response of the spine to different loading conditions, under both healthy and pathological conditions as well as after the virtual implantation of surgical devices. Several finite element models with particular relevance to specific topics are described in other chapters of this book; here we give only a brief overview from a historical perspective.

Several early models aimed to represent the whole thoracic or thoracolumbar spine, including the rib cage, in a simplified way by means of beam elements (Gilbertson et al., 1995; Roberts and Chen, 1970; Sundaram and Feng, 1977; Vanderby et al., 1986; Viviani et al., 1986). Such simplified solutions were also widely used in more recent years, because a detailed anatomical description of the spine may be deemed unnecessary for several applications, such as the study of spine deformities or sagittal balance (Stokes and Laible, 1990; Kiefer et al., 1997; Shirazi-Adl and Parnianpour, 1993; Galbusera et al., 2014).

One of the first attempts to model a functional spine unit with a higher degree of anatomical realism was published by Belytschko et al. (1974) (Fig. 5). The model was axisymmetric and included neither the posterior elements of the vertebrae nor any ligaments. Cortical and trabecular bone tissues were modeled as volumes with distinct material properties. Because orthotropic material properties taking into account the collagen fibers and their crisscross orientations were considered, the modeling approach for the intervertebral disc was rather sophisticated. The nucleus pulposus was modeled as an incompressible fluid. A validation against experimental results retrieved from the literature (Rolander, 1966) was provided, resulting in excellent agreement.

This work constituted the basis for further developments, which soon resulted in even more realistic three-dimensional models (Liu et al., 1975; Lin et al., 1978). Among the several research groups active in this field, Shirazi-Adl and colleagues are particularly worthy of mention. Starting in 1984 (Shirazi-adl et al., 1984, 1986), the group published several papers describing a model of a lumbar motion segment that is still a valuable reference today (Fig. 6). The model was fully three-dimensional and included posterior elements, ligaments, and articulating facet joints. To represent the annular fibers, the authors used nonlinear axial elements connecting the nodes of the solid hexahedral elements of the annulus, which were designed so that the resulting fiber angle matched the physiological fiber orientation. A comprehensive validation against literature in vitro data was also provided in terms of intradiscal pressure (Andersson and Schultz, 1979; Schultz et al., 1979), disc bulge (Brown et al., 1957), and ligament strain (Panjabi et al., 1982).

In more recent years, direct validation against in vitro experiments purposely designed and conducted was increasingly employed. A valuable example of such validation is reported in the studies of Schmidt, Heuer, and colleagues, who performed a stepwise reduction of lumbar motion segments to collect data describing the biomechanical role of each anatomical structure included in the model (Heuer et al., 2007a,b; Schmidt et al., 2006, 2007) (Fig. 7). The authors performed a comprehensive

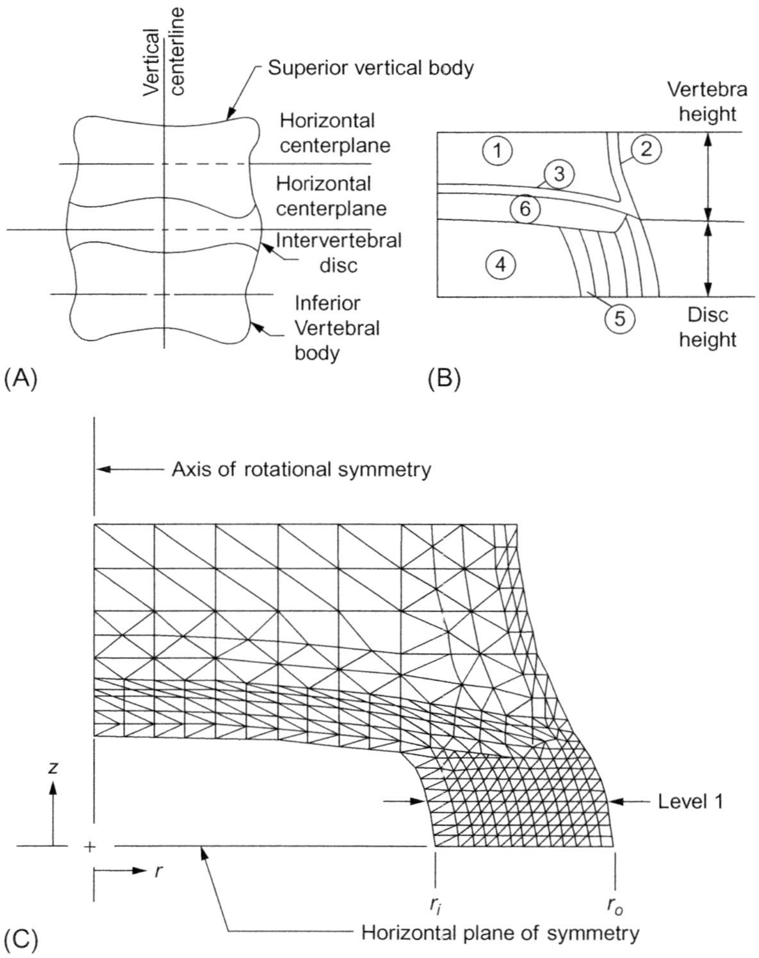

FIG. 5

Axisymmetric finite element model of a lumbar motion segment. (A) Sketch of the geometry of the motion segment. (B) Anatomical structures included in the model (1: trabecular bone; 2: cortical bone; 3: vertebral endplate; 4: nucleus pulposus; 5: annulus fibrosus; 6: cartilaginous endplate). (C) Finite element mesh.

Reproduced with permission from Belytschko, T., Kulak, R., Schultz. A., Galante, J., 1974. Finite element stress analysis of an intervertebral disc. J. Biomech. 7(3), 277–285.

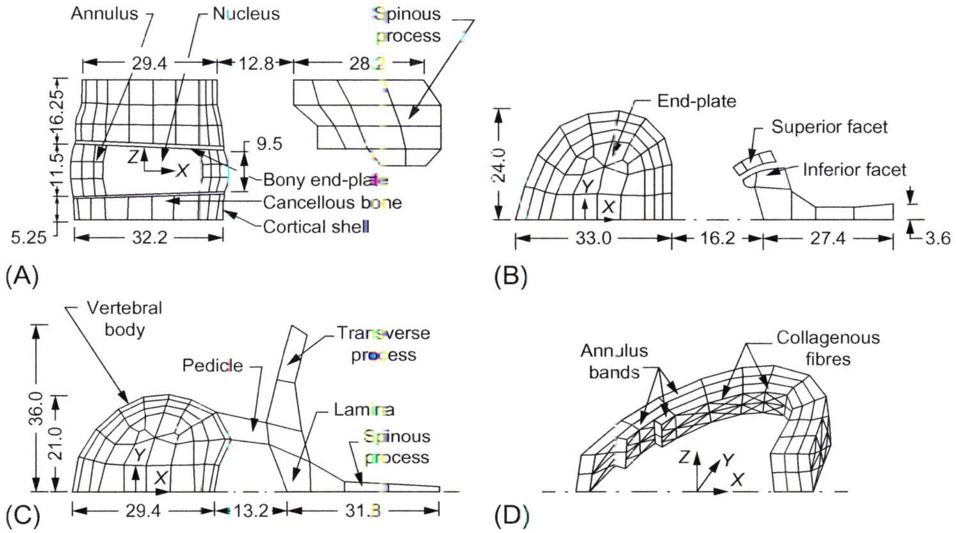

FIG. 6

Three-dimensional finite element model of a lumbar motion segment, including posterior elements, facet joints, and ligaments. (A) Sagittal view. (B) and (C) Axial views. (D) Modeling approach for the annulus fibrosus.

Reproduced with permission from Shirazi-Adl, A., Ahmed, A., Shrivastava, S., 1986. A finite element study of a lumbar motion segment subjected to pure sagittal plane moments. J. Biomech. 19(4) 331–350.

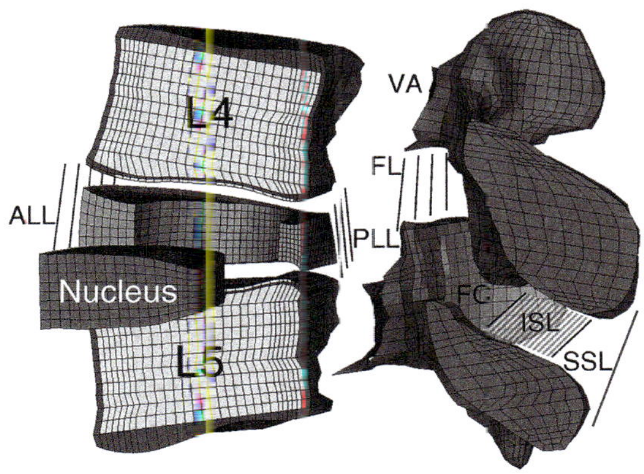

FIG. 7

Anatomically realistic three-dimensional finite element model of the L4-L5 motion segment.

Reproduced with permission from Schmidt, H., Heuer, F., Drumm, J., Klezl, Z., Claes, L., Wilke, H.J., 2007. Application of a calibration method provides more realistic results for a finite element model of a lumbar spinal segment. Clin. Biomech. 22(4), 377–384.

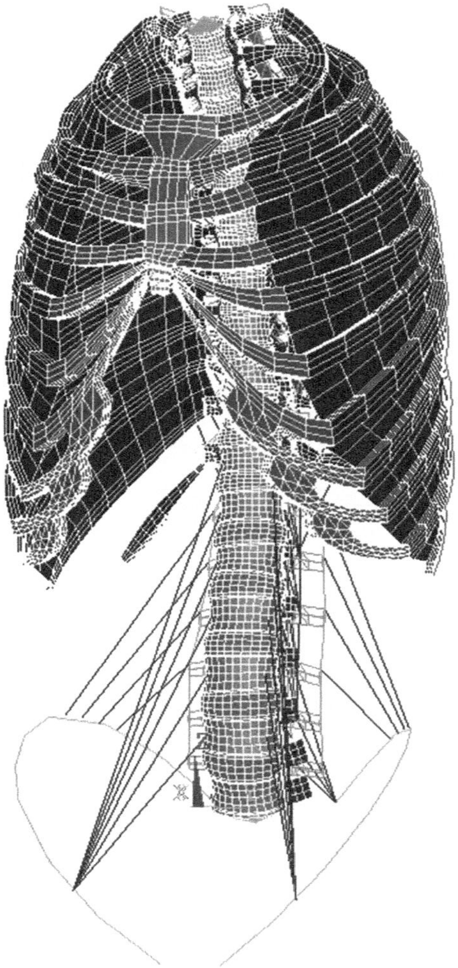

FIG. 8

Parametric finite element model of the thoracolumbar spine including the rib cage.

calibration of the model, which was shown to be able to provide realistic results also in defect conditions such as facetectomy and nucleotomy. This approach is now the foundation of several newer models that also consider longer spinal regions (Fig. 8), because in vitro flexibility data about stepwise reduction in the thoracic spine as well as in the rib cage is now available (Liebsch et al., 2017; Wilke et al., 2017).

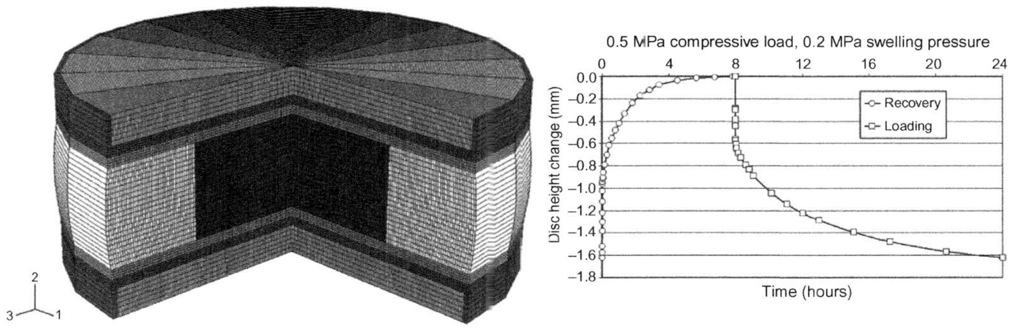

FIG. 9

Poroelastic finite element model of a lumbar intervertebral disc, which was used to investigate its response under cyclic loads and the transport of solutes.

Reproduced with permission from Ferguson, S.J., Ito, K., Nolte, L.P., 2004. Fluid flow and convective transport of solutes within the intervertebral disc. J. Biomech. 37(2), 213–221.

MULTIPHYSICAL NUMERICAL MODELS

Finite element modeling is increasingly employed to investigate not only the mechanical response of the spine but also other clinically relevant physical phenomena. One of the first multiphysical modeling applications in the spine field was the simulation of the intervertebral disc as a fluid-filled porous structure. In such models, the disc is modeled as a biphasic tissue in which a fluid permeates a porous solid matrix following the laws of Biot's poroelasticity model (Fig. 9). Several papers employed this approach, first by considering a simplified geometry of the disc and linear material properties (Laible et al., 1993; Simon et al., 1985), then by introducing a refined material model for the annulus (Shirazi-Adl, 1992), and finally by considering the full three-dimensional geometry of the motion segment and nonlinear strain-dependent tissue permeability (Argoubi and Shirazi-Adl, 1996). Poroelastic models have been widely used to investigate the creep behavior of the intervertebral disc (Argoubi and Shirazi-Adl, 1996; Ferguson et al., 2004) and its response under cyclic loading (Schmidt et al., 2010).

The availability of the fluid velocities as results of biphasic models was exploited in several papers in which the advection-diffusion of cell nutrients and waste products was investigated. Following pioneering papers studying the transport of solutes (Ferguson et al., 2004), more recent studies have modeled several coupled phenomena, such as the diffusion, advection, consumption, and production of several relevant substances, for example, oxygen, glucose, and lactate (Mokhbi Soukane et al., 2009; Soukane et al., 2007, 2005; Selard et al., 2003; Das et al., 2009). Cell viability criteria have also been introduced to investigate the effect of mechanical loading and its interaction with degenerative changes of the intervertebral disc, such as endplate calcification (Malandrino et al., 2011; Shirazi-Adl et al., 2010; Galbusera et al., 2013) (Fig. 10).

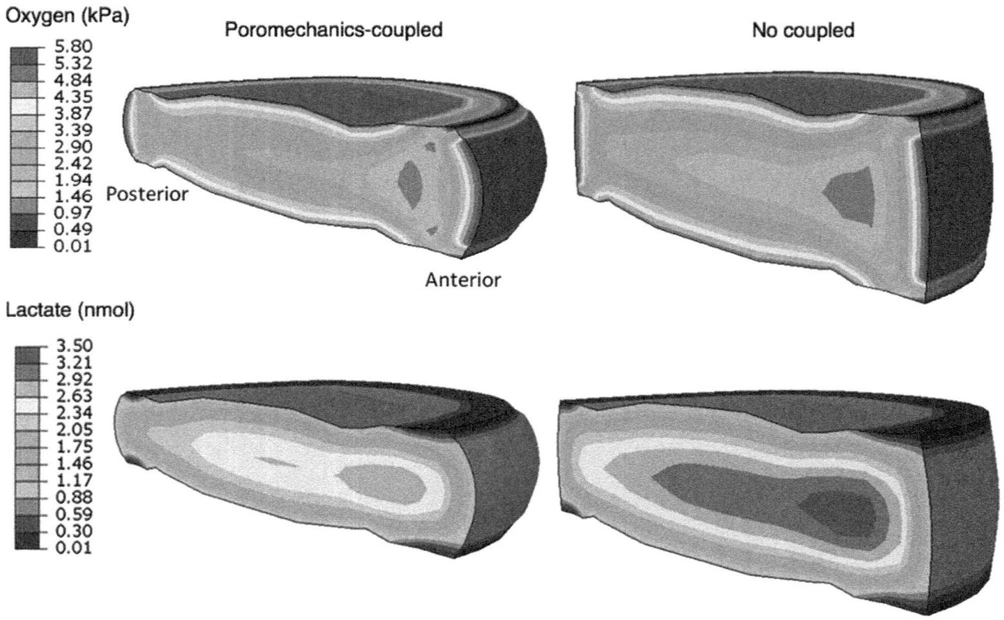

FIG. 10

Multiphysical model of a lumbar intervertebral disc, depicting the local concentrations of oxygen and lactate.

Reproduced from Malandrino, A., Noailly, J., Lacroix, D., 2011. The effect of sustained compression on oxygen metabolic transport in the intervertebral disc decreases with degenerative changes. PLoS Comput. Biol. 7(8), e1002112.

REFERENCES

Anderson, J.D., Wendt, J., 1995. Computational Fluid Dynamics. Springer, Heidelberg.

Andersson, G.B., Schultz, A.B., 1979. Effects of fluid injection on mechanical properties of intervertebral discs. J. Biomech. 12 (6), 453–458.

Argoubi, M., Shirazi-Adl, A., 1996. Poroelastic creep response analysis of a lumbar motion segment in compression. J. Biomech. 29 (10), 1331–1339.

Bangerth, W., Hartmann, R., Kanschat, G., 2007. Deal. II—a general-purpose object-oriented finite element library. ACM Trans. Math. Software (TOMS) 33 (4), 24.

Belytschko, T., Kulak, R., Schultz, A., Galante, J., 1974. Finite element stress analysis of an intervertebral disc. J. Biomech. 7 (3), 277–285.

Broberg, K.B., von Essen, H.O., 1980. Modeling of intervertebral discs. Spine 5 (2), 155–167.

Brown, T., Hansen, R.J., Yorra, A.J., 1957. Some mechanical tests on the lumbosacral spine with particular reference to the intervertebral discs: a preliminary report. JBJS 39 (5), 1135–1164.

Chew, L.P., 1989. Constrained delaunay triangulations. Algorithmica 4 (1–4), 97–108.

Courant, R., 1943. Variational methods for the solution of problems of equilibrium and vibrations. Bull. Am. Math. Soc. 49 (1), 1–23.

Das, D.B., Welling, A., Urban, J., Boubriak, O., 2009. Solute transport in intervertebral disc. Ann. N. Y. Acad. Sci. 1161 (1), 44–61.

Demers, S., Nadeau, S., Bouzid, A., 2016. Anisotropic multishell analytical modeling of an intervertebral disk subjected to axial compression. J. Biomech. Eng. 138 (4), 041004.
Dumont, E.R., 2010. Bone density and the lightweight skeletons of birds. Proc. Biol. Sci. 277 (1691), 2193–2198.
Ferguson, S.J., Ito, K., Nolte, L.P., 2004. Fluid flow and convective transport of solutes within the intervertebral disc. J. Biomech. 37 (2), 213–221.
Fourier, J., 1822. Theorie analytique de la chaleur. Chez Firmin Didot, père et fils, Paris, France.
Galbusera, F., Mietsch, A., Schmidt, H., Wilke, H.J., Neidlinger-Wilke, C., 2013. Effect of intervertebral disc degeneration on disc cell viability: a numerical investigation. Comput. Methods Biomech. Biomed. Engin. 16 (3), 328–337.
Galbusera, F., Brayda-Bruno, M., Costa, F., Wilke, H.J., 2014. Numerical evaluation of the correlation between the normal variation in the sagittal alignment of the lumbar spine and the spinal loads. J. Orthop. Res. 32 (4), 537–544.
Gilbertson, L.G., Goel, V.K., Kong, W.Z., Clausen, J.D., 1995. Finite element methods in spine biomechanics research. Crit. Rev. Biomed. Eng. 23 (5–6), 411–473.
Hecht, F., 2012. New development in FreeFem. J. Numer. Math. 20 (3–4), 251–266.
Henninger, H.B., Reese, S.P., Anderson, A.E., Weiss, J.A., 2010. Validation of computational models in biomechanics. Proc. Inst. Mech. Eng. H J. Eng. Med. 224 (7), 801–812.
Heuer, F., Schmidt, H., Claes, L., Wilke, H.J., 2007a. Stepwise reduction of functional spinal structures increase vertebral translation and intradiscal pressure. J. Biomech. 40 (4), 795–803.
Heuer, F., Schmidt, H., Klezl, Z., Claes, L., Wilke, H.J., 2007b. Stepwise reduction of functional spinal structures increase range of motion and change lordosis angle. J. Biomech. 40 (2), 271–280.
Hughes, T.J., Cottrell, J.A., Bazilevs, Y., 2005. Isogeometric analysis: CAD, finite elements, NURBS, exact geometry and mesh refinement. Comput. Methods Appl. Mech. Eng. 194 (39), 4135–4195.
Humphrey, J.D., 2003. Continuum biomechanics of soft biological tissues. Proc. Roy. Soc. Lond. A 459 (2029), 3–46.
Kiefer, A., Parnianpour, M., Shirazi-Adl, A., 1997. Stability of the human spine in neutral postures. Eur. Spine J. 6 (1), 45–53.
Kluess, D., 2010. Finite element analysis in orthopaedic biomechanics. In: Moratal, D. (Ed.), Finite Element Analysis. InTech, London.
Laible, J.P., Pflaster, D.S., Krag, M.H., Simon, B.R., Haugh, L.D., 1993. A poroelastic-swelling finite element model with application to the intervertebral disc. Spine 18 (5), 659–670.
Liebsch, C., Graf, N., Appelt, K., Wilke, H., 2017. The rib cage stabilizes the human thoracic spine: an in vitro study using stepwise reduction of rib cage structures. PLoS One 12 (6), e0178733.
Lin, H., Lui, Y.K., Ray, G., Nikravesh, P., 1978. Systems identification for material properties of the intervertebral joint. J. Biomech. 11 (1–2), 15–314.
Liu, Y.K., Ray, G., Hirsch, C., 1975. The resistance of the lumbar spine to direct shear. Orthop. Clin. North Am. 6 (1), 33–49.
Lorensen, W.E., Cline, H.E., 1987. Marching Cubes: A High Resolution 3D Surface Construction Algorithm, *SIGGRAPH '87*. ACM, New York, pp. 163–169.
Lyly, M., Ruokolainen, J., Järvinen, E., 1999. ELMER–a finite element solver for multiphysics. CSC-Report on Scientific Computing 2000, 156–159.
Maas, S.A., Ellis, B.J., Ateshian, G.A., Weiss, J.A., 2012. FEBio: finite elements for biomechanics. J. Biomech. Eng. 134 (1), 011005.
Malandrino, A., Noailly, J., Lacroix, D., 2011. The effect of sustained compression on oxygen metabolic transport in the intervertebral disc decreases with degenerative changes. PLoS Comput. Biol. 7 (8), e1002112.
McNally, D., Arridge, R., 1995. An analytical model of intervertebral disc mechanics. J. Biomech. 28 (1), 53–68.
Milne, W.E., Milne, W., 1953. Numerical solution of differential equations. Bull. Am. Math. Soc. 59 (6), 577–579.

Misra, J., Samanta, S., 1988. A mathematical model for the biomechanics of the body of a vertebra. Comput. Math. Appl. 15 (2), 85–96.

Mokhbi Soukane, D., Shirazi-Adl, A., Urban, J.P., 2009. Investigation of solute concentrations in a 3D model of intervertebral disc. Eur. Spine J. 18 (2), 254–262.

Pal, N.R., Pal, S.K., 1993. A review on image segmentation techniques. Pattern Recogn. 26 (9), 1277–1294.

Panjabi, M.M., Goel, V.K., Takata, K., 1982. Physiologic strains in the lumbar spinal ligaments: an in vitro biomechanical study. Spine 7 (3), 192–203.

Reddy, J.N., 1993. An Introduction to the Finite Element Method. McGraw-Hill, New York.

Reddy, J.N., Gartling, D.K., 2010. The Finite Element Method in Heat Transfer and Fluid Dynamics. CRC Press, Boca Raton.

Rho, J., Hobatho, M., Ashman, R., 1995. Relations of mechanical properties to density and CT numbers in human bone. Med. Eng. Phys. 17 (5), 347–355.

Ritz, W., 1909. Über eine neue Methode zur Lösung gewisser Variationsprobleme der mathematischen Physik. Journal für die reine und angewandte Mathematik (135), 1–61.

Roache, P.J., 1998. Verification and Validation in Computational Science and Engineering. Hermosa, Albuquerque.

Roberts, S., Chen, P., 1970. Elastostatic analysis of the human thoracic skeleton. J. Biomech. 3 (6), 527–545.

Rolander, S.D., 1966. Motion of the lumbar spine with special reference to the stabilizing effect of posterior fusion: an experimental study on autopsy specimens. Acta Orthop. Scand. 37 (sup90), 1–144.

Saad, Y., 2003. Iterative Methods for Sparse Linear Systems. SIAM, Philadelphia.

Schmidt, H., Heuer, F., Simon, U., Kettler, A., Rohlmann, A., Claes, L., Wilke, H.J., 2006. Application of a new calibration method for a three-dimensional finite element model of a human lumbar annulus fibrosus. Clin. Biomech. (Bristol, Avon) 21 (4), 337–344.

Schmidt, H., Heuer, F., Drumm, J., Klezl, Z., Claes, L., Wilke, H.J., 2007. Application of a calibration method provides more realistic results for a finite element model of a lumbar spinal segment. Clin. Biomech. (Bristol, Avon) 22 (4), 377–384.

Schmidt, H., Shirazi-Adl, A., Galbusera, F., Wilke, H.J., 2010. Response analysis of the lumbar spine during regular daily activities—a finite element analysis. J. Biomech. 43 (10), 1849–1856.

Schreiber, J.J., Anderson, P.A., Rosas, H.G., Buchholz, A.L., Au, A.G., 2011. Hounsfield units for assessing bone mineral density and strength: a tool for osteoporosis management. J. Bone Joint Surg. Am. 93 (11), 1057–1063.

Schultz, A., Warwick, D., Berkson, M., Nachemson, A., 1979. Mechanical properties of human lumbar spine motion segments. J. Biomech. Eng. 101, 46–52.

Selard, E., Shirazi-Adl, A., Urban, J.P., 2003. Finite element study of nutrient diffusion in the human intervertebral disc. Spine 28 (17), 1945–1953. discussion 1953.

Shirazi-Adl, A., 1992. Finite-element simulation of changes in the fluid content of human lumbar discs. Mechanical and clinical implications. Spine 17 (2), 206–212.

Shirazi-Adl, A., Parnianpour, M., 1993. Nonlinear response analysis of the human ligamentous lumbar spine in compression. On mechanisms affecting the postural stability. Spine 18 (1), 147–158.

Shirazi-adl, S.A., Shrivastava, S.C., Ahmed, A.M., 1984. Stress analysis of the lumbar disc-body unit in compression a three-dimensional nonlinear finite element study. Spine 9 (2), 120–134.

Shirazi-Adl, A., Ahmed, A., Shrivastava, S., 1986. A finite element study of a lumbar motion segment subjected to pure sagittal plane moments. J. Biomech. 19 (4), 331–350.

Shirazi-Adl, A., Taheri, M., Urban, J.P., 2010. Analysis of cell viability in intervertebral disc: effect of endplate permeability on cell population. J. Biomech. 43 (7), 1330–1336.

Simon, B., Wu, J., Carlton, M., Evans, J., Kazarian, L., 1985. Structural models for human spinal motion segments based on a poroelastic view of the intervertebral disk. J. Biomech. Eng. 107 (4), 327–335.

Soukane, D.M., Shirazi-Adl, A., Urban, J.P., 2005. Analysis of nonlinear coupled diffusion of oxygen and lactic acid in intervertebral discs. J. Biomech. Eng. 127 (7), 1121–1126.

Soukane, D.M., Shirazi-Adl, A., Urban, J.P., 2007. Computation of coupled diffusion of oxygen, glucose and lactic acid in an intervertebral disc. J. Biomech. 40 (12), 2645–2654.

Stokes, I.A., Laible, J.P., 1990. Three-dimensional osseo-ligamentous model of the thorax representing initiation of scoliosis by asymmetric growth. J. Biomech. 23 (6), 589–595.

Sundaram, S., Feng, C., 1977. Finite element analysis of the human thorax. J. Biomech. 10 (8), 505–516.

Vanderby, R., Daniele, M., Patwardhan, A., Bunch, W., 1986. A method for the identification of in-vivo segmental stiffness properties of the spine. J. Biomech. Eng. 108 (4), 312–316.

Viceconti, M., Olsen, S., Nolte, L., Burton, K., 2005. Extracting clinically relevant data from finite element simulations. Clin. Biomech. 20 (5), 451–454.

Viviani, G., Ghista, D., Lozada, P., Subbaraj, K., Barnes, G., 1986. Biomechanical analysis and simulation of scoliosis surgical correction. Clin. Orthop. Relat. Res. (208), 40–47.

Wilke, H., Herkommer, A., Werner, K., Liebsch, C., 2017. In vitro analysis of the segmental flexibility of the thoracic spine. PLoS One 12 (5), e0177823.

Young, T., 1809. I. The Croonian lecture. On the functions of the heart and arteries. Philos. Trans. R. Soc. Lond. 99, 1–31.

Zienkiewicz, O.C., Taylor, R.L., Zienkiewicz, O.C., Taylor, R.L., 1977. The Finite Element Method. McGraw-Hill, London.

Zitova, B., Flusser, J., 2003. Image registration methods: a survey. Image Vis. Comput. 21 (11), 977–1000.

CHAPTER 15

MUSCULOSKELETAL MODELING

Tito Bassani, Fabio Galbusera
IRCCS Galeazzi Orthopedic Institute, Milan, Italy

THE MUSCULOSKELETAL MODELING APPROACH

When focusing on the characterization of the human spine, investigating the relationship between body motion and biomechanical loads is of prime importance for determining optimal postures and exercises during physiological activities (e.g., ergonomics, training, and rehabilitation), for designing implants, and for preventing, evaluating, and treating spinal disorders in pathological scenarios (e.g., spine deformities and surgical fixation strategies). The evaluation of the intervertebral loads during physiological and pathological conditions has been carried out in vivo by means of instrumented spinal internal fixators (Rohlmann et al., 1995, 2008) and indirectly by measuring intradiscal pressure using needles equipped with pressure transducer tips (Meir et al., 2007; Polga et al., 2003; Wilke et al., 2001; Sato et al., 1999; Schultz et al., 1982). However, the invasiveness of these methods makes them inapplicable mainly for assessing spinal loads in both healthy subjects and subjects with pathological conditions.

A useful alternative is musculoskeletal modeling, which can be exploited as a noninvasive way to investigate the relationship between body motion and internal biomechanical loads in a wide range of conditions. This section presents in detail the principal aspects that characterize the musculoskeletal modeling approach: (i) constitutive elements, (ii) model properties, (iii) motion-solving approaches, (iv) motion definition, and (v) model validation.

(i) *Constitutive elements:* The musculoskeletal modeling approach describes the human body by characterizing bones as rigid segments connected by joints and muscles as tensile actuator elements connecting the bony segments. The constitutive elements of musculoskeletal models are thus segments, joints, muscles, and ligaments.

- *Segments:* Musculoskeletal modeling allows the characterization of segments (e.g., a single vertebra or entire spinal sections such as the whole thoracic region) by accounting for total mass, dimensions, and inertial properties, but neglects deformability as well as morphological properties, including interaction by means of contact surfaces. Indeed, unlike other methods (e.g., finite element analysis), musculoskeletal modeling is not aimed at describing stress-displacement relations and shear distribution inside bony and soft tissues, but is limited to describing the motion and loads acting on the segments as a whole. Segments are also used in musculoskeletal modeling to model cranks, pedals, handles, tools, sports equipment, tables, chairs, and all other physical objects that may be connected to the body.

  When modeling the thoracic spine, most authors chose to simplify their models by representing the whole thoracic region (including the rib cage) as a simple lumped segment,

Biomechanics of the Spine. https://doi.org/10.1016/B978-0-12-812851-0.00015-X

thus ignoring its deformability (Christophy et al., 2012; Han et al., 2012; de Zee et al., 2007a,b; El-Rich et al., 2004; Stokes and Gardner-Morse, 1995). This assumption is motivated by the intrinsic high stiffness of the thoracic chest that, in principle, greatly limits the intervertebral range of motion. However, some authors attempted to differentiate the thoracic section in single vertebral levels (Fig. 1) in order to refine the spine models (Bassani et al., 2017a; Ignasiak et al., 2016a,b). Ignasiak et al. (2016a) found that a rigid thorax assumption, in comparison with an articulated thorax, affected model loading predictions at the upper but not the lower lumbar levels.

- *Joints:* Different joint types (i.e., universal, revolute, and spherical joints) can be used to connect segments, depending on the articulation being modeled. In the spine, the intervertebral discs are usually characterized by spherical joints that by definition can provide reaction forces but not reaction moments (Bassani et al., 2017a; de Zee et al., 2007a,b). In order to overcome this limitation, musculoskeletal modeling allows adding passive elastic elements, such as springs, to the specific intervertebral level. According to the definition of spring stiffness and force-deformation relation, these elements provide passive force values acting over the intervertebral disc and can thus be exploited to obtain a measurement of the total reaction moment at the specific intervertebral level. Alternatively, stiffness can be indirectly accounted for at the joint level by applying forces and torques whose force-displacement relation is

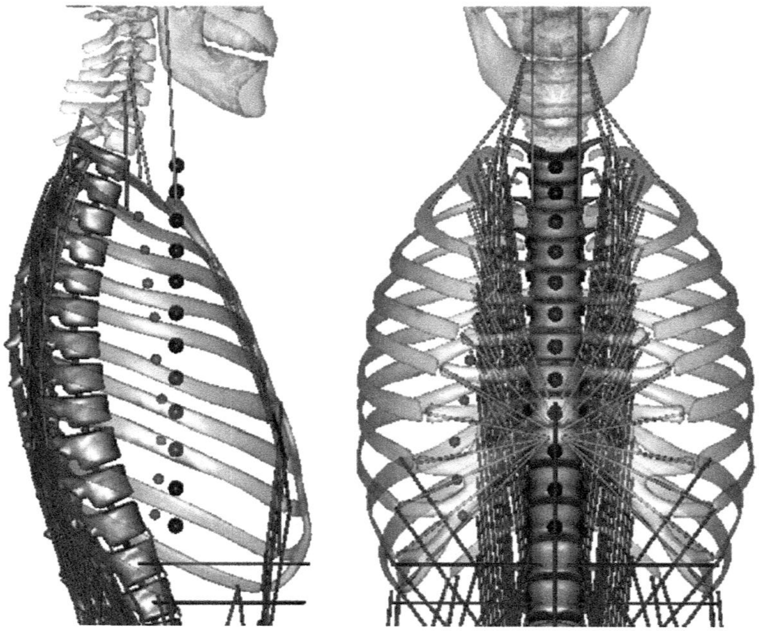

FIG. 1

Musculoskeletal model of the thoracolumbar spine with articulated rib cage developed by Ignasiak et al. (2016b) with AnyBody software.

Reproduced from the original paper with courteous publisher permission.

known a priori, which, for example, makes disc stiffness a weighting factor. In addition, some modeling software programs (e.g., AnyBody—AnyBody Technology, Aalborg, Denmark) account for the deformation of the intervertebral joint through the force-dependent kinematic technique, which assesses force-dependent translational displacements by setting joint stiffness (Bassani et al., 2017a; Kang et al., 2017).
Instead of using spherical joints, other authors have used deformable beams to represent the intervertebral discs (Fig. 2) (Ghezelbash et al., 2016; Hajihosseinali et al., 2014; Arjmand and Shirazi-Adl, 2006; Shirazi-Adl et al., 2005; El-Rich et al., 2004). In this approach, based on the finite element method (Kiefer et al., 1997), the beams modeled the overall nonlinear stiffness of the motion segments (i.e., vertebrae, disc, facets, and ligaments) and provided both moments and force reactions.

- *Muscles:* The muscles are modeled as lines going from origin to insertion points and can go through multiple points on the bones or wrap over geometrical surfaces. Skeletal muscles are unilateral elements that can exert only tensile forces (i.e., pull and not push), and their

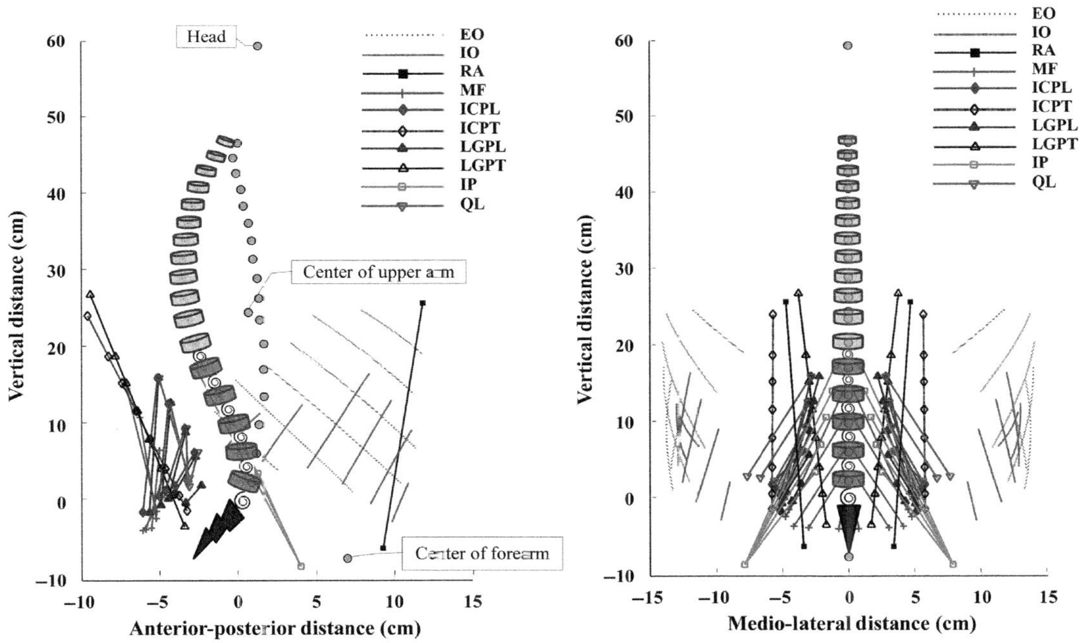

FIG. 2

Sagittally symmetric trunk model developed by Hajihosseinali et al. (2014) exploiting ABAQUS software. The model is depicted in sagittal and coronal planes with local and global muscles. *ICPL*, iliocostalis lumborum pars lumborum; *LGPL*, longissimus thoracis pars lumborum; *MF*, multifidus; *QL*, quadratus lumborum; *IP*, iliopsoas; *ICPT*, iliocostalis lumborum pars thoracis; *LGPT*, longissimus thoracispars thoracic; *IO*, internal oblique; *EO*, external oblique; *RA*, rectus abdominis.

Reproduced from the original paper with courteous publisher permission.

mechanical properties are typically modeled by use of a Hill-type muscle model, which includes elements such as force-length and force-velocity relationships as well as serial and passive elasticity (Zajac, 1989; Hill, 1938).

- *Ligaments:* These elements are passive, nonlinear elastic structures that connect articulating bones and keep joints assembled. Mechanically, they are much like muscles, but with no active contractile element. Ligaments can provide force only when stretched by the relative movement of the segments that they connect, and they are not considered as stretched until their length goes beyond an imposed slack length. When a ligament is stretched, it exerts a tensile force whose relationship with stretching depends on the imposed stiffness of the ligament (Putzer et al., 2016a; Mattucci and Cronin, 2015; Iwaskiw et al., 2012).

(ii) *Model properties:* The determination of the model properties (muscle attachment points, wrapping surfaces, bone lengths, joint centers, etc.) is typically based on reference values from anatomic and cadaver studies. To obtain accurate spine models, adequate values must be set to define the length of segments (Nissan and Gilad, 1986), the muscles' insertion points (Bayoglu et al., 2017; Phillips et al., 2008; Hansen et al., 2006; Stokes and Gardner-Morse 1999; Bogduk et al., 1992), the intervertebral disc stiffness (Gardner-Morse and Stokes, 2004; Patwardhan et al., 2003; Panjabi et al., 1976), the joints' centers of rotation (Pearcy and Bogduk, 1988), and the distribution of the body mass weight along the vertebral levels (Kiefer et al., 1997; Pearsall, 1994).

Nonetheless, the intrinsic limitation associated with the use of reference anatomic values is that they generate only general template models (e.g., male adult subject with average weight and height) that must be adapted and scaled to guarantee subject-specificity. Han et al. (2013) and Hajihosseinali et al. (2015) reported that the biomechanical loads change in relation to a subject's height and weight. Moreover, spine loading was found to be related to anatomical parameters such as the dimensions of the vertebral body, disc, and posterior parts of the vertebrae (Putzer et al., 2016b). Accordingly, to properly match a single individual or a specific population, an adequate model must account for gender, segment dimensions and mass properties, muscle properties, and ability and disability. To this end, the segment models are usually scaled according to different scaling methods (Lund et al., 2015), that is, uniform scaling (use the same ratio to scale everything in all dimensions, thus potentially creating unrealistic models), length mass scaling (scale the length according to external dimension and diameter consistent with the segment mass), and length, mass, fat scaling (try to modulate strength by fat percentage). Moreover, muscle properties must be adjusted (Heinen et al., 2016).

(iii) *Motion-solving approaches:* The motion of the constrained segments is determined by means of equations that basically result from Newton's second law of motion. The equations are written for general motion of the single segments accounting for mass and moments of inertia, with the addition of constraints (e.g., applied loads, ground and environment reaction forces). Depending on how the equations of motion are solved, two different approaches to analysis can be conducted: inverse and forward dynamics.

- *Inverse dynamics:* This approach allows calculating muscle activations and the joint reaction forces acting during the execution of specifically imposed kinematics and under the action of known external loads. Kinematic and kinetic analyses are treated separately. Kinematic analysis is performed to obtain the position, velocity, and acceleration of each segment model. Subsequently, together with the constraint conditions, the computed accelerations are used to form the equations of motion (Rasmussen et al., 2001). The unknown muscle and joint reaction forces (i.e., kinetic) are calculated by solving the muscle recruitment problem (Rasmussen et al., 2001; Dul et al., 1984; Crowninshield, 1978). Specifically, the muscle recruitment algorithms are based on the assumption that the muscle forces are distributed according to some optimal criterion, for instance, minimum fatigue. The inclusion of such assumption is necessary, because of the overdetermined structure of the modeling equations. The number of muscles is indeed greater than that of the degrees of freedom, thus leading to infinite solutions to the equations of motion.

- *Forward dynamics:* This approach allows estimating segment motions and internal reactions from known joint torques and applied forces. It involves predicting the muscle forces without assuming a known kinematic. This can be achieved either by deriving a criterion from which both the internal forces and the kinematic model can be calculated (Anderson and Pandy, 1999, 2001) or by measuring the state of the muscles, typically by using EMG and integrating the equations of motion to obtain the kinematic model (Gagnon et al., 2011; Marras et al., 2009; Buchanan et al., 2004; Lloyd and Besier, 2003; Granata and Marras, 1995; Cholewicki and McGill, 1994). Whereas in inverse dynamics, the kinematic model is directly used in the equations, in forward dynamics, it can be used as a reference trajectory that the model should follow as closely as possible.

  An aspect that can affect motion-solving approaches is the modeling of muscle coactivation (Le et al., 2017a; Rashedi et al., 2010; Hughes et al., 1995), which is generally understood as the synergy of antagonist and agonist activations and which represents the physiological/metabolic cost of inefficient muscular loading patterns onto a joint. Le et al. (2017b) recently published an exhaustive review that provided an understanding of the factors that may influence coactivation; the review documented the metrics used to assess coactivity, assessed the feasibility of those metrics, and defined the necessary variables for a coactivation index that can be used for a variety of tasks. Unfortunately, not all available musculoskeletal modeling software allows accounting for muscle coactivation, which is seen as an intrinsic limitation.

(iv) *Motion definition:* Kinematic models can be set by imposing assigned translational and rotational motions to segments and joints in order to achieve specific movements. Nevertheless, data acquired with motion capture techniques can be exploited to allow for the accurate characterization of the motion of each body segment during movements. Segment kinematics can be recorded through an optoelectronic camera system by monitoring the trajectories of active LED markers or passive reflective markers placed on a subject's skin (Fig. 3). Moreover, information about the forces exchanged between the subject and the ground can be acquired through force plates integrated into walkways. The acquired data are then processed by means of musculoskeletal modeling (Fig. 4) to achieve motion and parameter optimization before computing muscle activations and internal reaction forces. Concerning the definition of the spine kinematic model, it is worth noting that although some authors have exploited active LED markers placed along the thoracic and lumbar spines on the tip of the spinous processes (Arjmand and Shirazi-Adl, 2006), the common motion capture protocols (e.g., VICON's Plug-in-Gait) deal only with thoracic and pelvis passive markers and neglect the lumbar region (Fig. 3B). In the lumbar region, the parameter optimization procedures arrange the distribution of the vertebral motion angles in the lumbar joints (i.e., the lumbar spine rhythm) based on

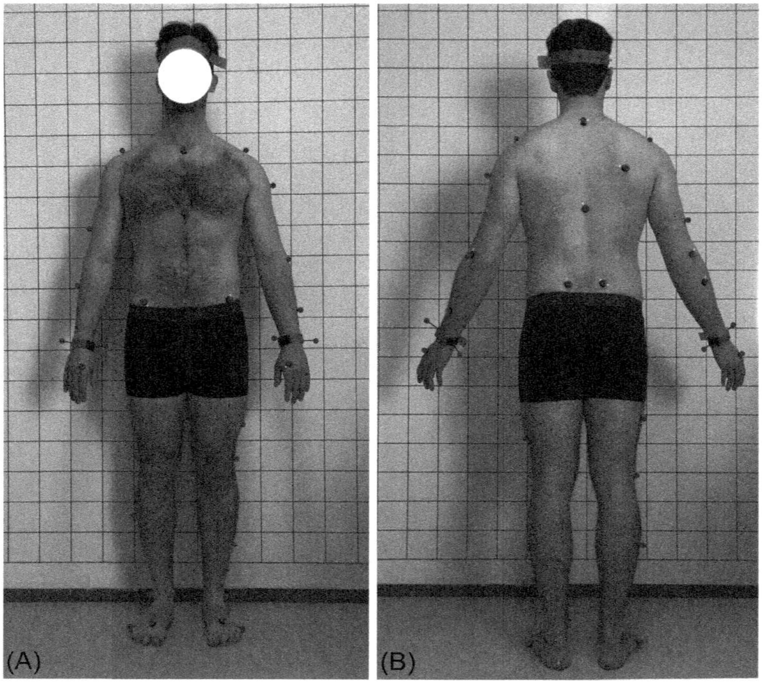

FIG. 3

Front (A) and back (B) view of the 41 passive markers placed on the skin of the subject evaluated by Bassari et al. (2017b), according to VICON Plug-in-Gait protocol.

Reproduced from the original paper with courteous publisher permission.

reference points for dynamic motion measurements (Wong et al., 2006; Lee et al., 2002; Dvorak et al., 1991; Pearcy, 1985). In lower to upper vertebral levels, different authors have identified the increasing contribution of the individual vertebral segments to the overall lumbar motion during flexion-extension movements (Aiyangar et al., 2014; Han et al., 2012; Gercek et al., 2008). Nevertheless, the use of literature values to define lumbar spine rhythm could present a potential limitation because, in principle, the real vertebral motion pattern of the evaluated subject may not be interpreted. It was also demonstrated that different lumbar spine rhythm assumptions can affect the distribution of the predicted lumbar loads (Arshad et al., 2016).

(v) *Model validation:* After a novel musculoskeletal model has been defined, it becomes essential to verify its accuracy and its suitability in describing the simulated system. Model validation is thus a fundamental and obligatory step aimed at supporting the model and promoting its use to simulate protocol conditions. The common model validation techniques consist of comparing the model outcomes (i.e., internal reactions and muscle forces) with that of previously validated models and with reference measurements obtained from in vivo and in vitro studies (Hicks et al., 2015; Lund et al., 2012). In spine modeling, the intersegmental reactions provided by the intervertebral joints can be compared with the loading values gathered in vivo with spinal

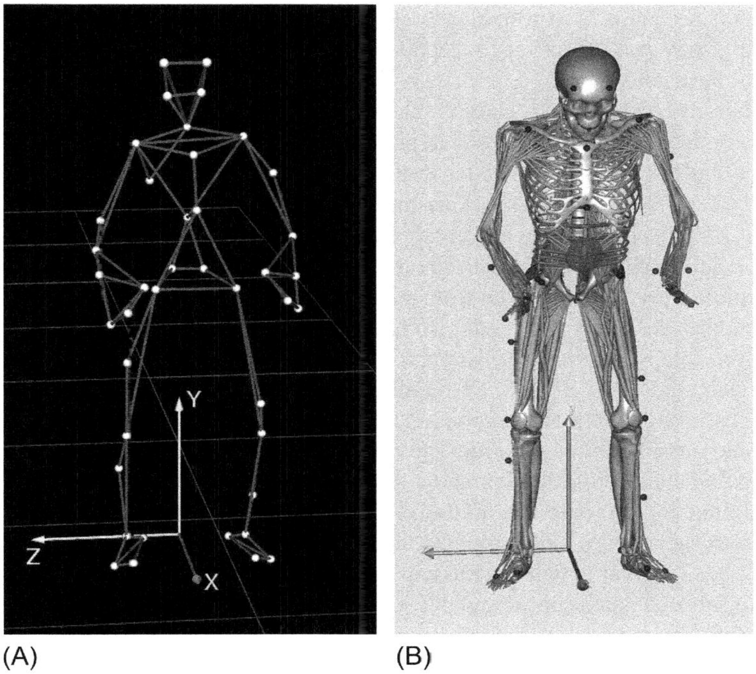

FIG. 4

Markers positions in the 3D space (A) acquired with motion capture system by Bassani et al. (2017b) and subsequently imported to set the AnyBody full-body kinematic model (B).

Reproduced from the original paper with courteous publisher permission.

internal fixators (Rohlmann et al., 1995, 2003). Furthermore, the joint reaction forces can be processed to obtain measurements of the intradiscal pressure (Ghezelbash et al., 2016; Nachemson, 1960), which can be compared with reference values evaluated in vivo in the thoracic section (Meir et al., 2007; Polga et al., 2003) and at lumbar levels (Wilke et al., 2001; Sato et al., 1999). Muscle activation can be assessed by comparing the model outcomes with EMG-based reference values (Arjmand and Shirazi-Adl, 2006; El-Rich et al., 2004; Schultz et al., 1982) and with evaluations of the muscle moment arms obtained in vivo (Phillips et al., 2008; Jorgensen et al., 2001; Macintosh et al., 1993; Bogduk et al., 1992).

MODELING SOFTWARE

In the 1990s the development of computational models aimed at characterizing the human spine was based mainly on in-house software. However, from the beginning of the 21st century, continual and exponential growth in computing capacity led to the development of novel software programs able to support graphic user interfaces, some of which were specifically devised for musculoskeletal modeling. This section provides an overview of many of the currently available tools. The section begins with a more extensive discussion of two software programs that are specifically aimed at musculoskeletal modeling (i.e., AnyBody and OpenSim), whereas additional software programs are briefly described in the latter part of the section.

AnyBody (AnyBody Technology, Aalborg, Denmark): This software is a powerful commercial tool that provides both the possibility of developing original models and a full-body musculoskeletal model (Fig. 4B) (AnyBody Managed Model Repository, AMMR). This model was originally developed by AnyBody Technology through collaboration in research projects with academic institutions (Damsgaard et al., 2006) and is increasingly utilized by numerous researchers worldwide. Indeed, more than 50 publication references were listed on the AnyBody Technology website for the year 2016, most of which exploit the full-body model from AMMR. Currently, this software uses only the inverse dynamic solving approach. The spine model is characterized as follows: the cervical section contains seven vertebrae connected with spherical joints from T1 to C2, a joint between C2 and the skull, and 136 muscle fascicles (de Zee et al., 2007a); the thoracic spine is defined as a single lumped segment from T1 to T12; the lumbar spine model contains five vertebrae connected with spherical joints, 188 muscle fascicles and intraabdominal pressure (de Zee et al., 2007b; Hansen et al., 2006), and the pelvis and sacrum. Specific details about lumbar spine rhythm, muscle parameter, facet joint definition, and exploitation of motion capture data can be found in the software's tutorial at the company's website (http://www.anybodytech.com). Concerning the ability of the AMMR full-body model to predict lumbar loads during physiological activities, Bassani et al. (2017b) recently performed a comprehensive validation analysis whose results support model suitability.

OpenSim (SimTK, Stanford, CA): This program is an open-source modeling and simulation framework of documented models that are freely available for download on the http://simtk.org platform (Seth et al., 2011; Delp et al., 2007). Researchers can contribute models, scripts, data, and plug-ins to enhance the software. Both inverse and forward- dynamic-solving approaches are allowed. In regard to the human spine, OpenSim provides a full-body musculoskeletal model (Fig. 5) based on the work of Christophy et al. (2012) and Vasavada et al. (1998). Briefly, the model consists of the following bodies: head and neck, single lumped segment of thoracic spine and rib cage, five lumbar vertebrae connected with spherical joints, rigid pelvis, and sacrum. The model also features 238 muscle fascicles associated with the eight main muscle groups of the lumbar spine. Moreover, a model of a fully articulated thoracolumbar spine (T1 through L5) and rib cage (24 individual ribs and a sternum) based on the work of Bruno et al. (2015) is available.

In addition to the preceding software, here is a list of modeling solutions utilized in other studies:

- *LifeMOD (LifeModeler, Inc., San Clemente, CA):* Discontinued commercial software that provided a basic whole body model consisting of 19 segments and 118 muscles, in which the spine can be refined into individual vertebral segments of cervical, thoracic, and lumbar regions (Huynh et al., 2015; Hajizadeh et al., 2013). The computer program has been used in numerous biomechanical studies before being withdrawn from the market in 2012.
- *ADAMS (MSC Software, Newport Beach, CA):* Multibody dynamics and motion analysis software used by several authors, for example, Nagavarapu et al. (2017), Abouhossein et al. (2011), Knapik and Marras (2009), Petit et al. (2004), and Aubin et al. (2003).
- *SIMM (Musculographics, Inc., Santa Rosa, CA)*: A toolkit that allows modeling, animation, and analysis of 3D musculoskeletal systems, exploited, for example, by Nevins et al. (2014) and Bae and Mun (2010).
- *ArtiSynth:* An open-source biomechanical modeling toolkit (Lloyd et al., 2012) freely available for download at http://artisynth.org (Malakoutian et al., 2016a).

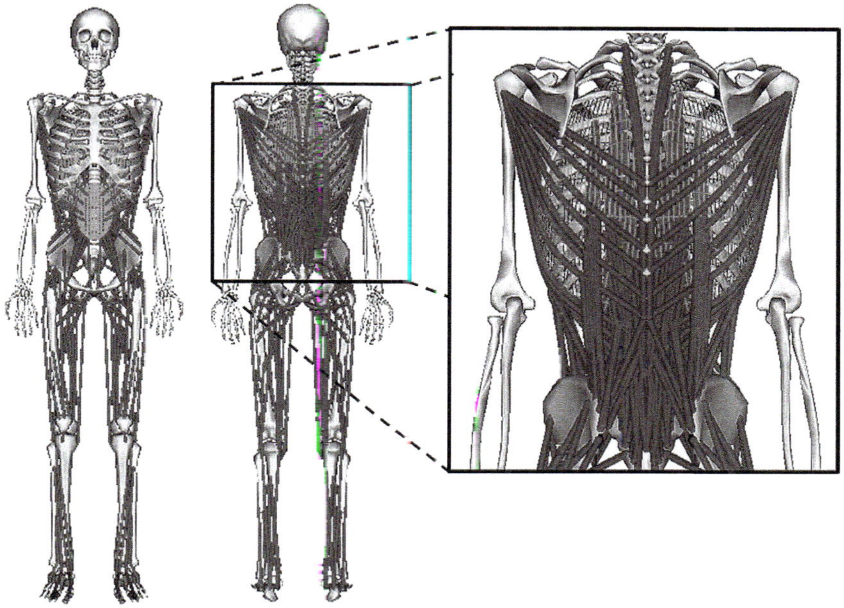

FIG. 5

Full-body lumbar spine model based on OpenSim software and developed by Raabe and Chaudhari (2016).

Reproduced from the original paper with courteous publisher permission.

Furthermore, some authors have created their own spine models using MATLAB software (MathWorks Inc., Natick, MA), a popular numerical computing environment and programming language (Curtin and Lowery, 2014; Van Toen et al., 2012; Iyer et al., 2010; Franklin and Granata, 2007; Keller et al., 2005). MATLAB provides an additional software package specifically devised for musculoskeletal modeling: Biomechanics of Bodies (BoB). This modeling package contains a human musculoskeletal model consisting of rigid segments representing the major anatomical components connected by physiologically appropriate joints and more than 600 locomotor muscles. Other authors have exploited the complementary use of MATLAB and ABAQUS (Dassault Systemes, France), a software suite for finite element analysis (Fig. 2) (Hajihosseinali et al., 2014; Arjmand and Shirazi-Adl, 2006; Shirazi-Adl et al., 2005; El-Rich et al., 2004).

MODELING THE SPINE IN PHYSIOLOGICAL CONDITIONS

This section provides an overview of the scientific works devised to exploit musculoskeletal modeling for the evaluation of spine biomechanics during physiological activities. Three main contexts, that is, ergonomics, training, and rehabilitation, are presented here:

Ergonomics: In terms of physiological conditions, the ergonomic field is indubitably the most investigated and explored field, with numerous contributions in the scientific literature. This area is so attractive because of the opportunities musculoskeletal modeling offers to assess spine loads in working conditions. Spine loads can be evaluated in specific working postures and during weight-lifting activities, providing measures aimed at decreasing the risk factors for spinal injuries and back pain. In order to avoid excessively extended presentations, the works considered as most representative are reported in detail in the following discussion, whereas other works are briefly cited.

Arjmand et al. (2015) used their previously validated trunk model (Fig. 2), implemented in ABAQUS software (Arjmand et al., 2011). The authors verified the estimated lifting equation for hands, proposed by the National Institute for Occupational Safety and Health. Moreover, they assessed the risk of spinal injury (Fig. 6). Their results showed that, in lifting activities involving moderate to large forward trunk flexion, the estimated recommended weight limit generated L5-S1 spine loads exceeding the safety limits. Thus they proposed a revised version of the equation to also account for the trunk flexion angle. Nimbarte et al. (2013) used AnyBody software to quantify the effect of dynamic cart-pushing exertions on the biomechanical loading of the lower back. Stambolian et al. (2016) developed and validated via AnyBody a 3D dynamic lifting model for lower back evaluation of careful box placement (Fig. 7). Kumar (1999) utilized a validated 3D model (Cheng and Kumar, 1991) to assess lumbar spine load in palletizing tasks with restriction to access and varying degrees of headroom. By breaking down the tasks' components, the authors found that the take-off stage was the most stressful, followed by the shelf stage. Headroom affected the biomechanical loads, and symmetry affected the disc compressive load during take off only, whereas clearance had no impact.

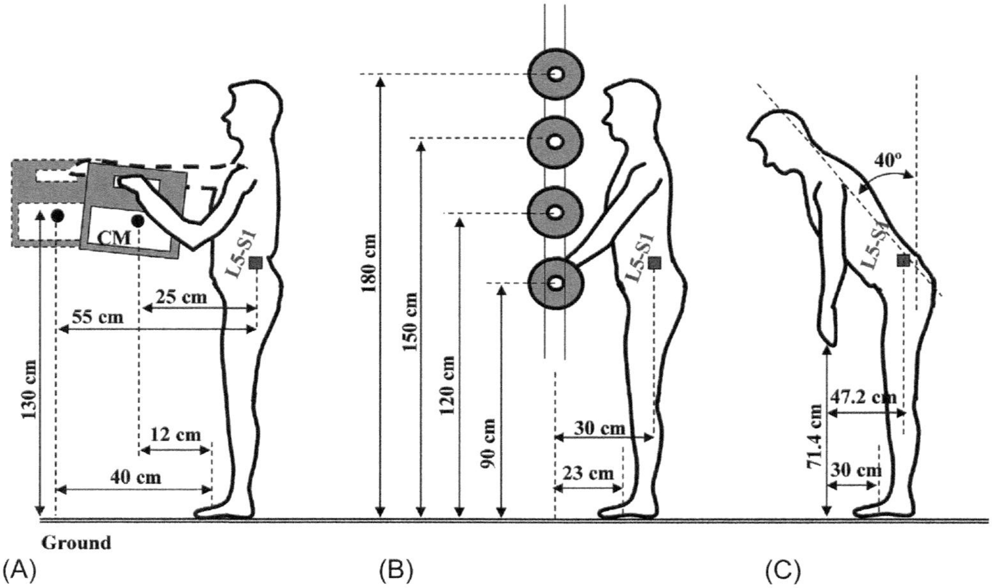

FIG. 6

Schematic presentation of the tasks performed by the subject in the in vivo study of Arjmand et al. (2015) to determine position of load and body posture.

Reproduced from the original paper with courteous publisher permission.

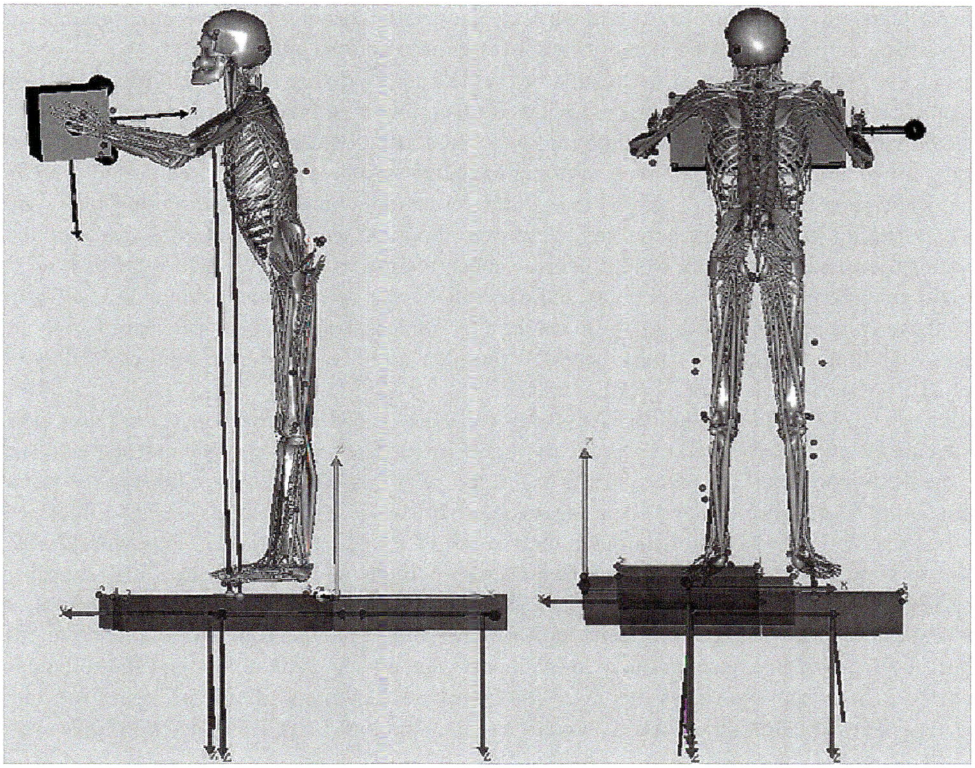

FIG. 7

The box lifting model developed exploiting AnyBody full-body model by Stambolian et al. (2016) for lower back evaluation for careful box placement. The blue vectors represent the ground reaction forces exchanged between feet and ground surface.

Reproduced from the original paper with courteous publisher permission.

Nussbaum et al. (1999) performed a biomechanical analysis of material handling manipulators in short distance transfers of moderate mass objects. The authors evaluated joint strength, spine forces, and muscular antagonism and found that strength requirements decreased significantly at the lower back when a manipulator was used in comparison with similar transfers performed manually. However, the overall results suggested that hoist-assisted transfers, although better in reducing spine compression forces, may impose relatively higher demands on coordination and/or stability at extreme heights or with torso twisting motions.

Marras et al. (2009) examined how different physical factors can influence spine loads during pushing tasks that workers might experience in industrial settings. Load magnitude, speed of push, required control, and handle height were varied while pushing both carts and overhead suspended loads. The authors exploited a validated EMG-assisted model to assess loads at the various levels of the lumbar spine. The findings revealed that spine shear forces result from a complex coactivation of trunk muscle activities and spine orientations that are influenced by several occupational factors. The authors interpreted the results as helping to explain why low back pain rates in some work environments associated with lifting may not be reduced even when lifting interventions are employed.

Rajaee et al. (2015) compared the results of six distinct modeling tools when predicting loads at lower lumbar levels in response to static activities with and without hand load. Differences were found in predicted spinal loads depending on the tool used. Accordingly, each tool was evaluated to identify its shortcomings and preferred application domains. Nagavarapu et al. (2017) exploited a previously validated EMG-assisted model (Theado et al., 2007) to evaluate the spine loading during the application and removal of lifting slings, assessing the effects of patient weight, bed height, and work method. They found that peak compression and anterior shear forces exceeded recognized thresholds when placing slings underneath heavier patients and that raising the bed to at least knuckle level helped to mitigate these spinal loads. The loading on the lumbar spine during the performance of several manual patient-handling activities was also assessed by other authors in previous studies (Gallagher and Marras, 2012; Marras et al. 1999; McGill, 1997).

Grujicic et al. (2010) exploited the AnyBody full-body model to investigate the main causes of long-distance fatigue experienced by vehicle drivers (Fig. 8). The results showed that seat adjustments, drivers' back supports, and the nature of the seat upholstery can have complex influences on: muscle activation, joint forces, normal and shear stresses on soft tissue. All of these aspects affect not only the drivers' perceptions of comfort but also their sense of fatigue. Mathys and Ferguson (2012) used OpenSim to simulate the effects of two different pilot helmets on the pilot's neck loading during air combat. The increased weight and the forward shift of the center of mass of the new generation helmet (with additional mounted equipment) led to higher muscle activations and higher joint reaction loads over a wide range of head and neck movements. The muscle activations required to balance the head and neck in extreme postures increased the compressive force at the T1-C7 level substantially, whereas in a neutral posture, the muscle activations remained low. Fritz (1998) used in-house software

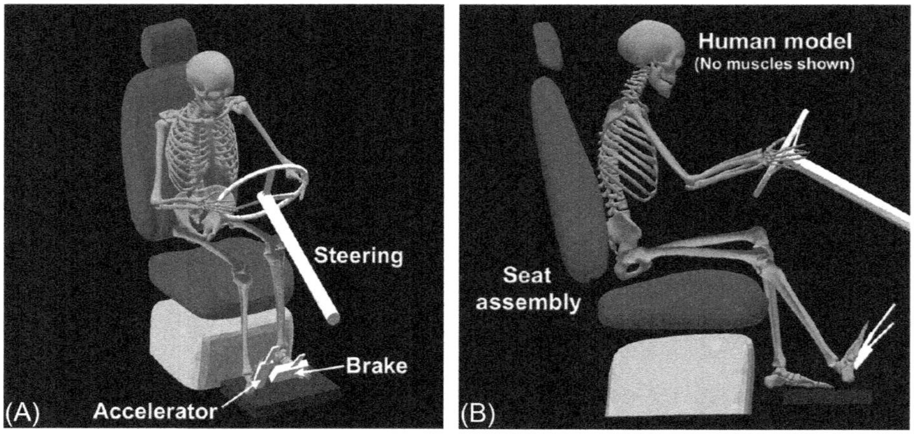

FIG. 8

Isometric (A) and side-view (B) of the car-driver-seat assembly evaluated by Grujicic et al. (2010) with AnyBody full-body model to assess long-distance driving fatigue. Muscles are not shown.

Reproduced from the original paper with courteous publisher permission.

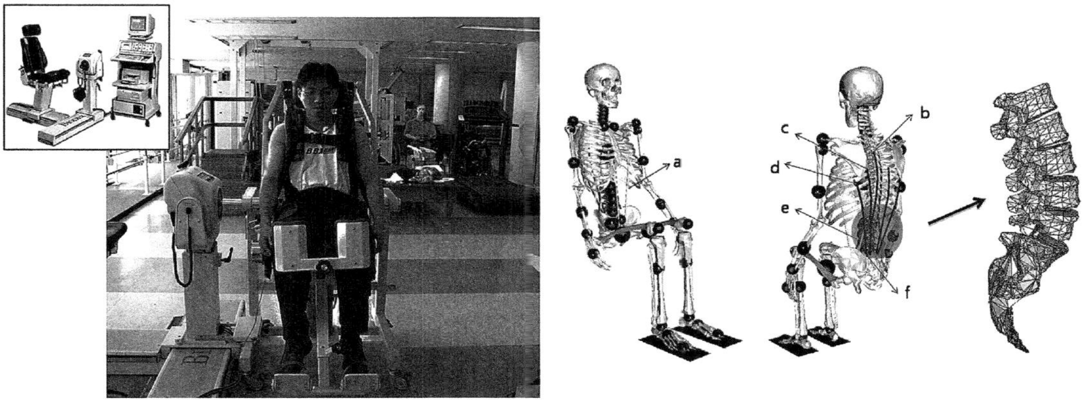

FIG. 9

Full-body model with back muscles and lumbosacral joint model (right panel) based on SIMM software, exploited by Bae and Mun (2010) to evaluate the effect of lumbar lordotic angle on lumbosacral joint during isokinetic exercise (left panel).

Reproduced from the original paper with courteous publisher permission.

to simulate the response of the spine and pelvis to vibration stress, under the proved assumption that long-term exposure to whole-body vibrations can induce low back pain.

Training: Holmberg and Lund (2008) used AnyBody software to assess motion angles and muscle activations during cross-country skiing. The authors found results in agreement with values reported in literature with similar kinematic and external kinetic values thus supporting the use of computer simulations to investigate training conditions. Hamner et al. (2010) utilized OpenSim software to evaluate lower limb muscle activations in relation to lumbar and pelvis motion angles during running. Using the same software, Raabe and Chaudhari (2016) investigated jogging biomechanics. They enhanced and validated a previous full-body model by comparing muscle moment arms in reference to experimental values. Bae and Mun (2010) used SIMM software to evaluate the effect of the lumbar lordotic angle on lumbosacral joint during isokinetic exercise (Fig. 9). Muscle joint torque, joint forces of each joint, system energy, and estimated muscle forces at the lumbosacral joint were estimated in cases of normal lordosis, excessive lordosis, lumbar kyphosis and hypolordosis. Results revealed that largest joint torque of excessive lordosis was greater than that of normal lordosis and that lumbar kyphosis was less than normal lordosis. Furthermore, lumbar kyphosis required the least and most energy during flexion and extension, respectively. Koizumi et al. (2011) utilized the same software to compare lumbar spine loads during Nordic walking with those during ordinary walking. They found lower spinal loads during Nordic walking when the subjects walked on a level surface and going upstairs, but not when going downstairs.

Furthermore, other authors have exploited musculoskeletal modeling to assess conditions characterized by altered gravity (i.e., microgravity), after recognizing that exposure to microgravity affects the human musculoskeletal system in various ways. For example, during spaceflight many astronauts experience low back pain, and the risk of spine injuries is significantly greater post-flight. With this assumption, Lindenroth et al. (2015) evaluated an exercise device developed to specifically counteract atrophy of deep spinal and postural muscles. Using AnyBody software, the authors tested the possibility of transferring this exercise concept from Earth to space. The results showed that when accounting microgravity modifications, the model led to increased muscle activation of deep spinal muscles and to decreased activation of trunk muscles which create superficial moment.

Rehabilitation: Drake et al. (2006) evaluated the advantage of training with an exercise ball for trunk extensor exercises. They recorded three-dimensional lumbar spine postures and upper-body kinematics and used an EMG-driven model to calculate spinal loading. Their results revealed that in a healthy, young population, no training advantages appeared when performing extensor exercises on a ball versus a mat. However, they proposed that in a rehabilitation scenario, when performed on a ball, these exercises could reduce low back loading and therefore reduce the potential for reinjury. Van Toen et al. (2012) used MATLAB software to implement a one-dimensional lumped parameter model of the upper body and pelvis, with the aim of simulating a backward fall onto the buttocks. The authors exploited the model to evaluate the effect of floor stiffness on spine forces during falls and to compare predicted impact forces with forces previously measured to fracture the spine. They found that lowering the floor stiffness can attenuate peak lumbosacral spine forces in a backward fall onto the buttocks from standing by 46% to values closer to the average tolerance of the spine to fracture. Han et al. (2015) investigated the effect of core muscle strengthening in the spine during spine stabilization exercises using a whole body tilt device. The authors used AnyBody software to replicate the input motion from the whole body tilting exercise and to estimate spine loads and muscle forces depending on the tilting angles. The results revealed that the activation of long and superficial back muscles was mainly affected by the forward direction of the tilt, whereas the front muscles were mainly affected by the backward tilting direction. Moreover, deep muscles, such as the multifidi, and short muscles were found to be activated in most directions of the rotation and tilt. The findings supported the authors in concluding that the backward directions of the tilt device could be carefully applied for the elderly and for rehabilitation patients who are expected to have less muscle strength.

MODELING THE SPINE IN PATHOLOGICAL SCENARIOS

When used for the characterization of the human spine in pathological conditions, musculoskeletal modeling was aimed at investigating two main topics: spine deformities (i.e., scoliosis) and effects of spine surgery.

Spine deformities: Curtin and Lowery (2014) used MATLAB software to investigate muscle activation in combination with applied external forces as a treatment for scoliosis. Various scoliotic curvatures were modelled, and optimization was applied to reduce the curves using combinations of both deep and superficial muscle activation and applied external forces. The results suggested that activation of superficial and deep muscles may be effective in reducing spinal curvature in the case of moderate scoliosis when muscle groups are selected for activation based on the curve type. Furthermore, the findings suggested the potential for a hybrid treatment involving combined muscle activation and applied external forces in the case of severe scoliosis. Bassani et al. (2017a) proposed a semiautomatic software approach to reconstruct a 3D subject-specific musculoskeletal model of the thoracolumbar spine and the pelvis from radiographic digitized images. They evaluated adolescents with mild idiopathic scoliosis and used AnyBody software to evaluate the relationship between intervertebral loads and anatomical parameters and the degree of scoliosis. The results showed that the intervertebral loads were significantly correlated with the anatomical parameters in the sagittal and axial planes.

Aubin et al. (2003) exploited ADAMS software to simulate surgical instrumentation of vertebral fusion in scoliosis. Their model was personalized according to the patient's spine geometry and binary joints, and kinematic constraints were used to represent the rod-implant-vertebra joints. Based on the results, the authors interpreted their tool as being able to provide realistic representation of the surgical maneuvers and to predict spinal geometry correction and assist in the preoperative planning of surgical instrumentation of the scoliotic spine. Petit et al. (2004) used the same piece of software to develop a flexible multibody model to characterize in vivo the mechanical properties of the scoliotic spine. The approach was applied to scoliotic patients undergoing spinal correction. The identification of the spine mechanical properties was interpreted by the authors as potentially helping clinicians in the planning of surgical instrumentation maneuvers.

Hajizadeh et al. (2013) exploited LifeMOD software to develop a 3D multibody model of the scoliotic spine. Rib cage intervertebral loads and motion angles and muscle activations were assessed during lateral bending. According to the results and to the qualitative experimental results from previous studies, the mobility of the rib cage in the scoliosis model was found lower than that of the normal (nonscoliotic) model with the same anthropometric data.

Effects of spine surgery: Bresnahan et al. (2010) evaluated how graded resection of the lumbar paraspinal muscles, as a result of posterior lumbar surgery, affects muscle activity for a variety of movement tasks. Specifically, the authors modified the lumbar spine model from the AnyBody repository to study the change in muscle activation as a result of posterior lumbar surgery at the L3-L4 and L4-L5 segments. The results showed that preservation of the paraspinal musculature associated with minimally invasive surgical approaches better preserved postoperative muscle activity. Furthermore, the study suggested a positive correlation between the reduction of the paraspinal muscle cross-sectional area following surgery and the alteration in trunk muscle activity.

Malakoutian et al. (2016b) used ArtiSynth software to assess the effects of posterior spinal surgery on spinal loading. The authors evaluated the effect of simulated muscle damage on postoperative spinal loading at the adjacent levels to a spinal fusion, during upright postures. The results pointed out an increased compressive load and shear forces at the adjacent levels after muscle damage, with bigger changes at the rostral level than at the caudal level. These changes were due to a redistribution of muscle activity from the local lumbar to the global spinal musculature.

Jamshidnejad and Arjmand (2015) investigated the effects of the iatrogenic injuries to paraspinal muscles during posterior lumbar surgery on postoperative patterns of muscle activations and spinal loads. They simulated various activities using combined modeling (exploiting the concomitant use of MATLAB and ABAQUS software) and in vivo MR imaging. Postoperatively, they found decreased trunk extension strength and decreased force in multifidus fascicles in flexion tasks and upright position. In contrast, the sum of the forces in all other intact and less injured extensor muscles slightly increased in the postoperative model. Furthermore, postoperative L5-S1 compressive and shear loads varied slightly. The authors concluded that intraoperative injuries induced a shift in load sharing from the most injured muscle (multifidus) toward other less injured and intact muscles during all simulated activities. They thus suggested that the postoperative rehabilitation programs should strengthen and facilitate not only the injured multifidus but also other intact and less injured trunk muscles that play a compensatory role after an operation.

REFERENCES

Abouhossein, A., Weisse, B., Ferguson, S.J., 2011. A multibody modelling approach to determine load sharing between passive elements of the lumbar spine. Comput. Methods Biomech. Biomed. Eng. 14 (6), 527–537.

Aiyangar, A.K., Zheng, L., Tashman, S., Anderst, W.J., Zhang, X., 2014. Capturing three-dimensional in vivo lumbar intervertebral joint kinematics using dynamic stereo-X-ray imaging. J. Biomech. Eng. 136 (1), 011004.

Anderson, F.C., Pandy, M.G., 2001. Dynamic optimization of human walking. J. Biomech. Eng. 123 (5), 381–390.

Anderson, F.C., Pandy, M.G., 1999. A dynamic optimization solution for vertical jumping in three dimensions. Comput. Methods Biomech. Biomed. Eng. 2 (3), 201–231.

Arjmand, N., Plamondon, A., Shirazi-Adl, A., Larivière, C., Parnianpour, M., 2011. Predictive equations to estimate spinal loads in symmetric lifting tasks. J. Biomech. 44 (1), 84–91.

Arjmand, N., Shirazi-Adl, A., 2006. Model and in vivo studies on human trunk load partitioning and stability in isometric forward flexions. J. Biomech. 39 (3), 510–521.

Arjmand, N., Amini, M., Shirazi-Adl, A., Plamondon, A., Parnianpour, M., 2015. Revised NIOSH lifting equation may generate spine loads exceeding recommended limits. Int. J. Ind. Ergon. 47, 1–8.

Arshad, R., Zander, T., Dreischarf, M., Schmidt, H., 2016. Influence of lumbar spine rhythms and intra-abdominal pressure on spinal loads and trunk muscle forces during upper body inclination. Med. Eng. Phys. 38 (4), 333–338.

Aubin, C.E., Petit, Y., Stokes, I.A., Poulin, F., Gardner-Morse, M., Labelle, H., 2003. Biomechanical modeling of posterior instrumentation of the scoliotic spine. Comput. Methods Biomech. Biomed. Eng. 6 (1), 27–32.

Bae, T.S., Mun, M., 2010. Effect of lumbar lordotic angle on lumbosacral joint during isokinetic exercise: a simulation study. Clin. Biomech. 25 (7), 628–635.

Bassani, T., Ottardi, C., Costa, F., Brayda-Bruno, M., Wilke, H.J., Galbusera, F., 2017a. Semiautomated 3D spine reconstruction from biplanar radiographic images: prediction of intervertebral loading in scoliotic subjects. Frontiers Bioeng. Biotechnol. 5, 1.

Bassani, T., Stucovitz, E., Qian, Z., Briguglio, M., Galbusera, F., 2017b. Validation of the AnyBody full body musculoskeletal model in computing lumbar spine loads at L4L5 level. J. Biomech. 58, 89–96.

Bayoglu, R., Geeraedts, L., Groenen, K.H.J., Verdonschot, N., Koopman, B., Homminga, J., 2017. Twente spine model: a complete and coherent dataset for musculo-skeletal modeling of the lumbar region of the human spine. J. Biomech. 53, 111–119.

Bogduk, N., Macintosh, J.E., Pearcy, M.J., 1992. A universal model of the lumbar back muscles in the upright position. Spine 17 (8), 897–913.

Bresnahan, L., Fessler, R.G., Natarajan, R.N., 2010. Evaluation of change in muscle activity as a result of posterior lumbar spine surgery using a dynamic modeling system. Spine 35 (16), E761–7.

Bruno, A.G., Bouxsein, M.L., Anderson, D.E., 2015. Development and validation of a musculoskeletal model of the fully articulated thoracolumbar spine and rib cage. J. Biomech. Eng. 137 (8), 081003.

Buchanan, T.S., Lloyd, D.G., Manal, K., Besier, T.F., 2004. Neuromusculoskeletal modeling: estimation of muscle forces and joint moments and movements from measurements of neural command. J. Appl. Biomech. 20 (4), 367–395.

Cheng, R.C.K., Kumar, S., 1991. A three dimensional biomechanical model for human back. Int. J. Ind. Ergon. 7, 327–339.

Cholewicki, J., McGill, S.M., 1994. EMG assisted optimization: a hybrid approach for estimating muscle forces in an indeterminate biomechanical model. J. Biomech. 27 (10), 1287–1289.

Christophy, M., Faruk Senan, N.A., Lotz, J.C., O'Reilly, O.M., 2012. A musculoskeletal model for the lumbar spine. Biomech. Model. Mechanobiol. 11 (1–2), 19–34.

Crowninshield, R.D., 1978. Use of optimization techniques to predict muscle forces. J. Biomech. Eng. 100 (2), 88–92.

Curtin, M., Lowery, M.M., 2014. Musculoskeletal modelling of muscle activation and applied external forces for the correction of scoliosis. J. Neuroeng. Rehabil. 11, 52-0003-11-52.

Damsgaard, M., Rasmussen, J., Christensen, S.T., Surma, E., de Zee, M., 2006. Analysis of musculoskeletal systems in the AnyBody modeling system. Simul. Model. Pract. Theory 14 (8), 1100–1111.

de Zee, M., Falla, D., Farina, D., Rasmussen, J., 2007a. A detailed rigid-body cervical spine model based on inverse dynamics. J. Biomech. 40 (Suppl. 2), S284.

de Zee, M., Hansen, L., Wong, C., Rasmussen, J., Simonsen, E.B., 2007b. A generic detailed rigid-body lumbar spine model. J. Biomech. 40 (6), 1219–1227.

Delp, S.L., Anderson, F.C., Arnold, A.S., Loan, P., Habib, A., John, C.T., Guendelman, E., Thelen, D.G., 2007. OpenSim: open-source software to create and analyze dynamic simulations of movement. IEEE Trans. Biomed. Eng. 54 (11), 1940–1950.

Drake, J.D., Fischer, S.L., Brown, S.H., Callaghan, J.P., 2006. Do exercise balls provide a training advantage for trunk extensor exercises? A biomechanical evaluation. J. Manip. Physiol. Ther. 29 (5), 354–362.

Dul, J., Johnson, G.E., Shiavi, R., Townsend, M.A., 1984. Muscular synergism—II. A minimum-fatigue criterion for load sharing between synergistic muscles. J. Biomech. 17 (9), 675–684.

Dvorak, J., Panjabi, M.M., Novotny, J.E., Chang, D.G., Grob, D., 1991. Clinical validation of functional flexion-extension roentgenograms of the lumbar spine. Spine 16 (8), 943–950.

El-Rich, M., Shirazi-Adl, A., Arjmand, N., 2004. Muscle activity, internal loads, and stability of the human spine in standing postures: combined model and in vivo studies. Spine 29 (23), 2633–2642.

Franklin, T.C., Granata, K.P., 2007. Role of reflex gain and reflex delay in spinal stability—a dynamic simulation. J. Biomech. 40 (8), 1762–1767.

Fritz, M., 1998. Three-dimensional biomechanical model for simulating the response of the human body to vibration stress. Med. Biol. Eng. Comput. 36 (6), 686–692.

Gagnon, D., Arjmand, N., Plamondon, A., Shirazi-Adl, A., Larivière, C., 2011. An improved multi-joint EMG-assisted optimization approach to estimate joint and muscle forces in a musculoskeletal model of the lumbar spine. J. Biomech. 44 (8), 1521–1529.

Gallagher, S., Marras, W.S., 2012. Tolerance of the lumbar spine to shear: a review and recommended exposure limits. Clin. Biomech. 27 (10), 973–978.

Gardner-Morse, M.G., Stokes, I.A.F., 2004. Structural behavior of human lumbar spinal motion segments. J. Biomech. 37 (2), 205–212.

Gercek, E., Hartmann, F., Kuhn, S., Degreif, J., Rommens, P.M., Rudig, L., 2008. Dynamic angular three-dimensional measurement of multisegmental thoracolumbar motion in vivo. Spine 33 (21), 2326–2333.

Ghezelbash, F., Shirazi-Adl, A., Arjmand, N., El-Ouaaid, Z., Plamondon, A., 2016. Subject-specific biomechanics of trunk: musculoskeletal scaling, internal loads and intradiscal pressure estimation. Biomech. Model. Mechanobiol. 15 (6), 1699–1712.

Granata, K.P., Marras, W.S., 1995. An EMG-assisted model of trunk loading during free-dynamic lifting. J. Biomech. 28 (11), 1309–1317.

Grujicic, M., Pandurangan, B., Xie, X., Gramopadhye, A.K., Wagner, D., Ozen, M., 2010. Musculoskeletal computational analysis of the influence of car-seat design/adjustments on long-distance driving fatigue. Int. J. Ind. Ergon. 40 (3), 345–355.

Hajihosseinali, M., Arjmand, N., Shirazi-Adl, A., 2015. Effect of body weight on spinal loads in various activities: a personalized biomechanical modeling approach. J. Biomech. 48 (2), 276–282.

Hajihosseinali, M., Arjmand, N., Shirazi-Adl, A., Farahmand, F., Ghiasi, M.S., 2014. A novel stability and kinematics-driven trunk biomechanical model to estimate muscle and spinal forces. Med. Eng. Phys. 36 (10), 1296–1304.

Hajizadeh, K., Gibson, I., Liu, G., 2013. Developing a 3D multi-body model of the scoliotic spine with lateral bending motion for comparison of ribcage flexibility. Int. J. Adv. Design Manuf. Technol. 6 (1).

Hamner, S.R., Seth, A., Delp, S.L., 2010. Muscle contributions to propulsion and support during running. J. Biomech. 43 (14), 2709–2716.

Han, K., Rohlmann, A., Zander, T., Taylor, W.R., 2013. Lumbar spinal loads vary with body height and weight. Med. Eng. Phys. 35 (7), 969–977.

Han, K., Zander, T., Taylor, W.R., Rohlmann, A., 2012. An enhanced and validated generic thoraco-lumbar spine model for prediction of muscle forces. Med. Eng. Phys. 34 (6), 709–716.

Han, K.S., Nam, H.D., Kim, K., 2015. Core muscle strengthening effect during spine stabilization exercise. J. Electr. Eng. Technol. 10 (6), 2413–2419.

Hansen, L., de Zee, M., Rasmussen, J., Andersen, T.B., Wong, C., Simonsen, E.B., 2006. Anatomy and biomechanics of the back muscles in the lumbar spine with reference to biomechanical modeling. Spine 31 (17), 1888–1899.

Heinen, F., Lund, M.E., Rasmussen, J., de Zee, M., 2016. Muscle-tendon unit scaling methods of Hill-type musculoskeletal models: an overview. Proc. Inst. Mech. Eng. H J. Eng. Med. 230 (10), 976–984.

Hicks, J.L., Uchida, T.K., Seth, A., Rajagopal, A., Delp, S.L., 2015. Is my model goodenough? Best practices for verification and validation of musculoskeletalmodels and simulations of movement. J. Biomech. Eng. 137 (2), 020905.

Hill, A.V., 1938. The heat of shortening and the dynamic constants of muscle. Proc. R. Soc. London B Biol. Sci. 126 (843), 136–195.

Holmberg, L.J., Lund, A.M., 2008. A musculoskeletal full-body simulation of cross-country skiing. Proc. Inst. Mech. Eng. P: J. Sports Eng. Technol. 222, 11–22.

Hughes, R.E., Bean, J.C., Chaffin, D.B., 1995. Evaluating the effect of co-contraction in optimization models. J. Biomech. 28 (7), 875–878.

Huynh, K.T., Gibson, I., Jagdish, B.N., Lu, W.F., 2015. Development and validation of a discretised multi-body spine model in LifeMOD for biodynamic behaviour simulation. Comput. Methods Biomech. Biomed. Eng. 18 (2), 175–184.

Ignasiak, D., Ferguson, S.J., Arjmand, N., 2016a. A rigid thorax assumption affects model loading predictions at the upper but not lower lumbar levels. J. Biomech. 49 (13), 3074–3078.

Ignasiak, D., Dendorfer, S., Ferguson, S.J., 2016b. Thoracolumbar spine model with articulated ribcage for the prediction of dynamic spinal loading. J. Biomech. 49 (6), 959–966.

Iwaskiw, A.S., Armiger, R.S., Ott, K.A., Wickwire, A.C., Merkle, A.C., 2012. Response of individual thoracolumbar spine ligaments under high-rate deformation. Biomed. Sci. Instrum. 48, 194–201.

Iyer, S., Christiansen, B.A., Roberts, B.J., Valentine, M.J., Manoharan, R.K., Bouxsein, M.L., 2010. A biomechanical model for estimating loads on thoracic and lumbar vertebrae. Clin. Biomech. 25 (9), 853–858.

Jamshidnejad, S., Arjmand, N., 2015. Variations in trunk muscle activities and spinal loads following posterior lumbar surgery: a combined in vivo and modeling investigation. Clin. Biomech. 30 (10), 1036–1042.

Jorgensen, M.J., Marras, W.S., Granata, K.P., Wiand, J.W., 2001. MRI-derived moment-arms of the female and male spine loading muscles. Clin. Biomech. 16 (3), 182–193.

Kang, K.T., Koh, Y.G., Jung, M., Nam, J.H., Son, J., Lee, Y.H., Kim, S.J., Kim, S.H., 2017. The effects of posterior cruciate ligament deficiency on posterolateral corner structures under gait- and squat-loading conditions: a computational knee model. Bone Joint Res. 6 (1), 31–42.

Keller, T.S., Colloca, C.J., Harrison, D.E., Harrison, D.D., Janik, T.J., 2005. Influence of spine morphology on intervertebral disc loads and stresses in asymptomatic adults: implications for the ideal spine. Spine J. 5 (3), 297–309.

Kiefer, A., Shirazi-Adl, A., Parnianpour, M., 1997. Stability of the human spine in neutral postures. Eur. Spine J. 6 (1), 45–53.

Knapik, G.G., Marras, W.S., 2009. Spine loading at different lumbar levels during pushing and pulling. Ergonomics 52 (1), 60–70.

Koizumi, T., Tsujiuchi, N., Takeda, M., Fujikura, R., Kojima, T., 2011. Load dynamics of joints in Nordic walking. Procedia Eng. 13, 544–551.

Kumar, S., 1999. Biomechanical load on human lumbar spine in palletizing tasks with restriction to access and varying headroom. Int. J. Ind. Ergon. 23 (4), 349–358.

Le, P., Aurand, A., Dufour, J.S., Knapik, G.G., Best, T.M., Khan, S.N., Mendel, E., Marras, W.S., 2017a. Development and testing of a moment-based coactivation index to assess complex dynamic tasks for the lumbar spine. Clin. Biomech. 46, 23–32.

Le, P., Best, T.M., Khan, S.N., Mendel, E., Marras, W.S., 2017b. A review of methods to assess coactivation in the spine. J. Electromyogr. Kinesiol. 32, 51–60.

Lee, S.W., Wong, K.W., Chan, M.K., Yeung, H.M., Chiu, J.L., Leong, J.C., 2002. Development and validation of a new technique for assessing lumbar spine motion. Spine 27 (8), E215–20.

Lindenroth, L., Caplan, N., Debuse, D., Salomoni, S.E., Evetts, S., Weber, T., 2015. A novel approach to activate deep spinal muscles in space—results of a biomechanical model. Acta Astron. 116, 202–210.

Lloyd, J.E., Stavness, I., Fels, S., 2012. ArtiSynth: a fast interactive biomechanical modeling toolkit combining multibody and finite element simulation. Soft Tissue Biomech. Model. Comput. Assisted Surg. 355–394.

Lloyd, D.G., Besier, T.F., 2003. An EMG-driven musculoskeletal model to estimate muscle forces and knee joint moments in vivo. J. Biomech. 36 (6), 765–776.

Lund, M.E., Andersen, M.S., de Zee, M., Rasmussen, J., 2015. Scaling of musculoskeletal models from static and dynamic trials. Int. Biomech. 2 (1), 1–11.

Lund, M.E., de Zee, M., Andersen, M.S., Rasmussen, J., 2012. On validation of multibody musculoskeletal models. Proc. Inst. Mech. Eng. H 226 (2), 82–94.

Macintosh, J.E., Bogduk, N., Pearcy, M.J., 1993. The effects of flexion on the geometry and actions of the lumbar erector spinae. Spine 18 (7), 884–893.

Malakoutian, M., Street, J., Wilke, H.-J., Stavness, I., Fels, S., Oxland, T., 2016a. A musculoskeletal model of the lumbar spine using ArtiSynth – development and validation. Comput. Methods Biomech. Biomed. Eng. Imaging Vis. 1–8.

Malakoutian, M., Street, J., Wilke, H.J., Stavness, I., Dvorak, M., Fels, S., Oxland, T., 2016b. Role of muscle damage on loading at the level adjacent to a lumbar spine fusion: a biomechanical analysis. Eur. Spine J. 25 (9), 2929–2937.

Marras, W.S., Davis, K.G., Kirking, B.C., Bertsche, P.K., 1999. A comprehensive analysis of low-back disorder risk and spinal loading during the transferring and repositioning of patients using different techniques. Ergonomics 42 (7), 904–926.

Marras, W.S., Knapik, G.G., Ferguson, S., 2009. Loading along the lumbar spine as influence by speed, control, load magnitude, and handle height during pushing. Clin. Biomech. 24 (2), 155–163.

Mathys, R., Ferguson, S.J., 2012. Simulation of the effects of different pilot helmets on neck loading during air combat. J. Biomech. 45 (14), 2362–2367.

Mattucci, S.F., Cronin, D.S., 2015. A method to characterize average cervical spine ligament response based on raw data sets for implementation into injury biomechanics models. J. Mech. Behav. Biomed. Mater. 41, 251–260.

McGill, S.M., 1997. The biomechanics of low back injury: implications on current practice in industry and the clinic. J. Biomech. 30 (5), 465–475.

Meir, A.R., Fairbank, J.C., Jones, D.A., McNally, D.S., Urban, J.P., 2007. High pressures and asymmetrical stresses in the scoliotic disc in the absence of muscle loading. Scoliosis 2, 4.

Nachemson, A., 1960. Lumbar intradiscal pressure. Experimental studies on post-mortem material. Acta Orthop. Scand., Suppl. 43, 1–104.

Nagavarapu, S., Lavender, S.A., Marras, W.S., 2017. Spine loading during the application and removal of lifting slings: the effects of patient weight, bed height and work method. Ergonomics 60 (5), 636–648.

Nevins, D.D., Zheng, L., Vasavada, A.N., 2014. Inter-individual variation in vertebral kinematics affects predictions of neck musculoskeletal models. J. Biomech. 47 (13), 3288–3294.

Nimbarte, A.D., Sun, Y., Jaridi, M., Hsiao, H., 2013. Biomechanical loading of the shoulder complex and lumbosacral joints during dynamic cart pushing task. Appl. Ergon. 44 (5), 841–849.

Nissan, M., Gilad, I., 1986. Dimensions of human lumbar vertebrae in the sagittal plane. J. Biomech. 19 (9), 753–758.

Nussbaum, M.A., Chaffin, D.B., Baker, G., 1999. Biomechanical analysis of materials handling manipulators in short distance transfers of moderate mass objects: joint strength, spine forces and muscular antagonism. Ergonomics 42 (12), 1597–1618.

Panjabi, M.M., Brand Jr., R.A., White 3rd, A.A., 1976. Mechanical properties of the human thoracic spine as shown by three-dimensional load-displacement curves. J. Bone Joint Surg.: Am. Vol. 58 (5), 642–652.

Patwardhan, A.G., Havey, R.M., Carandang, G., Simonds, J., Voronov, L.I., Ghanayem, A.J., Meade, K.P., Gavin, T.M., Paxinos, O., 2003. Effect of compressive follower preload on the flexion–extension response of the human lumbar spine. J. Orthop. Res. 21 (3), 540–546.

Pearcy, M.J., 1985. Stereo radiography of lumbar spine motion. Acta Orthop. Scand. Suppl. 212, 1–45.

Pearcy, M.J., Bogduk, N., 1988. Instantaneous axes of rotation of the lumbar intervertebral joints. Spine 13 (9), 1033–1041.

Pearsall, D.J., 1994. Segmental inertial properties of the human trunk as determined from computer tomography and magnetic resonance imagery (Ph.D. thesis). Queen's University, Kingston, Ontario.

Petit, Y., Aubin, C.E., Labelle, H., 2004. Patient-specific mechanical properties of a flexible multi-body model of the scoliotic spine. Med. Biol. Eng. Comput. 42 (1), 55–60.

Phillips, S., Mercer, S., Bogduk, N., 2008. Anatomy and biomechanics of quadratus lumborum. Proc. Inst. Mech. Eng. H J. Eng. Med. 222 (2), 151–159.

Polga, D., Beaubien, B., Kallemeier, P., Schellhas, K., Lew, W., Buttermann, G., Wood, K., 2003. In vivo measurement of intradiscal pressure in healthy thoracic intervertebral discs. Spine J. 3 (5 (Suppl.)), 83–84.

Putzer, M., Auer, S., Malpica, W., Suess, F., Dendorfer, S., 2016a. A numerical study to determine the effect of ligament stiffness on kinematics of the lumbar spine during flexion. BMC Musculoskelet. Disord. 17 (95).

Putzer, M., Ehrlich, I., Rasmussen, J., Gebbeken, N., Dendorfer, S., 2016b. Sensitivity of lumbar spine loading to anatomical parameters. J. Biomech. 49 (6), 953–958.

Raabe, M.E., Chaudhari, A.M.W., 2016. An investigation of jogging biomechanics using the full-body lumbar spine model: Model development and validation. J. Biomech. 49 (7), 1238–1243.

Rajaee, M.A., Arjmand, N., Shirazi-Adl, A., Plamondon, A., Schmidt, H., 2015. Comparative evaluation of six quantitative lifting tools to estimate spine loads during static activities. Appl. Ergon. 48, 22–32.

Rashedi, E., Khalaf, K., Nassajian, M.R., Nasseroleslami, B., Parnianpour, M., 2010. How does the central nervous system address the kinetic redundancy in the lumbar spine? Three-dimensional isometric exertions with 18 Hill-model-based muscle fascicles at the L4-L5 level. Proc. Inst. Mech. Eng. H J. Eng. Med. 224 (3), 487–501.

Rasmussen, J., Damsgaard, M., Voigt, M., 2001. Muscle recruitment by the min/max criterion—a comparative numerical study. J. Biomech. 34 (3), 409–415.

Rohlmann, A., Bergmann, G., Graichen, F., Mayer, H.M., 1995. Telemeterized load measurement using instrumented spinal internal fixators in a patient with degenerative instability. Spine 20 (24), 2683–2689.

Rohlmann, A., Graichen, F., Kayser, R., Bender, A., Bergmann, G., 2008. Loads on a telemeterized vertebral body replacement measured in two patients. Spine 33 (11), 1170–1179.

Sato, K., Kikuchi, S., Yonezawa, T., 1999. In vivo intradiscal pressure measurement in healthy individuals and in patients with ongoing back problems. Spine 24 (23), 2468–2474.

Schultz, A., Andersson, G., Ortengren, R., Haderspeck, K., Nachemson, A., 1982. Loads on the lumbar spine. Validation of a biomechanical analysis by measurements of intradiscal pressures and myoelectric signals. J. Bone Joint Surg. Am. 64 (5), 713–720.

Seth, A., Sherman, M., Reinbolt, J.A., Delp, S.L., 2011. OpenSim: a musculoskeletal modeling and simulation framework for in silico investigations and exchange. Proc. IUTAM 2, 212–232.

Shirazi-Adl, A., El-Rich, M., Pop, D.G., Parnianpour, M., 2005. Spinal muscle forces, internal loads and stability in standing under various postures and loads—application of kinematics-based algorithm. Eur. Spine J. 14 (4), 381–392.

Stambolian, D., Eltoukhy, M., Asfour, S., 2016. Development and validation of a three dimensional dynamic biomechanical lifting model for lower back evaluation for careful box placement. Int. J. Ind. Ergon. 54, 10–18.

Stokes, I.A., Gardner-Morse, M., 1995. Lumbar spine maximum efforts and muscle recruitment patterns predicted by a model with multijoint muscles and joints with stiffness. J. Biomech. 28 (2), 173–186.

Stokes, I.A.F., Gardner-Morse, M., 1999. Quantitative anatomy of the lumbar musculature. J. Biomech. 32 (3), 311–316.

Theado, E.W., Knapik, G.G., Marras, W.S., 2007. Modification of an EMG-assisted biomechanical model for pushing and pulling. Int. J. Ind. Ergon. 37 (11–12), 825–831.

Van Toen, C., Sran, M.M., Robinovitch, S.N., Cripton, P.A., 2012. Transmission of force in the lumbosacral spine during backward falls. Spine 37 (9), E519–27.

Vasavada, A.N., Li, S., Delp, S.L., 1998. Influence of muscle morphometry and moment arms on the moment-generating capacity of human neck muscles. Spine 23 (4), 412–422.

Wilke, H., Neef, P., Hinz, B., Seidel, H., Claes, L., 2001. Intradiscal pressure together with anthropometric data—a data set for the validation of models. Clin. Biomech. 16 (Suppl. 1), S111–S126.

Wong, K.W., Luk, K.D., Leong, J.C., Wong, S.F., Wong, K.K., 2006. Continuous dynamic spinal motion analysis. Spine 31 (4), 414–419.

Zajac, F.E., 1989. Muscle and tendon: properties, models, scaling, and application to biomechanics and motor control. Crit. Rev. Biomed. Eng. 17 (4), 359–411.

FURTHER READING

Holmes, J.W., 2006. Teaching from classic papers: Hill's model of muscle contraction. Adv. Physiol. Educ. 30 (2), 67–72.

CHAPTER

ANIMAL MODELS FOR SPINE BIOMECHANICS

16

Gloria Casaroli*, Kelly Wade†, Tomaso Villa*,‡, Hans-Joachim Wilke†

IRCCS Galeazzi Orthopedic Institute, Milan, Italy Institute of Orthopaedic Research and Biomechanics, Ulm University, Ulm, Germany† Politecnico di Milano, Milan, Italy‡*

WHY USE ANIMAL MODELS?

Animal models have played an important role in the investigation of spine biomechanics. Early experiments on vertebral and disc replacements were conducted on dogs, monkeys, and rabbits (Urbaniak et al., 1973; Schneider and Oyen, 1974; Stern and Coulson, 1976; Kahanovitz et al., 1985; Timmis et al., 1986; Moon et al., 1986; Hou et al., 1991). Other early studies on the development of spinal deformities (e.g., scoliosis) were conducted using different animal species such as rats, mice, and rabbits (Riggins et al., 1977; Rodriguez et al., 1965; Ponseti and Baird, 1952; Langenskiold and Michelsson, 1961; Miles and Heath, 1946). In conjunction with these early investigations, researchers have studied the anatomy, the biomechanical behavior, and the physiology of the animal models in order to find the optimal surrogate of the human spine for each specific purpose.

The use of animal samples is preferable to the use of human tissue for many reasons. First, human tissue has a high variability because of anatomical and pathological differences. The majority are gathered from aged donors who have degenerated discs, low bone quality, and other pathologies, all of which can influence test results in unexpected ways. Second, human specimens have a low availability and high associated expenses, making the collection of a large number of specimens impractical. Finally, certain ethical reasons favor the use of animal models. Indeed, biomechanical studies often require a relatively large number of samples to show significant trends; therefore, in the early stages of an investigation, it is not ethical to expend valuable healthy human samples to explore trends. Rather, the phenomena should be established by using a suitable animal model and then confirmed by using a small amount of suitable human tissue.

The use of animals for scientific purposes is strictly regulated by the European Parliament (Directive 2010/63/EU, 2010) and other such bodies worldwide. These regulations are intended to limit the use of animals and minimize the suffering of animals that are used in experiments. For conducting in vivo studies, permission must be obtained from an ethics committee, and the subjects' housing, feeding, and care must comply with the FELASA requirements (Federation of European Laboratory Animal Science Associations) and conform to ARRIVE guidelines (Animal Research: Reporting of In Vivo Experiments) (Kilkenny et al., 2012). Although the use of animal models is necessary, researchers have an ethical obligation to limit their use and to ensure that the optimal model is chosen to produce relevant results. Both large and small animal species have been studied and have

Biomechanics of the Spine. https://doi.org/10.1016/B978-0-12-812851-0.00016-1

been shown to exhibit similarities with the human spine. However, there is no single perfect surrogate for human specimens, but careful selection of an appropriate animal model can lead to useful, meaningful results from a study.

SPINAL LOADS

A common misconception is that quadrupedal spines are not subjected to axial load like the upright, bipedal human spine is. In reality, however, it has been demonstrated that quadruped spines not only experience axial loads but also that they may in fact be higher than in humans. Indeed, high muscle and ligament forces act on the quadrupedal spine to constrain its movement in the frontal and sagittal planes (Reitmaier et al., 2013). It is important to note that the trabeculae of the vertebral bodies of both bipeds and quadrupeds are oriented from endplate to endplate (Fig. 1), implying that they undergo axial loading (Smit, 2002).

Moreover, this implication is supported by the observation that the bone density of the vertebral bodies is two to four times higher in quadrupeds than in humans, suggesting that these bodies support higher axial stresses. Although there are no directly measured forces in the literature, this conclusion is supported by some experimental studies on sheep and cattle (runt cows) (Reitmaier et al., 2013; Buttermann et al., 2009). In these studies, intradiscal pressure (IDP) was measured in vivo during different activities, and the regression curve between the compressive force and the IDP was determined from in vitro testing (Fig. 2). This finding indicated that the discs of quadrupeds are generally subjected to higher pressure than those of humans. In sheep, the IDP level during static activities such as reclining and sleeping was four times that of humans, and two or three times higher during dynamic activities such as walking; whereas the values of cows were closer to humans. For standing, IDP in sheep and runt cows (0.70 and 0.80 MPa, respectively) was similar to that of humans (0.55–1.2 MPa). The regression curve in sheep was steeper than in humans or cattle, indicating that minor increases in compressive force caused a higher increase in pressure. This result is due to the lower cross-sectional area and height of the ovine disc, supporting the assertion that the different geometry is one of the main reasons for the

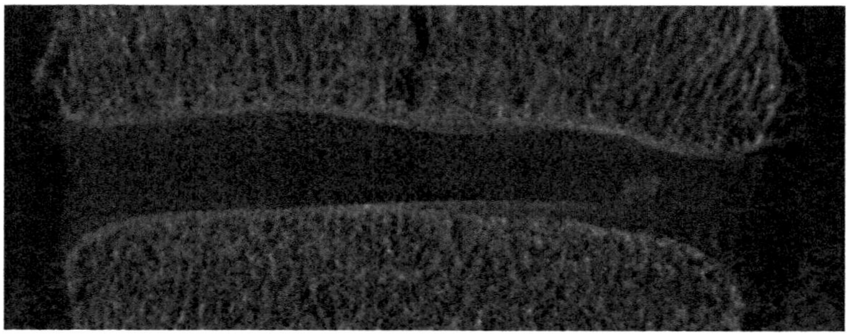

FIG. 1

Micro-CT scan of an ovine lumbar L2-L3 functional spinal unit. The CT scan clearly shows the cranio-caudal orientation of the trabeculae.

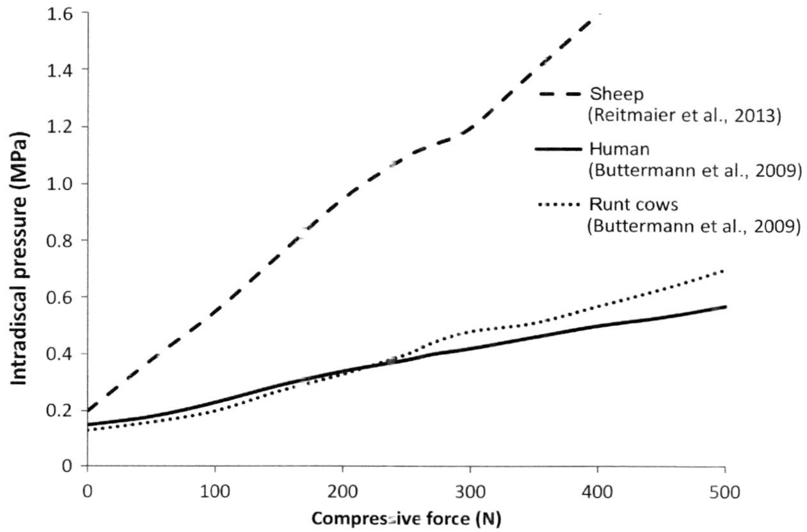

FIG. 2

Linear correlation between the compressive force and IDP measured ex vivo.

Adapted from Buttermann, G.R., Beaubien, B.P., Saeger, L.C., 2009. Mature runt cow lumbar intradiscal pressures and motion segment biomechanics. Spine J. 9(2), 105–114; Reitmaier, S., et al., 2013. Preliminary investigations on intradiscal pressures during daily activities: an in vivo study using the merino sheep. PLoS ONE 8(7), e69610.

different mechanical behavior to the human disc and such for differences in scale between the model systems need to be taken into account.

However, a direct comparison of in vitro measurements between human and animal specimens is complicated by the fact that the majority of animal specimens are healthy and thus well hydrated, whereas donated human discs often came from deceased elderly people with degeneration characterized by a lower IDP because of the reduced proteoglycan and water content of the nucleus (Urban and Roberts, 2003).

SPINAL FLEXIBILITY AND RANGE OF MOTION

One of the most important parameters used to compare the biomechanical behavior of the functional spinal units is flexibility in the three main anatomical planes. Many studies have been performed to measure the whole spine flexibility of common animal models to enable their comparison with the human spine (Figs. 3–5) (Wilke et al., 1996, 1997a, 2011). It has been found that in human and all animal species, the range of motion (ROM) in axial rotation is smallest in the lumbar region (Fig. 4). In addition, the absolute values of ROM in the lumbar region are smaller for bovine, ovine, and porcine, especially in flexion-extension (Fig. 3). Qualitative comparison among the different species indicates that the ovine and porcine spines have a behavior similar to that of the human spine in the lumbar

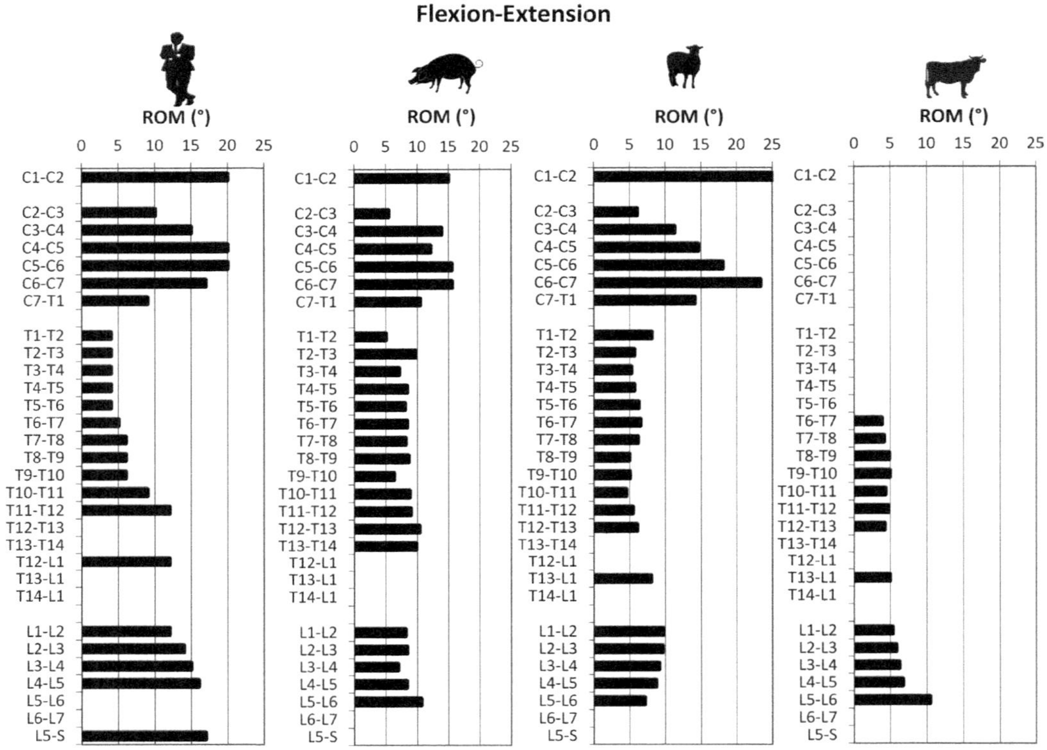

FIG. 3

Average range of motion (ROM) in flexion-extension for each single segment of human, pig, sheep and cow.

Adapted from Wilke, H.J., Krischak, S., Claes, L., 1996. Biomechanical comparison of calf and human spines. J. Orthop. Res. 14(3), 500–503; Wilke, H.-J., Kettler, A., Claes, L., 1997. Are sheep spines a valid biomechanical model for human spines? Spine 22(20), 2365–2374; Wilke, H.-J., Geppert, J., Kienle, A., 2011. Biomechanical in vitro evaluation of the complete porcine spine in comparison with data of the human spine. Eur. Spine J. 20(11), 1859–1868.

region, that the bovine spine has a similar flexibility to the human spine in the thoracic region, and that the porcine spine is most similar to the human spine in the cervical region.

STRUCTURAL AND BIOLOGICAL DIFFERENCES BETWEEN HUMANS AND COMMON ANIMAL MODELS

Animal models are widely used in biomechanical research to investigate herniation and degeneration processes and to evaluate new treatment methods for the intervertebral disc (IVD). Geometrical properties play a primary role in the determination of the mechanical behavior of the spine, and the selection of the most appropriate model depends on the particular process (herniation or degeneration) that will be investigated.

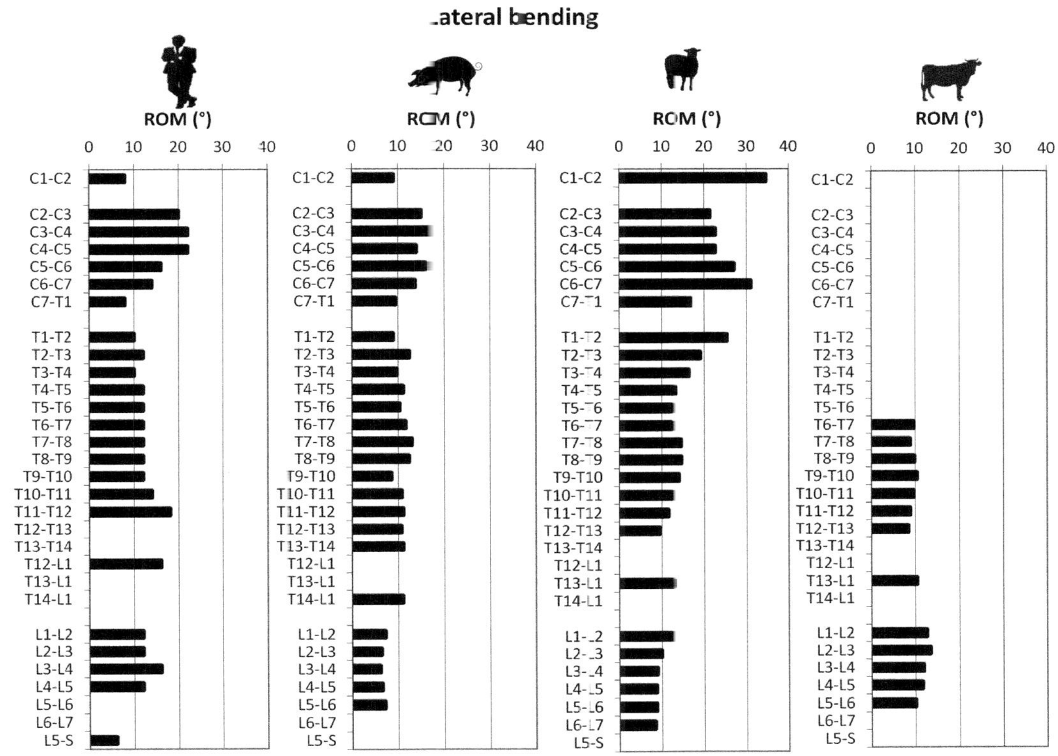

FIG. 4

Average range of motion (ROM) in right and left lateral bending for each single segment of human, pig, sheep and cow.

Adapted from Wilke, H.J., Krischak, S., Claes, L., 1996. Biomechanical comparison of calf and human spines. J. Orthop. Res. 14(3), 500–503; Wilke, H.-J., Kettler, A., Claes, L. 1997. Are sheep spines a valid biomechanical model for human spines? Spine 20(22), 2365–2374; Wilke, H.-J., Geppert, J., Kienle, A., 2011. Biomechanical in vitro evaluation of the complete porcine spine in comparison with data of the human spine. Eur. Spine J. 20(11), 1859–1868.

Therefore a sound knowledge of the geometrical properties helps researchers plan their experiments and adopt the optimal model for their specific purpose.

In fact, animal models with anatomical dimensions, spine flexibility, and bone density similar to that of humans are best suited for studying new implants (Kettler et al., 2007). On the other hand, investigation of the herniation process, degeneration, and mechanical behavior might be strictly related to the disc shape and the mechanical properties of the annulus fibrosus (Yates et al., 2010).

This section presents comparisons among the anatomical parameters of the animal models used in the literature and their suitability as models of the human spine.

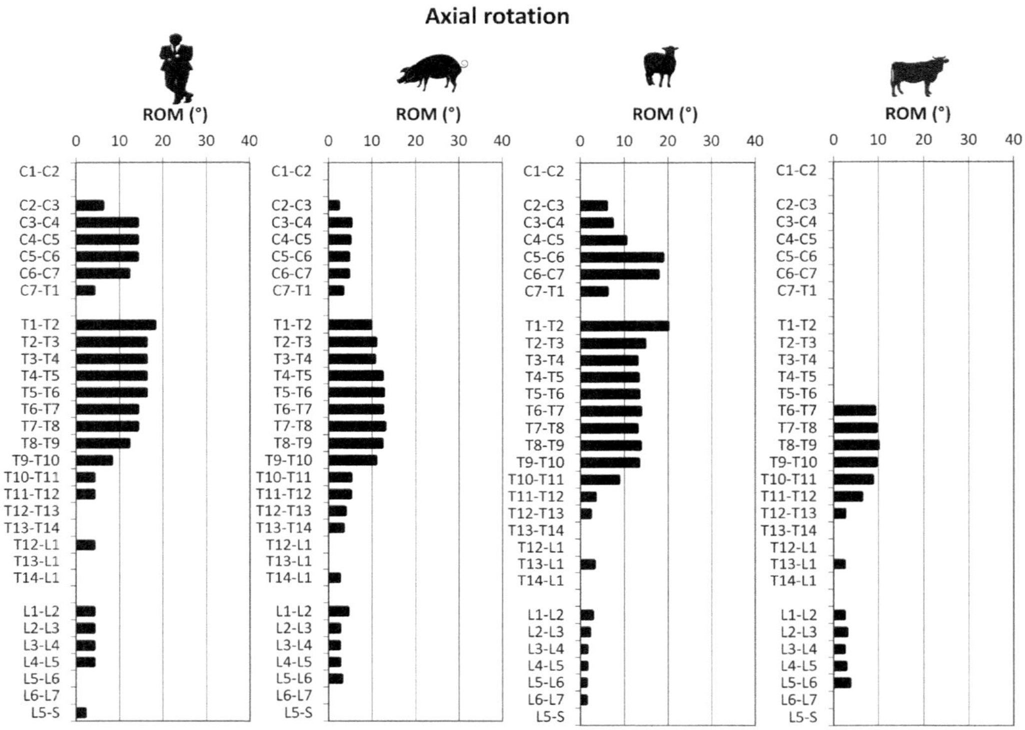

FIG. 5

Average range of motion (ROM) in right and left axial rotation for each single segment of human, pig, sheep and cow.

Adapted from Wilke, H.J., Krischak, S., Claes, L., 1996. Biomechanical comparison of calf and human spines. J. Orthop. Res. 14(3), 500–503; Wilke, H.-J., Kettler, A., Claes, L., 1997. Are sheep spines a valid biomechanical model for human spines? Spine 20(22), 2365–2374; Wilke, H.-J., Geppert, J., Kienle, A., 2011. Biomechanical in vitro evaluation of the complete porcine spine in comparison with data of the human spine. Eur. Spine J. 20(11), 1859–1868.

DISC ANATOMY

Considerable variability of the anatomical properties of the IVD exists across species (Fig. 6). The first difference between the human and the animal disc concerns their size. In general, human discs are bigger than animal discs (Table 1). Among large animal species, baboons have the disc transversal area closest to that of humans; however, few studies have been conducted on primates such as baboons and monkeys because of their low availability and the high cost of these species. In contrast, some small animal species are appropriate for investigating surgical techniques because their discs are shaped similarly to human discs; for example, when taking into account normalized height, anteroposterior width, and nucleus pulposus area, the mouse tail appears to be most representative of the human disc (O'Connell et al., 2007). Because it influences the mechanism of failure, the selection of a disc model with geometrical parameters similar to the human one is very important in biomechanical experiments.

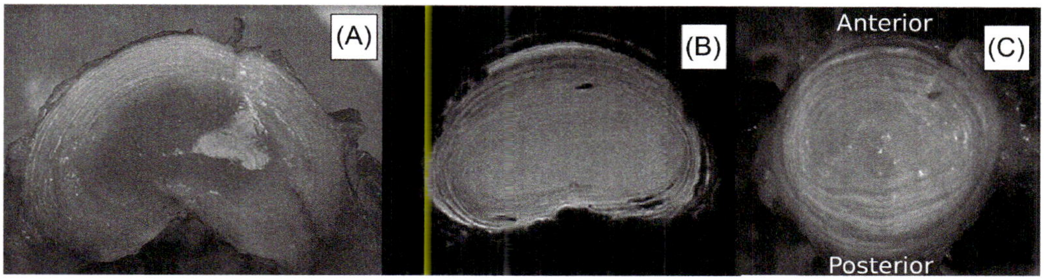

FIG. 6

Transversal shape of the lumbar intervertebral disc (A in pig (Mizrahi et al., 2013), (B) ovine, and (C) bovine tail (Michalek and Iatridis, 2012).

Table 1 Anatomical Properties of the Intervertebral Disc in the Main Animal Models

	DH (mm)	MLW (mm)	APW (mm)	DA (mm^2)	NPA(mm^2)	NPA:DA
Human	11.3[a] (0.3)	55.9 (9.4)[a]	37.[illegible] (4.7)[a]	1727 (550)[a]	479 (110)	0.28
Baboon	4.45[a] (1.39)	35.4 (2.3)[a]	22.[illegible] (0.8)[a]	749 (82)[a]	242 (50)	0.32
Ovine	3.93[a] (0.07)- 4.9[b,c] (0.4)	24.9 (1.7)[b]– 34.5 (2.9)[a]	18.[illegible] (1.8)[b]– 21.6 (2.1)[a]	676 (122)[a]	267 (79)	0.39
Bovine (T)	6.9 (0.35)[c]– 12.9 (0.4)[b]	23.7 (1.7)[b]– 28.9 (2.0)[a]	24.[illegible] (1.2)[b]– 27.[illegible] (1.3)[a]	622 (71)[a]	176 (22)	0.28
Porcine (C)	7.8 (1.3)[b]	33.3 (0.02)[d]	17.[illegible] (0.1)[d]			
Rat (L)	0.93 (0.24)[a]	5.79 (0.29)[a]	3.4[illegible] (0.6)[a]	20.4 (2.1)[a]	5.00 (2.06)	0.24
Mouse (L)	0.31 (0.03)[a]	1.84 (0.03)[a]	1.2[illegible] (0.1)[a]	1.81 (0.14)[a]	0.33 (0.07)	0.18

DH, disc height; MLW, mediolateral width; APW, anteroposterior width; DA, disc area; NPA, nucleus pulposus area.
[a]*O'Connell et al. (2007).*
[b]*Monaco et al. (2016).*
[c]*Wilke et al. (1997a,b).*
[d]*Sheng et al. (2016).*

For example, a limacon-shaped disc, such as in humans and sheep, is most likely to suffer herniation in the posterolateral part of the disc (van Heeswijk et al., 2017; Yates et al., 2010).

One approach to the use of animal models with a transversal area smaller than the human one is to scale the applied load to induce an IDP equal to the human one (Reitmaier et al., 2013; Alini et al., 2008). Indeed, the fact that the animal discs have a different size does not mean that generated stresses and strains are different from the human ones. The geometrical properties of the discs of each species are related to the loads that they have to support; therefore, in small animal species such as mice or rabbits, the forces will be lower than in large animal species such as sheep or cows, but the stresses within the discs could be similar. It is reasonable to conclude that if the IDP is similar, the stress state of the disc will be comparable.

Depending on the investigation, it may be important to consider how the disc anatomical parameters vary along the whole spine. For example, in humans the anteroposterior diameter and disc height decrease from the lumbar to the cervical region, whereas they are more constant in large animals such as pigs, ovine, and calves.

Therefore, compared to the human spine, these species have larger discs in the cervical region but smaller discs in the lumbar region (Sheng et al., 2016; Alini et al., 2008).

In addition to geometrical differences, differences in IVD behavior are determined by structural characteristics such as the annulus lamellar structure and thickness; if these differences did not exist among species, only the geometrical parameters would need to be taken into account in comparisons with the human disc. It has been demonstrated that the ovine lumbar disc is a reasonable model of the human lumbar IVD in terms of water content, collagen content, and collagen fiber orientation. The water content in sheep was found to be about 80%, as well as in the normal human disc (Reid et al., 2002; Lyons et al., 1981), and the collagen content was in reasonable agreement with the 33% found in the human disc (Lyons et al., 1981). In contrast, the collagen fibers' orientation in sheep was found to be between 20 and 30 degrees, with a slightly lower inclination in the posterior annulus (Reid et al., 2002; Casaroli et al., 2017); whereas in the human disc, the posterior fiber reached a double value of the anterior one (Holzapfel et al., 2005).

VERTEBRAL BODY ANATOMY

Large animal models are often used for preclinical tests to investigate the effect of new spinal implants and surgical situations. Obviously, an optimal surrogate of the human spine should have similar anatomical parameters, size, and mechanical properties.

In the lumbar region, sheep and deer are considered good alternatives to human specimens. These species have generally shown trends similar to those of humans, except in the vertebral body height, which in sheep is greater than the width (Table 2). However, in both species, the vertebrae have larger width than depth, giving the vertebral bodies an oval shape. It may also be important to note that sheep spines are slightly kyphotic, rather than lordotic. In the cervical region, porcine vertebrae have a very similar anatomy to the human vertebrae. Calf cervical vertebrae are larger and have thicker pedicles than humans do, but the spinal canal dimensions of the two exhibit only a few differences (Sheng et al., 2016) (Table 2). The main limitation of immature calves is that they have growth plates that could lead to inaccurate results in traumatic studies, whereas using adult specimens could occur in oversize errors (Sheng et al., 2016).

An important difference between the endplate region of human spines and those of almost all other animals (except for some large primates) (Ogden et al., 1994) relates to their development. As previously mentioned, in humans the growth of the vertebrae proceeds from the cartilaginous endplate and a ring of cartilage around the edge of the vertebral body, which is known as the ring apophysis (Bick and Copel, 1951; Peacock, 1952). In most other animals, the vertebrae grow from a separate growth plate that fuses at maturity (Hasler et al., 2010). The effect of fusion on the appearance of the ovine vertebrae is shown in Fig. 7A–C in which the demarcation between the growth plate and the rest of the vertebra is clearly visible in the immature subject. In the image from the mature ewe, the growth plate can be seen to have fused because of the breakdown of this line of demarcation. However, the approximate location of the growth plate is still visible in this image.

Table 2 Anatomical Properties of the Vertebral Bodies in the Main Animal Models

	EPW	EPD	VBH	SCW	SCD	PW	PH
Human (L)[a]	45.75	34.62	23.78	24.86	18.6	11.48	16.08
Human (C)[b]	19.1 (1.2)	15.6 (1.6)		26.1 (1.8)	27.3 (1.5)	5.5 (0.3)	11.2 (0.5)
Ovine (L)[c]	30.7	19.65	38.3	13.8	9.2	10	34.3
Porcine (L)[a]	37.48	23.32	35.32	17.98	11.78	12.13	21.20
Porcine (C)[b]	33.3 (2.1)	17.8 (0.6)		22.1 (2.2)	28.2 (1.5)	11.1 (0.7)	17.9 (1.3)
Bovine (L)[d]	31.2 (3.5)	24.8 (1.5)	26.4 (2.1)	19.6 (1.4)	15.7 (1.1)	8.5 (1.6)	
Baboon (C)[b]	10.6	10.6	10	14.8	10.5	2.8	7
Deer (L)[e]	30.8 (0.5)	19.8 (0.8)	43.7 (1.5)				

EPW, endplate width; EPD, endplate depth; VBH, vertebral body height; SCW, spinal canal width; SCD, spinal canal depth; PW, pedicle width; PH, pedicle height; L, lumbar; C, cervical.
[a]*Dath et al. (2007).*
[b]*Sheng et al. (2016).*
[c]*Wilke et al. (1997a,b).*
[d]*Cotterill et al. (2017).*
[e]*Wang et al. (2015).*

It is possible that this difference in structure may lead to differences between the behavior of human and animal motion segments especially in cases where the animal motion segments have unfused growth plates. Conversely, using adult specimens could lead to errors if the bovine model is used because the adult bovine vertebra is considerably larger than that of the human vertebra (Sheng et al., 2016). These factors should be considered when designing testing protocols and interpreting experimental results to ensure that the growth plate status is not producing unintended sample behavior and potentially misleading findings.

We conclude this section by saying that, as previously mentioned, the bone density is higher in large quadrupeds than in humans, which can significantly affect results. Therefore applying a scaling factor based on the morphological parameters does not necessarily guarantee the same mechanical effect on human and animal specimens, especially where vertebral fractures are concerned.

NATURE OF THE NUCLEUS-CHONDRODYSTROPHOID HUMAN (OR OVINE, BOVINE) VERSUS NON-CHONDRODYSTROPHOID (PIGS, RODENTS, SOME DOGS)

The nature of the nucleus varies between species. Although it is commonly described as an amorphous gelatinous structure possessing little or no connectivity with its surroundings, which could give the impression that the disc is a binary structure and that the nucleus could be easily separated from the annulus and endplates, we emphasize the points presented in Chapter 8. In humans, there is a gradual transition between the annulus and endplates so that, although the nucleus is highly mobile and hydrated, which enables its load-bearing properties, it does contain a network of fine fibers that support its cells and proteoglycans. Although this factor enables its hydrostatic behavior, we emphasize that the nucleus is not actually a liquid in its mature and nondegenerate state—considerable effort requiring cutting or tearing is needed to separate it from its surroundings!

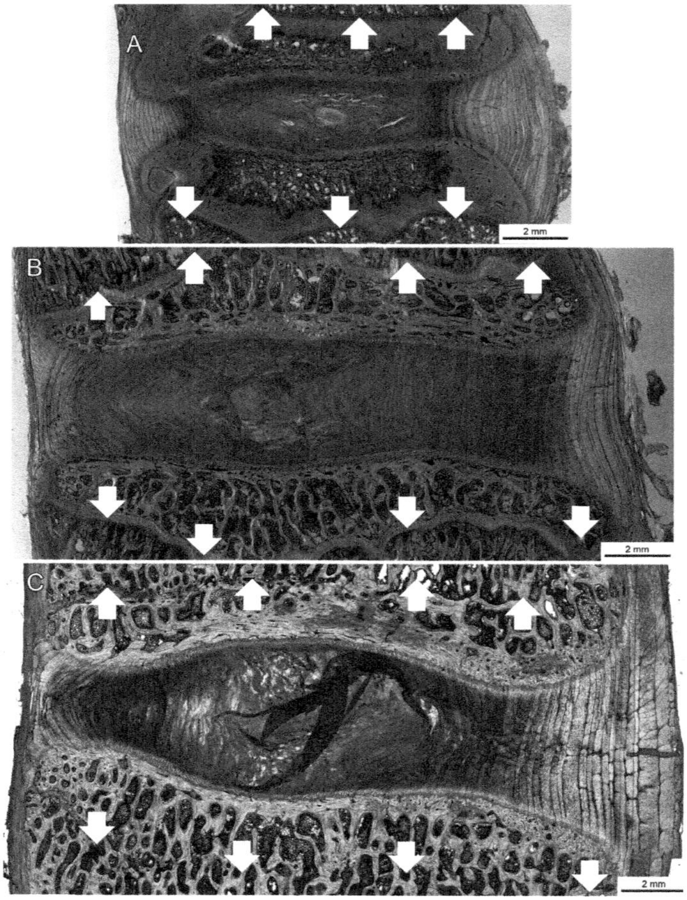

FIG. 7

Macro images from ovine lumbar motion segments of increasing levels of maturity. (A) and (B) are from newborn and three-month-old lambs, respectively, and in these images, the demarcation between the growth plate and the rest of the vertebra is clearly visible, as indicated by the red arrows. (C) is from a mature ewe, and in this image, there is no distinct boundary between the growth plate and the rest of the vertebra in the arrowed region. However, the approximate location of the growth plate can still be seen, as shown by the white arrows.

Of the common large-animal model systems, the ovine and bovine nuclei are similar in nature to the human nucleus, but it is notable that the porcine nucleus is quite different. Indeed, it is truly a viscous fluid that can flow under its own weight. This difference in properties should be considered both when choosing model systems for biomechanical testing and when interpreting results in the literature.

Recently a comparative study was conducted on cadaveric human tissue, which revealed that these structural features are also present in human discs (Brown et al., 2017). It was also found that degeneration leads to alteration or destruction of these features, particularly in the nucleus-endplate region, which indicates that, although the overall dimensions of discs may vary between species, the same or very similar mechanisms (or principles) operate within them. As has been noted elsewhere (Alini et al., 2008), it is quite reasonable to conclude that the aforementioned difference in structure will lead to differences in behavior of the inner annulus and nucleus when exposed to loading. Put another way, it is reasonable to expect a liquid nucleus to flow through smaller defects in a different manner to a fibrous semi-gel. It has therefore been suggested that the porcine model may approximate the behavior of the human disc in the outer annulus but not necessarily the inner annulus.

These similarities in alteration of the structural features also further strengthen the premise that animal models are a suitable surrogate for healthy human tissue in preliminary studies and can thus be used to explore disc behavior in detail before difficult-to-obtain human tissue is used to confirm this behavior.

BIOLOGICAL COMPOSITION

The IVD is populated with cells whose function is to produce and maintain the extracellular matrix of the disc, and thus they are crucial to maintaining the functional properties of the disc. Because the disc is fibrocartilage, these cells are thought to be similar to those found in articular cartilage. Indeed, the cell populations have been described in the past as being chondrocyte-like in the nucleus and fibroblast-like in the annulus (Johnson and Roberts, 2003; Errington et al., 1998). It should, however, be noted that the cells of the disc are in fact somewhat different from those found in articular cartilage (Johnson and Roberts, 2003). Specifically, the cell phenotypes have been shown to be different (Clouet et al., 2009), and the products of the cells are also different between the cell types (Cui et al., 2010; Melrose et al., 2003).

The nature of the cell population of the nucleus is ambiguous and the source of much debate. Broadly speaking, the nucleus contains two major cell types: chondrocyte-like cells (which are sometimes referred to as mature nucleus pulposus cells) and notochordal cells. In humans and many large animals, the notochordal cells are thought to be replaced by mature nucleus pulposus cells after skeletal maturity is reached (Hunter et al., 2004). In other animals, this is not necessarily the case; mice (McCann et al., 2012), pigs (Guehring et al., 2009), cats, and non-chondrodystryphoid dogs retain their notochordal cells into maturity. Furthermore, some species, such as rabbits, are considered to be skeletally mature by some researchers at the age of six months (Guehring et al., 2010), at which point their disc nuclei mostly consists of notochordal cells. However, it has been shown in rabbits that the notochordal cells are in fact largely replaced by chondrocyte-like cells by the age of 24 months (Souter and Taylor, 1970). Therefore age is important when considering the methods and findings of cell population studies especially in animals that may not be truly mature at the time of sacrifice.

ANIMAL MODELS OF SPINE DEFORMITIES AND DEGENERATIVE PROCESS

Since the beginning of the 20th century, animal models have been used to study the etiology and development of spinal deformities, pathologies, and new surgical techniques. One of the most investigated spinal deformities is idiopathic scoliosis. Invasive surgical techniques are required to reduce the deformity and to avoid progression of the disease (Odent et al., 2015; Bobyn et al., 2015; Wimmer et al., 2015). Idiopathic scoliosis has been observed to spontaneously occur only in humans and, in some limited cases, in chickens (Rigdon and Mack, 1968). Therefore many researchers have developed techniques to induce human scoliosis in animals. The main methods to induce scoliosis can be divided into five categories: prenatal induction (e.g., exposing the mother to hypoxia in mice or to insulin injection) (Ingalls and Curley, 1957; Duraiswami, 1951), dietary deficiency (e.g., feeding rats of *Lathyrus odoratus*) (Yamamoto, 1966; Barrow et al., 1974), immobilization techniques (in vivo experiments conducted on cats, dogs, and rabbits) (Poussa et al., 1991; Wullstein, 1992), systematic methods such as removing pineal glands in chickens (Coillard and Rivard, 1996; Wang et al., 1997), and local surgical procedures (An and Friedman, 1999).

Both biped and quadruped animals have been involved in in vivo and ex vivo investigations of scoliosis. Biped animals, especially chickens, are considered to have the advantage of being subjected to axial forces upon the spine other than gravity. However, the spine of these species is obviously smaller than the human one, with the consequence that they are not suitable for full-scale implant testing. Large animal models like pigs, sheep, goats, and calves are more suitable in this regard because the size of their vertebrae is closer to the human one.

Many surgical techniques have been developed to induce scoliosis in quadrupeds. The primary techniques include the excision or denervation of the local musculature, the resection of ribs, and the fixation of vertebrae or ribs to each other (Langenskiold and Michelsson, 1961). One of the most common recent techniques to induce scoliosis is unilateral tethering (Fig. 8). This surgical technique has the potential to be particularly successful for spinal growth modulation to manage early onset scoliosis (Thompson et al., 2007). Primary experiments have been conducted on calves (Newton et al., 2002), small pigs (Schwab et al., 2009), and goats (Braun et al., 2006). The limitation of the spine of goats is that it is too rigid and significantly smaller than the spine of humans, and thus the implants need to be scaled down. Bogie et al. suggested the use of a radiopaque polymethylmethacrylate tether to successfully induce a progressive curvature in three planes in a young porcine model (Bogie et al., 2017).

A reproducible progressive curvature of the lordo-scoliotic spine has also been replicated in immature ovine spines using a unilateral laminar tether (Burke et al., 2015). The spine of calves has been used to reproduce an in vitro model of idiopathic scoliosis (Wenger et al., 1982; Shono et al., 1991; Wilke et al., 2015). Recently, a scoliotic curvature was obtained through a complex surgical procedure (Fig. 9) in which the nuclei pulposi were substituted with silicone and the vertebral bodies were cut and fixed to obtain a wedge profile (Wilke et al., 2015).

Animal models have also been used to study the degenerative process. Probably the most famous of these is the ovine model developed by Fraser and Osti in which an incision was made in the outer annulus of the disc and the disc histologically examined up to 18 months following the procedure. This study revealed that, although the outer annulus tended to heal, tears and clefts progressed from the site of initial damage into the inner annulus (Osti et al., 1990). Other investigators have used models of rabbit and rat disc injuries (Zhang et al., 2009; Cunha et al., 2017; Kim et al., 2005) to investigate the progression of herniation, but these results should be interpreted with caution for reasons outlined previously in this chapter; most notably, the different nature of the nucleus in terms of structure cell populations may lead to differences in both the initial mechanical response of the disc to injury and its healing response.

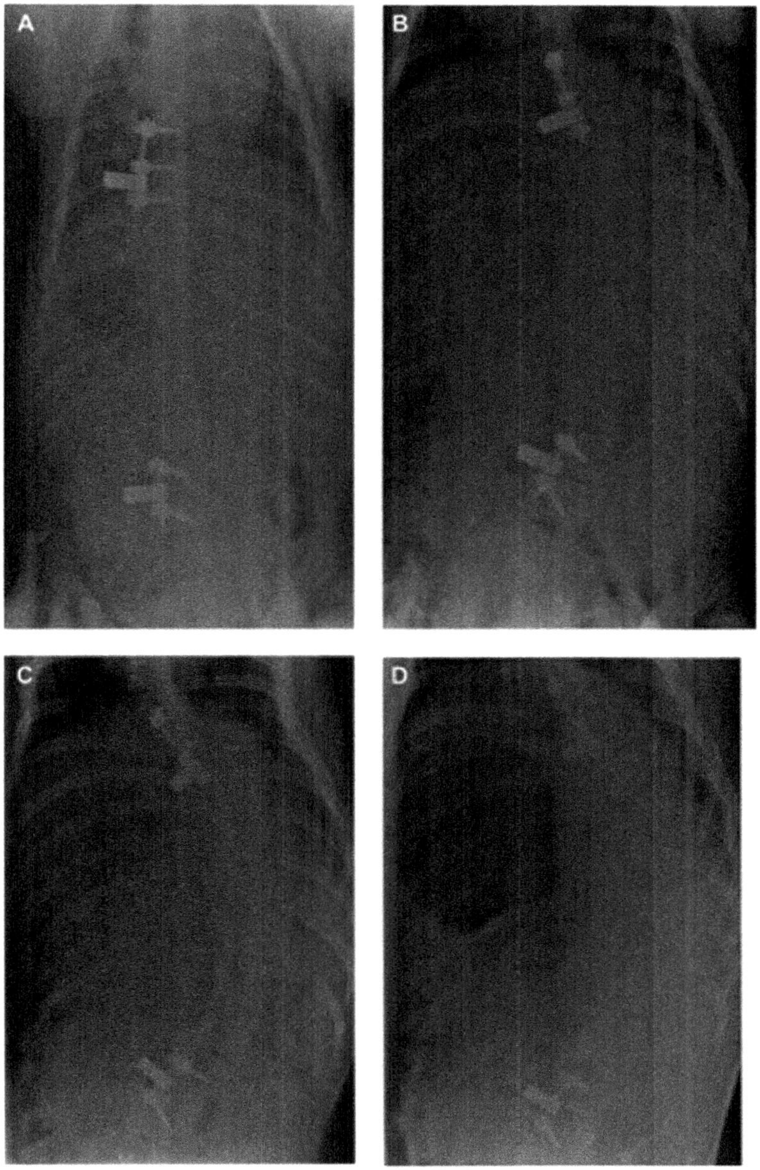

FIG. 8

Scoliotic curvature obtained by Bogie and colleagues on porcine spine through posterior spinal tether. (A) Instrumentation with straight spine direct postoperatively. (B) Four-week follow-up with the development of the curvature with thoracic asymmetry. (C) Ten-week follow-up before tether release. (D) Twenty-week follow-up with slight loss of curvature after tether release.

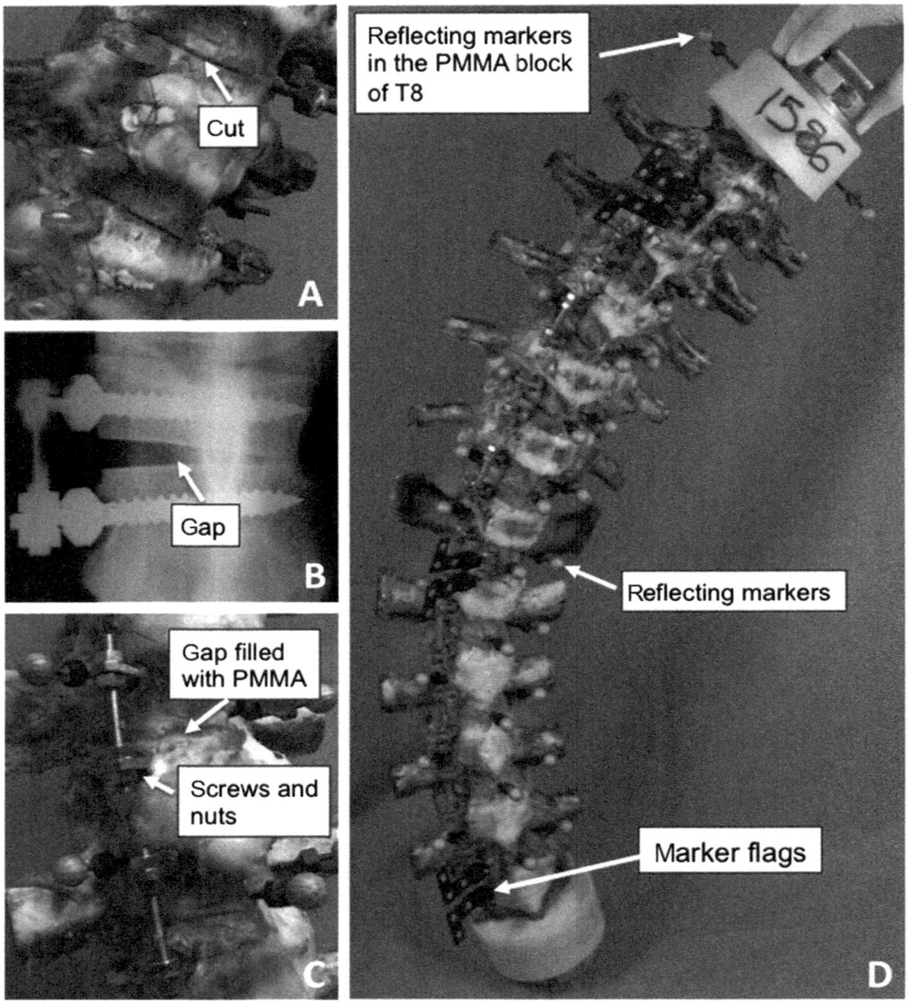

FIG. 9

Technical procedure proposed by Wilke to develop a scoliotic curvature of the bovine spine for in vitro test. Horizontal cuts (A+B) generated the wedge shape of the vertebral bodies. Then screws, thread rods, and nuts were used to spread and fix each in place, and resulting gaps were filled with polymethylmethacrylate (C). Reflecting markers were used to measure the intersegmental motion and to monitor the rigidity of the gaps (D).

The preceding studies are only few of the many attempts to reproduce clinical scenarios in animal models. There is no single optimal model for the human spine because the appropriate model depends on the purpose of the experiment. In general, large species are more suitable for investigating mechanical factors, preclinical tests on implants, and surgical techniques. Small species are often better suited to answering biological questions. It is the researcher's responsibility to select the optimal model for reaching the best result with a clinical relevance for humans.

REFERENCES

Alini, M., et al., 2008. Are animal models useful for studying human disc disorders/degeneration? Eur. Spine J. 17 (1), 2–19.

An, Y.H., Friedman, R.J., 1999. Animal Models in Orthopaedic Research. CRC Press, Taylor & Francis Group, Boca Raton, Florida.

Barrow, M.V., Simpson, C.F., Miller, E.J., 1974. Lathyrism: a review. Q. Rev. Biol. 49 (2), 101–128. Available from: http://www.jstor.org/stable/2820941.

Bick, E.M., Copel, J.W., 1951. The ring apophysis of the human vertebra; contribution to human osteogeny. II. J. Bone Joint Surg. Am. Vol. 33A (3), 783–787.

Bobyn, J.D., et al., 2015. Animal models of scoliosis. J. Orthop. Res. 33 (4), 458–467.

Bogie, R., et al., 2017. The development of a representative porcine early-onset scoliosis model with a standalone posterior spinal tether. Spine Deform. 5 (1), 2–10. Available from: http://linkinghub.elsevier.com/retrieve/pii/S2212134X16301113.

Braun, J.T., et al., 2006. Creation of an experimental idiopathic-type scoliosis in an immature goat model using a flexible posterior asymmetric tether. Spine 31 (13), 1410–1414.

Brown, S., et al., 2017. Staying connected: structural integration at the intervertebral disc – vertebra interface of human lumbar spines. Eur. Spine J. 26, 248–258.

Burke, J.G., et al., 2015. Creation of an ovine model of progressive structural lordo-scoliosis using a unilateral laminar tether. Eur. Spine J. 24 (7), 1382–1390.

Buttermann, G.R., Beaubien, B.P., Saeger, L.C., 2009. Mature runt cow lumbar intradiscal pressures and motion segment biomechanics. Spine J. 9 (2), 105–114. https://doi.org/10.1016/j.spinee.2007.09.006.

Casaroli, G., et al., 2017. A novel finite element model of the ovine lumbar intervertebral disc with anisotropic hyperelastic material properties. PLoS ONE 12 (5), e0177088.

Clouet, J., et al., 2009. Identification of phenotypic discriminating markers for intervertebral disc cells and articular chondrocytes. Rheumatology 48 (11), 1447–1450. Available at https://doi.org/10.1093/rheumatology/kep262.

Coillard, C., Rivard, C.H., 1996. Vertebral deformities and scoliosis. Eur. Spine J: Off. Publ. Eur. Spine Soc. Eur. Spinal Deform. Soc. Eur. Sec. Cervical Spine Res. Soc. 5 (2), 91–100.

Cotterill, P., et al., 2017. An anatomical comparison of the human and bovine thoracolumbar spine. J. Orthop. Res.: Off. Publ. Orthop. Res. Soc. 4 (3).

Cui, Y., et al., 2010. Differential gene expression profiling of metalloproteinases and their inhibitors: a comparison between bovine intervertebral disc nucleus pulposus cells and articular chondrocytes. Spine 35 (11), 1101–1108.

Cunha, C., et al., 2017. Joint analysis of IVD herniation and degeneration by rat caudal needle puncture model. J. Orthop. Res.: Off. Publ. Orthop. Res. Soc. 35 (2), 258–268.

Dath, R., et al., 2007. Anatomical measurements of porcine lumbar vertebrae. Clin. Biomech. 22 (5), 607–613.

Directive 2010/63/EU, 2010. Directive 2010/63/EU of the European Parliament and of the Council of 22 September 2010 on the protection of animals used for scientific purposes. Off. J. Eur. Union 33–79.

Duraiswami, P.K., 1951. Experimental causation of congenital skeletal defects and its significance in orthopaedic surgery. J. Bone Joint Surg. 34B (4), 646–698.

Errington, R.J., et al., 1998. Characterisation of cytoplasm-filled processes in cells of the intervertebral disc. J. Anat. 192 (Pt 3), 369–378.

Guehring, T., et al., 2009. Notochordal intervertebral disc cells: sensitivity to nutrient deprivation. Arthritis Rheum. 60 (4), 1026–1034.

Guehring, T., et al., 2010. Sensitivity of notochordal disc cells to mechanical loading: an experimental animal study. Eur. Spine J.: Off. Publ. Eur. Spine Soc. Eur. Spinal Deform. Soc. Eur. Sect. Cervical Spine Res. Soc. 19 (1), 113–121.

Hasler, C., Sprecher, C.M., Milz, S., 2010. Comparison of the immature sheep spine and the growing human spine: a spondylometric database for growth modulating research. Spine 35 (23), E1262–72.

van Heeswijk, V.M., et al., 2017. Posterolateral disc prolapse in flexion initiated by lateral inner annular failure: an investigation of the herniation pathway. Spine 42 (21), 1604–1613.

Holzapfel, G.A., et al., 2005. Single lamellar mechanics of the human lumbar anulus fibrosus. Biomech. Model. Mechanobiol. 3, 125–140.

Hou, T.S., et al., 1991. Lumbar intervertebral disc prosthesis. An experimental study. Chin. Med. J. 104 (5).

Hunter, C.J., Matyas, J.R., Duncan, N.A., 2004. Cytomorphology of notochordal and chondrocytic cells from the nucleus pulposus: a species comparison. J. Anat. 205 (5), 357–362.

Ingalls, H., Curley, J., 1957. Principles governing the genesis of congenital malformations induced in mice by hypoxia. N. Engl. J. Med. 257 (23), 1121–1127. https://doi.org/10.1056/NEJM195712052572306.

Johnson, W.E.B., Roberts, S., 2003. Human intervertebral disc cell morphology and cytoskeletal composition: a preliminary study of regional variations in health and disease. J. Anat. 203 (6), 605–612.

Kahanovitz, N., Arnoczky, S.P., Kummer, F., 1985. The comparative biomechanical, histologic, and radiographic analysis of canine lumbar discs treated by surgical excision or chemonucleolysis. Spine 10 (2).

Kettler, A., et al., 2007. Are the spines of calf, pig and sheep suitable models for pre-clinical implant tests? Eur. Spine J. 16, 2186–2192.

Kilkenny, C., et al., 2012. Improving bioscience research reporting: the arrive guidelines for reporting animal research. Osteoarthr. Cartil. 20, 256–260. Available at https://doi.org/10.1016/j.joca.2012.02.010.

Kim, K.S., et al., 2005. Disc degeneration in the rabbit: a biochemical and radiological comparison between four disc injury models. Spine 30 (1), 33–37.

Langenskiold, A., Michelsson, J., 1961. Experimental progressive scoliosis in the rabbit. J. Bone Joint Surg. 43B, 116–120.

Lyons, G., Eisenstein, S.M., Sweet, M.B., 1981. Biochemical changes in intervertebral disc degeneration. Biochim. Biophys. Acta 673 (4), 443–453.

McCann, M.R., et al., 2012. Tracing notochord-derived cells using a Noto-cre mouse: implications for intervertebral disc development. Dis. Model. Mech. 5 (1), 73–82. Available from: https://doi.org/10.1242/dmm.008128.

Melrose, J., Smith, S., Ghosh, P., 2003. Assessment of the cellular heterogeneity of the ovine intervertebral disc: comparison with synovial fibroblasts and articular chondrocytes. Eur. Spine J.: Off. Publ. Eur. Spine Soc. Eur. Spinal Deform. Soc. Eur. Sect. Cervical Spine Res. Soc. 12 (1), 57–65.

Michalek, A.J., Iatridis, J.C., 2012. Height and torsional stiffness are most sensitive to annular injury in large animal intervertebral discs. Spine J. 12 (5), 425–432.

Miles, M., Heath, R.D., 1946. Experimental production of scoliosis in rats. Anat. Rec. 94, 482.

Mizrahi, O., et al., 2013. Nucleus pulposus degeneration alters properties of resident progenitor cells. Spine J. 13 (7), 803–814. https://doi.org/10.1016/j.spinee.2013.02.065.

Monaco, L.A., Dewitte-Orr, S.J., Gregory, D.E., 2016. A comparison between porcine, ovine, and bovine intervertebral disc anatomy and single lamella annulus fibrosus tensile properties. J. Morphol. 277 (2), 244–251.

Moon, M.-S., Ok, I.-Y., Ha, K.-Y., 1986. The effect of posterior spinal fixation with acrylic cement on the vertebral growth plate and intervertebral disc in dogs. Int. Orthop. 10 (1).

Newton, P.O., et al., 2002. Asymmetrical flexible tethering of spine growth in an immature bovine model. Spine 27 (7), 689–693.

O'Connell, G.D.B., Vresilovic, E.J., Elliott, D.M., 2007. Comparison of animals used in disc research to human lumbar disc geometry [Miscellaneous Article]. Spine 32 (3), 328–333. Available from: http://graphics.tx.ovid.com/ovftpdfs/FPDDNCOBCCEFJA00/fs046/ovft/live/gv025/00007632/00007632-200702010-00008.pdf.

Odent, T., et al., 2015. Fusionless surgery in early-onset scoliosis. Orthop. Traumatol.: Surg. Res. 101 (6), S281–S288. https://doi.org/10.1016/j.otsr.2015.07.004.

Ogden, J.A., et al., 1994. Development and Maturation of the Axial Skeletoned. Raven Press, New York.

Osti, O.L., Vernon-Roberts, B., Fraser, R.D., 1990. Anulus tears and intervertebral disc degeneration. An experimental study using an animal model. Spine 15, 762–767.

Peacock, A., 1952. Observations on the postnatal structure of the intervertebral disc in man. J. Anat. 86 (2), 162–179.
Ponseti, I.V., Baird, W.A., 1952. Scoliosis and dissecting aneurysm of the aorta in rats fed with *Lathyrus odoratus* seeds. Am. J. Pathol. 28 (6).
Poussa, M., Schlenzka, D., Ritsila, V., 1991. Scoliosis in growing rabbits induced with an extension splint. Acta Orthop. Scand. 62 (2), 136–138.
Reid, J.E., et al., 2002. Sheep lumbar intervertebral discs as models for human discs. Clin. Biomech. 17, 312–314.
Reitmaier, S., et al., 2013. Preliminary investigations on intradiscal pressures during daily activities: an in vivo study using the merino sheep. PLoS ONE 8 (7), e69610. Available from: http://www.pubmedcentral.nih.gov/articlerender.fcgi?artid=3722231&tool=pmcentrez&rendertype=abstractAccessed November 18, 2014.
Rigdon, R.H., Mack, J., 1968. Spontaneous occurrence of scoliosis in the chicken. Avian Dis. 12 (3), 530–543.
Riggins, R.S., et al., 1977. Scoliosis in chickens. J. Bone Joint Surg. – Ser. A 59 (8).
Rodriguez Sr., R.M., Bailey, R.W., Rodriguez Jr., R.P., 1965. Skeletal lesions of lathyrism and effects of bipedalism on spine development. Clin. Orthop. Relat. Res. 41.
Schneider, P.G., Oyen, R., 1974. Surgical replacement of the intervertebral disc. I. Replacement of lumbar discs with silicon rubber. Theoretical and experimental investigation I PLASTISCHE BANDSCHEIBENCHIRURGIE. I. MITTEILUNG: BANDSCHEIBENERSATZ IM LUMBALEN BEREICH MIT SILIKONKAUTSCHUK. Z. Orthop. Ihre Grenzgeb. 112 (5).
Schwab, F., et al., 2009. A porcine model for progressive thoracic scoliosis. Spine 34 (11), E397–404.
Sheng, S.R., et al., 2016. Comparison of cervical spine anatomy in calves, pigs and humans. PLoS ONE 11 (2), 1–10.
Shono, Y., Kaneda, K., Yamamoto, I., 1991. A biomechanical analysis of Zielke, Kaneda, and Cotrel-Dubousset instrumentations in thoracolumbar scoliosis. A calf spine model. Spine 16 (11), 1305–1311.
Smit, T.H., 2002. The use of a quadruped as an in vivo model for the study of the spine – biomechanical considerations. Eur. Spine J. 11 (2), 137–144.
Souter, W.A., Taylor, T.K., 1970. Sulphated acid mucopolysaccharide metabolism in the rabbit intervertebral disc. J. Bone Joint Surg. Br. Vol. 52 (2), 371–384. Available from: http://www.ncbi.nlm.nih.gov/pubmed/4192265.
Stern, W.E., Coulson, W.F., 1975. Effects of collagenase upon the intervertebral disc in monkeys. J. Neurosurg. 44 (1).
Thompson, G.H., Akbarnia, B.A., Campbell, R.M.J., 2007. Growing rod techniques in early-onset scoliosis. J. Pediatr. Orthop. 27 (3), 354–361.
Timmis, D.P., et al., 1986. Comparative study of alloplastic materials for temporomandibular joint disc replacement in rabbits. J. Oral Maxillofac. Surg. 44 (7).
Urban, J.P.G., Roberts, S., 2003. Degeneration of the intervertebral disc. Arthritis Res. Ther. 5 (3), 120–130.
Urbaniak, J.R., Bright, D.S., Hopkins, J.E., 1973. Replacement of intervertebral discs in chimpanzees by silicone-Dacron implants: a preliminary report. J. Biomed. Mater. Res. 7 (3).
Wang, X., et al., 1997. Characterization of the scoliosis that develops after pinealectomy in the chicken and comparison with adolescent idiopathic scoliosis in humans. Spine 22 (22), 2626–2635.
Wang, Y., et al., 2015. Anatomical characteristics of deer and sheep lumbar spines: comparison to the human lumbar spine. Int. J. Morphol. 33 (1), 105–112.
Wenger, D.R., et al., 1982. Laboratory testing of segmental spinal instrumentation versus traditional Harrington instrumentation for scoliosis treatment. Spine 7 (3) 265–269.
Wilke, H.-J., Geppert, J., Kienle, A., 2011. Biomechanical in vitro evaluation of the complete porcine spine in comparison with data of the human spine. Eur. Spine J.: Off. Publ. Eur. Spine Soc. Eur. Spinal Deform. Soc. Eur. Sect. Cervical Spine Res. Soc. 20 (11), 1859–1868. Available from: http://www.pubmedcentral.nih.gov/articlerender.fcgi?artid=3207338&tool=pmcentrez&rendertype=abstract.
Wilke, H.-J., Kettler, A., Claes, L., 1997a. Are Sheep Spines a Valid Biomechanical Model for Human Spines? 2365–2374. Available from: http://www.ncbi.nlm.nih.gov/pubmed/9355217.
Wilke, H.J., et al., 1997b. Anatomy of the sheep spine and its comparison to the human spine. Anat. Rec. 247, 542–555.

Wilke, H.J., et al., 2015. Development of a scoliotic spine model for biomechanical in vitro studies. Clin. Biomech. 30 (2), 182–187. Available from: https://doi.org/10.1016/j.clinbiomech.2014.12.002.

Wilke, H.J., Krischak, S., Claes, L., 1996. Biomechanical comparison of calf and human spines. J. Orthop. Res.: Off. Publ. Orthop. Res. Soc. 14 (3), 500–503. Available from: http://www.ncbi.nlm.nih.gov/pubmed/8676264.

Wimmer, C., Siam, A.E., Pfandlsteiner, T., 2015. Operative Behandlung von Skoliosen: praeoperative planung, intraoperatives monitoring und postoperatives management. Orthopade 44 (11), 859–868.

Wullstein, K.L., 1992. Die Skoliose in ihrer Behandlung und Entstehung nech Klinischen und experimentallen Studien. Z. Orthop. Chir. 10 (177).

Yamamoto, H., 1966. Experimental scoliosis in rachitic bipedal rats. Tokushima J. Exp. Med. 13, 1–34.

Yates, J.P., Giangregorio, L., McGill, S.M., 2010. The influence of intervertebral disc shape on the pathway of posterior/posterolateral partial herniation. Spine 35 (7), 734–739. Available from: http://www.ncbi.nlm.nih.gov/pubmed/20357638.

Zhang, H., et al., 2009. Developing consistently reproducible intervertebral disc degeneration at rat caudal spine by using needle puncture. J. Neurosurg. Spine 10 (6), 522–530.

SECTION

SPINAL DISORDERS AND SPINE SURGERY

4

DEGENERATIVE DISORDERS

CHAPTER

FIXATION AND FUSION

17

Luigi La Barbera[*,†]

Politecnico di Milano, Milan, Italy[*] *IRCCS Galeazzi Orthopedic Institute, Milan, Italy*[†]

INTRODUCTION

Spinal fixation and fusion represent the gold standard for the surgical treatment of a variety of spinal diseases including degenerative disorders. The overall aims of fixation are to stop the degenerative process and to provide pain relief through stabilization. Fixation techniques historically developed after fusion. Fusion represented the first attempt to achieve a solid arthrodesis after intervertebral disc removal and replacement with stable spacers (i.e., cages, bone grafts).

Today fixation and fusion are achieved through a variety of approaches and devices. The growth of new techniques attempts to shorten operation times and hospital stays and to achieve faster recoveries with reduced postoperative complications.

SPINAL FIXATION WITH PEDICLE SCREWS

Spinal fixation can be achieved using different anchorage components (e.g., screws, hooks, and cables) connected to longitudinal (e.g., plates, rods) and transversal elements (e.g., connectors).

TECHNIQUE

Pedicle screw fixation was developed in 1963 by Roy-Camille (Roy-Camille et al., 1986). It was first used for the treatment of fractures and then extended to tumors, high-grade spondylolisthesis, and malunions (Magerl, 1982; Harrington and Tullos, 1969; Steffee et al., 1986; Dick et al., 1985). This technique, which was initially based on the use of pedicle screws and plates, currently is based on the use of pedicle screws coupled to spinal rods. This technique is now the most widely used one to solidly anchor the instrumentation to the vertebrae's strongest fixation point.

Screws can be placed on the posterior section of the vertebrae using a posterior approach with minimally invasive techniques. Pedicle screw placement is usually performed in a stepwise manner. A careful exposure of the posterior elements enables cleaning of the facet joints removal of soft tissues and the identification of important anatomical landmarks. Pilot hole drilling, pedicle wall checking, tapping, and screw insertion are then performed. The entry point at different spinal levels depends on the particular pedicle morphology and vertebral body's anatomy—pedicle orientation and size play a major role (Fig. 1). After screws are in place, a longitudinal element (e.g., plate, spinal rod) is connected to each screw.

Biomechanics of the Spine. https://doi.org/10.1016/B978-0-12-812851-0.00017-3

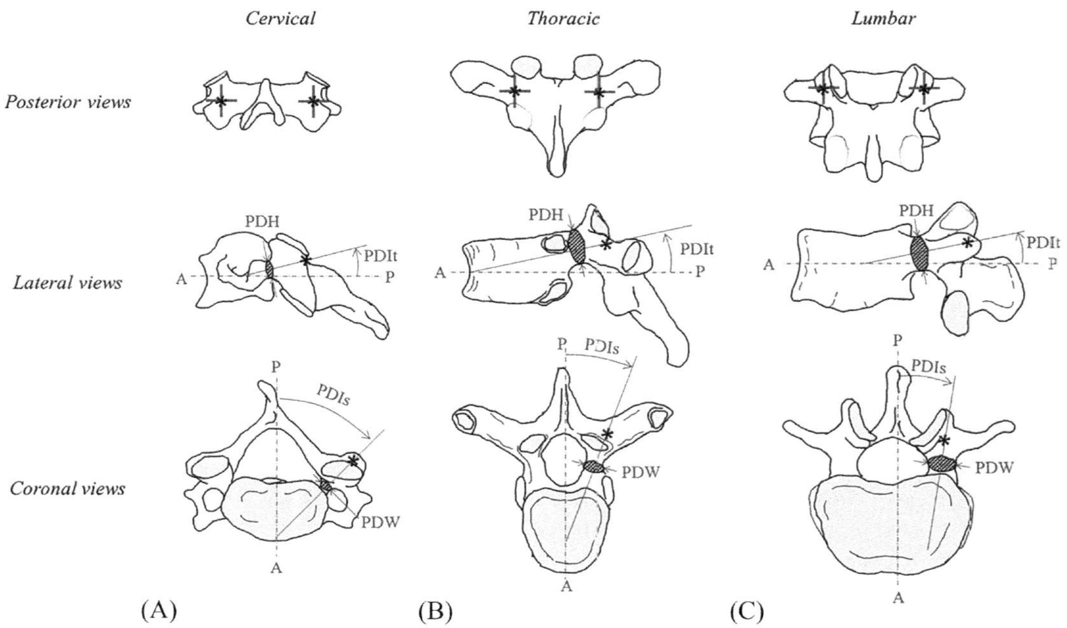

FIG. 1

Schematic drawings of the vertebrae on a posterior *(top)*, lateral *(middle)*, and top *(bottom)* views. Detail of pedicle screws entry point *(top)* and orientation of pedicle axis on the sagittal plane at cervical (A), thoracic (B), and lumbar (C) levels. PDIs/t: pedicle inclination referred to sagittal/transverse planes; *PDW/H*, pedicle width/height; *P*, posterior; *A*, anterior (Panjabi et al., 1991a, b, 1992).

Minimally invasive transcutaneous access minimizes the high morbidity rate typical of more invasive open access approaches. Pedicle screw technology applies simple intraoperative correction maneuvers to short fractured segments (Schmoelz, 2018), as well as more complex derotation maneuvers for long deformities (Schlager et al., 2018). Pedicle screw insertion has always been associated with specific complications. Screw misplacement may cause transient or permanent injuries, such as radicular pain, weakness or sensory loss, paralysis, and major vascular damage. Intraoperative radiographic control or navigation promise more accurate screw positioning within the pedicles (Gelalis et al., 2012).

Pedicle screws have been used to routinely treat degenerative spinal pathologies at thoracic and lumbar levels, including disc herniation, spinal stenosis, degenerative spondylolisthesis, and severe discogenic low back pain (Mobbs et al., 2011).

Abumi et al. (1994) were first to propose the use of pedicle screw fixation for traumatic cervical spine injuries at the C3 to C6 vertebrae. This technique allows sagittal alignment reconstruction without requiring stabilization to the lamina and adequate decompression. Morphometric measurements taken on human cadaveric vertebrae revealed significant variability in the critical pedicle dimensions. Because pedicle entry point, size, and orientation cannot be visualized directly from a posterior approach, intraoperative radiography, fluoroscopy, and navigation are extremely useful to avoid catastrophic complications. Successful pedicle screw placement requires a thorough knowledge of cervical spine anatomy (Coe et al., 1989; Pech et al., 1985). Intraoperative risks are related to screw malpositioning involving violation of the pedicle cortex, nerve root damage, or injury to the vertebral arteries (Reinhold et al., 2007; Abumi et al., 2000).

Transpedicular screw fixation of the cervical spine is indicated in cases ranging from degenerative arthritis and spondylolisthesis to rheumatoid instabilities and posterior fusion for anterior pseudarthrosis (Ludwig et al., 1999). Reconstruction of the anterior column is recommended for instability caused by the destruction of the vertebral bodies.

BIOMECHANICS

Spinal fixation, by definition, is aimed at providing immediate stability through rigid immobilization. In the case of degenerative disorders, posterior support facilitates the restoration of disc height, which is needed to avoid compression of the nerve roots, to unload the intervertebral disc, and to prevent further degeneration.

A discussion of the biomechanics of spinal fixation begins with the fact that the spine is a two-column structure, where the anterior and middle columns, typical of a three-column theory (Denis, 1983), are collapsed together (Fig. 2).

Posterior spinal fixation increases the stiffness of the posterior spine by introducing elements that constrain the overall kinematics and resist the complex loads occurring during everyday activities (Fig. 3).

KINEMATICS

Implant stiffness contributes to an increase of the overall stiffness of the instrumented spine segment, as well as to a reduction of its motion. Traditional in vitro flexibility tests on human and animal specimens demonstrated that rigid instrumentation, typically in titanium and stainless steel, provides effective short-term multidirectional stability to unstable and destabilized spine segments (Bruner et al., 2010; Gornet et al., 2011; Ponnappan et al., 2009; Cripton et al., 2000). Several finite element studies demonstrated that reductions in range of motion are related to spinal rod material and its geometrical features (i.e., diameter, cross-sectional shape) (Galbusera et al., 2011; Jahng et al., 2013; Rohlmann et al., 2012; Gornet et al., 2011; Ahn et al., 2008 Rohlmann et al., 2007; Rohlmann et al., 1999a) or,

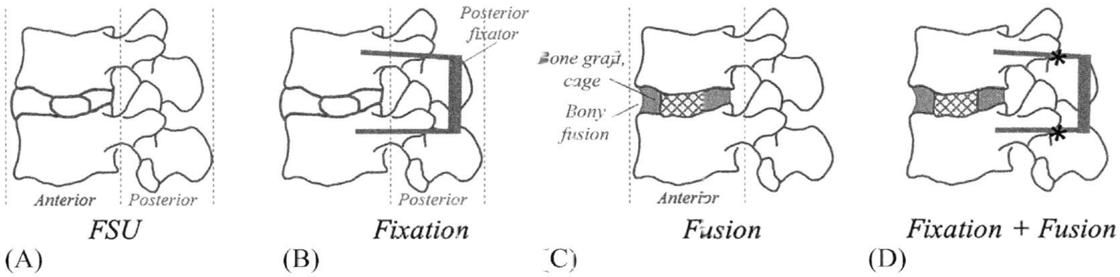

FIG. 2

Simplification of a functional spinal unit (FSU) as a two-column structure (A); treatment of a degenerated FSU with posterior fixation (B); anterior fusion (C); or, more often, both (D).

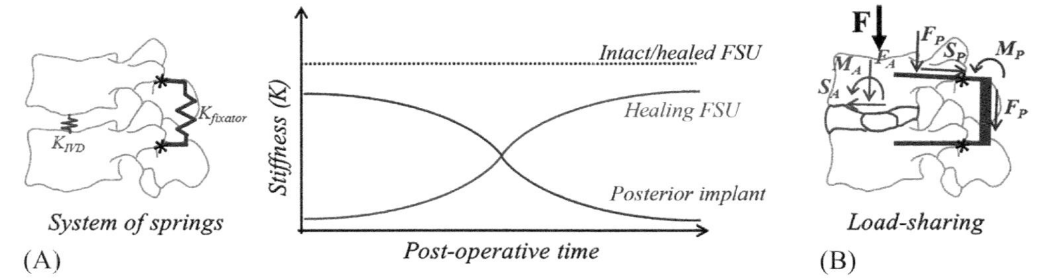

FIG. 3

Simplification of a functional spinal unit instrumented with posterior pedicle screw fixator as a system of springs with specific spring stiffness. (A) The posterior implant is expected to provide intrinsic primary stability contributing most in the short operative time, when the degenerated FSU provides incomplete anterior support; ideally, the stiffness of the implant will gradually disappear as the FSU heals. Load-sharing mechanism across the anterior and posterior spine. (B) A simple vertical force may contribute to bending moment and axial and shear force components both anteriorly and posteriorly.

in general, to its axial, bending, and torsional stiffness (Rohlmann et al., 2012; Schmidt et al., 2009). Working with calf specimens in vitro, Oda and colleagues found that pedicle screw fixation alone is adequate to limit spinal instability with preserved anterior load sharing, but may not provide enough stability to prevent high implant strains when the anterior column is damaged by a facetectomy or a discectomy (Oda et al., 2003).

Experimental in vitro tests demonstrated that, compared to other anterior and posterior internal fixation constructs, posterior fixation provided adequate spinal stability when two- and three-column and multilevel instability at the cervical level are addressed (Kotani et al., 1994). More recent studies confirmed this result, especially when multilevel instrumentation was used without excessive three-column instability (Espinoza-Larios et al., 2007). Anterior cervical plating in combination with posterior pedicle screw fixation offers increased stability compared to other conventional fixation systems, particularly in multiple level constructs and in the case of severe three-column instability (Kotani et al., 1994). Pedicle screw fixation demonstrated comparable kinematics when compared to lateral mass screws in isolated cervical vertebrae tested in vitro (Fensky et al., 2014).

The altered kinematics of an instrumented segment can also be investigated in terms of instantaneous center of rotation (CoR). Schmidt and colleagues found that the CoR is located in the center of the disc nucleus and that it moves slightly posteriorly or anteriorly, respectively, in flexion and extension (Schmidt et al., 2008). Posterior fixation produces a shift of the CoR toward the spinal canal (Jahng et al., 2013), but results may be controversial (Chun et al., 2015).

Primary Stability

Spinal stability is achieved through a solid anchorage to the vertebra and optimal bending stiffness characteristics of the implant. Pedicle screw triangulation was found to contribute significantly to the stability of the fixation construct (Carson et al., 1990; Ruland et al., 1991). Pedicle screws are designed to provide adequate resistance to axial force, which contributes to pull-out; shear forces, which contribute to loosening; and bending moments, which produce high bending stresses (Fig. 3B).

Alternative anchorage systems, designs, materials, and hardware constructs have been compared in the past. Pedicle screw fixation provided superior anchorage capabilities compared to laminar hooks and wires (Coe et al., 1990; Ruland et al., 1991), but in vitro tests reported that they could behave equally well in patients with osteopenia (Wilke et al., 2016). Slightly different techniques have been proposed to promote fixation through bicortical engagement, modified screw trajectory, and use of bone cement or expandable screws, particularly in osteoporotic bone (Demir, 2015; Kueny et al., 2014; Shea et al., 2014; Chao et al., 2013).

Pedicle screws experience a very complex combination of shear forces and bending moments in different anatomical planes (Freeman et al., 2012). Primary stability depends on screw-bone interaction, static mechanical response of the implants, and factors related to the surgical technique. Uncomplicated mechanical tests are usually adopted to evaluate screw resistance under a simple axial load (pull-out), a simple shear load (toggle), or a combination of both. Bone mineral density is known to influence pull-out resistance (Johnston et al., 2006; Pfeifer et al., 1996; Coe et al., 1990), with pedicle cortical bone not contributing to an increase of screw purchase (Defino and Vendrame, 2001). Among the design parameters playing a biomechanical role against pull-out, the insertion length, the outer diameter of the screw, and the thread type are important (Bianco et al., 2017; Crawford III et al., 2011). Other studies found that thread depth, pitch, and shape were related to the bone volume inside the threads, thus influencing the pull-out strength (Crawford III et al., 2011). In regard to the surgical technique, relevant factors contributing to pull-out were found to be correct pedicle-screw positioning (Costa et al., 2013), an undersized pilot hole compared to screw diameter (Silva et al., 2013), a screw trajectory (Bianco et al., 2017; Ruland et al., 1991), and insertion depth of pedicle screws (Krag et al., 1988). Templier and colleagues found that more rigid posterior fixators induced high push-in/pull-out force concentrations at the screw-bone interface, possibly related to screw mobilization (Templier et al., 1998).

Long-Term Stability

Several studies highlighted that nonaxial shear forces induce screw plowing, leading to bone compaction, subsequent screw loosening, or bone failure, thus reducing the pedicle screw fixation strength (Bianco et al., 2016; Kueny et al., 2014). Higher toggle resistance is commonly related to a more rigid screw-bone interface with increasing bone mineral density, rather than pilot-hole overdrilling (Tan et al., 2003), and screw trajectory (Newcomb et al., 2017). It was found that loosening could be reduced by using higher screw diameters (Kiner et al., 2008), with augmentation particularly in osteoporotic bone (Hirano et al., 1997), and increasing the surface area of the screw with hydroxyapatite coating or rough surface finishing (Dakhil-Jerew et al., 2009). Pedicle screws demonstrated higher toggle resistance than lateral mass screws did in isolated cervical vertebrae tested in vitro, whereas their kinematics was similar (Fensky et al., 2014).

A recent experimental study on isolated vertebrae proposed that bending moments coupled with shear-force components may be more realistic in the screw-loosening process (Liebsch et al., 2017). Moreover, in vitro tests on instrumented cadavers demonstrated that bone mineral density is a determinant factor for the long-term stability of pedicle screws (Wilke et al., 2016).

LOAD SHARING

Posterior fixation produces a shift of the spinal loads from the anterior spine to the instrumentation, according to a well-known load-sharing mechanism (Fig. 3B).

In vitro tests on lumbar specimens demonstrated that posterior fixation significantly decreases intradiscal pressure in axial compression, flexion-extension, and lateral bending compared to intact and destabilized segments (Cripton et al., 2000; Cunningham et al., 1997; Abe et al., 1998). Cripton and colleagues used a semi-analytical approach based on in vitro measurements on cadavers to estimate the bending moments supported by the spinal structures and the fixator according to a destabilization protocol including removal of the posterior elements and subsequent discectomy (Cripton et al., 2000). Numerical simulations of instrumented lumbar disc degeneration found a reduction in the facet joint force and intradiscal pressure in extension and standing, with a further reduction when distraction was simulated (Rohlmann et al., 2007); the spinal rod diameter was found to play an important role in reducing the stresses in the bridged disc (Rohlmann et al., 1999a). Further in silico studies significantly correlated the effects of rod stiffness characteristics with intradiscal pressure, posterior disc bulging, and facet joint loads (Chun et al., 2015; Rohlmann et al., 2012). Other models found a clear correlation between the load shared by the posterior fixator—as well as facet joint forces—and its elastic modulus (Jahng et al., 2013; Gornet et al., 2011; Ahn et al., 2008). Such an effect may help in providing recovery to a degenerated disc, but may pose further problems in terms of stress-shielding in the case of anterior fusion.

Melnyk and colleagues investigated in vitro the effect of shear forces on an unstable model of degenerative spondylolisthesis in a human cadaver spine (Melnyk et al., 2012a). They found that the posterior fixator supported a significantly greater shear force as the specimen was progressively destabilized (facet and disc destabilization) and the implant stiffness increased (Melnyk et al., 2012b, 2015). Rohlmann and colleagues (Rohlmann et al., 2002a) predicted an increase of both axial and shear forces in the implant with increasing distraction.

The internal mechanical loads acting on the implants have been measured in vivo by using instrumented posterior fixators. Several studies investigated the effect of different body positions and tasks on the loading components (Rohlmann et al., 2000). Axial forces are mainly compressive and depend on bodyweight, the surgical procedure (i.e., compression, distraction of the bridged region), and the level of fusion (higher at lumbar levels); whereas the bending moments act mainly in flexion, being dependent on the indication for surgery and the surgical procedure (distraction implies higher moments) (Rohlmann et al., 1999b). Among the common life activities, walking is considered the most critical loading condition for implant fatigue, more because of the high number of repetitions performed every day than for the intensity of the bending moments achieved (Rohlmann et al., 1997).

Bending moments are also considered the primary cause of fatigue failure due to high repeated stresses on the posterior spinal fixator during clinical use. Based on an estimate of the bending moment on the posterior instrumentation, Cripton warned that, in the case of severe anterior column injuries, the implant may be at risk of fracture, and secondary to reversed bending moments (Cripton et al., 2000). Recent numerical studies highlighted the significant role of the anterior support in reducing the internal loads and the maximum stresses on different components of the posterior fixator (La Barbera et al., 2016a; La Barbera and Villa, 2016; La Barbera et al., 2014; Villa et al., 2014a, b). With an increasing degree of anterior spine instability, the spinal implant could reach much higher loads compared to an intact disc, even during noncritical everyday activities (e.g., standing) (La Barbera et al., 2016a, b). The predictions of such models agree with the failure modes met during clinical experience (La Barbera et al., 2017).

Newcomb and colleagues determined with a combined in vitro and in silico approach that pedicle screw trajectory may be optimized to decrease the stress, thus avoiding screw breakage (Newcomb et al., 2017). Of course, design-dependent parameters may also play an important role. Several numerical models correlated the high internal loads to implant stresses through the elastic modulus of the longitudinal element (Jahng et al., 2013; Gornet et al., 2011; Ahn et al., 2008) and to its diameter

(Rohlmann et al., 1999a). Shih and colleagues compared conical and cylindrical screw designs using bench tests coupled with simple finite element simulations (Shih et al., 2015); they found that a conical design exhibits superior bending performances under both static and fatigue-loading conditions. Amaritsakul performed a similar study showing that a conical design could be further optimized using a dual-core thread shape to increase bending strength (Amaritsakul et al., 2014).

POSTEROLATERAL FUSION (PLF) IN THORACOLUMBAR SPINE

TECHNIQUE

Fusion of the lumbar spine with a posterior technique using bone grafting was first described by Albee (1911). Posterolateral fusion (PLF) was first introduced in 1953 by Watkins to achieve fusion of the lumbar and lumbosacral spine (Watkins, 1953, 1959). In this procedure, a posterior access is used with the patient in a prone position. A midline incision is performed and the paravertebral fascia is opened bilaterally to expose the neural arc. The facet joints, laminae, and transverse processes are decorticated. Laminectomy, medial facetectomy, and foraminotomy are usually performed to provide canal decompression.

Bone chips can be obtained from the same patient (autologous graft); for instance, a significant amount of spongious bone can be harvested from the ilium. However, allograft from humans and animals as well as bioactive substances (e.g., bone morphogenetic protein and osteoinductive factor) can also be used (Strömqvist, 1993). The graft material is then placed along the transverse processes and the lateral part of the laminae to stimulate bone growth. Chronic pain at the donor site is a common complication and is often indicative of an unsatisfactory result. This outcome seems to be avoided by using the posterior iliac crest as a donor site (Strömqvist, 1993).

PLF alone is indicated for degenerative disc disease requiring a wide decompression with an inherently stable spine (e.g., isthmic spondylolisthesis). Orthotic treatment is required after this procedure. PLF must also be supplemented with posterior fixation in cases of severe translational instability.

Common complications are superficial infection, nerve root injury, and dural tears (Wu et al., 2011). Despite lengthening the operation time and the risk of complications, posterior pedicle screw fixation is often used to provide primary stability until solid fusion is achieved (Thomsen et al., 1997). A satisfactory outcome is achieved in terms of pain relief and fusion rate, both with and without posterior fixation, but the latter seems to reduce the occurrence of pseudarthrosis (Schwab et al., 1995; Narayan et al., 2002). PLF supplemented with pedicle screw fixation has demonstrated clinical and functional outcomes similar to those of other posterior fusion techniques (i.e., PLIF) but higher complications related to hardware biomechanics (Wu et al., 2011).

BIOMECHANICS

Kinematics

PLF represents a first attempt to achieve fusion of a spine segment by constraining the motions of the facet joints and by reinforcing the posterior lamina with high stiffness material (in the long term). In vitro tests on cadaveric specimens demonstrated that PLF alone is a valuable method providing good stabilization to the fused segment and having the least effect on the adjacent unfused segment (Lee and Langrana, 1984).

Primary Stability and Long-Term Fusion

From a mechanical perspective, constraining the posterior spine with PLF alone may not be the best strategy to achieve fusion. In fact, because of the loads acting on the anterior spine, the lamina and the facet joints may experience some relatively high motions, while needing a relatively longer time to fuse. This may explain the pseudarthrosis rate seen in clinical practice. Moreover, the use of external orthosis may not be as effective as other internal fixation implants. This second strategy has all the mechanical advantages of posterior fixation alone, providing primary stability in the very short-term after surgery, while shielding the loads on the hardware, particularly after solid fusion is achieved.

McAfee and colleagues demonstrated that bilateral posterolateral bone grafting without instrumentation may lead to lower fusion rates, compared to more stable instrumented PLF, mainly because of the intrinsic superior stability of the implant in flexion and torsion (McAfee et al., 1989, 1991).

Chen and colleagues warned that bone graft union in PLF does not guarantee the prevention of screw breakage (Chen et al., 2005); they used simple finite element models to demonstrate that the instrumentation may experience very high stresses, supporting the high revision rate often met in clinical practice.

PLF demonstrated superior stability when coupled with anterior interbody fusion (Lee and Langrana, 1984), thus reducing the motions of the anterior spine and the loads on posterior hardware. Another study confirmed the higher complications related to pedicle screw fracture and loosening without adequate anterior support (Wu et al., 2011). Moreover, despite the fact it is not associated with a poorer clinical outcome, PLF is less effective than anterior fusion (e.g., PLIF, TLIF) in increasing and maintaining the disc height and avoiding anterior translational/angular slippage of the superior vertebral body relative to the lower vertebra (Campbell et al., 2017; Lidar et al., 2005).

INTERBODY THORACOLUMBAR FUSION

TECHNIQUES

Interbody fusion (IF) is an established technique for the treatment of a variety of spinal disorders. It involves the use of cages, spacers, or bone grafts that are positioned within the intervertebral space following discectomy and after a careful endplate preparation. IF can be used alone or, more commonly, in combination with posterior pedicle screw fixation. Today a surgeon can choose from a variety of surgical approaches to achieve a satisfactory IF, with each approach presenting advantages and disadvantages, from both a clinical and a biomechanical point of view (Fig. 4). In general, surgical options can be classified in two categories taking the transverse process as a reference (Mobbs et al., 2015): anterior approaches (ALIF, LLIF) imply a retroperitoneal corridor to reach the lumbar spine, whereas posterior techniques (PLIF/TLIF/MI-TLIF) necessarily traverse the spinal canal or the foramen to access the intervertebral disc.

Clinical evidence did not demonstrate that one specific approach is to be preferred or avoided over the others in terms of fusion rate or clinical outcomes. However, IF is considered superior to PLF because of the lower postoperative complication rate and the lower occurrence of pseudarthrosis (Eck et al., 2007; Wu et al., 2011), as well as maintenance of the intervertebral disc height (Cloward, 1953).

Anterior Lumbar Interbody Fusion (ALIF)

The anterior retroperitoneal approach provides direct access to the ventral aspect of the intervertebral disc. The patient is lying in supine position during surgery. Incision and approaches are performed at

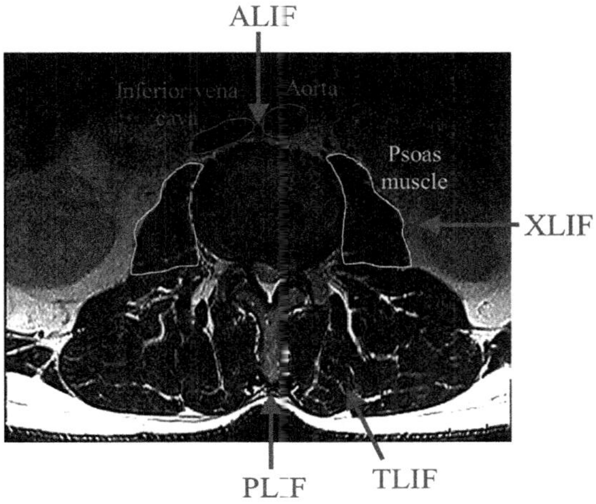

FIG. 4

Schematic identification of alternative surgical approaches to reach the anterior lumbar spine: anterior (ALIF), lateral or extreme (LLIF or XLIF), transforaminal (TLIF), and posterior (PLIF) lumbar interbody fusions. The presence of nerves, muscles fasciae, and great vessels determines the feasibility of each approach at different spinal levels.

midline, paramedian, or mini-Pfannenstiel (at L5-S1) planes, thus allowing a retroperitoneal corridor with vascular mobilization and dissection (Mobbs et al., 2015).

Vascular anatomy makes ALIF suitable for L4/L5 and L5/S1. However, it is limited for L2/L3 and L3/L4 in cases of extensive peritoneal and kidney retraction and the risk of superior mesenteric artery thrombosis. The ALIF approach is suitable for degenerative disc disease, discogenic disease, and revision of previously failed posterior fusion (Rao et al., 2015a). Contraindications include high-grade degenerative spondylolisthesis without posterior fusion (Malham et al., 2014), as well as isthmic spondylolisthesis at L5/S1 (Rao et al., 2015b) where ALIF should be combined with posterior fixation.

The ALIF approach facilitates efficient discectomy with rapid endplate preparation and the insertion of an interbody implant. The anterior access allows maximization of the implant size and surface area, which optimizes the fusion rate (Phan et al., 2015a, b; Malham et al., 2014) and makes ALIF suitable for foraminal height restoration and high lordosis correction. This approach helps to spare spinal muscles and to reduce postoperative pain and disability. Surgery-related complications may include retrograde ejaculation and visceral and vascular injuries (Phan et al., 2015a, b; Malham et al., 2014; Mobbs et al., 2015).

The anterior technique is also used at the thoracic level. The techniques to access the thoracic disc space include thoracoabdominal, open transthoracic, and thoracoscopic approaches (Patel et al., 2010).

Posterior Lumbar Interbody Fusion (PLIF)

Briggs and Milligan were first to describe surgical posterior access to the intervertebral disc via a posterior approach in 1944 (Briggs and Milligan, 1944). This approach is well known by the majority of spinal surgeons, who are well trained to perform it. The patient is placed in the prone position. An open midline approach with bilateral strip dissection or MIS paramedian Wiltse muscle splitting is used to reach the posterior aspect of the vertebral body. A laminotomy is performed medial to the facets, and the disc space is exposed upon dura retraction. Finally, endplates and disc space are prepared to allow the insertion of an interbody spacer.

The posterior lumbar interbody fusion (PLIF) approach is indicated for degenerative indications requiring fusion. It can also be used for selected patients with segmental instability, symptomatic stenosis, recurrent disc herniation, and pseudarthrosis (Mobbs et al., 2015).

The posterior approach is a valuable technique for treatment of the thoracic spine (Smith et al., 2008). Posterior thoracic cage interbody fusion (PTCIF) can be considered at the thoracic and thoracolumbar junction (Shin et al., 2016).

Posterior exposure ensures excellent visualization of the nerve roots and surrounding structures. PLIF allows for adequate interbody height restoration and effective neural decompression, while preserving posterior structures (Lestini et al., 1994). The posterior approach facilitates pedicle screw insertion and 360 degree fusion through a single incision. Among the disadvantages of this approach is prolonged paraspinal muscle retraction, which may provoke iatrogenic injury with delayed recovery and mobilization (Fan et al., 2010). Compared to the ALIF approach, with PLIF, endplate preparation, correction of coronal imbalance, and lordosis restoration may be more difficult to achieve (Landham et al., 2017).

Transforaminal Lumbar Interbody Fusion (TLIF)

The transforaminal route was initially described by Harms and Rolinger in 1982 to achieve the insertion of an interbody cage packed with bone graft (Harms and Rolinger, 1982). transforaminal lumbar interbody fusion (TLIF) involves posterior unilateral access to the intervertebral foraminal space from L1 to S1. A midline, bilateral paramedian incision is usually performed with the patient lying in the prone position. A unilateral laminectomy and inferior facetectomy facilitate bone graft positioning. The access on one side guarantees limited damage to bony structures and nerve roots, ligamentum flavum preservation, as well as limited dura retraction compared to that of PLIF. In PLIF, a prolonged paraspinal muscle retraction may produce an iatrogenic injury. Access using a MIS mini-open incision and microscopy may also reduce muscle injury, minimizing bleeding and improving postoperative recovery. Despite endplate preparation may be more difficult compared to ALIF approach, MIS-TLIF is equally adequate compared to PLIF (Rihn et al., 2014).

TLIF is indicated for treating degenerative lumbar diseases following failed conservative treatment, including pseudarthrosis and spondylosis (Mobbs et al., 2015). However, it does not allow for coronal imbalance correction and lordosis restoration (McAfee et al., 2005; Humphreys et al., 2001).

Extreme Lateral (or Transpsoas) Interbody Fusion (XLIF)

The direct (DLIF) or lateral (LLIF) or extreme lateral interbody fusion (XLIF) technique was first introduced by Ozgur et al. (2006). With this technique, the patient is laterally positioned, and a small incision is performed. The XLIF approach involves the use of a lateral retroperitoneal transpsoas corridor to access the disc space from T12 to L5. Neuromonitoring is considered essential to prevent damage to the lumbar plexus and the psoas muscle and bowel injury at lower lumbar levels (Malham et al., 2012). The lateral iliac crest, the vascular anatomy, and the course of the anterior lumbar plexus do not make XLIF technique suitable for the L5-S1 level. The lateral approach guarantees minimal muscle splitting with rapid postoperative mobilization, despite the fact that transient neurological complications have been reported. A high fusion rate can be achieved easily thanks to comprehensive clearance of the disc space (Phan et al., 2015a, b).

XLIF is suitable for degenerative indications, adjacent level disease from prior fusions, and thoracic disk herniations (Karikari et al., 2011). It allows for an effective sagittal and coronal deformity correction, particularly for lumbar degenerative scoliosis with laterolisthesis (Arnold et al., 2012).

Stand-alone XLIF without posterior instrumentation is not recommended in cases of severe instability, facet arthropathy and central canal stenosis, and high-grade spondylolisthesis (Malham et al., 2015, 2017).

BIOMECHANICS

The biomechanics of spinal fusion involve a two-column structure (Fig. 2), where the anterior spine is stiffened by a very rigid cage/bone graft working in parallel with the remaining disc structures. Fusion is often supplemented with a posterior fixator stiffening the posterior column, thus providing primary and long-term stability (Fig. 5).

An exhaustive review of the existing literature dealing with fixation and fusion is a challenging task.

To the best of our knowledge, very few biomechanical studies have conducted extensive comparisons of the biomechanics of alternative fusion techniques. Moreover, many surgical techniques fall within the same category, even if they exhibit specific peculiarities in terms of implant design and surgical techniques. Therefore it is often very hard to make direct comparisons or to generalize about their biomechanical performances.

KINEMATICS AND PRIMARY STABILITY

Biomechanical studies on stand-alone cages showed controversial effects on kinematics, often affected by a lack of primary stability leading to cage dislocation.

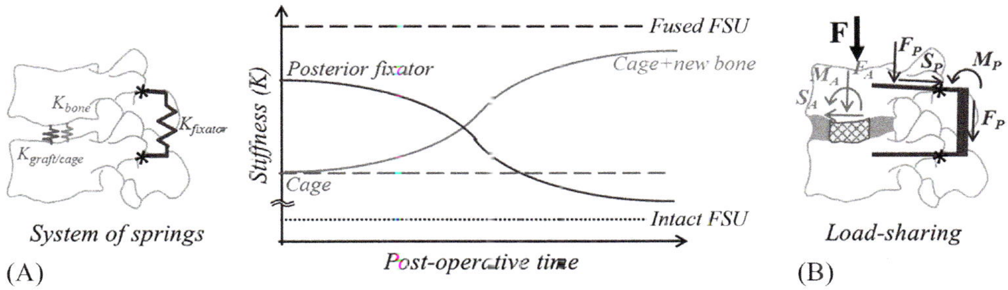

FIG. 5

Simplification of a functional spinal unit instrumented with an intervertebral spacer (e.g., cage/bone graft) and a posterior pedicle screw fixator, as a system of springs with specific spring stiffness (A). The posterior fixator is expected to contribute most in the short operative time, whereas the anterior fusion mass (e.g., cage, graft) provides incomplete anterior support; the anterior support stiffness (cage and newly formed bone) is expected to increase its contribution with the bone graft increasing its bone mineral density over time. It is important to notice that the stiffness of the implants alone is much higher than the stiffness of the intact FSU, providing intrinsic primary stability. Load-sharing mechanism across the anterior and posterior spine (B): a simple vertical force may contribute to bending moment and axial and shear force components, both anteriorly and posteriorly.

A review by Schmoelz reported a relatively low stabilizing effect of stand-alone cages, particularly in extension and axial rotation (Schmoelz and Keiler, 2015). Goh and colleagues compared in vitro the effect of PLIF cage sizes (Goh et al., 2000); they noted that a stand-alone cage must provide sufficient vertebral distraction and adequate tension in the annulus to restore stability without any additional instrumentation. If a small and a medium cage could restore the stiffness (i.e., stability) of the facetectomized FSU only in extension and lateral bending, a larger cage could restore torsional stiffness. However, flexural stiffness could not be recovered, regardless of the cage size (Goh et al., 2000). Kettler and colleagues found in vitro that an ALIF cage guaranteed significantly higher stabilization than an TLIF cage did but that both were insufficient in axial rotation (Kettler et al., 2006). Harris observed the same effect for a lumbar spine segment implanted with TLIF cages (Harris et al., 2004). Oxland highlighted that different ALIF designs did not stabilize the lumbar spine in extension (Oxland et al., 2000). Buttermann and colleagues performed in vitro tests on alternative ALIF cage designs (Buttermann et al., 2009). Cages with spikes or threads provided substantial motion segment rigidity compared to that of an intact state in bending, whereas endplate engagement with spikes limited torsion, which was difficult to achieve with stand-alone cages (Buttermann et al., 2009). Choi and colleagues demonstrated that an ALIF cage alone was effective in restoring the stability of a destabilized lumbar spine segment, especially when screws were secured to the endplates (Choi et al., 2013). Kim found a similar trend using two threaded cylindrical ALIF cages in a lumbar model (Kim, 2007). Other numerical studies have highlighted that stand-alone ALIF provides acceptable stability without dislocations only with locking screws and coronal wings compared to other less effective designs (Chen et al., 2013). Multidirectional in vitro flexibility tests coupled with numerical simulations found slightly different stiffening effects with similar ALIF cages, while confirming the holding power of screws anchoring through the endplates (Schleicher et al., 2008). Jost and Lund found that the insertion depth and holding power were quite different between plates, screws, and cylinders, potentially affecting the stabilizing ability of the holding mechanisms of interbody devices (Jost et al., 1998; Lund et al., 1998). Nagaraja and colleagues performed in vitro tests on cadaveric spine segments in pseudo-physiologic fatigue conditions (Nagaraja and Palepu, 2017; Palepu et al., 2017); they observed that an anterior plate may be more effective in decreasing the risk of screw loosening compared to ALIF with integrated anchoring systems (Nagaraja and Palepu, 2017; Palepu et al., 2017). Choi and colleagues used a numerical lumbar model to predict that a more secure cage fixation in the disc space with screw systems and plates may contribute to reducing the stress components on the adjacent endplates (Choi et al., 2013). Karim and colleagues found a comparable stability for cadaveric specimens implanted with a ALIF supplemented with anterior or posterior pedicle screw fixation (Karim et al., 2006).

In vitro tests on human specimens by Fogel and colleagues determined that XLIF cages alone significantly increased spinal stability compared to an intact condition, whereas a lateral plate may further reduce the kinematics in lateral bending and axial rotation (Fogel et al., 2014). Flexibility tests on synthetic and human FSU confirmed the same trends and demonstrated comparable effects between XLIF and ALIF supplemented with lateral and anterior plates (Wang et al., 2014; Cappuccino et al., 2010; Le Huec et al., 2002). Another in vitro study suggested that wider intervertebral cages are biomechanically more stable and could be used with less or even without supplemental posterior fixation (Pimenta et al., 2012); therefore wider XLIF cages may be more advantageous than conventional TLIF cages are.

Several biomechanical studies have highlighted the advantages of supplementing anterior support with posterior instrumentation. Buttermann and colleagues observed an increased stability with different ALIF cages when supplemented with anterior plating and even greater rigidity with posterior fixation (Buttermann et al., 2009). Cook confirmed that an ALIF spacer alone may need posterior stabilization to effectively reduce the intervertebral relative rotations and translations in vitro (Cook et al., 2015). Rapoff and colleagues measured a significant increase in stiffness compared to the intact spines and the resected spines, with higher stiffness in compression and in extension, respectively, for a BAK cage and a threaded interbody fusion device (Rapoff et al., 1997). Kettler and colleagues found that an ALIF cage guaranteed significantly higher stabilization only in flexion-extension compared to TLIF after posterior fixation (Kettler et al., 2006). Harris reported instead that TLIF with bilateral pedicle screws closely approximated the flexibility of the intact spine (Harris et al., 2004). Oda and colleagues found that posterior fixation with an additional PLIF cage significantly increased the stiffness of an instrumented construct and minimized the implant strains compared to fixation alone (Oda et al., 2003). Lund et al. determined that posterior fixation significantly increased the stabilization of lumbar spine segments in vitro after the insertion of PLIF cages (Lund et al., 1998). Posterior fixation in the XLIF approach increased spine stability significantly only in flexion-extension, as compared with a stand-alone cage (Fogel et al., 2014). Several finite element studies found a significant reduction in intervertebral motions following lumbar fixation and fusion with a TLIF cage with both spinal rods and plates (Chen et al., 2012; Li et al., 2015), with a big ALIF cage (Choi et al., 2013), and with two threaded cylindrical ALIF cages (Kim, 2007). Rohlmann and colleagues found a similar stabilization effect with two intervertebral cages on consecutive levels, with a reduction in facet joint forces (Rohlmann et al., 2002b, 2005). Moreover, they did not find a clear biomechanical effect depending on monosegmental or bisegmental stabilization and bone graft positioning (Zander et al., 2002). Boustani and colleagues simulated different scenarios from early-to-late fusion phases associated with instrumented lumbar fusion, predicting a gradual increase in stabilization potential, particularly in extension, where a perfect bond was assumed at cage-endplate interfaces (Boustani et al., 2012). Such an approach led to comparable effects with anterior fusion supplemented by an anterior plate system (Rohlmann et al., 2005). Kim highlighted that pedicle screw fixation increased segmental stiffness compared to a standalone ALIF cage, while reducing relative motions and bone stresses at the interface with ALIF cages. Therefore they speculated that it may be beneficial to promote bone fusion (Kim and Vanderby, 2000; Kim, 2007). Sim and colleagues found that PLIF provides a higher immediate stability than TLIF does, especially in lateral bending (Sim et al., 2010).

SUBSIDENCE

Interbody fusion has been claimed to increase or at least to maintain the height of a degenerated disc better than stabilization alone does. Graft or cage design, the quality of the surrounding bone, and the presence of supplementary posterior fixation are considered key factors for avoiding cage penetration in the endplates (i.e., subsidence) and to stimulate fusion.

Closkey and colleagues performed compressive in vitro tests on thoracic vertebrae (Closkey et al., 1993). They found a correlation between the bone mineral density and the PLIF bone graft compression strength of decorticated trabecular bone. Moreover, they found that endplate coverage by a bone graft of at least 30% of the endplate area was necessary to reach high failure loads and to prevent subsidence (Closkey et al., 1993). Jost and colleagues found a similar increase of compressive failure loads with increasing bone mineral density in lumbar specimens, but they did not find a significant correlation with PLIF cage design and posterior instrumentation (Jost et al., 1998). Brantigan and colleagues compared a carbon fiber cage to a bone graft and reported similar failure loads, with or without posterior fixation

(Brantigan et al., 1991). Kettler and Krammer reported a comparable response using threaded, bullet-shaped PEEK and rectangular titanium PLIF cages after cyclic axial loading without a correlation to bone quality (Kettler et al., 2002; Krammer et al., 2001). Other in vitro studies on TLIF cage design and positioning did not find significant correlation to endplate failure or stiffness in compression (Lam et al., 2012); however, it was found that bone mineral density significantly affected both failure forces and stiffness, but was not dependent on the positioning or design of the cage.

Polikeit and colleagues simulated ALIF cage usage in a lumbar FSU, demonstrating a stress concentration on the peripheral part of the endplate (Polikeit et al., 2003a, b); they highlighted the important contribution of the elastic modulus (i.e., bone quality) of the endplate and the underlying trabecular bone in reducing peak stresses, potentially related to subsidence or failure of adjacent vertebrae (Polikeit et al., 2003a, b).

De Beer and colleagues investigated the use of rapid prototyping to produce patient-specific cages with a conformal shape (de Beer and Scheffer, 2012); static mechanical tests confirmed that the optimized contact area could lead to higher failure loads and a decreased subsidence rate.

Endplate preparation is considered crucial for ensuring and promoting interbody fusion. Tatsumi and colleagues found a gradual increase in the endplate preparation area with ALIF, TLIF, PLIF, and XLIF (Tatsumi et al., 2015). Endplate damage area was highest in the TLIF group, whereas the XLIF group appeared to provide a greater area and more complete endplate preparation, thus minimizing damages (Tatsumi et al., 2015). Kim predicted that a larger disc area and a higher bone quality (i.e., elastic modulus) are related to an increased stability of threaded cages, which minimizes the risk of micromotions at the cage-endplate interface (Kim, 2001).

LOAD SHARING

The load-sharing effect across the anterior and posterior spine columns may pose challenging issues, depending on the presence of a fusion mass alone or with supplementary devices (Fig. 3). Stress-shielding, caused by the presence of supplementary posterior instrumentation, is a very common problem affecting the maturation of the anterior fusion mass. Asher and colleagues used a canine model to study in vivo the contribution of posterior implants' stiffness on fusion. They demonstrated how a higher rod diameter (i.e., implant stiffness) can be associated with a delayed fusion, despite the fact that the fusion rate is satisfactory (Asher et al., 2007). Other studies have discussed "device-related osteoporosis," reporting a drop in mineral bone quality and the apposition rate of the trabeculae in the region spanned by a rigid implant, as compared with an intact segment in vivo (Farey et al., 1989; McAfee et al., 1989). McAfee noted that the stiffness of the posterior instrumentation more than compensated for this effect, leading to a significantly more rigid construct in flexion and torsion, as well as to a more successful fusion rate, compared with bilateral posterolateral bone grafting without instrumentation (McAfee et al., 1989, 1991). Conversely, iliac crest grafting without fixation at the lumbar and sacral levels was related to a general increase in bone density on patients with painful disc disease (Lipscomb et al., 1989). Goel and colleagues were first to perform an experimental and numerical biomechanical study demonstrating how to reduce the rigidity of traditional implants to achieve a satisfactory primary stability and maturation of the fusion mass (Goel et al., 1991). The effectiveness of such an approach was confirmed in vivo with a canine model. Haher and colleagues proposed a technique to determine the amount of load transferred from the fusion mass to the instrumentation in a synthetic construct; they found that posterior fixation systems alone contributed to increasing the axial stiffness of the construct, thus providing sufficient stability in compression but shielding the anterior support, much more than anterior plates did (Haher et al., 2001). It has also been found that, contrary to a posterior placement, anterior cage positioning increases the overall stiffness of a synthetic construct, while increasing anterior load sharing (i.e., fusion) and minimizing posterior rod strains (i.e., fatigue failure) (Polly et al., 2000). Two anteriorly positioned side-by-side cages (i.e., TLIF) were found to be biomechanically equivalent or superior to a single large cage (i.e., ALIF, PLIF, XLIF). Cook and colleagues found that rigid posterior fixation may shield the compressive loads acting on an ALIF spacer instrumented with a pressure film transducer, especially in flexion (Cook et al., 2015).

The complex loads acting on interbody cages and bone grafts at different phases of maturation of the fusion mass are not yet fully understood. In vivo studies made the first attempt to understand the complex internal loads acting on a telemeterized anterior vertebral body replacement coupled with posterior fixation in patients with degenerative disorders (Dreischarf et al., 2015; Rohlmann et al., 2008, 2014). The loads depended on the task, with relatively low loads kept in the first postoperative time (Rohlmann et al., 2008); however, very high loads (up to 1650 N, with relatively low shear forces components) and consistent bending moments (bending and torsional moments beyond 6 Nm and 3.5 Nm, respectively) could be reached afterward (Rohlmann et al., 2014). A numerical study demonstrated that the elastic modulus of fusion bone considerably increased the compressive force in the vertebral body replacement where it was inserted, but reduced the internal loads on the ventral implant in loading conditions other than extension (Rohlmann et al., 2006). Pretensioning (shortening) of the posterior longitudinal rods was also found to increase the load on the bone graft (Zander et al., 2002). More promising finite element studies using mechano-regulation algorithms investigated the effect of stand-alone cage designs on tissue formation (Postigo et al., 2014; Bashkuev et al., 2015). Cage material stiffness and shape were found to play a key role in determining the mechanical conditions within the fusion region and regulating the course of the fusion process (Postigo et al., 2014; Bashkuev et al., 2015). Despite the important simplifications made (i.e., axisymmetry, pure compression, no validation), mechano-biological models may be very promising tools for interpreting in vivo observations.

Rohlmann and colleagues found that the internal loads acting in vivo on a posterior fixator, counterintuitively, often did not decrease after bony fusion occurred, but remained relatively high (Rohlmann et al., 2000). However, an anterior bony bridge may be useful for reducing pedicle screw fractures caused by bending loads (Rohlmann et al., 1998). La Barbera and colleagues found that posterior load sharing significantly decreased when a very rigid anterior body cage was used in a lumbar spine finite element model simulating a long-term fusion condition (La Barbera et al., 2016b, 2017). The same qualitative trend was found in a standardized synthetic construct by increasing the anterior support stiffness (La Barbera et al., 2016b, 2017). Similarly, using a pressure film transducer, Sengupta et al. (2013) found that the load shared by the anterior cage may decrease to 55% after posterior fixation compared to a more homogeneous 80% met in a physiological spine (White and Panjabi, 1990). An in vitro study by Kettler and colleagues did not find significant differences between the strain measurements on ALIF and TLIF cages, nor on the posterior instrumentation. Moreover, the estimated loads on the spinal rods were much lower than the load applied, suggesting that the remaining structures may still play an important role (Kettler et al., 2006). Chen and colleagues simulated alternative TLIF cage designs supplemented by pedicle screw fixation in a lumbo-sacral model, highlighting the effect of a stiff anterior support to decrease the stress on the instrumentation (Chen et al., 2012).

INTERBODY CERVICAL FUSION

ANTERIOR CERVICAL DISCECTOMY AND FUSION

Because of cervical anatomy, the intervertebral disc can be accessed easily via an anterior approach. Upon disc excision, an autogenous iliac bone graft or a cage are introduced in the disc space (Smith and Robinson, 1958). To minimize anterior graft pull-out, an anterior plate is usually fixed to the anterior aspects of the adjacent vertebrae, and to avoid cage migration, oblique fixation screws are often used to secure a stand-alone cage to the adjacent endplates. Anterior cervical discectomy and fusion (ACDF) with plate fixation has been associated with significantly higher vertebral fusion and lower subsidence rates, but it has also been associated with morbidity (Oliver et al., 2017; Njoku et al., 2014; Fujibayashi et al., 2008). However, a stand-alone cage is considered a valid alternative, and although it leads to comparable perioperative outcomes and complications, it significantly lowers blood loss and minimizes the morbidity rate related to dysphagia (Tabaraee et al., 2017; Chin et al., 2007). Zero-profile stand-alone cages were successfully implanted at contiguous (Moon et al., 2011) and non-contiguous levels (Wang et al., 2013) for cervical disc degenerative disease (Wu et al., 2012). Even if using a rigid anterior plate may shield the anterior bone graft from mechanical load, delaying its maturation, clinical results demonstrate a comparable fusion rate (Samartzis et al., 2004).

BIOMECHANICS

The biomechanics of spinal fixation at the cervical level demonstrate many similarities to the concepts already introduced for other spinal levels in terms of stability, cage dislocation, and subsidence.

The aim of spinal fixation is to promote bony fusion by restricting the intervertebral motions where the bone graft is implanted and thus avoid its migration (Ulrich et al., 2001).

A numerical study by Galbusera and colleagues demonstrated that a stand-alone cage led to the lowest stabilization among the configurations investigated. This scenario is also associated with the highest risk of subsidence because of the high stress levels reached at the bone-cage interface (Galbusera et al., 2008); using an anterior plate would reduce such occurrences and also further increase the stiffness of the spine segment. Scholz and colleagues found that a stand-alone PEEK cage significantly decreased the RoM in multilevel constructs in all motions but torsion (Scholz et al., 2015); the stabilizing effect was lower for anchored spacers and PEEK cages with a locking plate. It has also been found that posterior fixation supplemented with anterior bone graft significantly decreased intersegmental rotation in a cervical finite element model (Hussain et al., 2013); adding an anterior plate can further enhance such an effect. Cunningham used a hybrid in vitro protocol to test cadaveric segments and reported that, even when coupled with anterior plating, a spacer was not enough to achieve sufficient stability in all motion directions (Cunningham et al., 2010). In vitro flexibility tests on human cadaveric spines have demonstrated that spacers/cages with integrated screws guaranteed a comparable stabilizing effect to spacers with anterior plating in the case of single-level fixation (Stein et al., 2014; Paik et al., 2014). If anterior plating in multilevel reconstruction provided similar results compared to circumferential fusion (Setzer et al., 2012), lateral mass fixation may be another strategy to further increase spine stability (Paik et al., 2014). Bozkus and colleagues determined that both posterior fixation with pedicle screws and anterior plating may effectively reduced the kinematics of a cervical segment in vitro (Bozkus et al., 2005); however, a combined fixation further improved stability.

CONCLUSION

Spinal fixation and fusion are very well established surgical techniques useful in the treatment of a variety of diseases, often including diseases that are not limited to degenerative changes.

The biomechanics of both approaches can be interpreted as a system of springs representing the anterior and posterior columns. When the goal of surgery is fixation, a very stiff spine fixator is introduced posteriorly, which reduces the overall motions and shifts the loads on the instrumentation. When the aim is fusion, a very stiff anterior cage or bone graft is introduced anteriorly, which reduces the motions and shifts the loads anteriorly. More often, fixation and fusion are used together; therefore a careful balance in terms of load sharing must be established in order to promote bony fusion anteriorly and avoid mechanical issues related to excessive loads on the instrumentation or at the interface with the bone. In vitro experimental tests and numerical models are decisive tools that can be used to better investigate, understand, and control the biomechanics of fixation and fusion.

REFERENCES

Abe, E., Nickel, T., Buttermann, G.R., Lewis, J.L., Transfeldt, E.E., 1998. Lumbar intradiscal pressure after posterolateral fusion and pedicle screw fixation. Tohoku J. Exp. Med. 186 (4), 243–253.

Abumi, K., Itoh, H., Taneichi, H., Kaneda, K., 1994. Transpedicular screw fixation for traumatic lesions of the middle and lower cervical spine: description of the techniques and preliminary report. J. Spinal Disord. 7 (1), 19–28.

Abumi, K., Shono, Y., Ito, M., Taneichi, H., Kotani, Y., Kaneda, K., 2000. Complications of pedicle screw fixation in reconstructive surgery of the cervical spine. Spine 25 (8), 962–969.

Ahn, Y.H., Chen, W.M., Lee, K.Y., Park, K.W., Lee, S.J., 2008. Comparison of the load-sharing characteristics between pedicle-based dynamic and rigid rod devices. Biomed. Mater. 3 (4), 044101https://doi.org/10.1088/1748-6041/3/4/044101.

Albee, F.H., 1911. Transplantation of a portion of the tibia into the spine for Pott's disease. JAMA 57 (1), 885–886.

Amaritsakul, Y., Chao, C.K., Lin, J., 2014. Biomechanical evaluation of bending strength of spinal pedicle screws, including cylindrical, conical, dual core and double dual core designs using numerical simulations and mechanical tests. Med. Eng. Phys. 36 (9), 1218–1223. https://doi.org/10.1016/j.medengphy.2014.06.014.

Arnold, P.M., Anderson, K.K., RA Jr., M.G., 2012. The lateral transpsoas approach to the lumbar and thoracic spine: a review. Surg. Neurol. Int. 3, S198–215.

Asher, M.A., Carson, W.L., Hardacker, J.W., Lark, R.G., Lai, S.M., 2007. The effect of arthrodesis, implant stiffness, and time on the canine lumbar spine. J. Spinal Disord. Tech. 20, 549–559.

Bashkuev, M., Checa, S., Postigo, S., Duda, G., Schmidt, H., 2015. Computational analyses of different intervertebral cages for lumbar spinal fusion. J. Biomech. 48 (12), 3274–3282. https://doi.org/10.1016/j.jbiomech.2015.06.024.

Bianco, R.J., Aubin, C.E., Mac-Thiong, J.M., Wagnac, E., Arnoux, P.J., 2016. Pedicle screw fixation under nonaxial loads: a cadaveric study. Spine 41 (3), E124–30. https://doi.org/10.1097/BRS.0000000000001200.

Bianco, R.J., Arnoux, P.J., Wagnac, E., Mac-Thiong, J.M., Aubin, C.É., 2017. Minimizing pedicle screw pullout risks: a detailed biomechanical analysis of screw design and placement. Clin. Spine Surg. 30 (3), E226–E232. https://doi.org/10.1097/BSD.0000000000000151.

Boustani, H.N., Rohlmann, A., van der Put, R., Burger, A., Zander, T., 2012. Which postures are most suitable in assessing spinal fusion using radiostereometric analysis? Clin. Biomech. (Bristol, Avon). 27 (2), 111–116. https://doi.org/10.1016/j.clinbiomech.2011.08.012.

Bozkus, H., Ames, C.P., Chamberlain, R.H., Nottmeier, E.W., Sonntag, V.K., Papadopoulos, S.M., Crawford, N.R., 2005. Biomechanical analysis of rigid stabilization techniques for three-column injury in the lower cervical spine. Spine 30 (8), 915–922.

Brantigan, J.W., Steffee, A.D., Geiger, J.M., 1991. A carbon fiber implant to aid interbody lumbar fusion. Mechanical testing. Spine 16 (6 Suppl), S277–82.

Briggs, H., Milligan, P.R., 1944. Chip Fusion of the low back following exploration of the spinal canal. J. Bone Joint Surg. Am. 26, 125–130.

Bruner, H.J., Guan, Y., Yoganandan, N., Pintar, F.A., Maiman, D.J., Slivka, M.A., 2010. Biomechanics of polyaryletherketone rod composites and titanium rods for posterior lumbosacral instrumentation. J. Neurosurg. Spine 13 (6), 766–772. https://doi.org/10.3171/2010.5.SPINE09948.

Buttermann, G.R., Beaubien, B.P., Freeman, A.L., Stoll, J.E., Chappuis, J.L., 2009. Interbody device endplate engagement effects on motion segment biomechanics. Spine J. 9 (7), 564–573. https://doi.org/10.1016/j.spinee.2009.03.014.

Campbell, R.C., Mobbs, R.J., Lu, V.M., Xu, J., Rao, P.J., Phan, K., 2017 Aug. Posterolateral fusion versus interbody fusion for degenerative spondylolisthesis: systematic review and meta-analysis. Global Spine J. 7 (5), 482–490. https://doi.org/10.1177/2192568217701103.

Cappuccino, A., Cornwall, G.B., Turner, A.W., Fogel, G.R., Duong, H.T., Kim, K.D., Brodke, D.S., 2010. Biomechanical analysis and review of lateral lumbar fusion constructs. Spine 35 (26 Suppl), S361–S367. https://doi.org/10.1097/BRS.0b013e318202308b.

Carson, W.L., Duffield, R.C., Arendt, M., et al., 1990. Internal forces and moments in transpedicular spine instrumentation. Effect of pedicle screw angle and transfixation—the 4R-4bar linkage concept. Spine 15 (9), 893–901.

Chao, K.H., Lai, Y.S., Chen, W.C., Chang, C.M., McClean, C.J., Fan, C.Y., Chang, C.H., Lin, L.C., Cheng, C.K., 2013. Biomechanical analysis of different types of pedicle screw augmentation: a cadaveric and synthetic bone sample study of instrumented vertebral specimens. Med. Eng. Phys. 35 (10), 1506–1512. https://doi.org/10.1016/j.medengphy.2013.04.007.

Chen, C.S., Chen, W.J., Cheng, C.K., Jao, S.H., Chueh, S.C., Wang, C.C., 2005. Failure analysis of broken pedicle screws on spinal instrumentation. Med. Eng. Phys. 27 (6), 487–496. https://doi.org/10.1016/j.medengphy.2004.12.007.

Chen, S.H., Lin, S.C., Tsai, W.C., Wang, C.W., Chao, S.H., 2012. Biomechanical comparison of unilateral and bilateral pedicle screws fixation for transforaminal lumbar interbody fusion after decompressive surgery—a finite element analysis. BMC Musculoskelet. Disord. 13, 72https://doi.org/10.1186/1471-2474-13-72.

Chen, S.H., Chiang, M.C., Lin, J.F., Lin, S.C., Hung, C.H., 2013. Biomechanical comparison of three stand-alone lumbar cages—a three-dimensional finite element analysis. BMC Musculoskelet. Disord. 14, 281https://doi.org/10.1186/1471-2474-14-281.

Chin, K.R., Eiszner, J.R., Adams, S.B., 2007. Role of plate thickness as a cause of dysphagia after anterior cervical fusion. Spine 32 (23), 2585–2590.

Choi, K.C., Ryu, K.S., Lee, S.H., Kim, Y.H., Lee, S.J., Park, C.K., 2013. Biomechanical comparison of anterior lumbar interbody fusion: stand-alone interbody cage versus interbody cage with pedicle screw fixation—a finite element analysis. BMC Musculoskelet. Disord. 14, 220https://doi.org/10.1186/1471-2474-14-220.

Chun, K., Yang, I., Kim, N., Cho, D., 2015. Effect of device rigidity and physiological loading on spinal kinematics after dynamic stabilization: an in-vitro biomechanical study. J. Korean Neurosurg. Soc. 58 (5), 412–418. https://doi.org/10.3340/jkns.2015.58.5.412.

Closkey, R.F., Parsons, J.R., Lee, K.C., Blacksin, M.F., Zimmermann, M.C., 1993. Mechanics of interbody spinal fusion analysis of critical bone graft area. Spine 18, 1011–1015.

Cloward, R.B., 1953. The treatment of ruptured lumbar intervertebral discs by vertebral body fusion indications, operation technique, after care. J. Neurosurg. 10, 154–168.

Coe, J.D., Warden, K.E., Sutterlin III, C.E., McAfee, P.C., 1989. Biomechanical evaluation of cervical stabilization methods in a human cadaveric model. Spine 14 (10), 1122–1131.

Coe, J.D., Warden, K.E., Herzig, M.A., McAfee, P.C., 1990. Influence of bone mineral density on the fixation of thoracolumbar implants. A comparative study of transpedicular screws, laminar hooks, and spinous process wires. Spine 15 (9), 902–907.

Cook, D.J., Yeager, M.S., Thampi, S.S., Whiting, D.M., Cheng, B.C., 2015. Stability and load sharing characteristics of a posterior dynamic stabilization device. Int. J. Spine Surg. 9, 9. https://doi.org/10.14444/2009.

Costa, F., Villa, T., Anasetti, F., Tomei, M., Ortolina, A., Cardia, A., La Barbera, L., Fornari, M., Galbusera, F., 2013. Primary stability of pedicle screws depends on the screw positioning and alignment. Spine J. 13 (12), 1934–1939. https://doi.org/10.1016/j.spinee.2013.03.046.

Crawford III, C., O'Shaughnessy, B.A., Kuklo, T.R., 2011. Basic pedicle screw and construct biomechanics. In: Bridwell, K.H., Dewald, R.L. (Eds.), The Textbook of Spinal Surgery. vol. 1. Lippincott, Philadelphia, pp. 74–85.

Cripton, P.A., Jain, G.M., Wittenberg, R.H., Nolte, L.P., 2000. Load-sharing characteristics of stabilized lumbar spine segments. Spine 25 (2), 170–179.

Cunningham, B.W., Kotani, Y., PS, M.N., Cappuccino, A., McAfee, P.C., 1997. The effect of spinal destabilization and instrumentation on lumbar intradiscal pressure: an in vitro biomechanical analysis. Spine 22 (22), 2655–2663.

Cunningham, B.W., Hu, N., Zorn, C.M., McAfee, P.C., 2010 Apr. Biomechanical comparison of single- and two-level cervical arthroplasty versus arthrodesis: effect on adjacent-level spinal kinematics. Spine J. 10 (4), 341–349. https://doi.org/10.1016/j.spinee.2010.01.006.

Dakhil-Jerew, F., Jadeja, H., Cohen, A., Shepperd, J.A.N., 2009. Inter-observer reliability of detecting Dynesys pedicle screw using plain X-rays: a study on 50 post-operative patients. Eur. Spine J. 18, 1486–1493.

de Beer, N., Scheffer, C., 2012. Reducing subsidence risk by using rapid manufactured patient-specific intervertebral disc implants. Spine J. 12 (11), 1060–1066. https://doi.org/10.1016/j.spinee.2012.10.003.

Defino, H.L., Vendrame, J.R., 2001. Role of cortical and cancellous bone of the vertebral pedicle in implant fixation. Eur. Spine J. 10, 325–333.

Demir, T., 2015. A new alternative to expandable pedicle screws: expandable poly-ether-ether-ketone shell. Proc. Inst. Mech. Eng. H 229 (5), 386–394. https://doi.org/10.1177/0954411915584430.

Denis, F., 1983. The three column spine and its significance in the classification of acute thoracolumbar spinal injuries. Spine 8 (8), 817–831.

Dick, W., Kluger, P., Magerl, F., Woersdörfer, O., Zäch, G., 1985. A new device for internal fixation of thoracolumbar and lumbar spine fractures: the 'fixateur interne'. Paraplegia 23 (4), 225–232.

Dreischarf, M., Albiol, L., Zander, T., Arshad, R., Graichen, F., Bergmann, G., Schmidt, H., Rohlmann, A., 2015. In vivo implant forces acting on a vertebral body replacement during upper body flexion. J. Biomech. 48 (4), 560–565. https://doi.org/10.1016/j.jbiomech.2015.01.007.

Eck, J.C., Hodges, S., Humphreys, S.C., 2007. Minimally invasive lumbar spinal fusion. J. Am. Acad. Orthop. Surg. 15, 321–329.

Espinoza-Larios, A., Ames, C.P., Chamberlain, R.H., Sonntag, V.K.H., Dickman, C.A., Crawford, N.R., 2007. Biomechanical comparison of two-level cervical locking posterior screw/rod and hook/rod techniques. Spine J. 7 (2), 194–204. https://doi.org/10.1016/j.spinee.2006.04.015.

Fan, S.W., Hu, Z.J., Fang, X.Q., et al., 2010. Comparison of paraspinal muscle injury in one-level lumbar posterior inter-body fusion: modified minimally invasive and traditional open approaches. Orthop. Surg. 2, 194–200.

Farey, I.D., McAfee, P.C., Gurr, K.R., Randolph, M.A., 1989. Quantitative histologic study of the influence of spinal instrumentation on lumbar fusions: a canine model. J. Orthop. Res. 7 (5), 709–722.

Fensky, F., Kueny, R.A., Sellenschloh, K., Püschel, K., Morlock, M.M., Rueger, J.M., Lehmann, W., Huber, G., Hansen-Algenstaedt, N., 2014. Biomechanical advantage of C1 pedicle screws over C1 lateral mass screws: a cadaveric study. Eur. Spine J. 23 (4), 724–731. https://doi.org/10.1007/s00586-013-3143-4.

Fogel, G.R., Parikh, R.D., Ryu, S.I., Turner, A.W., 2014. Biomechanics of lateral lumbar interbody fusion constructs with lateral and posterior plate fixation: laboratory investigation. J. Neurosurg. Spine 20 (3), 291–297. https://doi.org/10.3171/2013.11.SPINE13617.

Freeman, A.L., Fahim, M.S., Bechtold, J.E., 2012 Oct. Validation of an improved method to calculate the orientation and magnitude of pedicle screw bending moments. J. Biomech. Eng. 134 (10), 104502https://doi.org/10.1115/1.4007629.

Fujibayashi, S., Neo, M., Nakamura, T., 2008. Stand-alone interbody cage versus anterior cervical plate for treatment of cervical disc herniation: sequential changes in cage subsidence. J. Clin. Neurosci. 15 (9), 1017–1022. https://doi.org/10.1016/j.jocn.2007.05.011.

Galbusera, F., Bellini, C.M., Costa, F., Assietti, R., Fornari, M., 2008. Anterior cervical fusion: a biomechanical comparison of 4 techniques laboratory investigation. J. Neurosurg. Spine 9 (5), 444–449. https://doi.org/10.3171/SPI.2008.9.11.444.

Galbusera, F., Bellini, C.M., Anasetti, F., Ciavarro, C., Lovi, A., Brayda-Bruno, M., 2011. Rigid and flexible spinal stabilization devices: a biomechanical comparison. Med. Eng. Phys. 33 (4), 490–496. https://doi.org/10.1016/j.medengphy.2010.11.018.

Gelalis, I.D., Paschos, N.K., Pakos, E.E., Politis, A.N., Arnaoutoglou, C.M., Karageorgos, A.C., Ploumis, A., Xenakis, T.A., 2012. Accuracy of pedicle screw placement: a systematic review of prospective in vivo studies comparing free hand, fluoroscopy guidance and navigation techniques. Eur. Spine J. 21, 247–255.

Goel, V.K., Lim, T.H., Gwon, J., Chen, J.Y., Winterbottom, J.M., Park, J.B., Weinstein, J.N., Ahn, J.Y., 1991. Effects of rigidity of an internal fixation device. A comprehensive biomechanical investigation. Spine 16 (3 Suppl), S155–61.

Goh, J.C., Wong, H.K., Thambyah, A., Yu, C.S., 2000. Influence of PLIF cage size on lumbar spine stability. Spine 25 (1), 35–39. (discussion 40).

Gornet, M.F., Chan, F.W., Coleman, J.C., Murrell, B., Nockels, R.P., Taylor, B.A., Lanman, T.H., Ochoa, J.A., 2011. Biomechanical assessment of a PEEK rod system for semi-rigid fixation of lumbar fusion constructs. J. Biomech. Eng. 133 (8), 081009https://doi.org/10.1115/1.4004862.

Haher, T.R., Yeung, A.W., Ottaviano, D.M., Merola, A.A., Caruso, S.A., 2001. The inverse effects of load transfer and load sharing on axial compressive stiffness. Spine J. 1 (5), 324–329. (discussion 330).

Harms, J., Rolinger, H., 1982. A one-stager procedure in operative treatment of spondylolistheses: dorsal traction-reposition and anterior fusion (author's transl). Z. Orthop. Ihre Grenzgeb. 120, 343–347.

Harrington, P.R., Tullos, H.S., 1969. Reduction of severe spondylolisthesis in children. South. Med. J. 62, 1–7.

Harris, B.M., Hilibrand, A.S., Savas, P.E., Pellegrino, A., Vaccaro, A.R., Siegler, S., Albert, T.J., 2004. Transforaminal lumbar interbody fusion: the effect of various instrumentation techniques on the flexibility of the lumbar spine. Spine 29 (4), E65–70.

Hirano, T., Hasegawa, K., Takahashi, H.E., Uchiyama, S., Hara, T., Washio, T., Sugiura, T., Yokaichiya, M., Ikeda, M., 1997. Structural characteristics of the pedicle and its role in screw stability. Spine 22, 2504–2510.

Humphreys, S.C., Hodges, S.D., Patwardhan, A.G., et al., 2001. Comparison of posterior and transforaminal approaches to lumbar interbody fusion. Spine 26, 567–571.

Hussain, M., Nassr, A., Natarajan, R.N., An, H.S., Andersson, G.B., 2013. Biomechanics of adjacent segments after a multilevel cervical corpectomy using anterior, posterior, and combined anterior-posterior instrumentation techniques: a finite element model study. Spine J. 13 (6), 689–696. https://doi.org/10.1016/j.spinee.2013.02.062.

Jahng, T.A., Kim, Y.E., Moon, K.Y., 2013. Comparison of the biomechanical effect of pedicle-based dynamic stabilization: a study using finite element analysis. Spine J. 13 (1), 85–94. https://doi.org/10.1016/j.spinee.2012.11.014.

Johnston, T.L., Karaikovic, E.E., Lautenschlager, E.P., Marcu, D., 2006. Cervical pedicle screws vs. lateral mass screws: uniplanar fatigue analysis and residual pullout strengths. Spine J. 6 (6), 667–672.

Jost, B., Cripton, P.A., Lund, T., Oxland, T.R., Lippuner, K., Jaeger, P., Nolte, L.P., 1998. Compressive strength of interbody cages in the lumbar spine: the effect of cage shape, posterior instrumentation and bone density. Eur. Spine J. 7, 132–141.

Karikari, I.O., Nimjee, S.M., Hardin, C.A., Hughes, B.D., Hodges, T.R., Mehta, A.I., et al., 2011. Extreme lateral interbody fusion approach for isolated thoracic and thoracolumbar spine diseases: initial clinical experience and early outcomes. J. Spinal Disord. Tech. 24, 368–375.

Karim, A., Mukherjee, D., Ankem, M., Gonzalez-Cruz, J., Smith, D., Nanda, A., 2006. Augmentation of anterior lumbar interbody fusion with anterior pedicle screw fixation: demonstration of novel constructs and evaluation of biomechanical stability in cadaveric specimens. Neurosurgery 58 (3), 522–527. discussion 522–7 https://doi.org/10.1227/01.NEU.0000197322.52848.C8.

Kettler, A., Dietl, R., Krammer, M., Lumenta, C.B., Claes, L., Wilke, H.J., 2002. Dislocation tendency, stabilizing effect and sintering tendency of different lumbar vertebrae cages in an in vitro experiment. Orthopade 31 (5), 481–487.

Kettler, A., Niemeyer, T., Issler, L., Merk, U., Mahalingam, M., Werner, K., Claes, L., Wilke, H.J., 2006. In vitro fixator rod loading after transforaminal compared to anterior lumbar interbody fusion. Clin. Biomech. (Bristol, Avon). 21 (5), 435–442. https://doi.org/10.1016/j.clinbiomech.2005.12.005.

Kim, Y., 2001. Prediction of mechanical behaviors at interfaces between bone and two interbody cages of lumbar spine segments. Spine 26 (13), 1437–1442.

Kim, Y., 2007. Finite element analysis of anterior lumbar interbody fusion. Threaded cylindrical cage and pedicle screw fixation. Spine 32 (23), 2558–2568. https://doi.org/10.1097/BRS.0b013e318158cdd8.

Kim, Y., Vanderby, R., 2000. Finite element analysis of interbody cages in a human lumbar spine. Comput. Methods Biomech. Biomed. Eng. 3 (4), 257–272. https://doi.org/10.1080/10255840008915270.

Kiner, D.W., Wybo, C.D., Sterba, W., Yeni, Y.N., Bartol, S.W., Vaidya, R., 2008. Biomechanical analysis of different techniques in revision spinal instrumentation: larger diameter screws versus cement augmentation. Spine 33, 2618.

Kotani, Y., Cunningham, B.W., Abumi, K., McAfee, P.C., 1994. Biomechanical analysis of cervical stabilization systems. An assessment of transpedicular screw fixation in the cervical spine. Spine 19 (22), 2529–2539.

Krag, M.H., Beynnon, B.D., Pope, M.H., DeCoster, T.A., 1988. Depth of insertion of transpedicular vertebral screws into human vertebrae: effect upon screw-vertebra interface strength. J. Spinal Disord. 1 (4), 287–294.

Krammer, M., Dietl, R., Lumenta, C.B., Kettler, A., Wilke, H.J., Büttner, A., Claes, L., 2001. Resistance of the lumbar spine against axial compression forces after implantation of three different posterior lumbar interbody cages. Acta Neurochir. 143 (12), 1217–1222.

Kueny, R.A., Kolb, J.P., Lehmann, W., Püschel, K., Morlock, M.M., Huber, G., 2014. Influence of the screw augmentation technique and a diameter increase on pedicle screw fixation in the osteoporotic spine: pullout versus fatigue testing. Eur. Spine J. 23 (10), 2196–2202. https://doi.org/10.1007/s00586-014-3476-7.

La Barbera, L., Villa, T., 2016. ISO 12189 standard for the preclinical evaluation of posterior spinal stabilization devices—I: assembly procedure and validation. Proc. Inst. Mech. Eng. H 230 (2), 122–133. https://doi.org/10.1177/0954411915621587.

La Barbera, L., Galbusera, F., Villa, T., Wilke, H.-J., 2014. ASTM F1717 standard for the preclinical evaluation of posterior spinal fixators: can we improve it? Proc. Inst. Mech. Eng. H 228 (10), 1014–1026. https://doi.org/10.1177/0954411914554244.

La Barbera, L., Costa, F., Villa, T., 2016a. ISO 12189 standard for the preclinical evaluation of posterior spinal stabilization devices—II: a parametric comparative study. Proc. Inst. Mech. Eng. H 230 (2), 134–144. https://doi.org/10.1177/0954411915621588.

La Barbera, L., Galbusera, F., Wilke, H.-J., Villa, T., 2016b. Preclinical evaluation of posterior spine stabilization devices: can the current standards represent basic everyday life activities? Eur. Spine J. 25 (9), 2909–2918. https://doi.org/10.1007/s00586-016-4622-1.

La Barbera, L., Galbusera, F., Wilke, H.J., Villa, T., 2017. Preclinical evaluation of posterior spine stabilization devices: can we compare in vitro and in vivo loads on the instrumentation? Eur. Spine J. 26 (1), 200–209. https://doi.org/10.1007/s00586-016-4766-z.

Lam, F.C., Alkalay, R., Groff, M.W., 2012. The effects of design and positioning of carbon fiber lumbar interbody cages and their subsidence in vertebral bodies. J. Spinal Disord. Tech. 25 (2), 116–122. https://doi.org/10.1097/BSD.0b013e31820ef778.

Landham, P.R., Don, A.S., Robertson, P.A., 2017. Do position and size matter? An analysis of cage and placement variables for optimum lordosis in PLIF reconstruction. Eur. Spine J. https://doi.org/10.1007/s00586-017-5170-z.

Le Huec, J.C., Liu, M., Skalli, W., Josse, L., 2002. Lumbar lateral interbody cage with plate augmentation: in vitro biomechanical analysis. Eur. Spine J. 11 (2), 130–136. https://doi.org/10.1007/s005860100316.

Lee, C.K., Langrana, N.A., 1984. Lumhosacral spinal fusion. A biomechanical study. Spine 9, 574.

Lestini, W.F., Fulghum, J.S., Whitehurst, L.A., 1994. Lumbar spinal fusion: advantages of posterior lumbar interbody fusion. Surg. Technol. Int. 3, 577–590.

Li, J., Shang, J., Zhou, Y., Li, C., Liu, H., 2015. Finite element analysis of a new pedicle screw-plate system for minimally invasive transforaminal lumbar interbody fusion. PLoS One 10 (12), e0144637https://doi.org/10.1371/journal.pone.0144637.

Lidar, Z., Beaumont, A., Lifshutz, J., Maiman, D.J., 2005. Clinical and radiological relationship between posterior lumbar interbody fusion and posterolateral lumbar fusion. Surg. Neurol. 64, 303–308.

Liebsch, C., Zimmermann, J., Graf, N., Schilling, C., Wilke, H.J., Kienle, A., 2017. In vitro validation of a novel mechanical model for testing the anchorage capacity of pedicle screws using physiological load application. J. Mech. Behav. Biomed. Mater. 77, 578–585. https://doi.org/10.1016/j.jmbbm.2017.10.030.

Lipscomb, H.J., Grubb, S.A., Talmage, R.V., 1989. Spinal bone density following spinal fusion. Spine 14 (4), 477–479.

Ludwig, S.C., Kramer, D.L., Vaccaro, A.R., Albert, T.J., 1999. Transpedicle screw fixation of the cervical spine. Clin. Orthop. Relat. Res. 359, 77–88.

Lund, T., Oxland, T.R., Jost, B., Cripton, P., Grassmann, S., Etter, C., Nolte, L.P., 1998. Interbody cage stabilisation in the lumbar spine: biomechanical evaluation of cage design, posterior instrumentation and bone density. J. Bone Joint Surg. (Br.) 80 (2), 351–359.

Magerl, F., 1982. External skeletal fixation of the lower thoracic and the lumbar spine. In: Uhthoff, H. (Ed.), Current Concepts of External Fixation of Fractures. Springer, Berlin, pp. 353–366.

Malham, G.M., Ellis, N.J., Parker, R.M., et al., 2012. Clinical outcome and fusion rates after the first 30 extreme lateral interbody fusions. Sci. World J. 2012, 246989.

Malham, G.M., Parker, R.M., Ellis, N.J., et al., 2014. Anterior lumbar interbody fusion using recombinant human bone morphogenetic protein-2: a prospective study of complications. J. Neurosurg. Spine 21, 851–860.

Malham, G.M., Parker, R.M., Goss, B., et al., 2015. Clinical results and limitations of indirect decompression in spinal stenosis with laterally implanted interbody cages: results from a prospective cohort study. Eur. Spine J. 24 (Suppl. 3), 339–345.

Malham, G.M., Ellis, N.J., Parker, R.M., Blecher, C.M., White, R., Goss, B., Seex, K.A., 2017. Maintenance of segmental lordosis and disc height in standalone and instrumented extreme lateral interbody fusion (XLIF). Clin. Spine Surg. 30 (2), E90–E98. https://doi.org/10.1097/BSD.0b013e3182aa4c94.

McAfee, P.C., Farey, I.D., Sutterlin, C.E., Gurr, K.R., Warden, K.E., Cunningham, B.W., 1989. Volvo award in basic science. Device-related osteoporosis with spinal instrumentation. Spine 14 (9), 919–926.

McAfee, P.C., Farey, I.D., Sutterlin, C.E., Gurr, K.R., Warden, K.E., Cunningham, B.W., 1991. The effect of spinal implant rigidity on vertebral bonedensity. A canine model. Spine 16 (6 Suppl), S190–7.

McAfee, P.C., JG, D.V., Chaput, C.D., et al., 2005. The indications for interbody fusion cages in the treatment of spondylolisthesis: analysis of 120 cases. Spine 30, S60–5.

Melnyk, A.D., Chak, J.D., Cripton, P.A., Dvorak, M.F., Oxland, T.R., 2012a. Shear force measurements on low- and high-stiffness posterior fusion devices. Med. Eng. Phys. 34 (9), 1260–1267. https://doi.org/10.1016/j.medengphy.2011.12.016.

Melnyk, A.D., Wen, T.L., Kingwell, S., Chak, J.D., Singh, V., Cripton, P.A., Fisher, C.G., Dvorak, M.F., Oxland, T.R., 2012b. Load transfer characteristics between posterior spinal implants and the lumbar spine under anterior shear loading: an in vitro investigation. Spine 37 (18), E1126–E1133. https://doi.org/10.1097/BRS.0b013e31825b528d.

Melnyk, A.D., Chak, J.D., Singh, V., Kelly, A., Cripton, P.A., Fisher, C.G., Dvorak, M.F., Oxland, T.R., 2015. Characterization of the behavior of a novel low-stiffness posterior spinal implant under anterior shear loading on a degenerative spinal model. Eur. Spine J. 24 (4), 775–782. https://doi.org/10.1007/s00586-014-3735-7.

Mobbs, R.J., Sivabalan, P., Li, J., 2011. Technique, challenges and indications for percutaneous pedicle screw fixation. J. Clin. Neurosci. 18 (6), 741–749. https://doi.org/10.1016/j.jocn.2010.09.019.

Mobbs, R.J., Phan, K., Malham, G., Seex, K., Rao, P., 2015. Lumbar interbody fusion: techniques, indications and comparison of interbody fusion options including PLIF, TLIF, MI-TLIF, OLIF/ATP LLIF and ALIF. J. Spine Surg. 1 (1), 2–18. https://doi.org/10.3978/j.issn.2414-469X.2015.10.05.

Moon, H.J., Kim, J.H., Kim, J.H., Kwon, T.H., Chung, H.S., Park, Y.K., 2011. The effects of anterior cervical discectomy and fusion with stand-alone cages at two contiguous levels on cervical alignment and outcomes. Acta Neurochir. 153 (3), 559–565. https://doi.org/10.1007/s00701-010-0879-z.

Nagaraja, S., Palepu, V., 2017. Integrated fixation cage loosening under fatigue loading. Int. J. Spine Surg. 11, 20. https://doi.org/10.14444/4020.

Narayan, P., Haid, R.W., Subach, B.R., et al., 2002. Effect of spinal disease on successful arthrodesis in lumbar pedicle screw fixation. J. Neurosurg. 97 (3 Suppl), 277–280.

Newcomb, A.G., Baek, S., Kelly, B.P., Crawford, N.R. 2017. Effect of screw position on load transfer in lumbar pedicle screws: a non-idealized finite element analysis. Comput. Methods Biomech. Biomed. Eng. 20 (2), 182–192. https://doi.org/10.1080/10255842.2016.1209187.

Njoku I Jr, Alimi M, Leng LZ, Shin BJ, James AR, Bhangoo S, Tsiouris AJ, Härtl R. Anterior cervical discectomy and fusion with a zero-profile integrated plate and spacer device: a clinical and radiological study: clinical article. J. Neurosurg. Spine 2014;21(4):529–37. doi: https://doi.org/10.3171/2014.6.SPINE12951.

Oda, I., Abumi, K., Yu, B.S., et al., 2003. Types of spinal instability that require interbody support in posterior lumbar reconstruction: an in vitro biomechanical investigation. Spine 28, 1573–1580.

Oliver, J.D., Goncalves, S., Kerezoudis, P., Alvi, M.A., Freedman, B.A., Nassr, A., Bydon, M., 2017. Comparison of outcomes for anterior cervical discectomy and fusion with and without anterior plate fixation: a systematic review and meta-analysis. Spine https://doi.org/10.1097/BRS.0000000000002441.

Oxland, T.R., Hoffer, Z., Nydegger, T., Rathonyi, G.C., Nolte, L.P., 2000. A comparative biomechanical investigation of anterior lumbar interbody cages: central and bilateral approaches. J. Bone Joint Surg. Am. 82 (3), 383–393.

Ozgur, B.M., Aryan, H.E., Pimenta, L., et al., 2006. Extreme lateral interbody fusion (XLIF): a novel surgical technique for anterior lumbar interbody fusion. Spine J. 6, 435–443.

Paik H, Kang DG, Lehman RA Jr, Cardoso MJ, Gaume RE, Ambati DV, Dmitriev AE. Do stand-alone interbody spacers with integrated screws provide adequate segmental stability for multilevel cervical arthrodesis? Spine J. 2014;14(8):1740–7. doi: https://doi.org/10.1016/j.spinee.2014.01.034.

Palepu, V., Peck, J.H., Simon, D.D., Helgeson, M.D., Nagaraja, S., 2017. Biomechanical evaluation of an integrated fixation cage during fatigue loading: a human cadaver study. J. Neurosurg. Spine 26 (4), 524–531. https://doi.org/10.3171/2016.9.SPINE16650.

Panjabi, M.M., Takata, K., Goel, V., Federico, D., Oxland, T., Duranceau, J., Krag, M., 1991a. Thoracic human vertebrae. Quantitative three-dimensional anatomy. Spine 16 (8), 888–901.

Panjabi, M.M., Durenceau, J., Goel, V., Oxland, T., Takata, K., 1991b. Cervical human vertebrae. Quantitative three-dimensional anatomy of the middle and lower regions. Spine 16 (8), 861–869.

Panjabi, M.M., Goel, V., Oxland, T., Takata, K., Duranceau, J., Krag, M., Price, M., 1992. Human Lumbar vertebrae. Quantitative three-dimensional anatomy. Spine 17 (3), 299–306.

Patel, N.P., Birch, B.D., Dement, S.E., Elbert, G.A., 2010. The mini-open anterolateral approach for degenerative thoracolumbar disease. Clin. Neurol. Neurosurg. 112 (10), 853–857. https://doi.org/10.1016/j.clineuro.2010.07.008.

Pech, P., Daniels, D.L., Williams, A.L., Haughton, V.M., 1985. The cervical neural foramina: correlation of microtomy and CT anatomy. Radiology 155 (1), 143–146. https://doi.org/10.1148/radiology.155.1.3975392.

Pfeifer, M., Gilbertson, L.G., Goel, V.K., Griss, P., Keller, J.C., Ryken, T.C., Hoffman, H.E., 1996. Effect of specimen fixation method on the pullout tests of pedicle screws. Spine 21, 1037–1044.

Phan, K., Thayaparan, G.K., Mobbs, R.J., 2015a. Anterior lumbar interbody fusion versus transforaminal lumbar interbody fusion—systematic review and meta-analysis. Br. J. Neurosurg. 29, 705–711.

Phan, K., Rao, P.J., Scherman, D.B., et al., 2015b. Lateral lumbar interbody fusion for sagittal balance correction and spinal deformity. J. Clin. Neurosci. 22, 1714–1721.

Pimenta, L., Turner, A.W., Dooley, Z.A., Parikh, R.D., Peterson, M.D., 2012. Biomechanics of lateral interbody spacers: going wider for going stiffer. Sci.World J. 2012, 381814. https://doi.org/10.1100/2012/381814.

Polikeit, A., Ferguson, S.J., Nolte, L.P., Orr, T.E., 2003a. The importance of the endplate for interbody cages in the lumbar spine. Eur. Spine J. 12 (6), 556–561. https://doi.org/10.1007/s00586-003-0556-5.

Polikeit, A., Ferguson, S.J., Nolte, L.P., Orr, T.E., 2003 Augb. Factors influencing stresses in the lumbar spine after the insertion of intervertebral cages: finite element analysis. Eur. Spine J. 12 (4), 413–420. https://doi.org/10.1007/s00586-002-0505-8.

Polly DW Jr, Klemme WR, Cunningham BW, Burnette JB, Haggerty CJ, Oda I. The biomechanical significance of anterior column support in a simulated single-level spinal fusion. J. Spinal Disord. 2000;13(1):58–62.

Ponnappan, R.K., Serhan, H., Zarda, B., Patel, R., Albert, T., Vaccaro, A.R., 2009. Biomechanical evaluation and comparison of polyetheretherketone rod system to traditional titanium rod fixation. Spine J. 9 (3), 263–267. https://doi.org/10.1016/j.spinee.2008.08.002.

Postigo, S., Schmidt, H., Rohlmann, A., Putzier, M., Simón, A., Duda, G., Checa, S., 2014. Investigation of different cage designs and mechano-regulation algorithms in the lumbar interbody fusion process—a finite element analysis. J. Biomech. 47 (6), 1514–1519. https://doi.org/10.1016/j.jbiomech.2014.02.005.

Rao, P.J., Loganathan, A., Yeung, V., et al., 2015a. Outcomes of anterior lumbar interbody fusion surgery based on indication: a prospective study. Neurosurgery 76, 7–23. (discussion 23–4).

Rao, P.J., Ghent, F., Phan, K., et al., 2015b. Stand-alone anterior lumbar interbody fusion for treatment of degenerative spondylolisthesis. J. Clin. Neurosci. 22, 1619–1624.

Rapoff, A.J., Ghanayem, A.J., Zdeblick, T.A., 1997. Biomechanical comparison of posterior lumbar interbody fusion cages. Spine 22 (20), 2375–2379.

Reinhold, M., Magerl, F., Rieger, M., Blauth, M., 2007. Cervical pedicle screw placement: feasibility and accuracy of two new insertion techniques based on morphometric data. Eur. Spine J. 16 (1), 47–56. https://doi.org/10.1007/s00586-006-0104-1.

Rihn, J.A., Gandhi, S.D., Sheehan, P., Vaccaro, A.R., Hilibrand, A.S., Albert, T.J., Anderson, D.G., 2014. Disc space preparation in transforaminal lumbar interbody fusion: a comparison of minimally invasive and open approaches. Clin. Orthop. Relat. Res. 472 (6), 1800–1805. https://doi.org/10.1007/s11999-014-3479-z.

Rohlmann, A., Bergmann, G., Graichen, F., 1997. Loads on an internal spinal fixation device during walking. J. Biomech. 30 (1), 41–47.

Rohlmann, A., Bergmann, G., Graichen, F., Mayer, H.M., 1998. Placing a bone graft more posteriorly may reduce the risk of pedicle screw breakage analysis of an unexpected case of pedicle screw breakage. J. Biomech. 31 (8), 763–767.

Rohlmann, A., Calisse, J., Bergmann, G., Weber, U., 1999a. Internal spinal fixator stiffness has only a minor influence on stresses in the adjacent discs. Spine 24 (12), 1192–1195. (discussion 1195–6).

Rohlmann, A., Bergmann, G., Graichen, F., 1999b. Loads on internal spinal fixators measured in different body positions. Eur. Spine J. 8, 354–359.

Rohlmann, A., Graichen, F., Weber, U., Bergmann, G., 2000. Volvo award winner in biomechanical studies: monitoring in vivo implant loads with a telemeterized internal spinal fixation device. Spine 25 (23), 2981–2986.

Rohlmann, A., Graichen, F., Bergmann, G., 2002a. Loads on an internal spinal fixation device during physical therapy. Phys. Ther. 82 (1), 44–52.

Rohlmann, A., Zander, T., Fehrmann, M., Klöckner, C., Bergmann, G., 2002b. Effect of a double intervertebral cage on the mechanical behavior of the lumbar spine. Biomed Tech (Berl) 47 (5), 124–129.

Rohlmann, A., Zander, T., Bergmann, G., 2005. Comparison of the biomechanical effects of posterior and anterior spine-stabilizing implants. Eur. Spine J. 14 (5), 445–453. https://doi.org/10.1007/s00586-004-0784-3.

Rohlmann, A., Zander, T., Bergmann, G., 2006. Effects of fusion-bone stiffness on the mechanical behavior of the lumbar spine after vertebral body replacement. Clin. Biomech. (Bristol, Avon). 21 (3), 221–227. https://doi.org/10.1016/j.clinbiomech.2005.10.012.

Rohlmann, A., Burra, N.K., Zander, T., Bergmann, G., 2007. Comparison of the effects of bilateral posterior dynamic and rigid fixation devices on the loads in the lumbar spine: a finite element analysis. Eur. Spine J. 16 (8), 1223–1231.

Rohlmann, A., Graichen, F., Bender, A., Kayser, R., Bergmann, G., 2008. Loads on a telemeterized vertebral body replacement measured in three patients within the first postoperative month. Clin. Biomech. 23, 147–158. https://doi.org/10.1016/j.clinbiomech.2007.09.011.

Rohlmann, A., Zander, T., Bergmann, G., Boustani, H.N., 2012. Optimal stiffness of a pedicle-screw-based motion preservation implant for the lumbar spine. Eur. Spine J. 21 (4), 666–673. https://doi.org/10.1007/s00586-011-2047-4.

Rohlmann, A., Pohl, D., Bender, A., Graichen, F., Dymke, J., et al., 2014. Activities of everyday life with high spinal loads. PLoS One 9 (5), e98510https://doi.org/10.1371/journal.pone.0098510.

Roy-Camille, R., Saillant, G., Mazel, C., 1986. Internal fixation of the lumbar spine with pedicle screw plating. Clin. Orthop. Relat. Res. 7–17.

Ruland, C., McAfee, P., Warden, K., Cunningham, B., 1991. Triangulation Of pedicular instrumentation. A biomechanical analysis. Spine 16 (6 Suppl), S270–6.

Samartzis, D., Shen, F.H., Lyon, C., Phillips, M., Goldberg, E.J., An, H.S., 2004. Does rigid instrumentation increase the fusion rate in one-level anterior cervical discectomy and fusion? Spine J. 4 (6), 636–643.

Schlager, B., Ismael Aguirre, M.F., Galbusera, F., 2018. Scoliosis. In: Biomechanics of the Spine, Section 4: Spinal Disorders and Spine Surgery, Spinal Deformities. Elsevier Inc., Cambridge, MA. Chapter 20.

Schleicher, P., Gerlach, R., Schär, B., Cain, C.M., Achatz, W., Pflugmacher, R., Haas, N.P., Kandziora, F., 2008. Biomechanical comparison of two different concepts for stand alone anterior lumbar interbody fusion. Eur. Spine J. 17, 1757–1765.

Schmidt, H., Heuer, F., Claes, L., Wilke, H.J., 2008. The relation between the instantaneous center of rotation and facet joint forces—a finite element analysis. Clin. Biomech. (Bristol, Avon). 23 (3), 270–278. https://doi.org/10.1016/j.clinbiomech.2007.10.001.

Schmidt, H., Heuer, F., Wilke, H.J., 2009. Which axial and bending stiffnesses of posterior implants are required to design a flexible lumbar stabilization system? J. Biomech. 42 (1), 48–54. https://doi.org/10.1016/j.jbiomech.2008.10.005.

Schmoelz, W., 2018. Trauma and tumors. In: Biomechanics of the Spine, Section 4: Spinal Disorders and Spine Surgery, Degenerative Disorders. Elsevier Inc., Cambridge, MA. Chapter 23.

Schmoelz, W., Keiler, A., 2015. Intervertebral cages from a biomechanical point of view. Orthopade 44 (2), 132–137. https://doi.org/10.1007/s00132-014-3071-y.

Scholz, M., Schleicher, P., Pabst, S., Kandziora, F., 2015. A zero-profile anchored spacer in multilevel cervical anterior interbody fusion: biomechanical comparison to established fixation techniques. Spine 40 (7), E375–80. https://doi.org/10.1097/BRS.0000000000000768.

Schwab, F.J., Nazarian, D.G., Mahmud, F., Michelsen, C.B., 1995. Effects of spinal instrumentation on fusion of the lumbosacral spine. Spine 20 (18), 2023–2028.

Sengupta, D.K., Bucklen, B., McAfee, P.C., Nichols, J., Angara, R., Khalil, S., 2013. The comprehensive biomechanics and load-sharing of semirigid PEEK and semirigid posterior dynamic stabilization systems. Adv. Orthop. 2013, 745610. https://doi.org/10.1155/2013/745610.

Setzer, M., Eleraky, M., Johnson, W.M., Aghayev, K., Tran, N.D., Vrionis, F.D., 2012. Biomechanical comparison of anterior cervical spine instrumentation techniques with and without supplemental posterior fusion after different corpectomy and discectomy combinations: laboratory investigation. J. Neurosurg. Spine 16 (6), 579–584. https://doi.org/10.3171/2012.2.SPINE11611.

Shea, T.M., Laun, J., Gonzalez-Blohm, S.A., Doulgeris, J.J., Lee 3rd, W.E., Aghayev, K., Vrionis, F.D., 2014. Designs and techniques that improve the pullout strength of pedicle screws in osteoporotic vertebrae: current status. Biomed. Res. Int. 2014, 748393https://doi.org/10.1155/2014/748393.

Shih, K.S., Hsu, C.C., Hou, S.M., Yu, S.C., Liaw, C.K., 2015. Comparison of the bending performance of solid and cannulated spinal pedicle screws using finite element analyses and biomechanical tests. Med. Eng. Phys. 37 (9), 879–884. https://doi.org/10.1016/j.medengphy.2015.06.008.

Shin, H.K., Oh, S.K., Choi, I., Jeon, S.R., 2016. Posterior thoracic cage interbody fusion (PTCIF) as an alternative fusion technique after laminectomy in thoracic and thoracolumbar junctional spine. Nerve 2 (1), 5–9. https://doi.org/10.21129/nerve.2016.2.1.5.

Silva, P., Rosa, R.C., Shimano, A.C., Defino, H.L.A., 2013. Effect of pilot hole on biomechanical and in vivo pedicle screw–bone interface. Eur. Spine J. 22, 1829–1836. https://doi.org/10.1007/s00586-013-2810-9.

Sim, H.B., Murovic, J.A., Cho, B.Y., Lim, T.J., Park, J., 2010. Biomechanical comparison of single-level posterior versus transforaminal lumbar interbody fusions with bilateral pedicle screw fixation: segmental stability and the effects on adjacent motion segments. J. Neurosurg. Spine 12 (6), 700–708. https://doi.org/10.3171/2009.12.SPINE09123.

Smith, G.W., Robinson, R.A., 1958. The treatment of certain cervical-spine disorders by anterior removal of the intervertebral disc and interbody fusion. J. Bone Joint Surg. Am. 40-A (3), 607–624.

Smith, J.S., Ogden, A.T., Fessler, R.G., 2008. Minimally invasive posterior thoracic fusion. Neurosurg. Focus. 25 (2), E9. https://doi.org/10.3171/FOC/2008/25/8/E9.

Steffee, A.D., Biscup, R.S., Sitkowski, D.J., 1986. Segmental spine plates with pedicle screw fixation. A new internal fixation device for disorders of the lumbar and thoracolumbar spine. Clin. Orthop. Relat. Res. 45-53.

Stein MI, Nayak AN, Gaskins RB 3rd, Cabezas AF, Santoni BG, Castellvi A. Biomechanics of an integrated interbody device versus ACDF anterior locking plate in a single-level cervical spine fusion construct. Spine J. 2014;14(1):128–36. doi: https://doi.org/10.1016/j.spinee.2013.06.088.

Strömqvist, B., 1993. Posterolateral uninstrumented fusion. Acta Orthop. Scand. 64 (Suppl. 251), 97–99. https://doi.org/10.3109/17453679309160134.

Tabaraee, E., Ahn, J., Bohl, D.D., Collins, M.J., Massel, D.H., Aboushaala, K., Singh, K., 2017. Comparison of surgical outcomes, narcotics utilization, and costs after an anterior cervical discectomy and fusion: stand-alone cage versus anterior plating. Clin. Spine Surg. 30 (9), E1201–E1205. https://doi.org/10.1097/BSD.0000000000000341.

Tan, J.S., Kwon, B.K., Samarasekera, D., Dvorak, M.F., Fisher, C.G., Oxland, T.R., 2003. Vertebral bone density—a critical element in the performance of spinal implants. In: Melkerson, M.N., Griffith, S.L., Kirkpatrick, J.S. (Eds.), Spinal Implants: Are We Evaluating Them Appropriately ASTM STP 1431. ASTM International, West Conshohocken, PA.

Tatsumi, R., Lee, Y.P., Khajavi, K., Taylor, W., Chen, F., Bae, H., 2015. In vitro comparison of endplate preparation between four mini-open interbody fusion approaches. Eur. Spine J. 24 (Suppl 3), 372–377. https://doi.org/10.1007/s00586-014-3708-x.

Templier, A., Denninger, L., Mazel, C., Lavaste, F., Skalli, W., 1998. Comparison between two different concepts of lumbar posterior osteosynthesis implants a finite-element analysis. Eur. J. Orthop. Surg. Traumatol. 8, 27–36.

Thomsen, K., Christensen, F.B., Eiskjaer, S.P., Hansen, E.S., Fruensgaard, S., Bünger, C.E., 1997. Volvo award winner in clinical studies. The effect of pedicle screw instrumentation on functional outcome and fusion rates in posterolateral lumbar spinal fusion: a prospective, randomized clinical study. Spine 22, 2813–2822.

Ulrich, C., Arand, M., Nothwang, J., 2001. Internal fixation on the lower cervical spine—biomechanics and clinical practice of procedures and implants. Eur. Spine J. 10 (2), 88–100. https://doi.org/10.1007/s005860000233.

Villa, T., La Barbera, L., Galbusera, F., 2014a. Comparative analysis of international standards for the fatigue testing of posterior spinal fixation systems. Spine J. 14 (4), 695–704. https://doi.org/10.1016/j.spinee.2013.08.032.

Villa, T., La Barbera, L., Galbusera, F., 2014b. Reply to the letter to the editor entitled: response to "comparative analysis of international standards for the fatigue testing of posterior spinal fixation systems". Spine J. 14 (12), 3068. https://doi.org/10.1016/j.spinee.2014.08.008.

Wang, H.R., Li, X.L., Dong, J., Yuan, F.L., Zhou, J., 2013. Skip-level anterior cervical discectomy and fusion with self-locking stand-alone PEEK cages for the treatment of 2 noncontiguous levels of cervical spondylosis. J. Spinal Disord. Tech. 26 (7), E286–92. https://doi.org/10.1097/BSD.0b013e31828679b3.

Wang, T., Ball, J.R., Pelletier, M.H., Walsh, W.R., 2014. Biomechanical evaluation of a biomimetic spinal construct. J. Exp. Orthop. 1 (1), 3. https://doi.org/10.1186/s40634-014-0003-z.

Watkins, M.B., 1953. Posterolateral fusion of the lumbar and lumbosacral spine. J. Bone Joint Surg. Am. 35A (4), 1014–1018.

Watkins, M.B., 1959. Posterolateral bonegrafting for fusion of the lumbar and lumbosacral spine. J. Bone Joint Surg. Am. 41A (3), 388–396.

White, A., Panjabi, M., 1990. Physical properties and functional biomechanics of the spine. In: Clinical Biomechanics of the Spine. Lippincott Williams & Wilkins, Philadelphia, PA, USA. Chapter 1.

Wilke, H.J., Kaiser, D., Volkheimer, D., Hackenbroch, C., Püschel, K., Rauschmann, M., 2016. A pedicle screw system and a lamina hook system provide similar primary and long-term stability: a biomechanical in vitro study with quasi-static and dynamic loading conditions. Eur. Spine J. 25 (9), 2919–2928. https://doi.org/10.1007/s00586-016-4579-x.

Wu, Y., Tang, H., Li, Z., Zhang, Q., Shi, Z., 2011. Outcome of posterior lumbar interbody fusion versus posterolateral fusion in lumbar degenerative disease. J. Clin. Neurosci. 18, 780–783. https://doi.org/10.1016/j.jocn.2010.10.012.

Wu, W.J., Jiang, L.S., Liang, Y, Dai, L.Y., 2012 Jul. Cage subsidence does not, but cervical lordosis improvement does affect the long-term results of anterior cervical fusion with stand-alone cage for degenerative cervical disc disease: a retrospective study. Eur. Spine J. 21 (7), 1374–1382. https://doi.org/10.1007/s00586-011-2131-9.

Zander, T., Rohlmann, A., Klöckner, C., Bergmann, G., 2002. Comparison of the mechanical behavior of the lumbar spine following mono- and bisegmental stabilization. Clin. Biomech. (Bristol, Avon). 17 (6), 439–445.

CHAPTER

MOTION PRESERVATION 18

Fabio Galbusera*, Hans-Joachim Wilke†

IRCCS Galeazzi Orthopedic Institute, Milan, Italy Institute of Orthopaedic Research and Biomechanics, Ulm University, Ulm, Germany†*

RATIONALE FOR MOTION PRESERVATION SURGERY

In the previous chapter, instrumented fusion was presented as the gold standard for the surgical treatment of symptomatic degenerative disc disease in both the cervical and lumbar spines. In fact, a very good to excellent clinical outcome after fixation and fusion has been generally reported, and many surgeons do not consider other surgical options to treat the symptomatic degenerative spine. However, a non-negligible number of patients reported a low satisfaction after fusion surgery, and complications are not infrequent even after short fixations. Meta-analysis of the literature showed a rate of satisfactory outcomes of only 68% after lumbar fusion, whereas 14% of patients had pseudoarthrosis and 9% reported chronic pain at the graft donor site in the iliac crest (Turner et al., 1992). In the cervical spine, similar complication rates have been reported (Phillips et al., 1997). A large retrospective study described a non-union rate of 8.5% after anterior cervical discectomy, fixation with a locking plate, and fusion (Ning et al., 2008).

A frequently debated aspect of fixation and fusion concerns the risk of early degeneration of the segments adjacent to the fused ones, which is supposed to be due to the hypermobility and overload required to compensate for the lost flexibility of the treated segments. Intervertebral disc degeneration is the most commonly mentioned feature of adjacent segment disease (ASD), but abnormalities such as spondylolisthesis, instability, facet joint degeneration, and stenosis have also been reported (Lee et al., 2014; Park et al., 2004). Changes have been observed both at the immediately adjacent segments (Miyakoshi et al., 2000; Wiltse et al., 1999) and at the second adjacent segment (Hambly et al., 1998; Schulte et al., 2007). In the cervical spine, ASD has been associated with radiculopathy and myelopathy (Hilibrand et al., 1999). However, the available literature appears to be contradictory in this respect. It has been shown that the ASD rate was highly variable (between 11% and 100%) based on the definition applied (Lund and Oxland, 2011). In addition, it has been hypothesized that changes related to ASD were linked to a normal progression of the degenerative disease itself, and thus not linked to the fusion surgery (Lund and Oxland, 2011; Penta et al., 1995), although other authors disagree with this point of view (Lee et al., 2014; Kaito et al., 2010). A recent systematic revision of the literature showed that, in the lumbar spine, the studies failed to demonstrate a hypermobility of the adjacent segments and that their kinematic behavior is generally not affected by the surgery (Malakoutian et al., 2015). The authors, however, did not comment about possible overload of the adjacent segments, which cannot be estimated in vivo by means of radiological studies. In vitro studies do not help in this respect because

Biomechanics of the Spine. https://doi.org/10.1016/B978-0-12-812851-0.00018-5

they involve standardized loading conditions based on assumptions that are either false or unproven (Malakoutian et al., 2015; Volkheimer et al., 2015). There is, however, consensus that ASD exists to some degree, even if basic scientific investigations failed to identify and characterize its etiology.

Artificial discs emerged in the 1980s with the aim of treating degenerative disorders of the spine, while preserving its physiological mobility, with supposed beneficial effects on the clinical outcome in terms of both eliminating complications such as pseudoarthrosis and limiting the incidence of ASD. In parallel, implants to be positioned between the spinous processes were developed to treat discogenic low back pain without requiring fixation and fusion (Sénégas et al., 2007). Finally, the introduction of the Graf ligament (SEM, Créteil, France) in the early 1990s (Grevitt et al., 1995) opened the way to a further alternative strategy of motion preservation, that is, posterior dynamic stabilization.

CERVICAL ARTIFICIAL DISCS

Total disc replacement in the cervical spine was pioneered in the early days of motion-preserving surgery and is in widespread use today. Current indications for total cervical disc replacement are disc herniation and degenerative disc disease at one or two levels (Wu, 2014). Contraindications are kyphotic cervical alignment, facet joint degeneration, history of trauma, osteoporosis, and cervical ankylosis. The best candidates for cervical disc replacement are relatively young subjects with disc herniation and non-degenerated facet joints (Wu, 2014). Anterior discectomy and fusion remain the gold standard for the treatment of severe degenerative disorders involving the facet joints (Salari and McAfee, 2012).

In 2017 six implants approved by the FDA for cervical disc arthroplasty became available: the PRESTIGE Cervical Disc System (Medtronic Inc., Minneapolis, Minnesota), the ProDisc-C Total Disc Replacement (Synthes Spine, Inc., West Chester, Pennsylvania), the BRYAN Cervical Disc (Medtronic Inc., Minneapolis, Minnesota), the SECURE-C Cervical Artificial Disc (Globus Medical, Audubon, Pennsylvania), the PCM Cervical Disc System (NuVasive Manufacturing LLC, Fairborn, Ohio), and the Mobi-C Cervical Disc (Zimmer Biomet, Warsaw, Indiana), whereas several other models are available or under investigation in other countries. These devices are made of two or three articulating components and can be classified in two broad categories: semi-constrained implants, which guide the center of rotation to a specific point defined by the geometry of the articulating surfaces, and unconstrained implants, which leave the center of rotation free to accommodate the spine kinematics, usually by employing two sets of articulating surfaces (Fig. 1) (Galbusera et al., 2008b, 2008c). The implants are made of various materials and coatings designed to optimize the wear properties of the coupling and to improve osseointegration. Keels are commonly used to improve the primary stability of the implants; notably, some designs, such as the PCM and the BRYAN, do not have keels to preserve bone integrity.

A typical example of a semi-constrained device is the ProDisc-C, in which a spherical uncaptured joint (the "ball-in-socket") allows for pure rotations around axes passing through the center of the joint, which has been designed to be closest to the physiological center of rotation in flexion-extension of the lower cervical spine (Penning and Wilmink, 1987; Amevo et al., 1991) (Fig. 2). The PCM is based on a similar principle, but employs a saddle design like the shape of cervical intervertebral discs, which is supposed to guide the center of rotation in a more physiological way. An in vitro study showed the effectiveness of the PCM in replicating a physiological motion pattern, although only flexion-extension was tested (Cunningham et al., 2010) (Fig. 3).

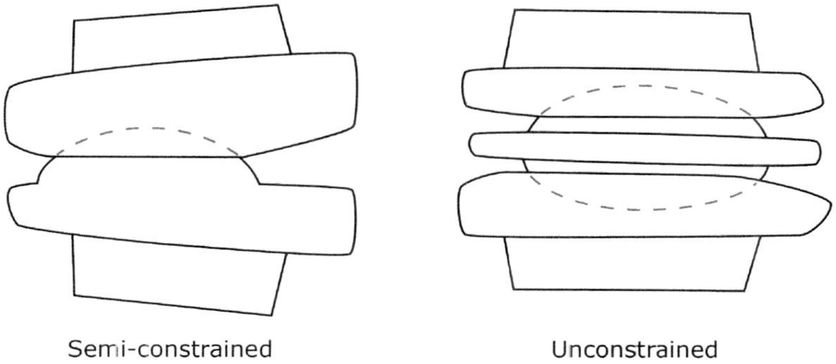

FIG. 1

Schematic representation of a semi-constrained artificial disc *(left)*, which guides the center of rotation to a specific point defined by the geometry of the articulating surfaces, and an unconstrained implant *(right)*, which leaves the center of rotation free to accommodate the spine kinematics by employing two sets of articulating surfaces.

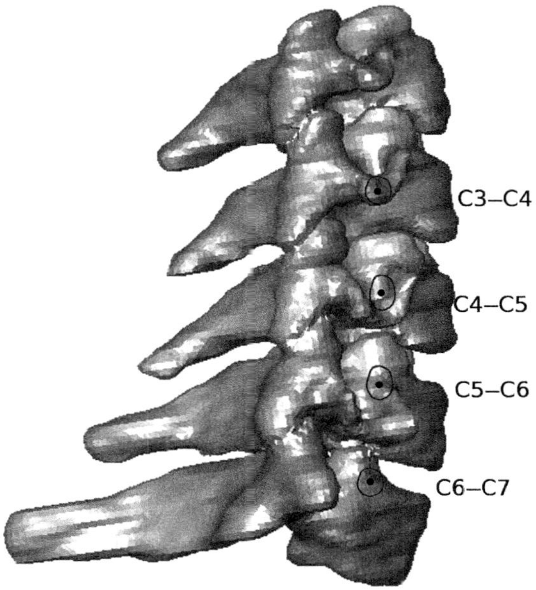

FIG. 2

Average location of the center of rotation during flexion-extension motion in the lower cervical spine.

Reproduced with permission from Galbusera, F., Bellini, C.M., Brayda-Bruno, M., Fornari, M., 2008b. Biomechanical studies on cervical total disc arthroplasty: a literature review. Clin. Biomech. (Bristol, Avon) 23(9), 1095–1104.

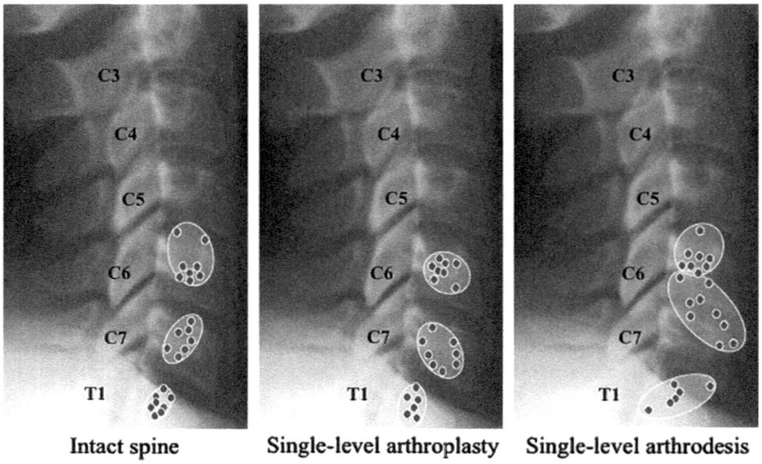

FIG. 3

Location of the centers of rotation in flexion-extension for the operated (C6–C7) and adjacent levels (C5–C6 and C7–T1) in specimens implanted with the PCM artificial disc *(center)* and after rigid fixation *(right)*, as compared with the centers of rotation before the implantation *(left)*. After arthroplasty, the point locations closely resembled those of the intact spine, whereas fixation induced more diffuse centers of rotation.

Reproduced with permission from Cunningham, B.W., Hu, N., Zorn, C.M., McAfee, P.C., 2010. Biomechanical comparison of single- and two-level cervical arthroplasty versus arthrodesis: effect on adjacent-level spinal kinematics. Spine J. 10(4), 341–349.

Most available unconstrained devices are based on two couplings of articulating surfaces with various shapes. In the BRYAN disc, the upper and lower couplings are symmetrical and have a spherical shape; a central post limits the maximal rotation to physiological values, as proven by radiological studies (Ryu et al., 2013). A finite element investigation calculated marginal alterations of the centers of rotation in flexion-extension (Galbusera et al., 2008a), which was confirmed by radiological analysis (Pickett et al., 2005); however, a recent in vitro study showed that the axis of rotation is non-negligibly affected (Wachowski et al., 2017). The prosthesis was shown not to be able to preserve the local lordosis of the segment, which decreased in average by 6 degree in a cohort of 14 patients (Pickett et al., 2004).

SECURE-C and Mobi-C have rather similar designs specifically developed to accommodate distinct mobilities in flexion-extension rather than in the other motions. In both prostheses, the upper coupling is spheroidal, whereas the lower articulating surfaces are flat in the Mobi-C and cylindrical in the SECURE-C. In both cases, a higher range of motion with limited constraint on the center of rotation is allowed in flexion-extension. Both designs also permit a pure anteroposterior shear motion. However, biomechanical studies quantitatively showing the motion of these two models have not been published yet.

Two designs substantially differ from the previously described ones. The PRESTIGE artificial disc and its updated version, PRESTIGE LP, feature a reversed ball-in-socket design with the spherical surface on the cranial side rather than on the caudal side. The socket is, however, oblong in the anteroposterior direction, thus providing little constraint to the movement in flexion-extension. Its first version was developed in 1989 (Cummins et al., 1998), and thus a wide range of literature about its clinical success and biomechanics is available. The prosthesis was shown to be able to preserve a physiological range of motion (Gornet et al., 2015), but radiological studies reported a superior shift of the center of rotation as well as an increase of the disc height and loss of the segmental lordosis (Kowalczyk et al., 2013). The M6-C Artificial Cervical Disc (Spinal Kinetics, Sunnyvale, California) features a deformable polymeric core surrounded by a fiber jacket resembling the annulus fibrosus, enclosed in a polymer sheath and two keeled titanium endplates for bone fixation. Its peculiar structure gives the prosthesis a shock absorption capability, in contrast with the other models. Although preliminary clinical results and biomechanical testing are encouraging (Thomas et al., 2016; Patwardhan et al., 2012), longer follow-up and further biomechanical studies are required to prove its possible superiority to the more conventional designs.

In general, clinical results after cervical disc replacement show that it is an acceptable alternative to anterior discectomy and fusion in selected patients, with lower reoperation rates and incidence of ASD (Buckland et al., 2016). Complications are, however, rather common and include subsidence, migration, and heterotopic ossification (Murtagh and Castellvi, 2014). The latter one is the most frequently observed complication, especially for some designs (Yi et al., 2010) and might be related to a decreased postoperative range of motion (Chang et al., 2017).

LUMBAR ARTIFICIAL DISCS

Despite being more challenging from a surgical point of view than cervical disc replacement is, the implantation of artificial discs in the lumbar spine was actually pioneered long before cervical disc replacement. In the 1950s, Nachemson implanted a testicular prosthesis into lumbar discs, which systematically failed because of mechanical fatigue (Nachemson, 1962). The first relatively large-scale trial was performed by Fernstrom, who implanted steel balls in the intervertebral space (Fernstrom, 1966), which subsided in the large majority of patients. The first modern total disc replacement (TDR) is considered to be the Charité Artificial Disc (DePuy Spine, Inc., Raynham, Massachusetts), which is now in its third version and still in widespread use.

Lumbar TDR is mainly indicated for the treatment of degenerative disc disease (Buttner-Janz et al., 2014). Other indications include spondylolisthesis, stenosis, degenerative scoliosis, segmental instability, discogenic pain, and failed back surgery (Bambakidis et al., 2005). Contraindications include facet joint degeneration, osteopenia and osteoporosis, and low disc height. All designs, with the notable exception of the XL TDR (NuVasive Manufacturing LLC, Fairborn, Ohio), require an invasive anterior surgical approach, which involves a substantial risk of complications in the immediate postoperative time, such as vascular injury and peritoneal entry (Murtagh and Castellvi, 2014). Even with the XL TDR, which is implanted through a far-lateral approach, peritoneal complications cannot be excluded. In male patients, retrograde ejaculation is another relatively frequent complication (4% in a group of 147 male patients (Holt et al., 2007)). Other reported complications include subsidence, heterotopic ossification, and ASD (Murtagh and Castellvi, 2014). The most common one is subsidence, which in many cases is attributed to an undersized implant that is not able to achieve purchase on the outer ring apophysis (Murtagh and Castellvi, 2014).

Three lumbar TDRs are currently approved by the FDA for use in the United States: the Charité, the ProDisc-L Total Disc Replacement (Synthes Spine, Inc., West Chester, Pennsylvania), and the activL Artificial Disc (Aesculap Implant Systems, LLC., Center Valley, Pennsylvania). Similar to cervical artificial discs, several non-FDA-approved implants are available in other countries.

The three preceding devices impose different degrees of kinematic constraint to the operated segment. The ProDisc-L is a typical ball-in-socket semi-constrained design that guides the axis of rotation to pass through the center of its spherical articulating surfaces. The stiffness and allowed range of motion is therefore the same in flexion-extension and lateral bending, whereas axial rotation is not constrained. The postoperative range of motion in flexion-extension measured on radiographic projections was reported in several studies and was similar to the preoperative one (Zigler et al., 2007) or slightly reduced (Kim et al., 2007; Yaszay et al., 2008). A cadaveric study demonstrated that the loss in mobility was attributed to an excessive implant height, which also determined a significant increase in the segmental lordosis (Gaffey et al., 2010).

The activL artificial disc possesses an additional degree of freedom, the anteroposterior shear translation, which is achieved by allowing for a translational motion of the polymeric insert. The implant is relatively new and has been investigated in only a few studies (Garcia Jr et al., 2015; Yue et al., 2016; Ha et al., 2009). On the contrary, the Charité, which features a three-component unconstrained design, has been extensively researched. A recent large-scale study observed a global preservation of the range of motion, which initially decreased 3 months after surgery, followed by a gradual increase to physiological levels (Guyer et al., 2016). This finding confirmed previous in vivo observations (Blumenthal et al., 2005; Katsimihas et al., 2010).

Another model that has been frequently implanted is the Maverick, a ball-in-socket design featuring a posterior center of rotation mimicking the physiological one in flexion-extension (Aiyangar et al., 2017). The device was prevented from being marketed in the United States because of patent infringement litigation. Similar to its cervical counterpart, the M6-L Artificial Lumbar Disc (Spinal Kinetics, Sunnyvale, California) mimics the structure of a physiological intervertebral disc; it features a deformable polymer core surrounded by a fiber jacket that limits the core bulging. Clinical and biomechanical studies of the M6-L Artificial Lumbar Disc, however, are still scarce (Schatz et al., 2015).

Actual preservation of the range of motion has been the subject of a heated debate in the biomechanical community. Wilke and colleagues compared the Charité with the ProDisc-L in cadaveric specimens and found equivalent results among the two in terms of flexibility, which were significantly higher than the intact condition, likely due to the resection of the anterior longitudinal ligament (Wilke et al., 2012). Similar findings were obtained in other biomechanical studies (Zander et al., 2009; Chung et al., 2009a, 2009b), but they conflicted with the in vivo observations previously described, which generally showed a preservation of the mobility within physiological ranges, or even a small loss in flexibility. However, the mechanical destabilization may be responsible for the increased segmental lordosis at the disc replacement level, which has been frequently observed in clinical studies (Cakir et al., 2005; Kasliwal and Deutsch, 2012; Le Huec et al., 2005). To limit the occurrence of such effects, some prostheses, such as the Mobidisc Lumbar (Zimmer Biomet, Warsaw, Indiana), allow for the selection of endplates with various lordotic angles. The actual impact of a possible increase in the segmental lordosis on the clinical outcome remains unclear.

Charles and colleagues suggested including the evaluation of the spinopelvic parameters and the Roussouly's classification (see Chapter 21) in the preoperative planning of lumbar TDR (Charles et al., 2011). The authors suggested avoiding the use of TDR at L5-S1 in patients with high pelvic incidence because of the relevant anteroposterior shear forces. Whereas in this case an unconstrained implant might face a higher risk of mechanical failure or migration, a semi-constrained device might lead to an excessive segmental lordosis, with a risk of painful impingement on the facet joints (Fig. 4).

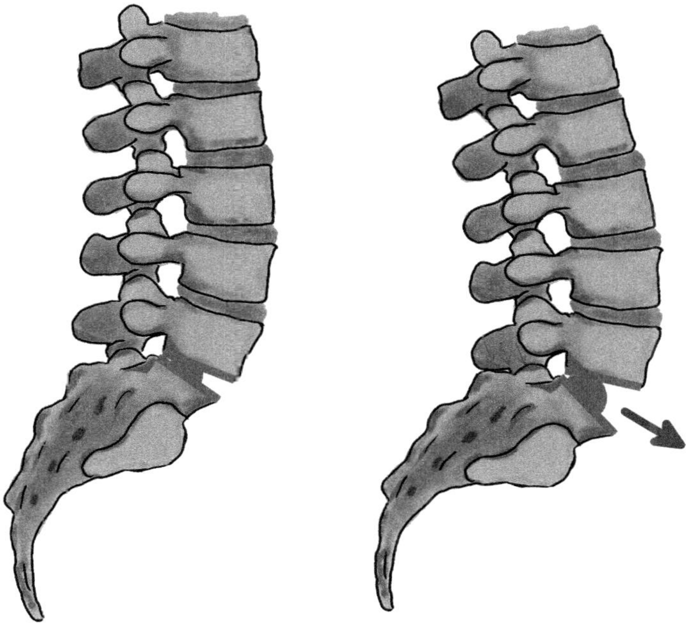

FIG. 4

Total disc replacement at L5-S1 in patients with high pelvic incidence and sacral slope *(right)* might induce high segmental lordosis, relevant anteroposterior shear forces, and impingement of the posterior elements (indicated by the arrows) as compared with patients with a lower sacral slope *(left)*.

PEDICLE-ANCHORED LUMBAR DYNAMIC STABILIZATION

The concept of dynamic stabilization arose from the description of the stages of lumbar spine degeneration proposed by Kirkaldy-Willis and Farfan (1982). The authors subdivided the degenerative process into three stages: a first temporary dysfunction that may respond to conservative treatment, followed by a phase of instability characterized by abnormal motion, and finally a spontaneous stabilization obtained by means of changes in the discal tissue composition, disc height loss, and osteophytosis. Although the concept of spinal instability in the early stages of degeneration was subsequently criticized based on investigations on cadaveric specimens (Kettler et al., 2011; Mimura et al., 1994), several implants have been developed with the aim of reducing the spinal flexibility in degenerated discs, with relatively good success from a clinical point of view.

Generally accepted indications for pedicle screw-based lumbar stabilization include symptomatic moderate disc degeneration, mild facet osteoarthritis, low-grade spondylolisthesis, and spinal stenosis (Prud'homme et al., 2015). Dynamic stabilization systems are also frequently used in hybrid constructs, including long fixations, in order to create a transition zone with a stiffness in between that of the instrumented region and the adjacent flexible segments. Contraindications include instability due to tumors and trauma, infections, degenerative scoliosis, and isthmic spondylolisthesis (Ozer, 2010).

The first device used clinically on a large scale was the Graf ligament, which consists of polymeric bands connecting the pedicle screws that are pre-tensioned in order to force the motion segment in a lordotic condition and to restrict flexion (Barrey et al., 2008). Cadaveric tests showed that the Graf ligament was effective in reducing the range of motion in flexion-extension and lateral bending, whereas the flexibility in axial rotation was only marginally affected (Strauss et al., 1994). Despite the fact that encouraging clinical results were achieved in the early days, longer follow-ups showed controversial results, including high reoperation rates and low patient satisfaction (Rigby et al., 2001).

The Dynesys Dynamic Stabilization System (Zimmer Biomet, Warsaw, Indiana) is approved by the FDA and, being first implanted in 1994, has a long clinical history. The device consists of a hollow polymeric spacer limiting the spinal flexibility in extension and a polymeric pre-tensioned cord tunneled through the spacer designed to limit the flexion range of motion. In vitro studies showed a rather high stiffness of the device, that is, a significant decrease in the range of motion in flexion-extension, by approximately 50% (Freudiger et al., 1999) or more (Niosi et al., 2006). After implantation in artificially destabilized spines, the stabilization with the Dynesys system showed a flexibility decrease in flexion and lateral bending with respect to the intact spines, an extension restored to physiological levels, and instability in axial rotation (Schmoelz et al., 2003). In spines with a less disruptive artificial destabilization, the implant induced a significant decrease in the range of motion in all directions and a marked posterior shift of the center of rotation in flexion-extension (Niosi et al., 2006). These findings may indicate a considerable effect on the intervertebral disc: With a posterior center of rotation, flexion, which is frequent in several postures and daily activities such as sitting, would result in an unphysiological loading of the disc in compression (Fig. 5). In extension, the disc would be tensioned, with possible facet overloading (Barrey et al., 2008). The clinical consequences of such alternations were, however, never extensively investigated.

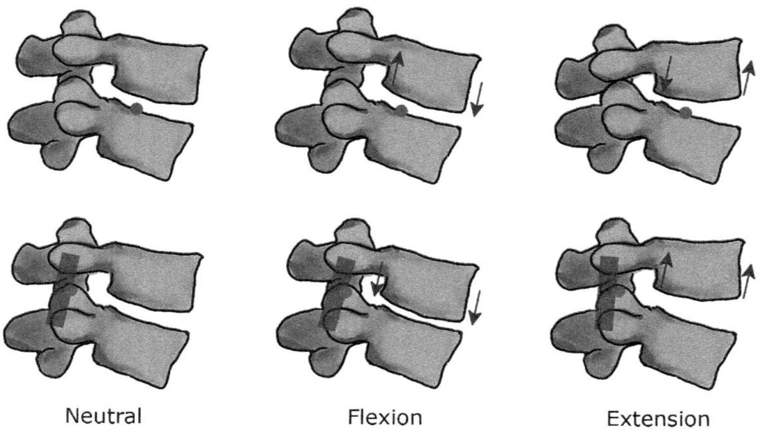

FIG. 5

In comparison with the intact spine *(top row)*, posterior dynamic stabilization *(bottom row)* can induce a posterior shift of the center of rotation in flexion-extension, which alters the strain patterns of the disc. With a posterior center of rotation, flexion, which is frequent in several postures and daily activities such as sitting, results in an unphysiological loading of the disc in compression, whereas extension determines a tensile strain with possible facet overloading.

Being designed not to achieve a fusion but instead to be subjected to the spinal loads for its whole life, the Dynesys system—as well as the other pedicle-anchored dynamic stabilization systems—is more at risk of screw loosening with respect to standard fixation. Indeed, literature reviews have reported loosening rates for the Dynesys system ranging between 0% and 17% (Kocak et al., 2010) and of 11.5% (Pham et al., 2016), whereas no literature exists regarding long-term follow-up of other implants. Implant loosening seems, however, to have a marginal effect on the clinical outcome (Fay et al., 2013).

Several other pedicle-anchored dynamic stabilization devices are currently available on the market, either with spring-like full metal designs (e.g., DSS Stabilization System, Paradigm Spine, LLC, New York, New York; AccuFlex, Globus Medical Inc., Audubon, Pennsylvania) or with integrating polymeric damping components (Isobar TTL, Scient'x Alphatec, Carlsbad, California; NFlex Dynamic Stabilization System, DePuy Synthes Spine, Raynham, Massachusetts). The clinical literature about these devices is growing and tends to indicate good outcomes in carefully selected patients, even if there is no clear evidence of superiority of dynamic stabilization to fusion in those patients (Prud'homme et al., 2015).

A related interesting aspect is the potential prevention of ASD when dynamic stabilization is used instead of rigid fixation. Large-scale clinical studies with follow-up long enough to properly address this issue are not currently available. In a smaller study comparing 15 patients treated with dynamic stabilization systems and 15 with rigid fixation, no signs of radiographic ASD were observed; however, the small size of the groups should be considered (Korovessis et al., 2002). In another prospective study including 32 patients treated with the Dynesys system or rigid fixation followed for 2 years, the authors observed that ASD did not appear to be influenced by the stiffness of the stabilization (Kumar et al., 2008). Another prospective trial on 26 elderly patients treated with the Dynesys system reported signs of radiographic ASD in seven patients (Schnake et al., 2006). In general, the available data seem therefore to indicate a limited effectiveness of dynamic stabilization in preventing ASD, but larger and longer clinical studies are required to draw a definitive conclusion.

OTHER MOTION-PRESERVING DEVICES

From the early days of disc replacement, the concept of replacing only the nucleus pulposus in order to restore disc height and to avoid inward bulging of the annulus fibrosus emerged as a viable alternative. The first device of this type implanted in human patients was the PDN (Prosthetic Disc Nucleus, Raymedica, Inc., Minneapolis, Minnesota), which consists of a polymeric nucleus surrounded by a permeable fiber jacket that is designed to minimize the risk of migration. Successive versions of the same device feature a softer core. Solutions based on injectable polymers were also implemented, such as the DASCOR Disc Arthroplasty System (Disc Dynamics, Inc., Eden Prairie, Minnesota) and the NuCore RPXL Injectable Nucleus (Spine Wave, Inc., Shelton, Connecticut), which both allow for a minimally invasive implantation. Although some positive clinical results have been reported (Di Martino et al., 2005; Bertagnoli et al., 2004; Bertagnoli and Vazquez, 2003), spine surgeons' acceptance of the concept of nucleus replacement is still rather limited. In fact, the indications for nucleus replacement described in the literature refer to low back pain not responding to conservative treatment but related degenerative changes at an early stage, with a good condition of the annulus fibrosus in order to achieve a proper containment of the device, and no severe facet arthrosis (Goins et al., 2005), which strongly restricts the number of eligible patients.

Interspinous implants have been used since the 1980s to treat spinal stenosis, generally in combination with decompression surgery (Machado et al., 2016). The most documented one is the X-Stop (Medtronic, Inc., Minneapolis, Minnesota), which is no longer available on the market; it consists of a titanium or PEEK spacer, which limits extension, contained in two metal wings designed to avoid migration. The implant can be disassembled and is implanted into the interspinous space from the lateral direction, thus preserving the supraspinous ligament that secures the X-Stop in place. The other device with a long clinical history, the Wallis Posterior Dynamic Stabilization System (Zimmer Biomet, Warsaw, Indiana), originally required the resection of the supraspinous ligament, whereas the current version overcomes this limitation. Nevertheless, supplementary fixation is granted by the bands that wrap around the spinous processes. These implants together with several others of note, including the DIAM (Medtronic, Inc., Minneapolis, Minnesota) and the FDA-approved coflex Interlaminar Stabilization (Paradigm Spine, LLC, New York, New York), began experiencing widespread use in the 2000s for the treatment of spinal stenosis. In recent years, several publications have highlighted the high reoperation rates caused by complications such as dislocation, fracture of the spinous processes, new onset of radiculopathy (Barbagallo et al., 2009; Bowers et al., 2010), and fracture of the facet joints (Chung et al., 2009a). Indeed, although the implantation of interspinous spacers can be an effective option in carefully selected patients, the high complication rates have resulted in a general decrease in their use.

Another the category of implants worthy of mention includes those designed to replace the function of the facet joints and posterior elements, which can be indicated in patients suffering from lumbar degenerative spondylolisthesis or spinal stenosis combined with severe facet arthritis (Anekstein et al., 2015; McAfee et al., 2007). A biomechanical study showed that the TOPS System (Premia Spine Ltd., Herzeila, Israel) is able to restore physiological flexibility of the lumbar spine in all motion directions (Wilke et al., 2006). The available data about the clinical results of facet joint arthroplasty are currently limited but are encouraging (Anekstein et al., 2015).

REFERENCES

Aiyangar, A., Zheng, L., Anderst, W., Zhang, X., 2017. Instantaneous centers of rotation for lumbar segmental extension in vivo. J. Biomech. 52, 113–121.

Amevo, B., Worth, D., Bogduk, N., 1991. Instantaneous axes of rotation of the typical cervical motion segments: a study in normal volunteers. Clin. Biomech. 6 (2), 111–117.

Anekstein, Y., Floman, Y., Smorgick, Y., Rand, N., Millgram, M., Mirovsky, Y., 2015. Seven years follow-up for total lumbar facet joint replacement (TOPS) in the management of lumbar spinal stenosis and degenerative spondylolisthesis. Eur. Spine J. 24 (10), 2306–2314.

Bambakidis, N.C., Feiz-Erfan, I., Klopfenstein, J.D., Sonntag, V.K., 2005. Indications for surgical fusion of the cervical and lumbar motion segment. Spine 30 (16S), S2–S6.

Barbagallo, G.M., Olindo, G., Corbino, L., Albanese, V., 2009. Analysis of complications in patients treated with the X-stop Interspinous process decompression system: proposal for a novel anatomic scoring system for patient selection and review of the literature. Neurosurgery 65 (1), 111–119 (discussion 119–20).

Barrey, C.Y., Ponnappan, R.K., Song, J., Vaccaro, A.R., 2008. Biomechanical evaluation of pedicle screw-based dynamic stabilization devices for the lumbar spine: a systematic review. SAS J. 2 (4), 159–170.

Bertagnoli, R., Vazquez, R.J., 2003. The anterolateral TransPsoatic approach (ALPA): a new technique for implanting prosthetic disc-nucleus devices. J. Spinal Disord. Tech. 16 (4), 398–404.

Bertagnoli, R., Schönmayr, R., Vazquez, R., 2004. Surgical and clinical results with the PDN® prosthetic disc-nucleus device. In: Arthroplasty of the Spine. Springer, Heidelberg, Germany, pp. 87–92.
Blumenthal, S., McAfee, P.C., Guyer, R.D., Hochschuler, S.H., Geisler, F.H., Holt, R.T., Garcia Jr., R., Regan, J.J., Ohnmeiss, D.D., 2005. A prospective, randomized, multicenter food and drug administration investigational device exemptions study of lumbar total disc replacement with the CHARITE™ artificial disc versus lumbar fusion: part I: evaluation of clinical outcomes. Spine 30 (14), 1565–1575.
Bowers, C., Amini, A., Dailey, A.T., Schmidt, M.H., 2010. Dynamic interspinous process stabilization: review of complications associated with the X-stop device. Neurosurg. Focus. 28 (6), E8.
Buckland, A.J., Baker, J.F., Roach, R.P., Spivak, J.M., 2016. Cervical disc replacement—emerging equivalency to anterior cervical discectomy and fusion. Int. Orthop. 40 (6), 1329–1334.
Buttner-Janz, K., Guyer, R.D., Ohnmeiss, D.D., 2014. Indications for lumbar total disc replacement: selecting the right patient with the right indication for the right total disc. Int. J. spine surg. 8. https://doi.org/10.14444/1012 (eCollection 2014).
Cakir, B., Richter, M., Käfer, W., Puhl, W., Schmidt, R., 2005. The impact of total lumbar disc replacement on segmental and total lumbar lordosis. Clin. Biomech. 20 (4), 357–364.
Chang, H., Chang, C., Tu, T., Wu, J., Huang, W., Fay, L., Chang, P., Wu, C., Cheng, H., 2017. Can segmental mobility be increased by cervical arthroplasty? Neurosurg. Focus. 42 (2), E3.
Charles, Y.P., Walter, A., Schuller, S., Steib, J.P., 2011. Dynamic instrumentation of the lumbar spine. Clinical and biomechanical analysis of success factors. Der Orthopade 40 (8), 703–712.
Chung, K.J., Hwang, Y.S., Koh, S.H., 2009a. Stress fracture of bilateral posterior facet after insertion of interspinous implant. Spine 34 (10), E380–3.
Chung, S.K., Kim, Y.E., Wang, K.C., 2009b. Biomechanical effect of constraint in lumbar total disc replacement: a study with finite element analysis. Spine 34 (12), 1281–1286.
Cummins, B.H., Robertson, J.T., Gill, S.S., 1998. Surgical experience with an implanted artificial cervical joint. J. Neurosurg. 88 (6), 943–948.
Cunningham, B.W., Hu, N., Zorn, C.M., McAfee, P.C. 2010. Biomechanical comparison of single-and two-level cervical arthroplasty versus arthrodesis: effect on adjacent-level spinal kinematics. Spine J. 10 (4), 341–349.
Di Martino, A., Vaccaro, A.R., Lee, J.Y., Denaro, V., Lim, M.R., 2005. Nucleus pulposus replacement: basic science and indications for clinical use. Spine 30 (16S), S16–S22.
Fay, L., Wu, J., Tsai, T., Wu, C., Huang, W., Cheng, H., 2013. Dynamic stabilization for degenerative spondylolisthesis: evaluation of radiographic and clinical outcomes. Clin. Neurol. Neurosurg. 115 (5), 535–541.
Fernstrom, U., 1966. Arthroplasty with intercorporal endoprothesis in herniated disc and in painful disc. Acta Chir. Scand. Suppl. 357, 154–159.
Freudiger, S., Dubois, G., Lorrain, M., 1999. Dynamic neutralisation of the lumbar spine confirmed on a new lumbar spine simulator in vitro. Arch. Orthop. Trauma Surg. 119 (3), 127–132.
Gaffey, J.L., Ghanayem, A.J., Voronov, M.L., Havey, R.M., Carandang, G., Abjornson, C., Patwardhan, A.G., 2010. Effect of increasing implant height on lumbar spine kinematics and foraminal size using the ProDisc-L prosthesis. Spine 35 (19), 1777–1782.
Galbusera, F., Bellini, C.M., Raimondi, M.T., Fornari, M., Assietti, R., 2008a. Cervical spine biomechanics following implantation of a disc prosthesis. Med. Eng. Phys. 30 (9), 1127–1133.
Galbusera, F., Bellini, C.M., Brayda-Bruno, M., Fornari, M., 2008b. Biomechanical studies on cervical total disc arthroplasty: a literature review. Clin. Biomech. (Bristol, Avon) 23 (9), 1095–1104.
Galbusera, F., Bellini, C.M., Zweig, T., Ferguson, S., Raimondi, M.T., Lamartina, C., Brayda-Bruno, M., Fornari, M., 2008c. Design concepts in lumbar total disc arthroplasty. Eur. Spine J. 17 (12), 1635–1650.
Garcia Jr., R., Yue, J.J., Blumenthal, S., Coric, D., Patel, V.V., Leary, S.P., Dinh, D.H., Buttermann, G.R., Deutsch, H., Girardi, F., Billys, J., Miller, L.E., 2015. Lumbar total disc replacement for discogenic low back pain: two-year outcomes of the activL multicenter randomized controlled IDE clinical trial. Spine 40 (24), 1873–1881.

Goins, M.L., Wimberley, D.W., Yuan, P.S., Fitzhenry, L.N., Vaccaro, A.R., 2005. Nucleus pulposus replacement: an emerging technology. Spine J. 5 (6), S317–S324.

Gornet, M.F., Burkus, J.K., Shaffrey, M.E., Argires, P.J., Nian, H., Harrell Jr., F.E., 2015. Cervical disc arthroplasty with PRESTIGE LP disc versus anterior cervical discectomy and fusion: a prospective, multicenter investigational device exemption study. J. Neurosurg. Spine 23 (5), 558–573.

Grevitt, M., Gardner, A., Spilsbury, J., Shackleford, I., Baskerville, R., Pursell, L., Hassaan, A., Mulholland, R., 1995. The Graf stabilisation system: early results in 50 patients. Eur. Spine J. 4 (3), 169–175.

Guyer, R.D., Pettine, K., Roh, J.S., Dimmig, T.A., Coric, D., McAfee, P.C., Ohnmeiss, D.D., 2016. Five-year follow-up of a prospective, randomized trial comparing two lumbar Total disc replacements. Spine 41 (1), 3–8.

Ha, S., Kim, S., Kim, D.H., Park, J., Lim, D., Lee, S., 2009. Biomechanical study of lumbar spinal arthroplasty with a semi-constrained artificial disc (Activ L) in the human cadaveric spine. J. Korean Neurosurg. Soc. 45 (3), 169–175.

Hambly, M.F., Wiltse, L.L., Raghavan, N., Schneiderman, G., Koenig, C., 1998. The transition zone above a lumbosacral fusion. Spine 23 (16), 1785–1792.

Hilibrand, A.S., Carlson, G.D., Palumbo, M.A., Jones, P.K., Bohlman, H.H., 1999. Radiculopathy and myelopathy at segments adjacent to the site of a previous anterior cervical arthrodesis. J. Bone Joint Surg. 81 (4), 519–528.

Holt, R.T., Majd, M.E., Isaza, J.E., Blumenthal, S.L., McAfee, P.C., Guyer, R.D., Hochschuler, S.H., Geisler, F.H., Garcia, R., Regan, J.J., 2007. Complications of lumbar artificial disc replacement compared to fusion: results from the prospective, randomized, multicenter US Food and Drug Administration investigational device exemption study of the Charité artificial disc. SAS J. 1 (1), 20–27.

Kaito, T., Hosono, N., Mukai, Y., Makino, T., Fuji, T., Yonenobu, K., 2010. Induction of early degeneration of the adjacent segment after posterior lumbar interbody fusion by excessive distraction of lumbar disc space: clinical article. J. Neurosurg. Spine 12 (6), 671–679.

Kasliwal, M.K., Deutsch, H., 2012. Effect of total lumbar disc replacement on lumbosacral lordosis. J. Spinal Disord. Tech. 25 (7), 370–373.

Katsimihas, M., Bailey, C., Issa, K., Rosas-Arellano, P., Bailey, S., Gurr, K., 2010. Prospective clinical and radiographic results of the Charite total disc replacement. Average 5 year follow-up. J. Bone Joint Surg. 92B, 380.

Kettler, A., Rohlmann, F., Ring, C., Mack, C., Wilke, H.J., 2011. Do early stages of lumbar intervertebral disc degeneration really cause instability? Evaluation of an in vitro database. Eur. Spine J. 20 (4), 578–584.

Kim, D., Ryu, K., Kim, M., Park, C., 2007. Factors influencing segmental range of motion after lumbar total disc replacement using the ProDisc II prosthesis. J. Neurosurg. Spine 7, 131–138.

Kirkaldy-Willis, W.H., Farfan, H.F., 1982. Instability of the lumbar spine. Clin. Orthop. Relat. Res. (165), 110–123.

Kocak, T., Cakir, B., Reichel, H., Mattes, T., 2010. Screw loosening after posterior dynamic stabilization—review of the literature. Acta Chir. Orthop. Traumatol. Cechoslov. 77 (2), 134–139.

Korovessis, P., Papazisis, Z., Lambiris, E., 2002. The role of rigid vs. dynamic instrumentation for stabilization of the degenerative lumbosacral spine. Stud. Health Technol. Inform. 91, 457–461.

Kowalczyk, I., Chaudhary, N., Duggal, N., 2013. Kinematic analysis following implantation of the PRESTIGE LP. Int. J. Spine Surg. 7 (1), e118-e122.

Kumar, A., Beastall, J., Hughes, J., Karadimas, E.J., Nicol, M., Smith, F., Wardlaw, D., 2008. Disc changes in the bridged and adjacent segments after Dynesys dynamic stabilization system after two years. Spine 33 (26), 2909–2914.

Le Huec, J., Basso, Y., Mathews, H., Mehbod, A., Aunoble, S., Friesem, T., Zdeblick, T., 2005. The effect of single-level, total disc arthroplasty on sagittal balance parameters: a prospective study. Eur. Spine J. 14 (5), 480–486.

Lee, J.C., Kim, Y., Soh, J.W., Shin, B.J., 2014. Risk factors of adjacent segment disease requiring surgery after lumbar spinal fusion: comparison of posterior lumbar interbody fusion and posterolateral fusion. Spine 39 (5), E339–45.

Lund, T., Oxland, T.R., 2011. Adjacent level disk disease—is it really a fusion disease? Orthop. Clin. N. Am. 42 (4), 529–541.

Machado, G.C., Ferreira, P.H., Yoo, R.I., Harris, I.A., Pinheiro, M.B., Koes, B.W., van Tulder, M.W., Rzewuska, M., Maher, C.G., Ferreira, M.L., 2016. Surgical options for lumbar spinal stenosis. Cochrane Database Syst. Rev. 11. The Cochrane Library.

Malakoutian, M., Volkheimer, D., Street, J., Dvorak, M.F., Wilke, H.J., Oxland, T.R., 2015. Do in vivo kinematic studies provide insight into adjacent segment degeneration? A qualitative systematic literature review. Eur. Spine J. 24 (9), 1865–1881.

McAfee, P., Khoo, L.T., Pimenta, L., Capuccino, A., Cacic, D., Hes, R., Conix, B., Asgarzadie, F., Hamzaoglu, A., Mirofsky, Y., 2007. Treatment of lumbar spinal stenosis with a total posterior arthroplasty prosthesis: implant description, surgical technique, and a prospective report on 29 patients. Neurosurg. Focus. 22 (1), 1–11.

Mimura, M., Panjabi, M.M., Oxland, T.R., Crisco, J.J., Yamamoto, I., Vasavada, A., 1994. Disc degeneration affects the multidirectional flexibility of the lumbar spine. Spine 19 (12), 1371–1380.

Miyakoshi, N., Abe, E., Shimada, Y., Okuyama, K., Suzuki, T., Sato, K., 2000. Outcome of one-level posterior lumbar interbody fusion for spondylolisthesis and postoperative intervertebral disc degeneration adjacent to the fusion. Spine 25 (14), 1837–1842.

Murtagh, R., Castellvi, A.E., 2014. Motion preservation surgery in the spine. Neuroimaging Clin. N. Am. 24 (2), 287–294.

Nachemson, A., 1962. Some mechanical properties of the lumbar intervertebral discs. Bull. Hosp. Jt Dis. 23, 130–143.

Ning, X., Wen, Y., Xiao-Jian, Y., Bin, N., De-Yu, C., Jian-Ru, X., Lian-Shun, J., 2008. Anterior cervical locking plate-related complications; prevention and treatment recommendations. Int. Orthop. 32 (5), 649–655.

Niosi, C.A., Zhu, Q.A., Wilson, D.C., Keynan, O., Wilson, D.R., Oxland, T.R., 2006. Biomechanical characterization of the three-dimensional kinematic behaviour of the Dynesys dynamic stabilization system: an in vitro study. Eur. Spine J. 15 (6), 913–922.

Ozer, A.F., 2010. Dynamic stabilization of the spine: a new classification system. Turk Neurosurg. 20 (2), 205–215.

Park, P., Garton, H.J., Gala, V.C., Hoff, J.T., McGillicuddy, J.E., 2004. Adjacent segment disease after lumbar or lumbosacral fusion: review of the literature. Spine 29 (17), 1938–1944.

Patwardhan, A., Tzermiadianos, M., Tsitsopoulos, P., Voronov, L., Renner, S., Reo, M., Carandang, G., Ritter-Lang, K., Havey, R., 2012. Primary and coupled motions after cervical total disc replacement using a compressible six-degree-of-freedom prosthesis. Eur. Spine J. 21 (5), 618–629.

Penning, L., Wilmink, J.T., 1987. Rotation of the cervical spine. A CT study in normal subjects. Spine 12 (8), 732–738.

Penta, M., Sandhu, A., Fraser, R.D., 1995. Magnetic resonance imaging assessment of disc degeneration 10 years after anterior lumbar interbody fusion. Spine 20 (6), 743–747.

Pham, M.H., Mehta, V.A., Patel, N.N., Jakoi, A.M., Hsieh, P.C., Liu, J.C., Wang, J.C., Acosta, F.L., 2016. Complications associated with the Dynesys dynamic stabilization system: a comprehensive review of the literature. Neurosurg. Focus. 40 (1), E2.

Phillips, F.M., Carlson, G., Emery, S.E., Bohlman, H.H., 1997. Anterior cervical pseudarthrosis: natural history and treatment. Spine 22 (14), 1585–1589.

Pickett, G.E., Mitsis, D.K., Sekhon, L.H., Sears, W.R., Duggal, N., 2004. Effects of a cervical disc prosthesis on segmental and cervical spine alignment. Neurosurg. Focus. 17 (3), 1–35.

Pickett, G.E., Rouleau, J.P., Duggal, N., 2005. Kinematic analysis of the cervical spine following implantation of an artificial cervical disc. Spine 30 (17), 1949–1954.

Prud'homme, M., Barrios, C., Rouch, P., Charles, Y.P., Steib, J., Skalli, W., 2015. Clinical outcomes and complications after pedicle-anchored dynamic or hybrid lumbar spine stabilization. J. Spinal Disord. Tech. 28 (8), E439–E448.

Rigby, M., Selmon, G., Foy, M., Fogg, A., 2001. Graf ligament stabilisation: mid-to long-term follow-up. Eur. Spine J. 10 (3), 234–236.

Ryu, W.H.A., Kowalczyk, I., Duggal, N., 2013. Long-term kinematic analysis of cervical spine after single-level implantation of Bryan cervical disc prosthesis. Spine J. 13 (6), 628–634.

Salari, B., McAfee, P.C., 2012. Cervical total disk replacement: Complications and avoidance. Orthop. Clin. N. Am. 43 (1), 97–107.

Schatz, C., Ritter-Lang, K., Gossel, L., Dressler, N., 2015. Comparison of single-level and multiple-level outcomes of Total disc Arthroplasty: 24-month results. Int. J. Spine Surg. 9, 14.

Schmoelz, W., Huber, J.F., Nydegger, T., Dipl-Ing, C., L. & Wilke, H.J., 2003. Dynamic stabilization of the lumbar spine and its effects on adjacent segments: an in vitro experiment. J. Spinal Disord. Tech. 16 (4), 418–423.

Schnake, K.J., Schaeren, S., Jeanneret, B., 2006. Dynamic stabilization in addition to decompression for lumbar spinal stenosis with degenerative spondylolisthesis. Spine 31 (4), 442–449.

Schulte, T.L., Leistra, F., Bullmann, V., Osada, N., Vieth, V., Marquardt, B., Lerner, T., Liljenqvist, U., Hackenberg, L., 2007. Disc height reduction in adjacent segments and clinical outcome 10 years after lumbar 360 fusion. Eur. Spine J. 16 (12), 2152–2158.

Sénégas, J., Vital, J., Pointillart, V., Mangione, P., 2007. Long-term actuarial survivorship analysis of an interspinous stabilization system. Eur. Spine J. 16 (8), 1279–1287.

Strauss, P.J., Novotny, J.E., Wilder, D.G., Grobler, L.J., Pope, M.H., 1994. Multidirectional stability of the Graf system. Spine 19 (8), 965–972.

Thomas, S., Willems, K., Van den Daelen, L., Linden, P., Ciocci, M.C., Bocher, P., 2016. The M6-C cervical disk prosthesis: first clinical experience in 33 patients. Clin. Spine Surg. 29 (4), E182–7.

Turner, J.A., Ersek, M., Herron, L., Haselkorn, J., Kent, D., Ciol, M.A., Deyo, R., 1992. Patient outcomes after lumbar spinal fusions. JAMA 268 (7), 907–911.

Volkheimer, D., Malakoutian, M., Oxland, T.R., Wilke, H.J., 2015. Limitations of current in vitro test protocols for investigation of instrumented adjacent segment biomechanics: critical analysis of the literature. Eur. Spine J. 24 (9), 1882–1892.

Wachowski, M.M., Weiland, J., Wagner, M., Gezzi, R., Kubein-Meesenburg, D. & Nägerl, H., 2017. "Kinematics of cervical segments C5/C6 in axial rotation before and after total disc arthroplasty". Eur. Spine J. (in press). https://doi.org/10.1007/s00586-017-5073-z.

Wilke, H.J., Schmidt, H., Werner, K., Schmolz, W., Drumm, J., 2006. Biomechanical evaluation of a new total posterior-element replacement system. Spine 31 (24), 2790–2796 (discussion 2797).

Wilke, H., Schmidt, R., Richter, M., Schmoelz, W., Reichel, H., Cakir, B., 2012. The role of prosthesis design on segmental biomechanics. Eur. Spine J. 21 (5), 577–584.

Wiltse, L.L., Radecki, S.E., Biel, H.M., DiMartino, P.P., Oas, R.A., Farjalla, G., Ravessoud, F.A., Wohletz, C., 1999. Comparative study of the incidence and severity of degenerative change in the transition zones after instrumented versus noninstrumented fusions of the lumbar spine. Clin. Spine Surg. 12 (1), 27–33.

Wu, J., 2014. Cervical total disc replacement. Formosan J. Surg. 47 (2), 49–52.

Yaszay, B., Bendo, J.A., Goldstein, J.A., Quirno, M., Spivak, J.M., Errico, T.J., 2008. Effect of intervertebral disc height on postoperative motion and outcomes after ProDisc-L lumbar disc replacement. Spine 33 (5), 508–512 (discussion 513).

Yi, S., Kim, K.N., Yang, M.S., Yang, J.W., Kim, H., Ha, Y., Yoon, D.H., Shin, H.C., 2010. Difference in occurrence of heterotopic ossification according to prosthesis type in the cervical artificial disc replacement. Spine 35 (16), 1556–1561.

Yue, J.J., Garcia Jr., R., Miller, L.E., 2016. The activL® artificial disc: a next-generation motion-preserving implant for chronic lumbar discogenic pain. Med. Devices (Auckl.) 9, 75.

Zander, T., Rohlmann, A., Bergmann, G., 2009. Influence of different artificial disc kinematics on spine biomechanics. Clin. Biomech. 24 (2), 135–142.

Zigler, J., Delamarter, R., Spivak, J.M., Linovitz, R.J., Danielson 3rd, G.O., Haider, T.T., Cammisa, F., Zuchermann, J., Balderston, R., Kitchel, S., Foley, K., Watkins, R., Bradford, D., Yue, J., Yuan, H., Herkowitz, H., Geiger, D., Bendo, J., Peppers, T., Sachs, B., Girardi, F., Kropf, M., Goldstein, J., 2007. Results of the prospective, randomized, multicenter Food and Drug Administration investigational device exemption study of the ProDisc-L total disc replacement versus circumferential fusion for the treatment of 1-level degenerative disc disease. Spine 32 (11), 1155–1162. discussion 1163.

SPINAL DEFORMITIES

CHAPTER

19 SCOLIOSIS

Benedikt Schlager*, Maryem-Fama Ismael Aguirre†, Hans-Joachim Wilke*, Fabio Galbusera†

*Institute of Orthopaedic Research and Biomechanics, Ulm University, Ulm, Germany**

IRCCS Galeazzi Orthopedic Institute, Milan, Italy†

OVERVIEW

Scoliosis is a three-dimensional deformity of the spine. The spinal deformity goes along with several other factors, such as deformity of the chest, reduction in lung capacity, modification of soft-tissue, changes in muscular architecture, as well as asymmetrical shoulder and head balance. Epidemiological studies have estimated that 1%–3% of adolescents are at risk of having a Cobb angle of >10 degrees (Lowe et al., 2000), while almost one person out of two who are >60 years of age develops degenerative scoliosis (Carter and Haynes, 1987; Schwab et al., 2005). Scoliosis is most common in girls, and a genetic disposition is very likely.

Scoliosis has different ages of onset, and it is commonly classified based on the age at which it is diagnosed:

- <3 years: Infantile scoliosis
- 3–9 years: Juvenile scoliosis
- 10–18 years: Adolescent scoliosis
- >18 years: Adult/degenerative scoliosis

Infantile and juvenile scoliosis share many characteristics and are commonly grouped under the name early onset scoliosis, to distinguish them from the more common adolescent type. Adult scoliosis generally is due to progression of idiopathic scoliosis or development of de novo scoliosis, which usually appears in adults after their fourth decade because of degenerative changes of the spine.

Scoliosis is commonly associated with disabilities, cosmetic deformity, pain, limitation of activity, quality of life issues, and breathing problems. In addition, health issues may remain throughout adulthood, such as increased risk of disc degeneration and curve progression, especially in the more severe curves.

Idiopathic scoliosis is initially treated conservatively with physical therapy and braces. Surgery is generally recommended when conservative treatment fails, the curvature progresses, and a Cobb angle >40 degrees is reached.

PATHOGENESIS

A slight deformity of the spine can be observed in almost every individual (Kouwenhoven et al., 2006).

The cause of scoliosis in children is in 85% of the cases unknown, namely idiopathic. The causes of the other 15% are explained either by congenital (from birth), myopathic (disease of muscles), or

Biomechanics of the Spine. https://doi.org/10.1016/B978-0-12-812851-0.00019-7

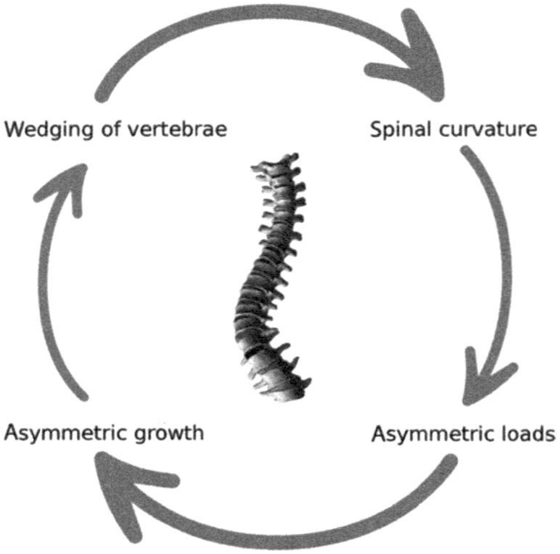

FIG. 1

Vicious cycle of the development of scoliosis.

neuropathic (neurological disease of the upper or lower motor neuron) disorders. Scoliosis diagnosed in the elderly is attributed to degenerative processes. This degenerative scoliosis can be further complicated by the presence of osteoporosis (reduce bone mineral density). In addition, scoliosis can develop secondary to interventions at the thorax, such as in pneumonectomy and rib resections.

Many reviews hypothesize that idiopathic scoliosis is multifactorial condition, as illustrated by Stokes et al. (1996a); thus the development of idiopathic scoliosis is often referred to as a "vicious cycle" (Fig. 1). Discussions on the pathogenesis of idiopathic scoliosis include genetic factors, growth effects, and structural causes (Kouwenhoven and Castelein, 2008; Hefti, 2013; Schlösser et al., 2014). However, no single causative factor has been identified (Kouwenhoven and Castelein, 2008). The main challenge is identifying the primary and secondary factors leading to the formation of scoliosis.

PATHOLOGY

Scoliosis is a three-dimensional deformity of the spine in all anatomical planes. The curvature with the highest Cobb angle within a scoliotic spine is defined as the main or primary curvature. Other concomitant curvatures are referred to as secondary or "compensatory" curvatures.

Morphological characteristics

From a biomechanical point of view, the shape of the bone and soft tissue within scoliosis is of particular interest. Because the biological tissue can adapt to the loads it is exposed to, the morphology can be used as an indicator for the stress distribution within the spine.

The loss of sagittal symmetry of the spine, or vertebral rotation, goes along with changes in the shape and composition of the disc, the vertebral body, the pedicle size, and the facets (Fig. 2).

Scoliotic spines exhibit a longer anterior height and a reduced spinal canal length compared to that of nonscoliotic spines (Fig. 3). This observation led to the theory that an anterior overgrowth might

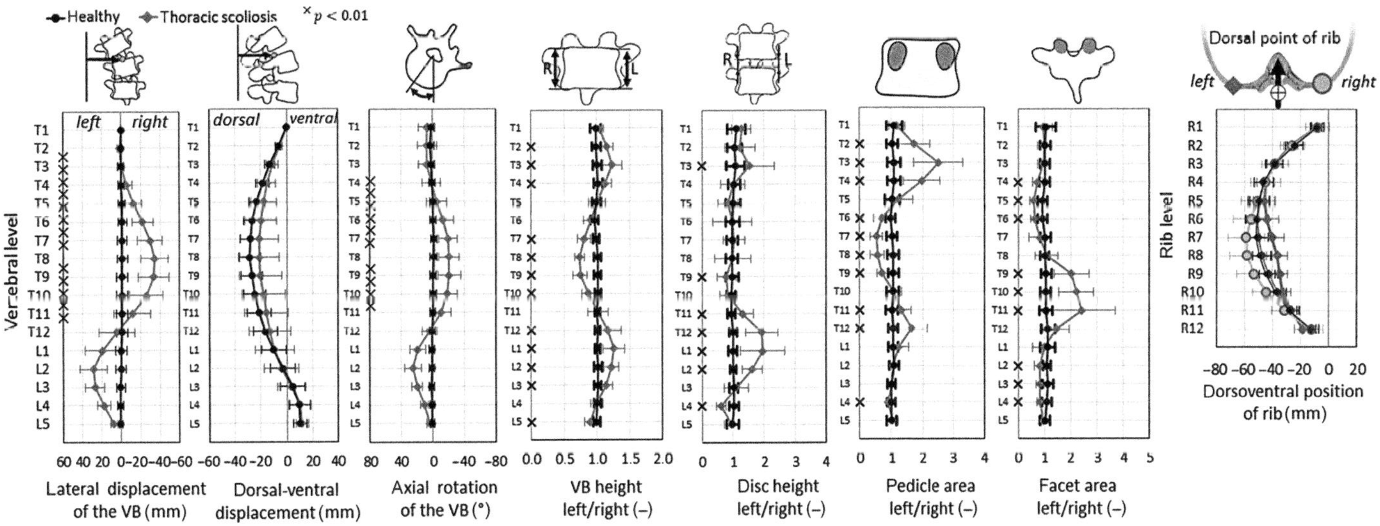

FIG. 2

Morphological characteristics of the vertebral body (VB) in scoliotic spines with a main thoracic curvature, compared to healthy nonscoliotic spines.

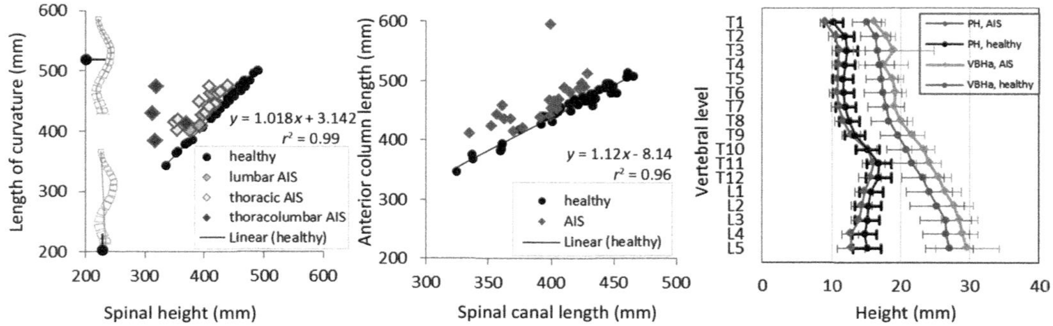

FIG. 3

Comparison of spinal parameters between scoliotic and healthy patient. The anterior vertebral column is generally longer in patients with scoliosis compared to patients with healthy straight spines. Also, the length of the posterior column (spinal canal) in relation to the anterior column is reduced in scoliotic patients. *PH*, pedicle height; *VBHa*, vertebral body height anterior.

be caused by tethering of the spinal canal (Porter, 2000, 2001). Yet evidence for this theory is lacking. However, the comparison of the spinal height and the length of the spinal curvature revealed that scoliotic spines are generally longer.

Wolff's law (Wolff, 1893) and the Hueter-Volkmann principle (Hueter, 1863; Volkmann, 1869) are often referred to when explaining the mechanics of bone adaption within scoliosis. According to the Hueter-Volkmann principle (Stokes et al., 1996b; Stokes, 2002), a decrease in the axial compression load will result in an increase in height. This finding agrees with findings of increased vertebral body height in scoliotic spines. The axial load component that acts in an erect posture along the spine should, therefore, be reduced.

Displacement and rotation of the thoracic vertebrae out of the sagittal symmetry plane alter the shape of the thorax. A rib hump usually appears on the convex side of the curvature, which is best visible during an Adam's forward bend test. On the other hand, a hump on the convex side of the lumbar curvature can usually be observed.

Further pathologies

The 3D deformity defining scoliosis also affects the thorax and the organs it contains and the muscle fiber distribution around the curve, and is characterized by a changing disc structure.

Scoliosis can lead to decreased respiratory function in the form of restrictive lung disease because of decreased lung volume, displacement of intrathoracic organs, and reduced mobility of the ribs affecting breathing. The degree of respiratory impairment is related to the severity of the curve, its location, and the amount of vertebral rotation. Curves with Cobb angles >90 degrees are a significant risk factor for cardiorespiratory failure, whereas patients with Cobb angles <70 degrees rarely present detectable respiratory signs and symptoms. The level of the curve is crucial because a more cranial curve can cause greater compromise of the lung on the convex side of the curve. Finally, vertebral rotation causes lateral and anterior displacement on the concave side and a posterior displacement on the convex side. In addition, the vertebral body is distorted toward the convex side causing a decrease in the sternum-to-vertebral body distance and an overall reduction of the lung capacity on the convex side. This leads to a decrease in the number of alveoli in infantile scoliosis, whereas it can lead to a normal number of reduced size in juvenile or adolescent scoliosis (Tsiligiannis and Grivas, 2012).

Histologic studies of idiopathic scoliosis reveal a muscle fiber asymmetry characterized by higher concentrations of type I fibers in the paraspinal muscle around the apex on the curve on the convex side. Type 1 muscle fibers are characterized by a greater concentration of mitochondria, aerobic respiration, lower contraction speeds, longer contraction times, and resistance to fatigue, and overall they produce and a more powerful contraction. This redistribution of muscle fibers may be a secondary remodeling to adapt to chronic asymmetric loads.

The vicious cycle of deformity and the asymmetrical load of the vertebrae and discs affect both the developing and the aging spine, especially toward the apex of the curve. In adolescent idiopathic scoliosis, vertebral wedging is more characteristic of the thoracic spine, whereas disc wedging is more likely when the curve affects the lumbar spine because of the increased flexibility of the lumbar intervertebral discs. In the aging spine, the asymmetrical load and reduced mobility of degenerative avascular discs may lead to curve development (Modi et al., 2008).

DIAGNOSIS AND CLASSIFICATION OF SCOLIOSIS

Scoliosis is diagnosed by the use of X-ray images in an upright standing position in the frontal and sagittal plane. The state-of-the-art method for the imaging of scoliosis is performed by the EOS Imaging System (EOS Imaging, Paris, France), which can capture simultaneous biplanar X-ray images with minimal distortion, low doses of radiation, and no magnification. The system allows for the determination of vertebral vectors that fully describe the size, position, and orientation of each vertebra in the three-dimensional space.

The severity of scoliosis is traditionally quantified by measuring the Cobb angle of each curvature within the frontal plane of the spine. The Cobb angle is defined as the curve angle formed between the endplates of the two vertebrae confining a curve. These are the superior endplate of the cranial vertebra and the inferior endplate of the caudal vertebra (Fig. 4). Scoliosis is defined by a Cobb angle >10 degrees. In addition, a variety of parameters were introduced to quantify the scoliotic curve, such as deviation from the C7 plumb line, the shoulder balance, rotation of the vertebrae, hip tilt, and the inclination angle of the rib cage asymmetry. The latter is quantified by using an inclinometer during forward bending. Which of the parameters are relevant generally depends on the specific question to be investigated. For example, explicit parameters are selected to create guidelines for the (surgical) treatment of the spine. The magnitudes of these parameters are used as a base for the classification of scoliosis.

One of the first classification systems that took into account the coronal appearance of the spine was defined by King et al. (1983) (Fig. 5). Lack of consideration of the sagittal profile and vertebral rotation and of some less common curve types (such as single thoracolumbar or lumbar curves) is considered to be a weaknesses of this classification system. A more comprehensive classification system was developed by Lenke et al. (2001, 2003). This system considered the scoliotic curvature in the frontal plane, the thoracic kyphosis, the lumbar curvature in standing X-rays, as well as the flexibility of the different curves by using lateral bending images in the supine position (Fig. 6). This later system also groups each curvature of a scoliotic spine into structural and nonstructural curvatures. This procedure attempts to address the self-restoration potential of the secondary curvature, allowing for "selective fusion" of only the main curve (see surgical treatment).

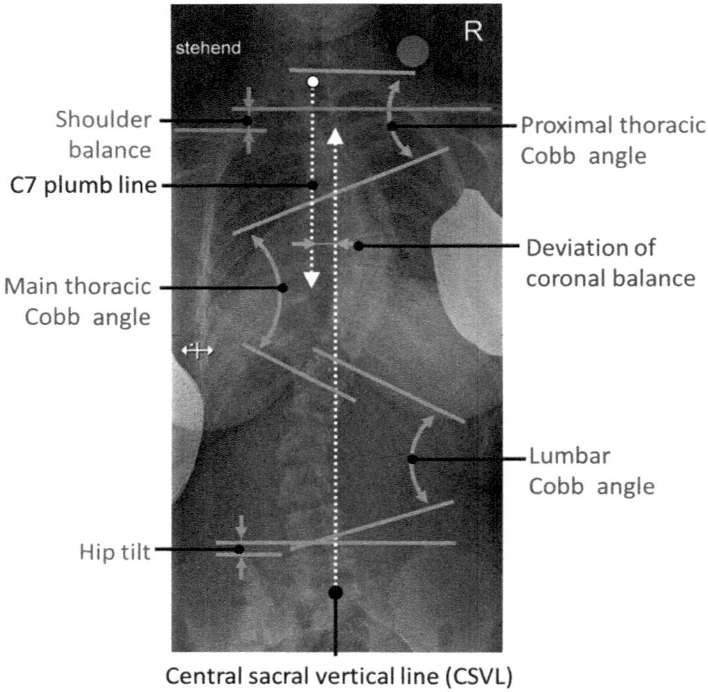

FIG. 4

Exemplary parameters to quantify the severity of the scoliotic curvature in the frontal plane.

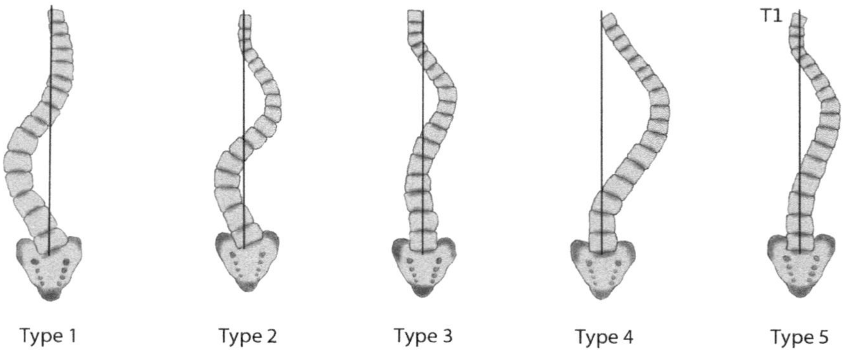

FIG. 5

King's classification of scoliosis, based on a single coronal X-ray image. In type 1, one thoracic and one lumbar curve are present, with the latter being larger and less flexible; both curves cross the midline. In type 2, the thoracic curve is equal or larger than the lumbar curve. In type 3, there is only a thoracic curve in which the lowest level does not cross the midline. In type 4, a single long thoracolumbar curve starting at L4 is present. Type 5 shows a double thoracic curve; T1 is involved in the upper curve.

	Type 1 Main thoracic	Type 2 Double thoracic	Type 3 Double major	Type 4 Triple major	Type 5 TL / lumbar	Type 6 TL / lumbar-main thoracic
PT	Non struct.	Struct.	Non struct.	Struct.	Non struct.	Non struct.
MT	Struct.	Struct.	Struct.	Struct.	Non struct.	Struct.
TL/L	Non struct.	Non struct.	Struct.	Struct.	Struct.	Struct.

The curve is defined as structural if the cobb angle inside bending is >25°, and the kyphosis angle >20°. The kyphosis angle is measured in PT region between T2 and T5, and in the MT and TL/L region between T10 and L2. PT, proximal thoracic; MT, main thoracic; TL, thoracolumbar; L, lumbar region

Lumbar modifier: location of the apical lumbar vertebrae relative to the vertical line from the center of the sacrum

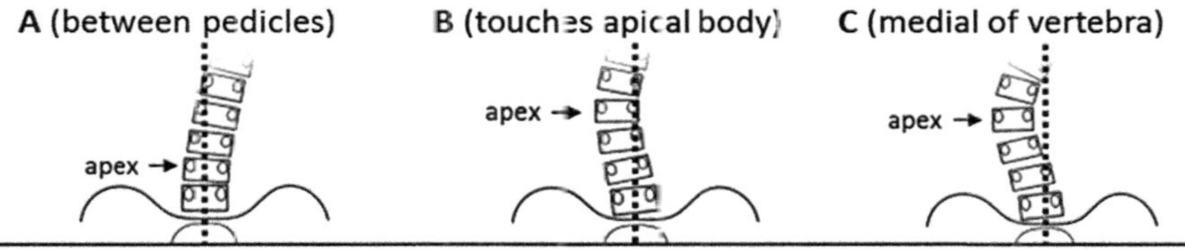

Thoracic sagittal modifier (angle measured between T5 and T12):

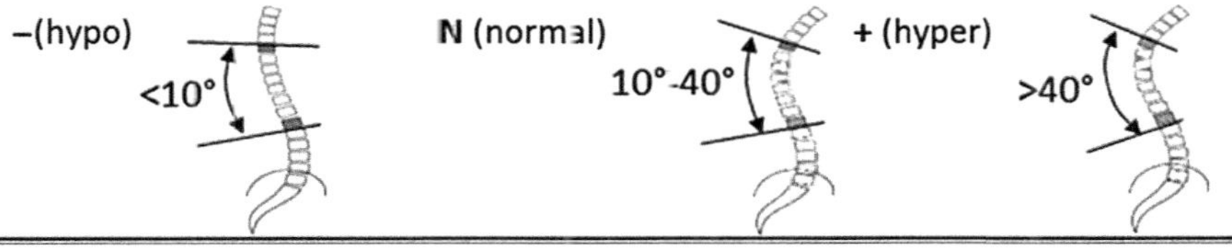

FIG. 6

Lenke classification system. The classification requires radiographic images in the frontal, sagittal plane, as well as lateral bending images to the right and left to determine the stiffness of the curve.

Different grading systems have been introduced regarding the rotation of the vertebra (Lam et al., 2008). The Nash-Moe grading system is currently the most popular one (Nash and Moe, 1969) (Fig. 7). The vertebral rotation may be a key indicator for curve progression, which is relevant for the prognosis and treatment of scoliosis.

A further classification system was developed by Suk to help determine the limits of a deformity and the extent of the necessary fusion (Suk et al., 2012). Fig. 8 shows an example of how a scoliotic spine can be classified.

In contrast to adolescent deformity, adult deformity is characterized by pain and dysfunction and is more commonly located in the lumbar area. Many classifications have been described, but Berjano and Lamartina described a treatment-based classification meant to define treatment based on the location of the symptomatic degeneration in relation to the apex of the curve (Berjano and Lamartina, 2014).

All the classification systems should be considered as guidelines that help predict the progression of the curve and therefore aid surgeons in their selection of a treatment, such as the selection of fusion

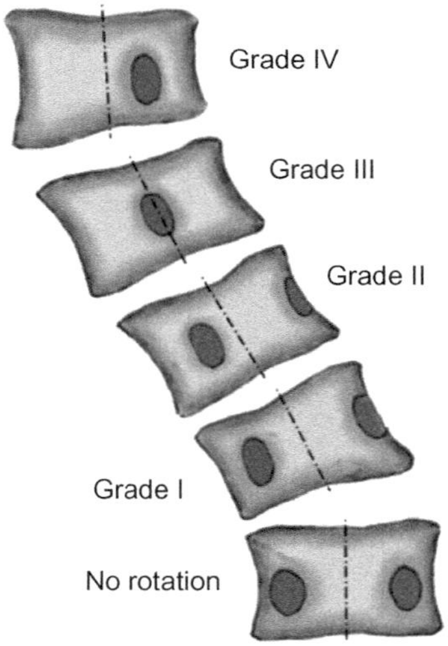

FIG. 7

Nash-Moe method for the assessment of the vertebral rotation. In grade I, the most lateral pedicle approaches the lateral margin of the vertebral projection, whereas in grade II it is only partially visible. In grade III, only one pedicle is visible, and touches the vertebral midline. In grade IV, the only visible pedicle is located beyond the vertebral midline.

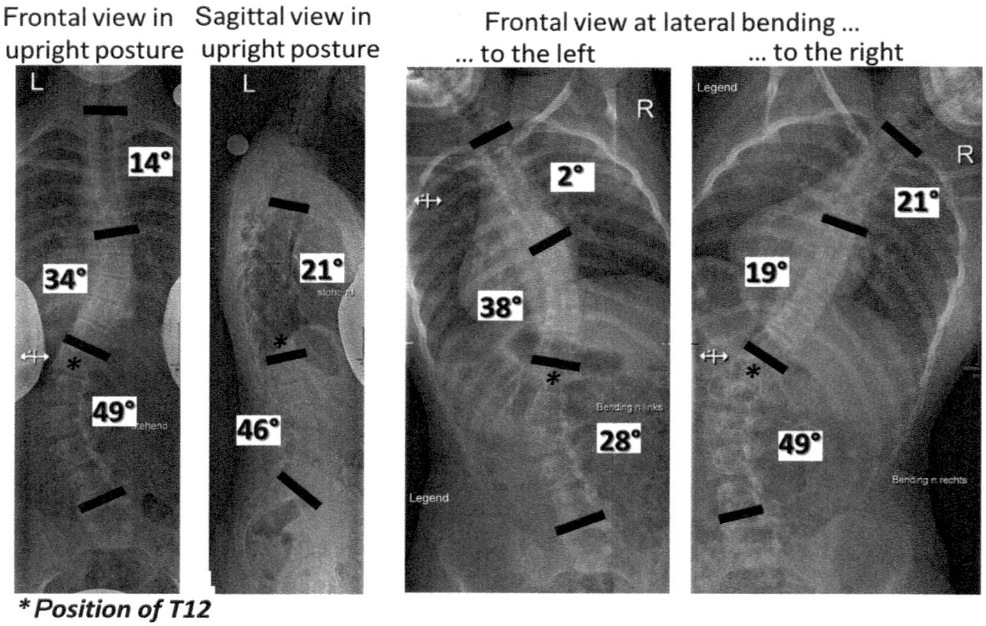

FIG. 8

Exemplary scoliosis case. This case is classified as a King type 1, Lenke-type 5cn, with a Nash-Moe grade 2 in the lumbar region.

levels. Despite numerous publications, there is no clear consensus on the optimal strategy for the correction of the spinal deformity, especially in idiopathic deformity. In particular, there are ongoing scientific debates, especially concerning selective thoracic/lumbar fusion, which deals with the added question of whether the secondary curvature needs to be fused as well.

Scoliosis surgery is generally based on empirical data, and the instrumentation technique depends especially on the preferences of the individual surgeon or institution.

TREATMENTS

The main treatment options for preventing curve progression include physical therapy, bracing, and surgery. Treatment options for idiopathic scoliosis depend on the severity of the scoliosis and the patient's stage of maturity. As a very general guideline, patients with curves smaller than 20 degrees are only monitored clinically, patients with curves between 20 and 40 degrees are treated with a brace, whereas surgery is recommended for patients with curves >40 degrees.

Physical therapy

In most countries, physical therapy is routinely recommended as treatment for patients with scoliosis. Although a strong musculoskeletal architecture is important for the patient, the literature's evidence on the role of exercise therapy in adolescent idiopathic scoliosis (AIS) is of poor quality.

Scoliosis specific exercises (SSEs), such as Bobath, consist of exercises that are individually adapted to the type of spinal curvature, while general physiotherapy (GPT) is more generic physical therapy that includes stretching and muscle-strengthening exercises. The therapeutic aim of SSE is to reduce the deformity by changing the spine's musculature and soft tissues. Furthermore, it is thought that SSE can alter the motor control, and thus the muscle activation of the spine.

However, the use of physical therapy to reduce curve progression and postpone or avoid more invasive treatments is controversial. Neither GPT (Negrini et al., 2012) nor SSE (Romano et al., 2013) have been proven to reduce curve progression, which is also due to a lack of evidence from high-level studies.

Brace treatment

Brace treatment is generally defined as the application of external corrective forces to the trunk. The main motivation behind brace treatment is to inhibit the progression of a deformity as well as to guide the growing spine toward an erect posture. There is little evidence in the literature to support the claim that braces help correct the deformity and reduce the magnitude of the curve, other than the application of consecutive bone casts, such as with the Lyon brace technique.

Therefore the effectiveness of brace treatment is controversial. Despite numerous research studies, mostly only low-level evidence studies, the efficacy of brace treatment has not been proven (Negrini et al., 2015).

Bracing is generally prescribed for (Risser sign 0–2) patients with immature skeletons with curves above a 20 ± 5 degrees Cobb angle, with a demonstrated progressive curve in consecutive X-ray controls (Negrini et al., 2012). It is recommended that patients should wear the brace at least 18 h a day for a minimum of 2–5 years until the end of bone growth.

Several aspects are considered for brace treatment:

- Design of the brace and pressure points should achieve effective correction of the spinal curvature and rib cage.
- The brace should enable the spine to develop toward a physiological three-dimensional shape during growth.

- It should allow flexibility of the spine and load-shearing between spine and brace. As movement is prohibited by the brace, the strength of back muscles, bones, and ligaments may decrease; therefore brace treatment is usually considered in combination with physical therapy.
- It should allow compliance and promote the patient's quality of life. An adolescent might not wear the brace because of discomfort and for social and cosmetic reasons. Brace treatment during childhood and adolescence is a life-changing treatment that can cause significant stress with both peers and parents, which can limit social inclusion and personal development.

The history of brace treatments has shown that an improper brace design can worsen the scoliosis. Although the frontal profile of the curve may be corrected by the brace, the vertebral rotation may increase or the thoracic kyphosis in the sagittal plane may be reduced. Identifying the right treatment for each patient remains a challenge.

Biomechanics of brace correction

The purpose of brace treatment is to treat scoliosis by applying specific pressure points on the trunk. The Society on Scoliosis Orthopedic and Rehabilitation Treatment (SOSORT) released a summary of how biomechanical correction is generally achieved (Rigo et al., 2006). A variety of brace types exist, and each is distinctive among the correction principles and building techniques (SRS, 2006; Zaina et al., 2014).

The main biomechanical aspects that are considered in the brace design are (Fig. 9):

- Thoracic curve correction:
  - Vector force level relative to the apex vertebra to correct the thoracic convexity
  - Direction of the vector force on the dorsal rib hump
  - Shape of the pad correcting the dorsal rib hump
  - Pad on the ventral rib hump
  - Derotation of the vertebrae

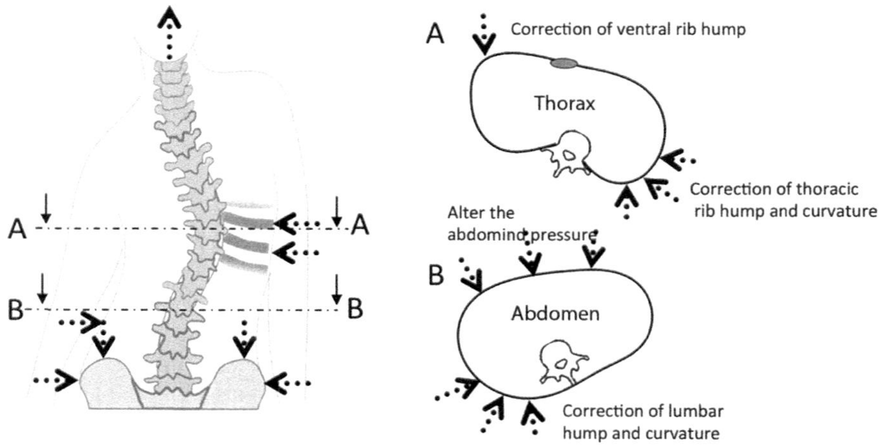

FIG. 9

Variety of possible load vectors that are applied during brace treatment (Rigo et al., 2006). The application and orientation of the load vectors depend on the selected brace type.

- Lumbar curve correction:
  - Vector force level relative to the apex vertebra to correct the lumbar convexity
  - Direction of vector force correcting the lumbar convexity
- Design of the abdominal pad to eventually alter the abdominal pressure
- Design of the brace in the pelvis region

The most efficient configuration of treatment has not yet been identified.

Improvement toward more sophisticated personalized brace designs using computer-aided design and manufacturing (CAD/CAM) is promising (Clin et al., 2010; Desbiens-Blais et al., 2012; Lebel et al., 2013; Cobetto et al., 2014) (Fig. 10). To reliably predict the performance of a brace by using numerical simulations, it is mandatory to create an adequate biomechanical model of the spine and chest.

Surgical treatment

Surgical treatment of a scoliotic deformity is performed by manually straightening and mechanically fusing the treated spinal section with implants. This is a very invasive procedure, commonly indicated for progressive curves with Cobb angles >40 degrees and at Risser signs >2.

The general goals of surgical treatment are to

- stop curve progression;
- correct deformity;
- restore cosmetic condition; and
- preserve flexibility and health of patient.

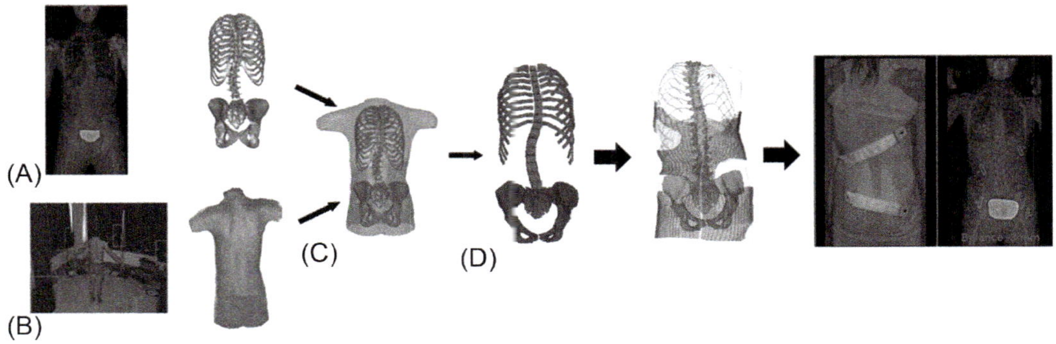

FIG. 10

Exemplary steps in the development of a patient-specific brace using CAD/CAM tools: (A) reconstruction of the patient-specific bone morphology, (B) scanning of the patients surface in an upright posture, (C) combination of inner morphology and outer surface, (D) development of a patient-specific finite-element model, following the simulation of different cage design and investigation of their impact on the spinal shape.

Reproduced with permission from Cobetto, N., Aubin, C.E., Clin, J., Le May, S., Desbiens-Blais, F., Labelle, H., Parent, S., 2014. Braces optimized with computer-assisted design and simulations are lighter, more comfortable, and more efficient than plaster-cast braces for the treatment of adolescent idiopathic scoliosis. Spine Deform. 2(4), 276–284. doi: 10.1016/j.jspd.2014.03.005. (Elsevier Inc).

Some desired goals can be contradictive, such as preserving curve flexibility yet restricting progression of the spinal curvature by fusion. Therefore the surgeon and patient need to strike a balance in the surgery's goal.

Surgical techniques

Surgical treatment for scoliosis is complex, and modern instrumentation techniques offer multiple possibilities. The choice of approach depends on the severity and type of scoliosis, the patient's symptoms and age, and the surgeon's experience.

There are several relevant biomechanical factors for surgery, including

- mechanical resistance of the spine for correction: functional/structural deformity, size of the curves, vertebral wedging, viscoelastic behavior;
- applied corrective forces: distraction, compression, translation, rotation;
- direction of approach: ventral, dorsal, or ventrodorsal;
- rigidity of the spine (immature curves are more flexible, whereas degenerative spines become very stiff);
- selection of fusion levels; and
- type and amount of fixations: various types of pedicle screws (monoaxial, uniaxial, polyaxial, multiple-degrees-of-freedom), hooks, wires, and bands).

To be able to perform significant curve corrections, the flexibility of the spine needs to be increased. This mobilization can be achieved by multiple resection steps, such as the resection of the articular processes, lamina, discs, as well as parts of the ribs in some cases. The resections are usually performed around the apex of the curve and can be differentiated, for example, by using the Schwab classification system (Schwab et al., 2014) (Fig. 11).

The materials used for fixation depend on the surgical needs. Monoaxial screws are more capable of maintaining the desired correction, although polyaxial screws are more easily adapted to the rod. A titanium alloy rod may be more easily adapted to the curve, although it is more frequently associated with rod failure (breakage) after fixation. Cobalt chrome alloy rods on the other hand are stiffer and are more resistant to loss of correction and fatigue, but may also be more frequently associated with proximal junctional kyphosis (failure of the proximal adjacent segment of the spine characterized by kyphosis, disc collapse, and/or fracture of the immediate cranial vertebra).

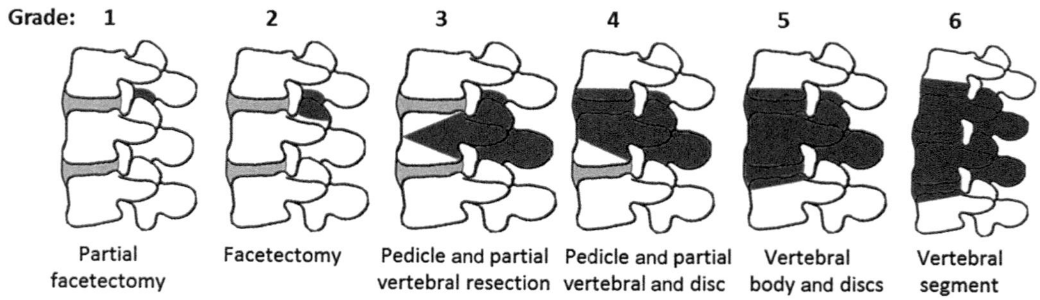

FIG. 11

Schwab classification of resection steps.

The mechanical correction of the spine depends on the curvature type, the instrumentation technique, and the direction of the approach. Some correction techniques using dorsal/posterior and/or ventral/anterior approaches follow adapted from Boos and Aebi (2008):

<table>
<tr><th colspan="2">Posterior approach</th></tr>
<tr><td>

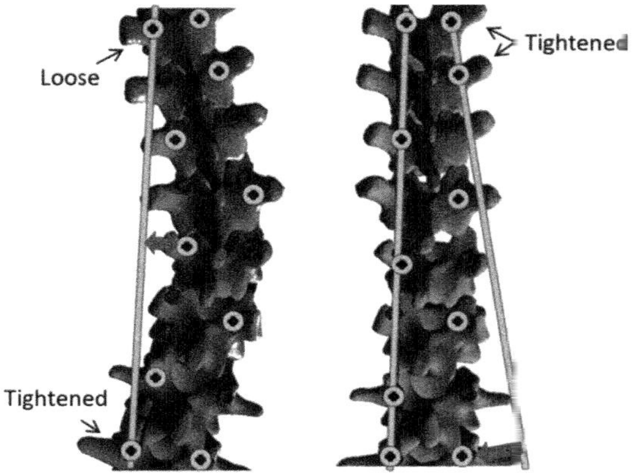

Posterior instrumentation

</td><td>

1. Pedicle screws and/or hooks are placed into carefully selected vertebrae.
2. A prebent rod is placed into the upper instrumented vertebra and lowest instrumented vertebra on the concave side.
3. Vertebrae are dragged using forceps toward the rod, vertebra by vertebra.
4. Second rod is inserted on the convex side and stepwise connected to the vertebrae.

</td></tr>
<tr><th colspan="2">Posterior approach using derotation maneuver</th></tr>
<tr><td>

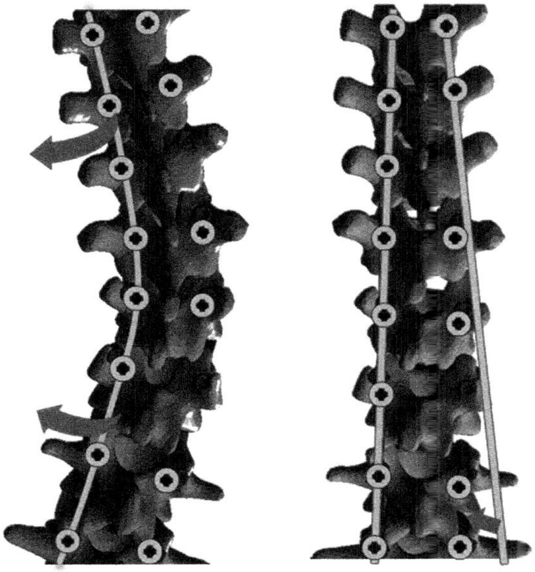

Posterior instrumentation using derotation maneuver

</td><td>

1. Pedicle screws are placed into the treated vertebrae.
2. A prebent rod is placed loose into pedicle screws on the concave side of the curvature.
3. The spine is derotated to a straight position, so that the scoliotic thoracic curve turns into the sagittal kyphotic curve and the scoliotic lumbar curve turns into the sagittal lordotic curve.
4. The screws are tightened, while adjusting level by level.
5. Second rod is inserted stepwise and fixed, providing additional stability.

</td></tr>
</table>

Anterior approach

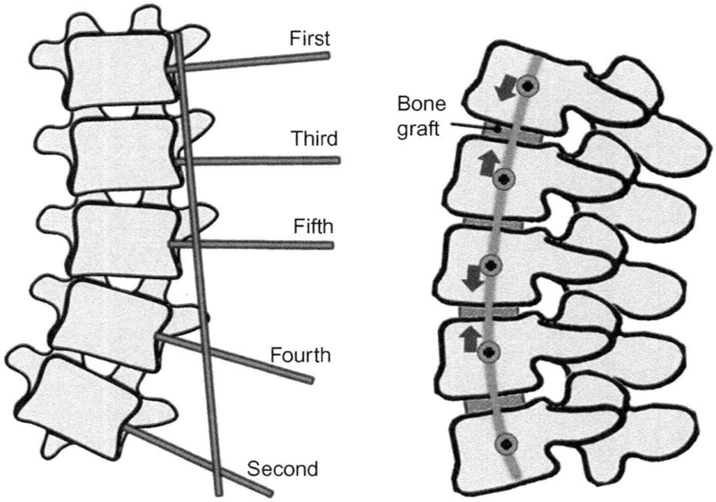

Anterior instrumentation

1. Discs are resected at the levels that will be fused.
2. Screws are placed laterally in the vertebral body.
3. A bent rod is inserted stepwise into the rods.
4. Before finally tightening the screws, the segments are contracted using forceps to achieve further correction.

Traditional surgical techniques are all varied techniques on fusion of the spine, which is an unphysiological and archaic solution, leading to a permanent fixation of the spine and its resulting biomechanical alterations to the spine. Alternative techniques, such as mini-invasive anterior vertebral body tethering, are continually being studied to reduce the deformity in the early stages of the deformity and to avoid the need for fusion.

Complications involved with surgery

Possible complications of a surgical treatment that relate to biomechanical modifications:

Complication	Description
General implant failure	Rod breakage caused by long instrumentation (occurs near thoracolumbar region, point of greatest lordosis due to significant bending of the rod, point of greatest implant concentration), screw/hook loosening (for example, at osteoporosis), implant abrasion.
Pseudarthrosis	When a vertebral segment involved in a surgical fusion does not heal/fuse together. This can cause chronic pain caused by instability. The spinal motion can also stress the metal hardware used to hold the fusion, causing rod and screws to break.
Elongation of spinal canal	A performed correction of a spinal deformity usually increases the length of the spinal cord. If the elongation of the spinal cord is too strong, the cord might suffer ischemia, causing severe temporary or permanent nerve damage and palsy.
Flat-back deformity	Loss of the sagittal thoracic kyphosis after correcting the frontal curve, especially with early surgical options such as the Harrington rod.

Complication	Description
Crankshaft phenomenon	Can occur in cases in patients with immature skeletons, despite posterior fusion. The posterior instrumentation acts as a posterior tethering bar, whereas the unfused anterior vertebral column continues to grow. This causes loss of correction and increased vertebral rotation (Dubousset et al., 1989).
Proximal/distal junctional kyphosis	Mechanical failure of the spine immediately above or below the instrumentation.
Coronal/sagittal imbalance	Deviation from the midline or a sagittal vertical axis (SVA) >5 cm.

REFERENCES

Berjano, P., Lamartina, C., 2014. Classification of degenerative segment disease in adults with deformity of the lumbar or thoracolumbar spine. Eur. Spine J. 23 (9), 1815–1824. https://doi.org/10.1007/s00586-014-3219-9.

Boos, N., Aebi, M., 2008. Spinal Disorder—Fundamentals of Diagnosis and Treatment. Springer, Berlin/Heidelberg.

Carter, O.D., Haynes, S.G., 1987. Prevalence rates for scoliosis in US adults: results from the first national health and nutrition examination survey. Int. J. Epidemiol. 16 (4), 537–544.

Clin, J., Aubin, C.E., Parent, S., Sangole, A., Labelle, H., 2010. Comparison of the biomechanical 3D efficiency of different brace designs for the treatment of scoliosis using a finite element model. Eur. Spine J. 19 (7), 1169–1178. https://doi.org/10.1007/s00586-009-1268-2.

Cobetto, N., Aubin, C.E., Clin, J., Le May, S., Desbiens-Blais, F., Labelle, H., Parent, S., 2014. Braces optimized with computer-assisted design and simulations are lighter, more comfortable, and more efficient than plaster-cast braces for the treatment of adolescent idiopathic scoliosis. Spine Deform. 2 (4), 276–284. https://doi.org/10.1016/j.jspd.2014.03.005.

Desbiens-Blais, F., Clin, J., Parent, S., Labelle, H., Aubin, C.E., 2012. New brace design combining CAD/CAM and biomechanical simulation for the treatment of adolescent idiopathic scoliosis. Clin. Biomech. 27 (10), 999–1005. Elsevier Ltd. https://doi.org/10.1016/j.clinbiomech.2012.08.006.

Dubousset, J., Herring, J.A., Shufflebarger, H., 1989. The crankshaft phenomenon. J. Pediatr. Orthop. 9 (5), 541–550. https://doi.org/10.1097/01241398-198,909,010-00008.

Hefti, F., 2013. Pathogenesis and biomechanics of adolescent idiopathic scoliosis (AIS). J. Child. Orthop. 7 (1), 17–24. https://doi.org/10.1007/s11832-012-0460-9

Hueter, C., 1863. Anatomische Studien an den Extremit{ä}tengelenken Neugeborener und Erwachsener. Archiv f{ü}r pathologische Anatomie und Physiologie und f{ü}r klinische Medicin 26 (5), 484–519. https://doi.org/10.1007/BF01878009.

King, H.A., Moe, J.H., Bradford, D.S., Winter, R.B., 1983. The selection of fusion levels in thoracic idiopathic scoliosis. J. Bone Joint Surg. Am. 65 (9), 1302–1313. https://doi.org/10.1097/01241398-198,405,000-00029.

Kouwenhoven, J.W., Castelein, R.M., 2008. The pathogenesis of adolescent idiopathic scoliosis: review of the literature. Spine (Phila Pa 1976) 33 (26), 2898–2908. https://doi.org/10.1097/BRS.0b013e3181891751.

Kouwenhoven, J.-W.M., Vincken, K.L., Bartels, L.W., Castelein, R.M., 2006. Analysis of preexistent vertebral rotation in the normal spine. Spine 31 (13), 1467–1472. United States https://doi.org/10.1097/01.brs.0000219938.14686.b3.

Lam, G.C., Hill, D.L., Le, L.H., Raso, J.V., Lou, E.H., 2008. Vertebral rotation measurement: a summary and comparison of common radiographic and CT methods. Scoliosis 3, 16. https://doi.org/10.1186/1748-7161-3-16.

Lebel, D.E., Al-Aubaidi, Z., Shin, E.J., Howard, A., Zeller, R., 2013. Three dimensional analysis of brace biomechanical efficacy for patients with AIS. Eur. Spine J. 22 (11), 2445–2448. https://doi.org/10.1007/s00586-013-2921-3.

Lenke, L.G., Betz, R.R., Harms, J., Bridwell, K.H., Clements, D.H., Lowe, T.G., Blanke, K., 2001. Adolescent idiopathic scoliosis: a new classification to determine extent of spinal arthrodesis. J. Bone Joint Surg. 83 (A(8)), 1169–1181. https://doi.org/10.1097/00007632-200,101,010-00009.

Lenke, L.G., Edwards, C.C., Bridwell, K.H., 2003. The Lenke classification of adolescent idiopathic scoliosis: how it organizes curve patterns as a template to perform selective fusions of the spine. Spine 28 (20), S199–207. https://doi.org/10.1097/01.BRS.0000092216.16155.33.

Lowe, T., Edgar, M., Margulies, J., 2000. Etiology of Idiopathic Scoliosis: Current Trends in Research. J. *Bone Surg.*. Available at: http://jbjs.org/article.aspx?articleID=24675Accessed August 6, 2013.

Modi, H.N., Suh, S.W., Song, H.R., Yang, J.H., Kim, H.J., Modi, C.H., 2008. Differential wedging of vertebral body and intervertebral disc in thoracic and lumbar spine in adolescent idiopathic scoliosis—a cross sectional study in 150 patients. Scoliosis 3, 11.

Nash, C.L.J., Moe, J.H., 1969. A study of vertebral rotation. J. Bone Joint Surg. 51 (2), 223–229.

Negrini, S., Aulisa, A.G., Aulisa, L., Circo, A.B., de Mauroy, J.C., Durmala, J., Grivas, T.B., Knott, P., Kotwicki, T., Maruyama, T., Minozzi, S., O'Brien, J.P., Papadopoulos, D., Rigo, M., Rivard, C.H., Romano, M., Wynne, J.H., Villagrasa, M., Weiss, H.-R., Zaina, F., 2012. 2011 SOSORT guidelines: orthopaedic and rehabilitation treatment of idiopathic scoliosis during growth. Scoliosis 7 (1), 3. https://doi.org/10.1186/1748-7161-7-3.

Negrini, S., Minozzi, S., Chockalingam, N., Grivas, T.B., Kotwicki, T., Maruyama, T., Romano, M., Zaina, F., 2015. Braces for idiopathic scoliosis in adolescents. Cochrane Database Syst. Rev. 6, https://doi.org/10.1002/14651858.CD006850.pub3. www.cochranelibrary.com.

Porter, R., 2000. Idiopathic scoliosis: the relation between the vertebral canal and the vertebral bodies. Spine 1360–1366. Available at http://journals.lww.com/spinejournal/Abstract/2000/06010/Idiopathic_Scoliosis__The_Relation_Between_the.7.aspx. (Accessed 31 January 2015).

Porter, R.W., 2001. Can a short spinal cord produce scoliosis? Eur. Spine J. 10 (1), 2–9. https://doi.org/10.1007/s005860000188.

Rigo, M., Negrini, S., Weiss, H.R., Grivas, T.B., Maruyama, T., Kotwicki, T., Sosort, 2006. SOSORT consensus paper on brace action: TLSO biomechanics of correction (investigating the rationale for force vector selection). Scoliosis 1, 11. https://doi.org/10.1186/1748-7161-1-11.

Romano, M., Minozzi, S., Zaina, F., Saltikov, J.B., Chockalingam, N., Kotwicki, T., Hennes, A.M., Negrini, S., 2013. Exercises for adolescent idiopathic scoliosis. Spine 38 (14), E883–E893. https://doi.org/10.1097/BRS.0b013e31829459f8.

Schlösser, T.P.C., Van Der Heijden, G.J.M.G., Versteeg, A.L., Castelein, R.M., 2014. How "idiopathic" is adolescent idiopathic scoliosis? A systematic review on associated abnormalities. PLoS One 9 (5), 1–10. https://doi.org/10.1371/journal.pone.0097461.

Schwab, F., Dubey, A., Gamez, L., El Fegoun, A.B., Hwang, K., Pagala, M., Farcy, J.P., 2005. Adult scoliosis: prevalence, SF-36, and nutritional parameters in an elderly volunteer population. Spine 30 (9), 1082–1085. https://doi.org/10.1097/01.brs.0000160842.43482.cd.

Schwab, F., Blondel, B., Chay, E., Demakakos, J., Lenke, L., Tropiano, P., Ames, C., Smith, J.S., Shaffrey, C.I., Glassman, S., Farcy, J.-P., Lafage, V., 2014. The comprehensive anatomical spinal osteotomy classification. Neurosurgery 74 (1), 112–120. discussion 120 https://doi.org/10.1227/NEU.0000000000000182o.

SRS, 2006. Brace Wear Compliance: Bracing Manual, Scoliosis Research Society. Available at: http://www.srs.org/professionals/education_materials/SRS_bracing_manual/.

Stokes, I.A.F., 2002. Mechanical effects on skeletal growth. J. Musculoskelet. Neuronal Interact. 2 (3), 277–280. Greece.

Stokes, I.A.F., Spence, H., Aronsson, D.D., Kilmer, N., 1996a. Mechanical modulation of vertebral body growth. Spine 1162–1167. https://doi.org/10.1097/00007632-199,605,150-00007.

Stokes, I.A., Spence, H., Aronsson, D.D., Kilmer, N., 1996b. Mechanical modulation of vertebral body growth. Implications for scoliosis progression. Spine 21 (10), 1162–1167. United States.

Suk, S.-I., Kim, J.-H., Kim, S.-S., Lim, D.-J., 2012. Pedicle screw instrumentation in adolescent idiopathic scoliosis (AIS). Eur. Spine J. 21 (1), 13–22. https://doi.org/10.1007/s00586-011-1986-0.

Tsiligiannis, T., Grivas, T., 2012. Pulmonary function in children with idiopathic scoliosis. Scoliosis 7 (7), https://doi.org/10.1186/1748-7161-7-7.

Volkmann, R., 1869. Die Krankheiten der Bewegungsorgane. Handbuch der allgemeinen und speziellen Chirurgie, Bd II, Abt 1, 350–351.

Wolff, J., 1893. Das Gesetz der Transformation der Knochen. Deutsche Medizinische Wochenschrift 19, 1222–1224. https://doi.org/10.1055/s-0028-1,144,106.

Zaina, F., De Mauroy, J.C., Grivas, T., Hresko, M.T., Kotwizki, T., Maruyama, T., Price, N., Rigo, M., Stikeleather, L., Wynne, J., Negrini, S., 2014. Bracing for scoliosis in 2014: state of the art. Eur. J. Phys. Rehabil. Med. 50 (1), 93–110.

FURTHER READING

Mordecai, S.C., Dabke, H.V., 2012. Efficacy of exercise therapy for the treatment of adolescent idiopathic scoliosis: a review of the literature. Eur. Spine J. 21 (3), 382–389. https://doi.org/10.1007/s00586-011-2063-4.

CHAPTER

NEUROMUSCULAR DISORDERS

20

Fabio Galbusera, Tito Bassani, Maryem-Fama Ismael Aguirre, Marco Brayda-Bruno
IRCCS Galeazzi Orthopedic Institute, Milan, Italy

INTRODUCTION

Neuromuscular disorders such as muscular dystrophy, spinal muscle atrophies, Parkinson's disease (PD), and Creutzfeldt-Jakob disease affect the muscles and the central nervous system's control over them. These disorders manifest symptoms such as alterations to muscle tone, rigidity, dystonia, and myalgia. Neuromuscular pathologies have various causes, including genetic and autoimmune disorders (Kraker, 2011), brain insults, and environmental factors. Because of the strong modifications to the muscle activity related to such diseases and therefore to the alterations of the biomechanical environment, spinal disorders such as scoliosis and sagittal imbalance are relatively frequent among patients with these diseases. The best documented disorders are neuromuscular scoliosis secondary to Duchenne and Becker muscular dystrophies, cerebral palsy, spinal muscular atrophy (Hart and McDonald, 1998), and misalignments in both the sagittal and coronal planes in patients suffering from Parkinson's disease (Doherty et al., 2011).

In this chapter, neuromuscular scoliosis and spinal disorders related to PD are described. The various forms of neuromuscular scoliosis are included first because of their wide similarities and because of their shared surgical principles and management options. PD is covered in a separate section because patients with PD are generally older than those suffering from neuromuscular scoliosis, and this factor presents distinct problems, treatment options, and complications.

In fact, biomechanical research on spinal disorders associated with neuromuscular diseases is almost nonexistent, likely because of their biomechanical complexity and their low prevalence, which in turn results in limiting the options available for funding research projects. Nevertheless, dedicated research and investment would benefit the knowledge about the potential role of neuromuscular factors in the initiation and progression of scoliosis as well as the high rate of complications after surgical treatment, and would have a direct positive impact on the patients' healthcare.

NEUROMUSCULAR SCOLIOSIS

The Scoliosis Research Society has classified neuromuscular scoliosis as two distinct types: those caused by neuropathic conditions such as cerebral palsy and spinal muscular atrophy, and those determined by myopathic conditions such as Duchenne and Becker muscular dystrophies. Both types frequently exhibit an increase in the thoracic kyphotic curve, which is likely caused by muscle weakness

Biomechanics of the Spine. https://doi.org/10.1016/B978-0-12-812851-0.00020-3

(Archer et al., 2016). Consequently, neuromuscular scoliosis is often referred to as "kyphoscoliosis." In addition to the spinal deformities, which are often severe, and to the neuromuscular symptoms, several patients suffer from associated pathologies such as reduced pulmonary function, cardiac arrhythmias, and malnutrition (Winter, 1994), making the clinical management of such patients complicated to the point that a multidisciplinary team is usually required. Indeed, these patients have a lower life expectancy as compared to the general population, especially due to the cardiopulmonary disorders secondary to the neuromuscular disease (Boos and Aebi, 2008).

SPINAL DEFORMITIES AND RISK OF PROGRESSION

As previously mentioned, the characteristic signs observed in radiographic evaluations are scoliosis and kyphosis. In addition, several patients present an increased lordosis of the lumbar spine and pelvic obliquity (Boos and Aebi, 2008; Tsirikos and Spielmann, 2007). (Lonstein and Akbarnia (1983) classified the appearance of neuromuscular scoliosis due to cerebral palsy based on coronal radiographic projections in two groups, either exhibiting (Group II) or not exhibiting (Group I) pelvic obliquity; and they further subdivided the two groups based on the location and size of the scoliotic curves (Fig. 1). For a comprehensive examination, Boos and Aebi (2008) further suggested evaluating whether the pelvic obliquity was secondary to the spinal deformity (suprapelvic obliquity), secondary to hip contractures (infrapelvic), or resulted from the morphology of the pelvis itself, which can be asymmetrical (intrapelvic) (Haas, 1942). Cerebral palsy is commonly associated with large thoracolumbar curves

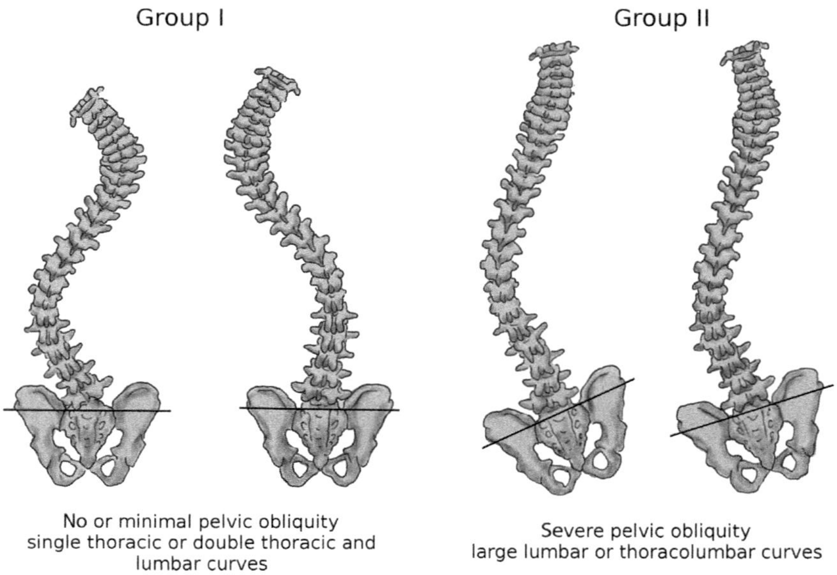

FIG. 1

Classification by Lonstein and Akbarnia (1983) of neuromuscular scoliosis due to cerebral palsy based on coronal radiographic projections.

Modified from Boos, N., Aebi, M., 2008. Spinal Disorders—Fundamentals of Diagnosis and Treatment, Springer, Heidelberg.

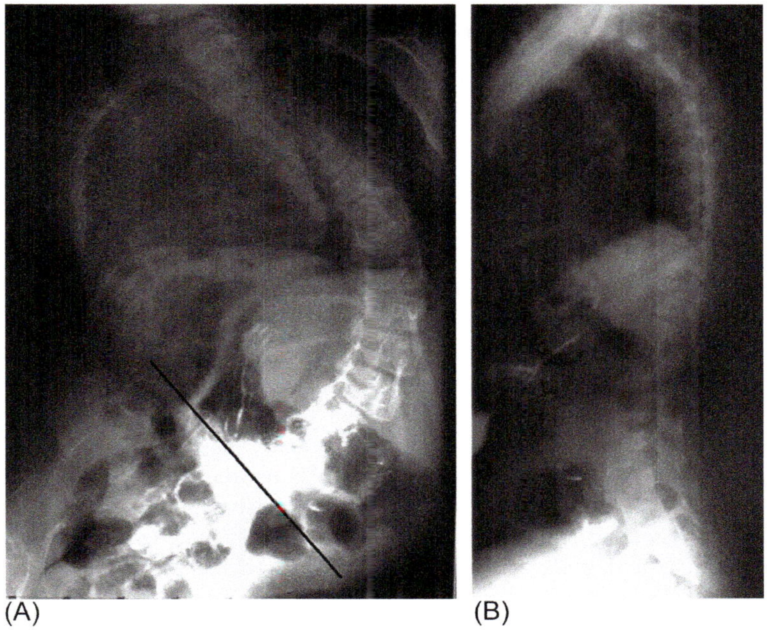

FIG. 2

Anteroposterior (A) and lateral (B) X-ray images of a patient suffering from cerebral palsy and exhibiting severe scoliosis with marked pelvic obliquity.

Reproduced with permission from Tsirikos, A.I., Spielmann, P., 2007. Spinal deformity in paediatric patients with cerebral palsy. Curr. Orthopaed. 21(2), 122–134.

with a pelvic obliquity exhibiting both suprapelvic and infrapelvic components (Cloake and Gardner, 2016; Winter and Pinto, 1986; Griffiths et al., 1977) (Fig. 2). The prevalence of scoliosis in cerebral palsy patients was reported to range between 15% and 80% (Koop, 2009), with a large variation due to the differences in the populations being studied. Oda et al. (2017) reported an average progression of the scoliosis between 0.1 and 2.7 degrees per year, depending on the severity of the pathology. Although these angles may not seem high in absolute terms, the progression of scoliosis in cerebral palsy is still a matter of concern because it does not respond well to conservative treatment and progresses even after skeletal maturity (Saito et al., 1998; Miller et al., 1996; Thometz and Simon, 1988).

Patients suffering from muscular dystrophy often show a more rapid curve progression, with a Cobb angle increase of 16–24 degrees yearly, with the fastest progression during the adolescent growth spurt (Archer et al., 2016). The coronal deformity often starts at the thoracolumbar junction, subsequently involving the entire thoracolumbar spine and in some cases the pelvis. In contrast with adolescent idiopathic scoliosis, which shows a generally lordotic sagittal alignment, in patients suffering from muscular dystrophy, the tendency to develop a thoracolumbar kyphosis is particularly accentuated (Archer et al., 2016) and can develop into a collapsing C-shaped curve pattern (Boos and Aebi, 2008). Oda et al. (1993) classified the risk of progression of scoliosis in patients with Duchenne muscular dystrophy in three categories: (i) fast-collapsing kyphoscoliotic curves including the pelvis,

(ii) hyperlordotic curves that either could be stable (if without pelvic obliquity) or could progress quickly (if with pronounced pelvic obliquity), and (iii) nonprogressive scoliotic curves with a Cobb angle lower than 30 degrees and with a minor curvature in the sagittal plane. Patients with Becker muscular dystrophy tend to be older and thus show a lower severity and higher stability of the curves in comparison with patients with Duchenne muscular dystrophy (Boos and Aebi, 2008; McDonald et al., 1995).

SURGICAL MANAGEMENT

Despite the high risks associated with major surgeries in young subjects with severe disorders and a complex clinical picture, there is a general consensus that spine surgery can significantly improve the quality of life in carefully selected patients. Indeed, neuromuscular scoliosis has a greater impact on the functional status of the patient than idiopathic scoliosis does because the former hinders the ability to perform daily activities such as standing, walking, and sitting.

In general, the same indications used for idiopathic scoliosis, that is, a curve greater than 50 degrees and the risk of progression, are considered valid for neuromuscular scoliosis (Boos and Aebi, 2008; Cloake and Gardner, 2016; Koop, 2009); however, a comprehensive multidisciplinary preoperative assessment is critical in order to correctly take into account the frequent concurrent pathologies. The aim of the surgery for the treatment of neuromuscular scoliosis is to correct the spinal deformity and restore a correct balance in both the coronal and sagittal planes. Patients with neuromuscular scoliosis have a limited capacity to activate compensatory strategies to achieve a balanced posture, and therefore special care should be taken in order to ensure that physiological alignment of the spine, including the pelvis, is achieved (Shapiro et al., 2003).

The first instrumentation used to correct neuromuscular deformities was the Harrington rod, which was employed with partial success in patients suffering from poliomyelitis in the 1950s (Harrington, 1988) and opened the way to the development of the modern implantable devices for the treatment of spinal deformities. Until recent years, posterior fixation with sublaminar wires and Luque rods (Luque, 1982) in combination with anterior releases was the most common solution for the correction of neuromuscular scoliosis, because of its capability of achieving a satisfactory but flexible local deformity correction without overloading the anchoring points (Boos and Aebi, 2008). Hooks and pedicle screws have also been used, but showed in the early stages a poorer clinical outcome and higher complication rates in comparison with wiring, mostly due to the complex insertion technique and inadequate purchase in the bone tissue, which is often osteopenic (Archer et al., 2016; Modi et al., 2010). Hybrid configurations including both wires and posterior instrumentation were then introduced and improved the outcomes (Modi et al., 2010; Arun et al., 2010) (Fig. 3). Solutions employing only pedicle screws are now gaining increasing interest because they reduce blood loss and surgical time (Archer et al., 2016; Arun et al., 2010).

The most common proximal fusion level for neuromuscular scoliosis is T2, even if the curve does not extend to such a level in a coronal radiograph (Boos and Aebi, 2008). In fact, the kyphotic deformity is frequently larger than the scoliosis, and a proximal extension ensures a proper correction in the sagittal plane. In order to minimize the risk of curve progression after fixation, fusion at all the involved levels is recommended (Boos and Aebi, 2008). If pelvic obliquity is present, its correction by means of sacral and pelvic fixation is mandatory. The classical technique employs the Luque-Galveston construct, in which the rods are inserted into the posterior superior iliac spine

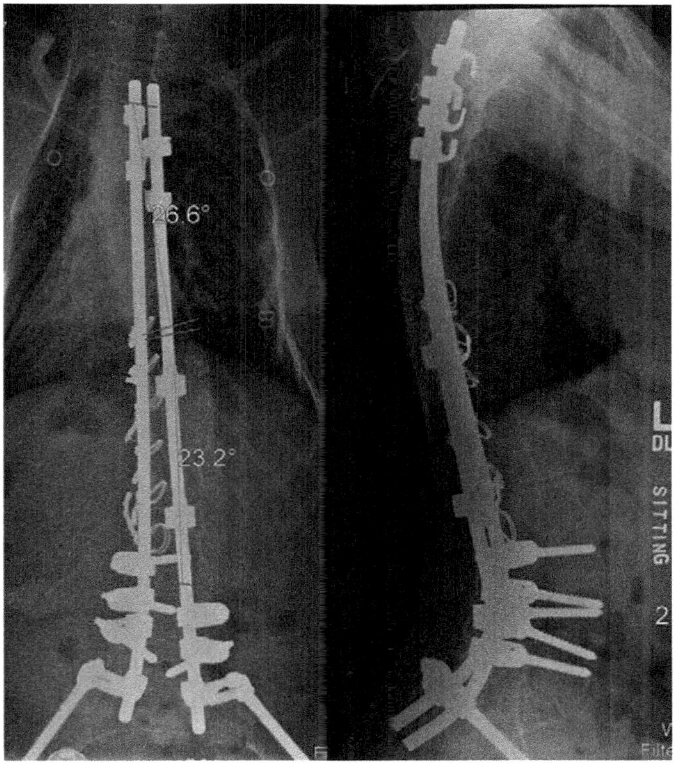

FIG. 3

Postoperative X-ray images of a patient suffering from spinal muscle atrophy, with a spinal deformity corrected by means of hybrid instrumentation comprising proximal hooks, distal pedicle screws, sublaminar wires, and pelvic fixation with screws.

Reproduced with permission from Maitra, S., Roberto, R.F., McDonald, C.M., Gupta, M.C., 2012. Treatment of spine deformity in neuromuscular diseases. Phys. Med. Rehabil.Clin. N. Am. 23(4), 869–883.

(Gau et al., 1991) and often additionally supported by S1 pedicle screws (Marchesi et al., 1997) (Fig. 4). The Luque-Galveston technique subsequently evolved into the unit-rod, and provided an improved potential to correct the pelvic obliquity (Bulman et al., 1996; Bell et al., 1989; Dias et al., 1996). In severe cases of rigid kyphosis, the use of technically demanding and risky procedures such as pedicle subtraction osteotomy and apical vertebral resection (Sharrard, 1968) has also been reported.

Complications during and after neuromuscular scoliosis corrective surgery are frequent and can be severe. Koop (2009) reported a complication rate of 40%–80% in patients with cerebral palsy, with an approximate death ratio of 1% (Lipton et al., 1999; Yazici et al., 2000). Similar figures were reported for patients with muscular dystrophy, with even higher occurrences of perioperative deaths

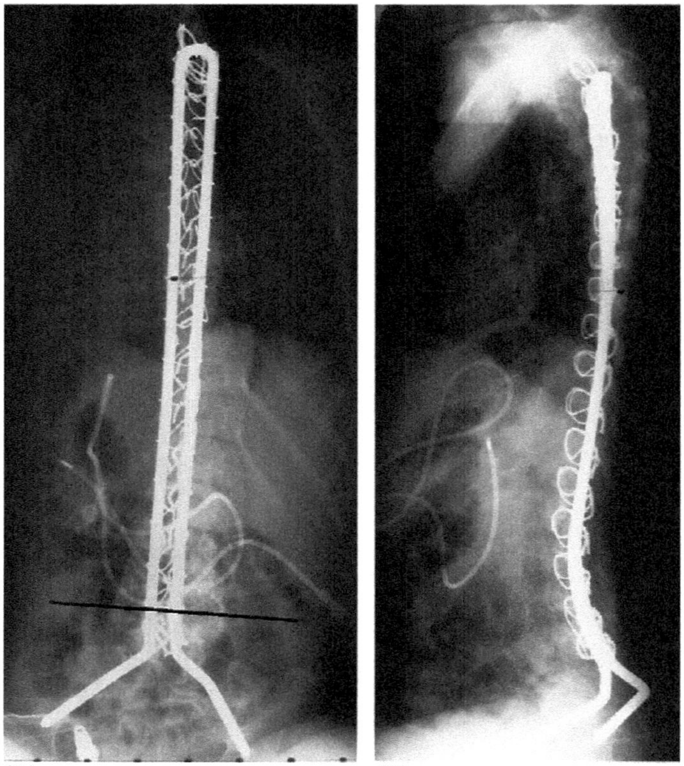

FIG. 4

Anteroposterior (A) and lateral (B) X-ray images of the same patient in Fig. 2 after correction of the spinal deformity by means of the Luque-Galveston technique and sublaminar wires.

Reproduced with permission from Tsirikos, A.I., Spielmann, P., 2007. Spinal deformity in paediatric patients with cerebral palsy. Curr. Orthopaed. 21(2), 122–134.

(Modi et al., 2010). Other reported major complications include respiratory failure, lung collapse, postoperative pneumonia, cardiac arrhythmias, and wound infections (Archer et al., 2016; Ramirez et al., 1997). Instrumentation-related complications such as pseudoarthrosis and cutout of the sublaminar wires are not common but have been reported as well (Tsirikos et al., 2008). Iliac loosening of the rods after fixation with the Luque-Galveston technique was found in approximately 30% of the patients, but with only a minor clinical impact and no need for re-operation (Nectoux et al., 2010). However, despite these complications and the burden for the patient as well as for the caregivers because of the complexity of the surgeries, a high degree of satisfaction has been reported (Jones et al., 2003; Tsirikos et al., 2004), thus testifying that the surgical correction is effective in improving the patients' quality of life. A key factor in satisfaction is improvement in the shape of the trunk, which facilitates sitting and other daily activities (Koop, 2009).

PARKINSON'S DISEASE

PD is characterized by a stooped posture in which the hips and knees are flexed and the head and trunk tend to be bent forward (Doherty et al., 2011). As the disease progresses, the posture's alterations can result in severe spinal deformities in the sagittal plane. A lateral bending of the trunk is also frequent, as well as true scoliosis (Fig. 5). The age of onset for PD is highly variable (Diederich et al., 2003);

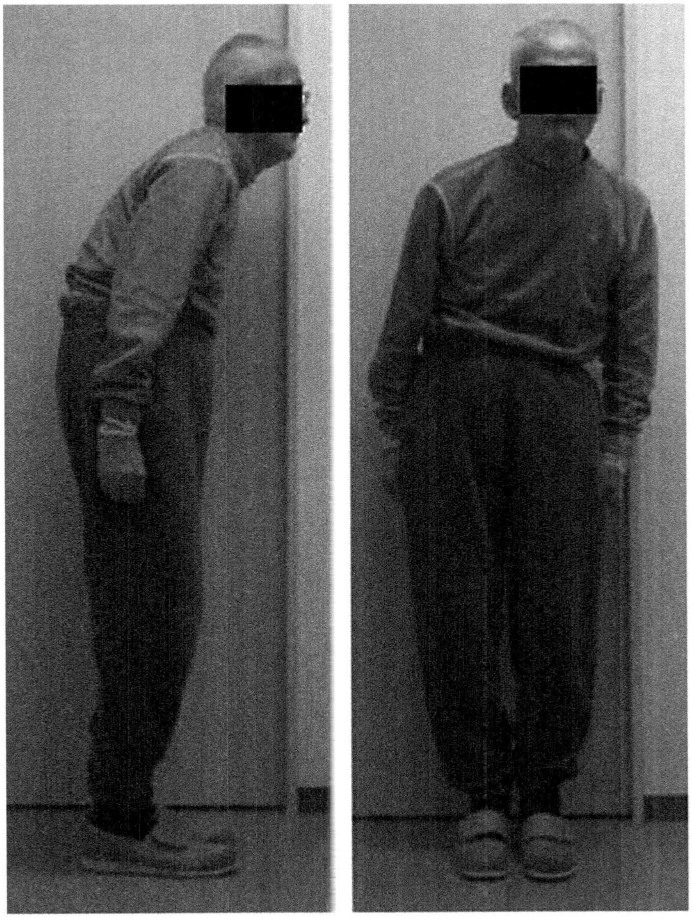

FIG. 5

Typical stooped posture of a patient suffering from Parkinson's disease, with lateral deviation of the trunk (Pisa syndrome).

Reproduced with permission from Furusawa, Y., Hanakawa, T., Mukai, Y., Aihara, Y., Taminato, T., Iawata, Y., Takei, T., Sakamoto, T., Murata, M., 2015. Mechanism of camptocormia in Parkinson's disease analyzed by tilt table-EMG recording. Parkinsonism Relat. Disord. 21(7), 765–770.

most patients are middle age or older when they start showing signs of PD, and the concomitant presence of degenerative signs in the spine is very frequent. The use of spine surgery aimed at correcting deformities as well as alleviating the symptoms of degenerative disorders such as spinal stenosis has become relatively common in recent years, but high rates of complications and re-operations are still reported in the literature.

The progression of PD is graded in stages that describe the symptoms and functional status of the patient with specific regard to autonomous capabilities. A widely used scale is the one published by Hoehn and Yahr (1967), which considers five stages from minimal or no functional disability to confinement to bed or wheelchair unless aided. The Unified Parkinson's Disease Rating Scale (UPDRS) includes the Hoehn and Yahr scale as a subsection and also covers behavior and mood, motor function, capability to carry out daily activities, complications, and evaluation of the autonomy of the patient (Goetz et al., 2008). None of these scales includes an evaluation of the posture or of the possible associated spinal deformities, but they take into account their effect on limiting the functional status of the patient.

SAGITTAL ALIGNMENT

As previously mentioned, patients with PD tend to show a forward bending of the thoracolumbar spine, which was found to be correlated with the stage of the disease (Watanabe et al., 2017; Oh et al., 2014). The sagittal vertical axis (SVA) increases with a progressing Hoehn and Yahr stage, and a kyphotic curvature of the lumbar spine can be observed in most patients (Watanabe et al., 2017; Oh et al., 2014). The thoracic kyphosis is also significantly higher than in age-matched non-PD subjects, and increases as the disease progresses. Advanced age and female gender are considered to be risk factors for forward bending (Oh et al., 2014).

Following the growing interest of pelvis morphology and orientation, radiographic investigations of the spinopelvic parameters were also carried out on PD patients. The available literature on this topic is still limited and rather contradictory, but evidence of a higher prevalence of spinopelvic disorders in these patients with respect to age-matched nonpathological subjects seems to have emerged (Watanabe et al., 2017; Oh et al., 2014; Bissolotti et al., 2014). Some authors have suggested that patients suffering from PD may have less ability to use pelvic retroversion as a compensatory mechanism to counterbalance the forward bending of the spine (Watanabe et al., 2017; Oh et al., 2014), whereas other studies reported that several PD patients had a rather low sacral slope with respect to their pelvic incidence, thus showing an ability to retrovert the pelvis to compensate for the sagittal imbalance (Bissolotti et al., 2014, 2017). Further research is needed to clarify these findings.

Camptocormia is a large flexion of the trunk observable in the standing posture, which is corrected mostly or completely if the patients lies supine or prone (Djaldetti et al., 1999; Tiple et al., 2009). Its prevalence in PD patients is around 7% (Tiple et al., 2009). The standardized diagnostic criterion defines camptocormia as a marked flexion originating in the thoracolumbar spine (minimum 45 degrees) that can be almost completely resolved in the supine position (Doherty et al., 2011). In comparison with degenerative sagittal imbalance, camptocormia is therefore substantially more flexible. In fact, most of the literature about PD does not employ the standard nomenclature used in orthopedics to describe the sagittal alignment because the diagnostic workup does not usually include the radiographic examination; therefore a direct comparison of the data gathered from orthopedic patients is not always straightforward.

Camptocormia is associated with an advanced age and severity of the PD symptoms (Djaldetti et al., 1999; Azher and Jankovic, 2005; Bloch et al., 2006; Spuler et al., 2010) and does not respond well to L-DOPA (Azher and Jankovic, 2005; Bloch et al., 2006). Based on this observation and on clear differences in the symptoms in comparison with PD patients not exhibiting camptocormia, some authors suggested that this pathology is likely to represent a specific form of PD (Bloch et al., 2006).

CORONAL ALIGNMENT

Many PD patients show a bending of the trunk to one side, which is referred to as the Pisa syndrome (Doherty et al., 2011). Its diagnostic definition describes a lateral flexion of at least 10 degrees that can be resolved completely by passive mobilization or in the supine position. Therefore, because there is no structural curve as well as axial rotation of the vertebrae, the Pisa syndrome should not be considered a true spinal deformity.

Scoliosis is a rather common finding in patients with PD, with a higher prevalence than in the general elderly population (Duvoisin and Marsden, 1975). Indeed, although epidemiological studies are rare, Baik and colleagues found scoliosis in more than 30% of the examined PD subjects. On the other hand, scoliosis was found in approximately 6%–30% in the elderly, non–PD population (Baik et al., 2009; Robin et al., 1982; Vanderpool et al., 1969). The Cobb angle of the scoliotic curve is associated with the severity of the PD symptoms, and a correlation between the curve progression and worsening of the pathology has been suggested (Choi et al., 2015). Scoliosis is more prevalent in female PD patients and does not respond well to L-DOPA treatment (Baik et al., 2009; Choi et al., 2015). Although many patients have exhibited concomitant sagittal misalignment and scoliosis, an association between the two could not be determined (Choi et al., 2015; Bissolotti et al., 2016).

An interesting aspect of scoliosis relates to a possible association between the laterality of the PD symptoms and the direction of the scoliosis. Early literature based on case reports showed a contralateral scoliosis with respect to the symptoms (Duvoisin and Marsden, 1975; Martin, 1965), whereas subsequent works conducted on human patients showed contrasting findings (Baik et al., 2009; Choi et al., 2015; Grimes et al., 1987; Di Matteo et al., 2011). Studies conducted on dogs (Martin, 1965) and rodents (Herrera-Marschitz et al., 1990; Lundblad et al., 2004) seemed to confirm an association between the laterality of PD symptoms and scoliosis, but in general the available results do not allow drawing a conclusion in this matter.

SURGICAL TREATMENT

Spine surgery is frequently performed on PD patients to treat degenerative disorders such as stenosis of the spinal canal and segmental instability, often in association with short fixation and fusion, with generally good clinical results (Koller et al., 2010). In recent years, the availability of novel surgical techniques and implantable devices have opened the way to the correction of the spinal deformities, in both the sagittal and the coronal planes, associated with PD, including severe deformities such as camptocormia (Fig. 6). Available data about such surgeries, however, are limited, and there is still no strong consensus about the techniques that should be used and the indications for surgery.

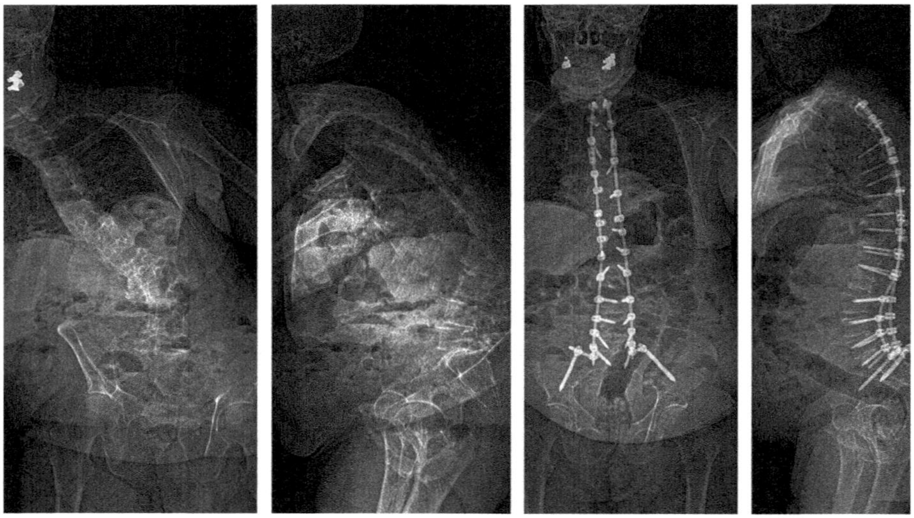

FIG. 6

Preoperative (left) and postoperative (right) X-ray images of a PD patient suffering from both sagittal and coronal imbalance, which was corrected by means of posterior instrumentation from T3 to the pelvis, including a pedicle subtraction osteotomy in L2 to reconstruct the lumbar lordosis.

In general, the correction of spinal deformities in PD patients suffers from a number of complications such as instrumentation loosening and loss of correction with progression of the deformity (Sapkas et al., 2014; Babat et al., 2004; Bouyer et al., 2017) (Fig. 7). The techniques used include fixation with pedicle screws and hooks, sacroiliac fixation with screws (Bouyer et al., 2017), as well as corpectomies and osteotomies in the case of stiff deformities (Koller et al., 2010; Bouyer et al., 2017). In all available studies, the rates of complications and re-operations were found to be substantial (Babat et al., 2004; Kaspar et al., 2005; Bourghli et al., 2012; Moon et al., 2012). Nevertheless, the majority of the patients reported positive satisfaction (Bouyer et al., 2017; Kaspar et al., 2005; Bourghli et al., 2012), thus indicating that the surgical treatment has the potential to improve the quality of life considerably. An aspect that was indicated as critical to achieving a good clinical outcome and limiting the complication rates was the restoration of a correct sagittal balance, including that of the pelvis, with a focus on the lumbar lordosis and the global alignment (Koller et al., 2010). Indeed, because PD patients might not have sufficient ability to activate compensatory mechanisms due to the muscular disorders, a correct surgical restoration has been hypothesized as a way to reduce muscle fatigue and therefore improve the perception of stability.

A factor implicated in the high risk of complications is the generally poor bone quality of patients with PD, which has been reported by several authors (Koller et al., 2010; Babat et al., 2004; Bouyer et al., 2017; Moon et al., 2012). Patients suffering from PD have a lower bone mineral density in comparison with the age-matched non-PD population (Torsney et al., 2014). In addition to a vitamin D deficiency secondary to hyperparathyroidism and reduced exposure to sunlight (Torsney et al., 2014), studies showed that L-DOPA might have a negative impact on the bone mineral density,

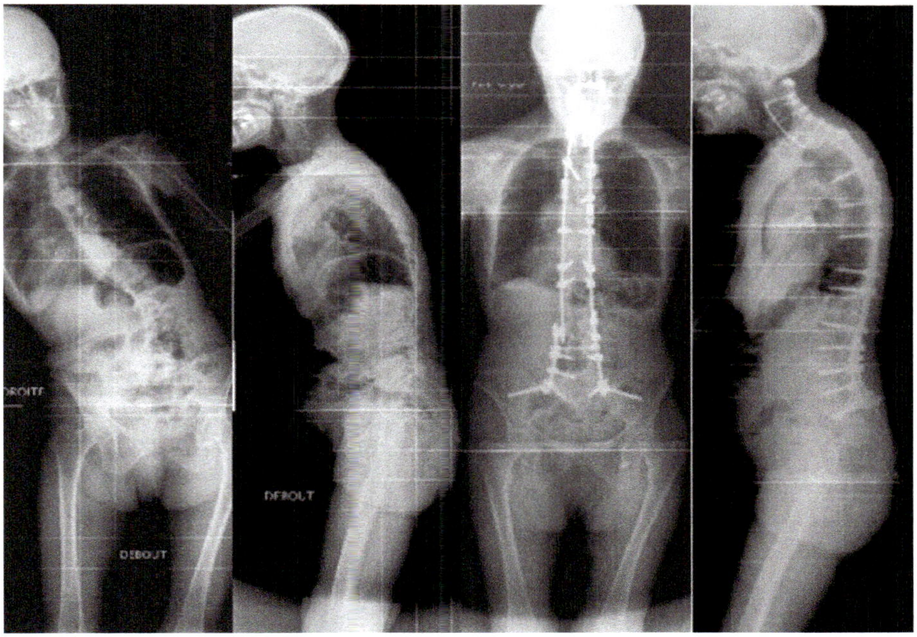

FIG. 7

Anteroposterior and lateral X-ray images before the surgical deformity correction (left) and after surgery and three re-operations for lumbosacral pseudoarthrosis and proximal junctional kyphosis in a patient suffering from Parkinson's disease. The high fusion length, from the cervical spine to the pelvis, should be noted.

Reproduced with permission from Bouyer, B., Scemama, C., Roussouly, P., Laouissat, F., Obeid, I., Boissière, L., Parent, F., Schuller, S., Steib, J., Pascal-Moussellard, H., 2017. Evolution and complications after surgery for spine deformation in patients with Parkinson's disease. Orthopaed. Traumatol. Surg. Res.

and the effect might be dose-dependent (Vestergaard et al., 2007). A higher fracture risk was also reported (Torsney et al., 2014), but this could be partially attributed to the higher risk of falls due to postural instability and motor disorders.

Other factors that are potentially responsible for complications have been identified. Sarkiss and colleagues found that a poor clinical outcome was associated with being older, having a high level of camptocormia, and having disorders of the hip joint (Sarkiss et al., 2015). Negative factors related to the surgical techniques were postoperative forward bending of the trunk with SVA higher than 5 cm, insufficient sacropelvic fixation, and inadequate selection of the fusion levels. In fact, typical features of PD, such as disorders of the muscle tone, inability to maintain a correct posture and to activate compensatory mechanisms, as well as poor bone quality, challenge even the most recent surgical techniques and devices; and further research is needed in order to improve surgical outcomes. The high rate of satisfaction reported by patients, even with the current techniques, should stimulate studies directed at gaining a better understanding of the biomechanics of spinal disorders related to PD, with the final aim of improving the quality of life of patients.

REFERENCES

Archer, J.E., Gardner, A.C., Roper, H.P., Chikermane, A.A., Tatman, A.J., 2016. Duchenne muscular dystrophy: the management of scoliosis. J. Spine Surg. 2 (3), 185.

Arun, R., Srinivas, S., Mehdian, S., 2010. Scoliosis in Duchenne's muscular dystrophy: a changing trend in surgical management. Eur. Spine J. 19 (3), 376–383.

Azher, S.N., Jankovic, J., 2005. Camptocormia: pathogenesis, classification, and response to therapy. Neurology 65 (3), 355–359.

Babat, L.B., McLain, R.F., Bingaman, W., Kalfas, I., Young, P., Rufo-Smith, C., 2004. Spinal surgery in patients with Parkinson's disease: construct failure and progressive deformity. Spine 29 (18), 2006–2012.

Baik, J.S., Kim, J.Y., Park, J.H., Han, S.W., Park, J.H., Lee, M.S., 2009. Scoliosis in patients with Parkinson's disease. J. Clin. Neurol. 5 (2), 91–94.

Bell, D.F., Moseley, C.F., Koreska, J., 1989. Unit rod segmental spinal instrumentation in the management of patients with progressive neuromuscular spinal deformity. Spine 14 (12), 1301–1307.

Bissolotti, L., Gobbo, M., Villafane, J.H., Negrini, S., 2014. Spinopelvic balance: new biomechanical insights with clinical implications for Parkinson's disease. Eur. Spine J. 23 (3), 576–583.

Bissolotti, L., Donzelli, S., Gobbo, M., Zaina, F., Villafane, J.H., Negrini, S., 2016. Association between sagittal balance and scoliosis in patients with Parkinson disease: a cross-sectional study. Am. J. Phys. Med. Rehabil. 95 (1), 39–46.

Bissolotti, L., Berjano, P., Zuccher, P., Zenorini, A., Buraschi, R., Villafane, J.H., Negrini, S., 2017. Sagittal balance is correlated with Parkinson's Disease clinical parameters: an overview of spinopelvic alignment on 175 consecutive cases. Eur. Spine J. 26 (Suppl. 4), 471–478.

Bloch, F., Houeto, J.L., Tezenas du Montcel, S., Bonneville, F., Etchepare, F., Welter, M.L., Rivaud-Pechoux, S., Hahn-Barma, V., Maisonobe, T., Behar, C., Lazennec, J.Y., Kurys, E., Arnulf, I., Bonnet, A.M., Agid, Y., 2006. Parkinson's disease with camptocormia. J. Neurol. Neurosurg. Psychiatry 77 (11), 1223–1228.

Boos, N., Aebi, M., 2008. Spinal Disorders—Fundamentals of Diagnosis and Treatment. Springer, Heidelberg.

Bourghli, A., Guerin, P., Vital, J.M., Aurouer, N., Luc, S., Gille, O., Pointillart, V., Obeid, I., 2012. Posterior spinal fusion from T2 to the sacrum for the management of major deformities in patients with Parkinson disease: a retrospective review with analysis of complications. J. Spinal Disord. Tech. 25 (3), E53–60.

Bouyer, B., Scemama, C., Roussouly, P., Laouissat, F., Obeid, I., Boissière, L., Parent, F., Schuller, S., Steib, J., Pascal-Moussellard, H., 2017. Evolution and complications after surgery for spine deformation in patients with Parkinson's disease. Orthop. Traumatol. Surg. Res. 103 (4), 517–522.

Bulman, W., Dormans, J.P., Ecker, M.L., Drummond, D.S., 1996. Posterior spinal fusion for scoliosis in patients with cerebral palsy: a comparison of Luque rod and unit rod instrumentation. J. Pediatr. Orthop. 16 (3), 314–323.

Choi, H.J., Smith, J.S., Shaffrey, C.I., Lafage, V.C., Schwab, F.J., Ames, C.P., Matsumoto, M., Baik, J.S., Ha, Y., 2015. Coronal plane spinal malalignment and Parkinson's disease: prevalence and associations with disease severity. Spine J. 15 (1), 115–121.

Cloake, T., Gardner, A., 2016. The management of scoliosis in children with cerebral palsy: a review. J. Spine Surg. 2 (4), 299.

Di Matteo, A., Fasano, A., Squintani, G., Ricciardi, L., Bovi, T., Fiaschi, A., Barone, P., Tinazzi, M., 2011. Lateral trunk flexion in Parkinson's disease: EMG features disclose two different underlying pathophysiological mechanisms. J. Neurol. 258 (5), 740–745.

Dias, R.C., Miller, F., Dabney, K., Lipton, G., Temple, T., 1996. Surgical correction of spinal deformity using a unit Rodin children with cerebral palsy. J. Pediatr. Orthop. 16 (6), 734–740.

Diederich, N.J., Moore, C.G., Leurgans, S.E., Chmura, T.A., Goetz, C.G., 2003. Parkinson disease with old-age onset: a comparative study with subjects with middle-age onset. Arch. Neurol. 60 (4), 529–533.

Djaldetti, R., Mosberg-Galili, R., Sroka, H., Merims D., Melamed, E., 1999. Camptocormia (bent spine) in patients with Parkinson's disease—characterization and possible pathogenesis of an unusual phenomenon. Mov. Disord. 14 (3), 443–447.

Doherty, K.M., van de Warrenburg, Bart, P., Peralta, M.C., Silveira-Moriyama, L., Azulay, J., Gershanik, O.S., Bloem, B.R., 2011. Postural deformities in Parkinson's disease. Lancet Neurol. 10 (6), 538–549.

Duvoisin, R.C., Marsden, C.D., 1975. Note on the scoliosis of parkinsonism. J. Neurol. Neurosurg. Psychiatry 38 (8), 787–793.

Gau, Y., Lonstein, J.E., Winter, R.B., Koop, S., Denis, F., 1991. Luque-Galveston procedure for correction and stabilization of neuromuscular scoliosis and pelvic obliquity: a review of 68 patients. Clin. Spine Surg. 4 (4), 399–410.

Goetz, C.G., Tilley, B.C., Shaftman, S.R., Stebbins, G.T., Fahn, S., Martinez-Martin, P., Poewe, W., Sampaio, C., Stern, M.B., Dodel, R., 2008. Movement disorder society-sponsored revision of the unified Parkinson's disease rating scale (MDS-UPDRS): scale presentation and clinimetric testing results. Mov. Disord. 23 (15), 2129–2170.

Griffiths, G., Evans, K., Roberts, G., Lloyd, K., 1977. The radiology of the hip joints and pelvis in cerebral palsy. Clin. Radiol. 28 (2), 187–191.

Grimes, J.D., Hassan, M.N., Thakar, J., 1987. Antioxidant therapy in Parkinson's disease. Can. J. Neurol. Sci. 14 (S3), 483–487.

Haas, S., 1942. Spastic scoliosis and obliquity of the pelvis. JBJS 24 (4), 774–780.

Harrington, P.R., 1988. The history and development of Harrington instrumentation. Clin. Orthop. Relat. Res. (227), 3–5.

Hart, D.A., McDonald, C.M., 1998. Spinal deformity in progressive neuromuscular disease. Natural history and management. Phys. Med. Rehabil. Clin. N. Am. 9 (1), 213–232 (viii).

Herrera-Marschitz, M., Utsumi, H., Ungerstedt, U., 1990. Scoliosis in rats with experimentally-induced hemiparkinsonism: dependence upon striatal dopamine denervation. J. Neurol. Neurosurg. Psychiatry 53 (1), 39–43.

Hoehn, M.M., Yahr, M.D., 1967. Parkinsonism: onset, progression and mortality. Neurology 17 (5), 427–442.

Jones, K.B., Sponseller, P.D., Shindle, M.K., McCarthy, M.L., 2003. Longitudinal parental perceptions of spinal fusion for neuromuscular spine deformity in patients with totally involved cerebral palsy. J. Pediatr. Orthop. 23 (2), 143–149.

Kaspar, S., Riley, L., Cohen, D., Long, D., Kostuik, J., Hassanzadeh, H., 2005. In: Spine surgery in Parkinsons. Orthopaedic Proceedings, p. 292.

Koller, H., Acosta, F., Zenner, J., Ferraris, L., Hitzl, W. Meier O., Ondra, S., Koski, T., Schmidt, R., 2010. Spinal surgery in patients with Parkinson's disease: experiences with the challenges posed by sagittal imbalance and the Parkinson's spine. Eur. Spine J. 19 (10), 1785–1794.

Koop, S.E., 2009. Scoliosis in cerebral palsy. Dev. Med. Child Neurol. 51 (s4), 92–98.

Kraker, J., 2011. Autoimmune neuromuscular disorders. Curr. Neuropharmacol. 9 (3), 400–408.

Lipton, G.E., Miller, F., Dabney, K.W., Altiok, H., Bachrach, S.J., 1999. Factors predicting postoperative complications following spinal fusions in children with cerebral palsy. Clin. Spine Surg. 12 (3), 197–205.

Lonstein, J.E., Akbarnia, A., 1983. Operative treatment of spinal deformities in patients with cerebral palsy or mental retardation. An analysis of one hundred and seven cases. JBJS 65 (1), 43–55.

Lundblad, M., Picconi, B., Lindgren, H., Cenci, M., 2004. A model of L-DOPA-induced dyskinesia in 6-hydroxydopamine lesioned mice: relation to motor and cellular parameters of nigrostriatal function. Neurobiol. Dis. 16 (1), 110–123.

Luque, E.R., 1982. Segmental spinal instrumentation for correction of scoliosis. Clin. Orthop. Relat. Res. (163), 192–198.

Marchesi, D., Arlet, V., Stricker, U., Aebi, M., 1997. Modification of the original Luque technique in the treatment of Duchenne's neuromuscular scoliosis. J. Pediatr. Orthop 17 (6), 743–749.

Martin, J.P., 1965. Curvature of the spine in post-encephalitic parkinsonism. J. Neurol. Neurosurg. Psychiatry 28 (5), 395–400.

McDonald, C.M., Abresch, R.T., Carter, G.T., Fowler Jr., W.M., Johnson, E.R., Kilmer, D.D., 1995. Profiles of neuromuscular diseases. Becker's muscular dystrophy. Am. J. Phys. Med. Rehabil. 74 (5 Suppl), S93–103.

Miller, A., Temple, T., Miller, F., 1996. Impact of orthoses on the rate of scoliosis progression in children with cerebral palsy. J. Pediatr. Orthop. 16 (3), 332–335.

Modi, H.N., Suh, S., Hong, J., Cho, J., Park, J., Yang, J., 2010. Treatment and complications in flaccid neuromuscular scoliosis (Duchenne muscular dystrophy and spinal muscular atrophy) with posterior-only pedicle screw instrumentation. Eur. Spine J. 19 (3), 384–393.

Moon, S.H., Lee, H.M., Chun, H.J., Kang, K.T., Kim, H.S., Park, J.O., Moon, E.S., Chong, H.S., Sohn, J.S., Kim, H.J., 2012. Surgical outcome of lumbar fusion surgery in patients with Parkinson disease. J. Spinal Disord. Tech. 25 (7), 351–355.

Nectoux, E., Giacomelli, M., Karger, C., Herbaux, B., Clavert, J., 2010. Complications of the Luque–Galveston scoliosis correction technique in paediatric cerebral palsy. Orthop. Traumatol. Surg. Res. 96 (4), 354–361.

Oda, T., Shimizu, N., Yonenobu, K., Ono, K., Nabeshima, T., Kyoh, S., 1993. Longitudinal study of spinal deformity in Duchenne muscular dystrophy. J. Pediatr. Orthop. 13 (4), 478–488.

Oda, Y., Takigawa, T., Sugimoto, Y., Tanaka, M., Akazawa, H., Ozaki, T., 2017. Scoliosis in patients with severe cerebral palsy: three different courses in adolescents. Acta Med. Okayama 71 (2), 119–126.

Oh, J.K., Smith, J.S., Shaffrey, C.I., Lafage, V., Schwab, F., Ames, C.P., Matsumoto, M., Baik, J.S., Ha, Y., 2014. Sagittal spinopelvic malalignment in Parkinson disease: prevalence and associations with disease severity. Spine 39 (14), E833–41.

Ramirez, N., Richards, B.S., Warren, P.D., Williams, G.R., 1997. Complications after posterior spinal fusion in Duchenne's muscular dystrophy. J. Pediatr. Orthop. 17 (1), 109–114.

Robin, G.C., Span, Y., Steinberg, R., Makin, M., Menczel, J., 1982. Scoliosis in the elderly: a follow-up study. Spine 7 (4), 355–359.

Saito, N., Ebara, S., Ohotsuka, K., Kumeta, H., Takaoka, K., 1998. Natural history of scoliosis in spastic cerebral palsy. Lancet 351 (9117), 1687–1692.

Sapkas, G., Lykomitros, V., Soultanis, K., Papadopoulos, E.C., Papadakis, M., 2014. Spinal surgery in patients with Parkinson's disease: unsatisfactory results, failure and disappointment. Open Orthop. J. 8, 264–267.

Sarkiss, C.A., Fogg, G.A., Skovrlj, B., Cho, S.K., Caridi, J.M., 2015. To operate or not? A literature review of surgical outcomes in 95 patients with Parkinson's disease undergoing spine surgery. Clin. Neurol. Neurosurg. 134, 122–125.

Shapiro, G.S., Taira, G., Boachie-Adjei, O., 2003. Results of surgical treatment of adult idiopathic scoliosis with low back pain and spinal stenosis: a study of long-term clinical radiographic outcomes. Spine 28 (4), 358–363.

Sharrard, W.J., 1968. Spinal osteotomy for congenital kyphosis in myelomeningocele. J. Bone Joint Surg. 50 (3), 466–471.

Spuler, S., Krug, H., Klein, C., Medialdea, I.C., Jakob, W., Ebersbach, G., Gruber, D., Hoffmann, K., Trottenberg, T., Kupsch, A., 2010. Myopathy causing camptocormia in idiopathic Parkinson's disease: a multidisciplinary approach. Mov. Disord. 25 (5), 552–559.

Thometz, J., Simon, S., 1988. Progression of scoliosis after skeletal maturity in institutionalized adults who have cerebral palsy. JBJS 70 (9), 1290–1296.

Tiple, D., Fabbrini, G., Colosimo, C., Ottaviani, D., Camerota, F., Defazio, G., Berardelli, A., 2009. Camptocormia in Parkinson disease: an epidemiological and clinical study. J. Neurol. Neurosurg. Psychiatry 80 (2), 145–148.

Torsney, K.M., Noyce, A.J., Doherty, K.M., Bestwick, J.P., Dobson, R., Lees, A.J., 2014. Bone health in Parkinson's disease: a systematic review and meta-analysis. J. Neurol. Neurosurg. Psychiatry 85 (10), 1159–1166.

Tsirikos, A.I., Spielmann, P., 2007. Spinal deformity in paediatric patients with cerebral palsy. Curr. Orthop. 21 (2), 122–134.

Tsirikos, A.I., Chang, W., Dabney, K.W., Miller, F., 2004. Comparison of parents' and caregivers' satisfaction after spinal fusion in children with cerebral palsy. J. Pediatr. Orthop. 24 (1), 54–58.

Tsirikos, A.I., Lipton, G., Chang, W.N., Dabney, K.W., Miller, F., 2008. Surgical correction of scoliosis in pediatric patients with cerebral palsy using the unit rod instrumentation. Spine 33 (10), 1133–1140.

Vanderpool, D., James, J., Wynne-davies, R., 1969. Scoliosis in the elderly. JBJS 51 (3), 446–455.

Vestergaard, P., Rejnmark, L., Mosekilde, L., 2007. Fracture risk associated with parkinsonism and anti-Parkinson drugs. Calcif. Tissue Int. 81 (3), 153–161.

Watanabe, K., Hirano, T., Katsumi, K., Ohashi, M., Shoji, H., Hasegawa, K., Yamazaki, A., Ishikawa, A., Koike, R., Endo, N., 2017. Characteristics of spinopelvic alignment in Parkinson's disease: comparison with adult spinal deformity. J. Orthop. Sci. 22 (1), 16–21.

Winter, S., 1994. Preoperative assessment of the child with neuromuscular scoliosis. Orthop. Clin. N. Am. 25 (2), 239–245.

Winter, R.B., Pinto, W.C., 1986. Pelvic obliquity: its causes and its treatment. Spine 11 (3), 225–234.

Yazici, M., Asher, M.A., Hardacker, J.W., 2000. The safety and efficacy of Isola-Galveston instrumentation and arthrodesis in the treatment of neuromuscular spinal deformities. JBJS 82 (4), 524.

FURTHER READING

Furusawa, Y., Hanakawa, T., Mukai, Y., Aihara, Y., Taminato, T., Iawata, Y., Takei, T., Sakamoto, T., Murata, M., 2015. Mechanism of camptocormia in Parkinson's disease analyzed by tilt table-EMG recording. Parkinsonism Relat. Disord. 21 (7), 765–770.

Maitra, S., Roberto, R.F., McDonald, C.M., Gupta, M.C., 2012. Treatment of spine deformity in neuromuscular diseases. Phys. Med. Rehabil. Clin. N. Am. 23 (4), 869–883.

CHAPTER

SAGITTAL IMBALANCE

21

Claudia Ottardi*, Andrea Luca†, Fabio Galbusera†

Politecnico di Milano, Milan, Italy IRCCS Galeazzi Orthopedic Institute, Milan, Italy†*

THE GLOBAL ALIGNMENT AND THE SPINO-PELVIC PARAMETERS

In healthy individuals, the spine is straight in the coronal plane, whereas it exhibits an S-shape in the sagittal plane. The combination the regional curvatures allows keeping the standing posture with minimal energy consumption (Le Huec et al., 2011b). Indeed, the vertical projection of the center of gravity of the trunk passes through the sacrum; the curved sagittal profile of the spine contributes to this equilibrium condition by avoiding buckling under gravity load. The basis of the cervical spine, that is, the C7 vertebra, is used clinically to measure a parameter to describe the global spine alignment, the sagittal vertical axis (SVA), defined as the distance in the horizontal direction between the plumb line conducted from the center of C7 to the posterosuperior corner of the sacral endplate (Fig. 1) (Jackson and McManus, 1994). In an ideal spinal alignment, SVA should be very low; a positive value (i.e., the plumb line more anterior than the sacral landmark) of 5 cm has been suggested as a threshold for spinal misalignment (Jackson and McManus, 1994). SVA is commonly used in the clinical setting to easily describe the spinal alignment with a single value however, it should be noted that SVA is dependent on the posture, and compensatory mechanisms may reduce its value even in subjects with alignment disorders.

In recent years, an approach emerged to characterize the global alignment of the spine starting from anatomical parameters that describe the pelvis emerged, and the approach has gained a general consensus. The key element of the approach is the pelvic incidence (PI) (Fig. 2), which is the angle measured in a sagittal projection between the line connecting the center of the femoral head to the center of the S1 endplate and the line orthogonal to the same endplate (Duval-Beaupere et al., 1992). This parameter is by definition purely morphological, because it does not vary when the hip joints are flexed or extended and is therefore commonly used to concisely describe the shape of the pelvis of each patient (Legaye et al., 1998). It should, however, be noted that several studies showed that the value of this parameter is not constant throughout life, but changes slightly during infancy and adolescence (Mangione et al., 1997; Mac-Thiong et al., 2004). Studies showed that the PI ranges between 35 degree and 85 degree and has an average value in asymptomatic adult subjects of 52 degree (Vialle et al., 2005).

PI is normally complemented by two other angles, the sacral slope (SS) and the pelvic tilt (PT), which are not morphological parameters but depend on the pelvis orientation. SS is defined as the angle between the sagittal projection of the S1 endplate and the horizontal line, whereas PT is defined as the angle between the vertical line and the line connecting the center of the S1 endplate and the center of the femoral heads. The values of the three parameters are interdependent:

Biomechanics of the Spine. https://doi.org/10.1016/B978-0-12-812851-0.00021-5

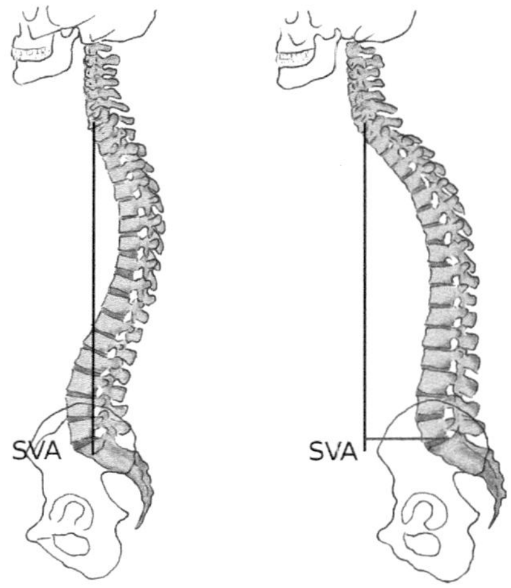

FIG. 1

Sagittal vertical axis (SVA) in a healthy subject *(left)* and in an imbalanced subject *(right)*. SVA is defined as the distance between the posterior margin of the S1 endplate and the C7 plumb line.

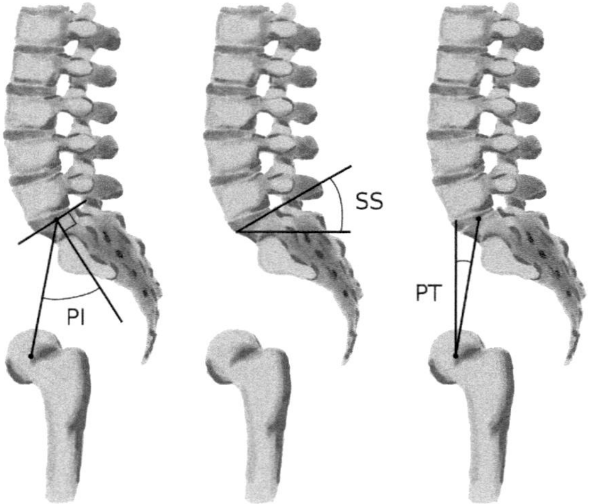

FIG. 2

Geometrical definitions of the pelvic incidence (PI), sacral slope (SS), and pelvic tilt (PT).

$$PI = PT + SS$$

Indeed, if the pelvis is retroverted, SS decreases and PT increases, and their sum remains constant and dependent on the pelvic morphology.

There is a general consensus that the value of PI is one of the main determinants of the global spine alignment (Legaye et al., 1998; Vialle et al., 2005). Arguably, patients with a high PI will also tend to have a high SS and PT; indeed, it has been shown that SS increases more than PT, and a high PI will therefore dictate a higher slope of the sacral endplate (Vialle et al., 2005). In order to keep a correct global alignment of the head over the sacrum (i.e , a low SVA), a pronounced lumbar lordosis will be necessary. The chain of correlation has also been shown to propagate to the thoracic and cervical regions, even with lower, or lack of, statistical significance (Ames et al., 2013b).

Roussouly and colleagues introduced a classification system for the alignment of the lumbosacral spine that described its normal variability in the adult asymptomatic population and that is commonly used today (Roussouly et al., 2005). The system is based on four radiological parameters: (i) SS, (ii) the positions of the apex of the lumbar lordosis, (iii) the inflection point between kyphosis and lordosis, and (iv) the lordosis tilt angle (defined between the vertical line and that connecting the inflection point and the anterior corner of the sacral endplate). The authors defined four categories to classify the spines of 160 asymptomatic volunteers (Fig. 3). They found that in type 1, a low SS as well as a short lordosis (low inflection point) determined a kyphotic curvature in the proximal lumbar spine, which propagated to the thoracic region. Subjects showing a type 2 alignment also had a low SS, but combined with a longer, flatter lordosis that generally gave a hypolordotic and hypokyphotic appearance to the spine. In type 3, SS ranged between 35 degree and 45 degree, and the inflection point reached L1, resulting in a harmonious spinal shape. Type 3 spines were found to be the most common types in the cohort (60 subjects). In type 4, SS was higher than 45 degree, generally because of a high PI. Also, both lumbar lordosis and thoracic kyphosis were pronounced for this type.

The strong correlation between PI and lumbar lordosis led to the widespread use of the former value, which can be easily measured on sagittal X-rays images, to determine the target lordosis to be achieved in the surgical treatment of sagittal spinal misalignment. Indeed, Schwab and colleagues suggested that if the difference between PI and lumbar lordosis (PI-LL) is around 10 degree after surgical correction, a correct alignment is achieved benefiting the clinical outcome (Schwab et al., 2010; Schwab et al., 2013). However, the general validity of such a simple approach has been questioned;

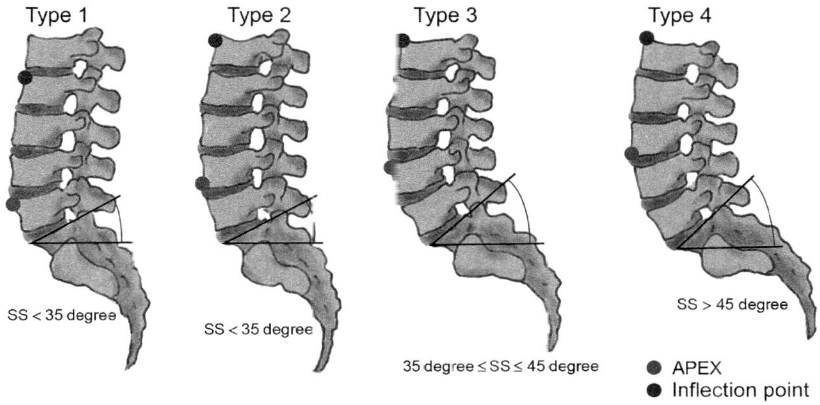

FIG. 3

The Roussouly classification of the physiological spinopelvic organization, based on the sacral slope (SS), the position of the apex and the inflection point, and the lordosis angles. In type 1, a low SS as well as a short lordosis are observable. In type 2, a longer, lower lordosis gives the spine a flatter appearance. In type 3, the spinal shape is harmonious. In type 4, both lumbar lordosis and thoracic kyphosis are pronounced.

patients with a high PI might require a smaller lumbar lordosis, and a pronounced thoracic kyphosis might need to be counterbalanced by a lower lordosis even in the case of a high PI (Schwab et al., 2014a). Nevertheless, it is generally believed that a target value of PI-LL around 10 degree is a good starting point when planning a surgical correction of the sagittal spine profile.

AGING AND PATHOLOGICAL CHANGES IN THE SAGITTAL ALIGNMENT AND COMPENSATORY MECHANISMS

In the course of the aging process, changes that alter the spine alignment may occur, such as degenerative disc disease, which generally leads to a loss of the lumbar lordosis and thoracic hyperkyphosis. Both these changes induce a forward shift of the trunk and a higher activation of the trunk muscles in order to maintain an erect posture. To keep the energy expenditure to a minimum and to reach an ergonomic posture, several compensation mechanisms, which might be active individually or in combination, are actuated to return the body's gravity line close to the feet. If the compensation mechanisms are effective and the patient can reach a satisfactory and nonsymptomatic posture and gait even in presence of loss of lordosis or hyperkyphosis (i.e., a small value of SVA), the spine alignment is categorized as "compensated imbalance" and in most cases does not require surgical treatment (Diebo et al., 2015).

An effective compensation mechanism is pelvic retroversion (Barrey et al., 2011; Barrey et al., 2007; Jackson et al., 1998). By rotating the pelvis backward (i.e., decreasing SS and increasing PT), the gravity line is shifted backward, and SVA is reduced. Indeed, some degrees of pelvic retroversion is observable in most elderly subjects, even asymptomatic ones (Jackson et al., 1998). This mechanism may be less effective in subjects with low PI, because their hip joints are in a more extended position in the standing posture, and therefore their range of motion in extension is more limited (Fig. 4).

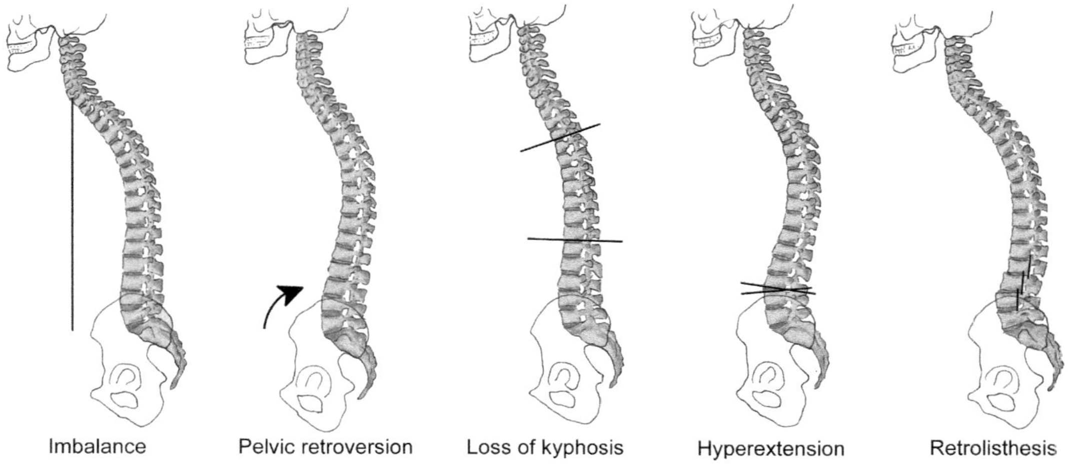

FIG. 4

Degenerative changes may lead to a forward shift of the trunk and a higher activation of the trunk muscles in order to maintain the erect posture (imbalance). Compensation mechanisms such as pelvic retroversion, loss of thoracic kyphosis, segmental lumbar hyperextension, and retrolisthesis may help in regaining an ergonomic posture.

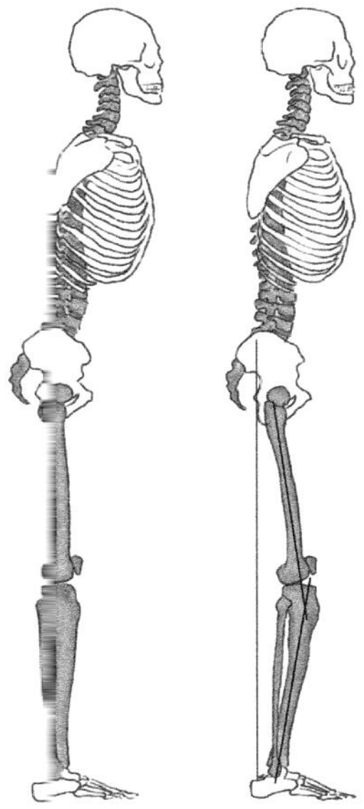

FIG. 5

Knee flexion can act as a compensatory mechanism in imbalanced subjects. The pelvic shift, defined as the distance between the sacrum plumb line and the ankle, is indicated by the arrow and is used to describe the combined action of pelvic retroversion and knee flexion.

As the imbalance increases and pelvic retroversion alone is not able to cope with the misalignment, other compensation mechanisms are recruited. The lumbar spine may increase its lordosis in the non-degenerated segments, which undergo hyperextension and in some cases retrolisthesis (Barrey et al., 2011; Jackson et al., 1998). This mechanism may, however, induce symptoms related to the overload of the posterior elements and stenosis of the spinal canal. In younger subjects with a flexible spine, the thoracic kyphosis may decrease, resulting in a generally flat sagittal profile, similar to the type 2 alignment in the Roussouly classification (Barrey et al., 2011; Barrey et al., 2007).

In severe cases, the lower limbs contribute to the chain of compensation (Fig. 5). Many elderly subjects adopt a standing posture in which the knees are flexed, which allows for a higher range for the realignment of the pelvis and therefore an effective compensation for the loss of lordosis or hyperkyphosis (Obeid et al., 2011). Ankle extension is also commonly observable in such patients. However, maintaining such a posture requires the activation of the quadriceps, gluteus maximus, and medius and is thus tiresome; the gait may be impaired and exhausting as well. The resulting combination of pelvic retroversion and knee flexion is commonly described by a single radiological parameter, the pelvic shift, which represents the position of the pelvis with respect to the feet (Schwab et al., 2006).

It should be noted that sagittal imbalance may also be initiated by a knee degenerative pathology with reduced range of motion in extension, which forces a neutral posture with flexed knees and lumbar flattening as a compensation mechanisms (Murata et al., 2003). However, in elderly patients showing multiple degenerative disorders, it may be difficult to determine which is the primary cause of the misalignment, and planning of the treatment may not be straightforward.

SURGICAL TREATMENT OF SAGITTAL IMBALANCE

Several studies reported a significant correlation between non-compensated sagittal imbalance and a decrease in the quality of life of the affected subject (Schwab et al., 2008; Glassman et al., 2005a; Glassman et al., 2005b). On the other hand, a successful surgical restoration of a correct sagittal alignment was linked to an improvement of patients' health. These studies therefore support the concept of realigning the spine by means of a surgical intervention, as well as of taking into account the sagittal alignment of the spine when surgery is needed for the treatment of other spinal disorders. For example, in planning the surgical correction of scoliosis, a correct sagittal alignment is currently considered of utmost importance (Ilharreborde et al., 2014).

The surgical correction of sagittal imbalance is based on posterior instrumentation of the spine with pedicle screws and rods. In the case of flexible curves, instrumentation alone that is eventually supplemented with anterior or posterior decompression and release is generally sufficient, whereas for stiff deformities, the use of osteotomies needs to be considered (Savage and Patel, 2014). Therefore one of the crucial steps in preoperative planning is to assess the flexibility of the deformity, which can be done by means of supine X-rays with fulcrums or bolsters, as well as with CT or MRI scans. With the latter two imaging methods, the condition of the intervertebral discs and the possible presence of osteophytes or ankylosis, which increase the spinal stiffness, can be assessed (Savage and Patel, 2014).

A precise preoperative plan based on lateral X-rays taken in a standing posture is currently considered mandatory before surgical correction of sagittal imbalance. First, radiographic parameters such as PI, SVA, lumbar lordosis, and thoracic kyphosis are measured on the image, either manually or with the assistance of a computer program (Lafage et al., 2015). These values are used to conduct an evaluation of the standing posture of the patient, including the possible presence of compensatory mechanisms. Because the degenerative loss of lumbar lordosis is a common driver of sagittal imbalance, its evaluation is critical and is usually performed by calculating the $PI-LL$ angle. As previously mentioned, a restoration of $PI-LL$ lower than 10 degree (typically 9 degree) is commonly considered as a desirable target after surgery. Therefore the target LL can be calculated as $LL=PI-9$ (Savage and Patel, 2014; Schwab et al., 2010).

Successful restoration of the sagittal balance should also take into account the compensatory mechanisms and should be aimed at minimizing them (Lafage et al., 2015). Pelvic retroversion is assessed by evaluating PT, which increases when this compensation mechanism is active. A value of PT lower than 20 degree is commonly assumed as a target for the surgery (Schwab et al., 2010). In addition, TK is measured to evaluate a possible loss of kyphosis; knee flexion is also assessed if a whole body X-ray image is available. The majority of spine surgeons accept the general rule that, after a successful surgery, SVA should be lower than 5 cm (Schwab et al., 2010). It should be noted that other, more sophisticated rules about preoperative planning have been presented and are currently in wide use. For example, the full body integrated (FBI) method (Le Huec et al., 2011a) is widely used and well accepted by the medical community.

VERTEBRAL OSTEOTOMIES

The correction of nonflexible deformities is commonly performed by means of vertebral osteotomies. Depending on the patient's age and condition, the type of deformity, and the surgeon's experience, several surgical techniques can be used and different outcomes can be reached. Recently, an anatomical

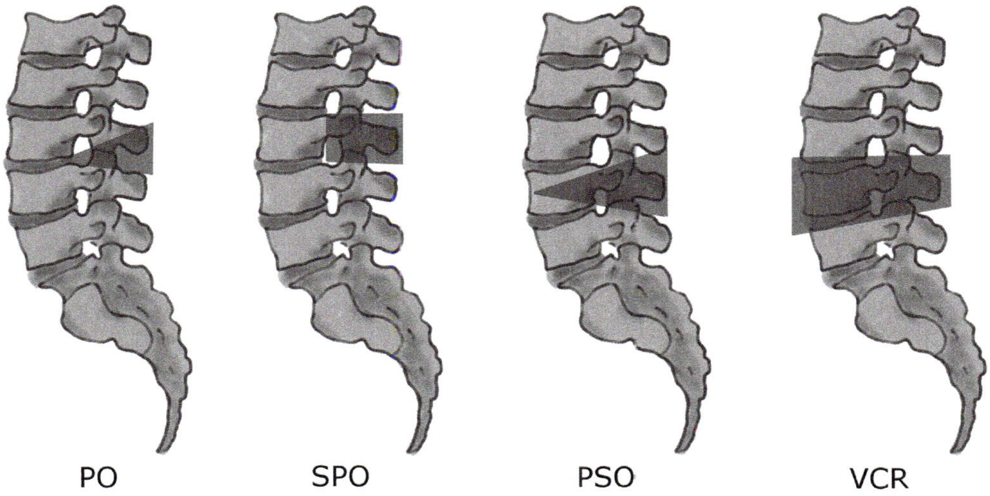

FIG. 6

Schematic representation of the most common osteotomies. From left to right: Ponte osteotomy (PO), Smith-Petersen osteotomy (SPO), pedicle subtraction osteotomy (PSO), and vertebral column resection (VCR).

classification of spinal osteotomies was proposed in order to standardize the description of the surgical interventions (Schwab et al., 2014b). Six grades of osteotomies were identified, which corresponded to different resected regions and to increasing destabilization of the spine: (1) partial and (2) complete facet joint removal; (3) partial and (4) complete pedicle and body resection (in some cases involving also the disc); (5) complete resection of one vertebra and the disc; and (6) removal of multiple vertebrae and discs. However, the surgical techniques most commonly employed are (Schwab et al., 2014b; Bergin et al., 2010; Bridwell, 2006) (Fig. 6) as follows:

- Ponte osteotomy (PO): Originally developed in 1984 to treat Scheuermann kyphosis (Ponte et al., 1984), this technique is performed by creating a posterior closing wedge on the articular processes of adjacent vertebrae, leading to a 5- to 15-degree correction. It is commonly performed on multiple levels.
- Smith-Petersen osteotomy (SPO): First described in 1969 (Smith-Petersen et al., 1969) and used to treat ankylosis spondylitis or Scheuermann kyphosis, this technique involves both the anterior and posterior columns. It allows for a correction up to 15 degree by shortening the posterior column and opening the anterior one.
- Pedicle subtraction osteotomy (PSO): First described in 1985 (Thomasen, 1985), this technique is now the most commonly used one for the treatment of fixed sagittal imbalance. It is a posterior closing wedge technique that achieves higher degree of correction in a single level than the other techniques achieve (up to 35 degree more). A single PSO can be an alternative to two or three SPOs.
- Vertebral column resection (VCR): This technique provides a very high degree of correction (up to 60 degree) by complete removal of one vertebra and the discs above and below it (Lenke et al., 2010). It is used only in severe cases and by expert hands and is associated with high morbidity and risk of complications.

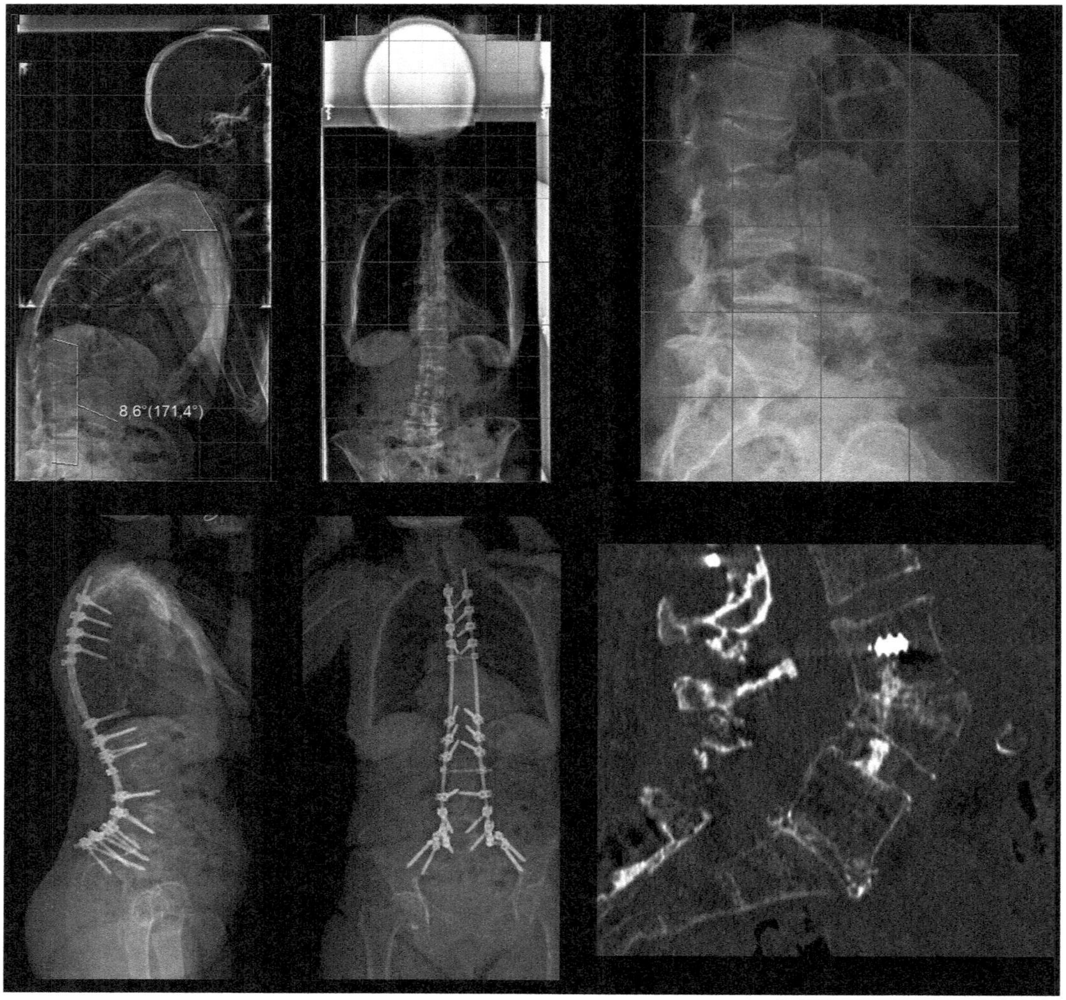

FIG. 7

Example of pedicle subtraction osteotomy to correct sagittal imbalance. First row, from left to right: pre-operative sagittal and coronal X-rays of the entire thoracolumbar spine as well as sagittal projection of the lumbar spine showing a major misalignment with loss of lumbar lordosis and increased thoracic kyphosis. Second row: postoperative sagittal and coronal X-rays and lumbar sagittal CT showing the correction by means of pedicle subtraction osteotomy at L4, posterior fixation from T4 to the sacroiliac joint with anterior support through an interbody cage at L4–L5.

PSO is generally performed when other less invasive techniques are not sufficient to achieve the targeted outcome (Berjano and Aebi, 2015). PSO is usually performed in the lumbar spine rather than in the thoracic spine, although a few clinical studies have described use of PSO on the thoracic vertebrae (Huang et al., 2015; Xia et al., 2017) (Fig. 7). Surgeons often prefer to operate on L2–L4 instead of other segments, even though studies have shown that performing a PSO in the lower lumbar spine

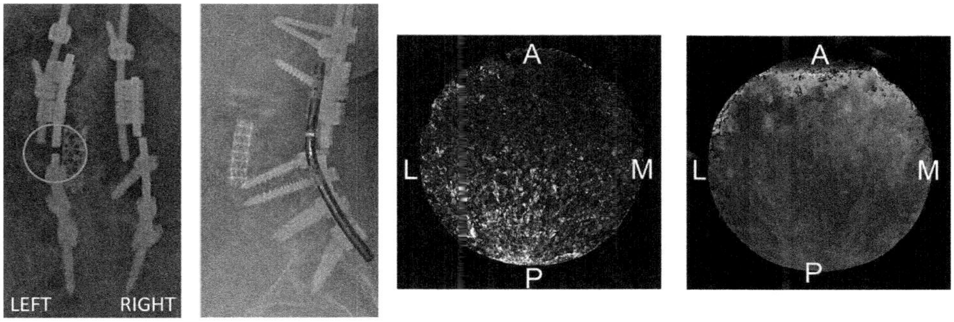

FIG. 8

Analysis of a retrieved rod broken at the level of a pedicle subtraction osteotomy. From left to right: (1) anteroposterior X-rays of the patient, showing both rods broken at the osteotomy site (the circle indicates the retrieved rod); (2) the rod superimposed to the lateral X-rays; (3) a confocal microscopy image of the fracture surface; (4) a height map of the fracture surface, showing a typical fatigue failure initiated on the posterior side. "A" indicates the anterior side, "P" the posterior one, "L" lateral, and "M" medial.

may have a higher risk of complications (Bridwell, 2006). Several surgeons perform the surgery on L4 whenever possible (Berjano and Aebi, 2015). It has been shown that PSO is linked to a non-negligible risk of major complications (e.g., excessive blood loss, hardware failure, neurological and motor deficits, cardiovascular problems, and infections), and minor complications (superficial infection, radicular pain, and sensory deficiency) were found in up to 90% of cases (Ames et al., 2013a). Hardware failure, typically rod breakage at the osteotomy level, is also very common (up to 39% of cases), thus making PSO a challenging operation from a biomechanical point of view (Enercan et al., 2013) (Fig. 8).

BIOMECHANICAL STUDIES OF SAGITTAL IMBALANCE AND REALIGNMENT

Most of the current knowledge about the sagittal alignment of the spine is based on clinical experience, that is, analysis of images of patients taken preoperatively, postoperatively, and at follow-up, as well as on the investigation of the clinical outcomes and complications. Bioengineering methods such as finite element modeling have been used in only a few studies. Keller and colleagues used a simple numerical model to study the relationship between sagittal alignment and spinal loads based on X-ray images of asymptomatic young subjects. They found that both compressive and shear spinal loads were strongly influenced by the sagittal parameters. Galbusera and colleagues built a more sophisticated finite element model that also included the trunk muscles and that quantified the correlations between the spinal loads and radiological parameters, such as LL and type following the Roussouly classification (Galbusera et al., 2013; Galbusera et al., 2014). Nevertheless, fundamental aspects such as the validity of the surgical targets (SVA less than 5 cm, PI − LL less than 10 degree, PT less than 20 degree) have never been investigated from a biomechanical point of view.

Despite the high clinical interest in the correction of sagittal imbalance, including the critical aspects related to instrumentation failure, only a few published biomechanical and computational studies have focused on the correction of sagittal imbalance. However, the related techniques are relatively new, and their mid-to-long-term clinical outcomes and complications have been available for only a few years. PO and discectomy were compared in a cadaveric study on the thoracic spine; this study looked at the ROM and the translations of the treated vertebrae with respect to the intact one (Wang et al., 2015). In another study, also on the thoracic spine, sequential POs were performed, and the flexibility of the spine was measured after each resection (Sangiorgio et al., 2013). A recent study focused on the effect of PSO and vertebral column resection (reduction of one or more motion segments) on spinal cord tension and dural buckling. The authors found that with both techniques the spinal cord tension can be significantly decreased but that the reduction of the spinal height causes dural buckling that can lead to compression of the neural structures (Safain et al., 2015). An interesting in vitro study was performed with the goal of comparing the stiffness and fatigue life of fixation constructs used to perform PSO, with or without interbody cages above and below the osteotomy site (Deviren et al., 2012). The authors performed a PSO at L3 on human specimens, followed by instrumentation with titanium polyaxial screws (6.0×45 mm) and Cr-Co rods with a diameter of 5.5 mm. In half of the specimens, an XLIF cage was inserted at the L2–L3 and L3–L4 levels. Then the authors tested the specimens by performing a multidirectional bending test and found that the anterior support can increase the construct rigidity. Consequently, they suggested using interbody cages associated with posterior fixation in order to avoid the need for revision surgeries. In another cadaver study, the stiffness of various configurations of devices used in the revision of PSOs were compared (Scheer et al., 2011). The use of supplementary rods at the osteotomy level was investigated on cadaveric specimens by Hallager and colleagues (Hallager et al., 2016), who found that the four-rod construct was effective in reducing the strains in the instrumentation, with a possible positive effect on the risk of rod fracture.

Concerning computational studies, Hato and colleagues studied a closing-opening correction osteotomy with a numerical model of the thoracolumbar spine, creating various models, varying the kyphotic angle, and considering different degree of osteoporosis (Hato et al., 2007). Charosky and colleagues developed a simplified model in order to reproduce a PSO on L4 (with three defect situations) and study loads and stresses in different configurations of spinal implants (Charosky et al., 2014). Ottardi and colleagues built a finite element model to investigate the effect of PSO without instrumentation (Ottardi et al., 2016), which was found to be major in all motion directions as well as greater if PSO was performed at L4 rather than at L3. The model was subsequently modified to study the effect of instrumentation, namely posterior fixation and fixation with supplementary rods at the level of osteotomy (Luca et al., 2017). Results indicated that stiffer configurations (double rods and rods with higher diameter) induced a marked reduction of the stresses on the instrumentation, but might slow down healing at the osteotomy site because of the decreased load transferred through the anterior column.

REFERENCES

Ames, C.P., Barry, J.J., Keshavarzi, S., Dede, O., Weber, M.H., Deviren, V., 2013a. Perioperative outcomes and complications of pedicle subtraction osteotomy in cases with single versus two attending surgeons. Spine Deform. 1 (1), 51–58.

Ames, C.P., Blondel, B., Scheer, J.K., Schwab, F.J., Le Huec, J.C., Massicotte, E.M., Patel, A.A., Traynelis, V.C., Kim, H.J., Shaffrey, C.I., Smith, J.S., Lafage, V., 2013b. Cervical radiographical alignment: comprehensive assessment techniques and potential importance in cervical myelopathy. Spine 38 (22 Suppl 1), S149–60.

Barrey, C., Jund, J., Noseda, O., Roussouly, P., 2007. Sagittal balance of the pelvis-spine complex and lumbar degenerative diseases. A comparative study about 85 cases. Eur. Spine J. 16 (9), 1459–1467.

Barrey, C., Roussouly, P., Perrin, G., Le Huec, J.C., 2011. Sagittal balance disorders in severe degenerative spine. Can we identify the compensatory mechanisms? Eur. Spine J. 20 (Suppl. 5), 626–633.

Bergin, P.F., O'Brien, J.R., Matteini, L.E., Yu, W.D., Kebaish, K.M., 2010. The use of spinal osteotomy in the treatment of spinal deformity. Orthopedics 33 (8), 586–594.

Berjano, P., Aebi, M., 2015. Pedicle subtraction osteotomies (PSO) in the lumbar spine for sagittal deformities. Eur. Spine J. 24 (Suppl. 1), S49–57.

Bridwell, K.H., 2006. Decision making regarding Smith-Petersen vs. pedicle subtraction osteotomy vs. vertebral column resection for spinal deformity. Spine 31 (Suppl 19), S171–8.

Charosky, S., Moreno, P., Maxy, P., 2014. Instability and instrumentation failures after a PSO: a finite element analysis. Eur. Spine J. 23 (11), 2340–2349.

Deviren, V., Tang, J.A., Scheer, J.K., Buckley, J.M., Pekmezci, M., McClellan, R.T., Ames, C.P., 2012. Construct rigidity after fatigue loading in pedicle subtraction osteotomy with or without adjacent interbody structural cages. Global Spine J. 2 (4), 213–220.

Diebo, B.G., Henry, J., Lafage, V., Berjano, P., 2015. Sagittal deformities of the spine: factors influencing the outcomes and complications. Eur. Spine J. 24 (Suppl. 1), S3–15.

Duval-Beaupere, G., Schmidt, C., Cosson, P., 1992. A Barycentremetric study of the sagittal shape of spine and pelvis: the conditions required for an economic standing position. Ann. Biomed. Eng. 20 (4), 451–462.

Enercan, M., Ozturk, C., Kahraman, S., Sarier, M., Hamzaoglu, A., Alanay, A., 2013. Osteotomies/spinal column resections in adult deformity. Eur. Spine J. 22 (Suppl. 2), S254–64.

Galbusera, F., Wilke, H.J., Brayda-Bruno, M., Costa, F., Fornari, M., 2013. Influence of sagittal balance on spinal lumbar loads: a numerical approach. Clin. Biomech. (Bristol, Avon) 28 (4), 370–377.

Galbusera, F., Brayda-Bruno, M., Costa, F., Wilke, H.J., 2014. Numerical evaluation of the correlation between the normal variation in the sagittal alignment of the lumbar spine and the spinal loads. J. Orthop. Res. 32 (4), 537–544.

Glassman, S.D., Berven, S., Bridwell, K., Horton, W., Dimar, J.R., 2005a. Correlation of radiographic parameters and clinical symptoms in adult scoliosis. Spine 30 (6), 682–688.

Glassman, S.D., Bridwell, K., Dimar, J.R., Horton, W., Berven, S., Schwab, F., 2005b. The impact of positive sagittal balance in adult spinal deformity. Spine 30 (18), 2024–2029.

Hallager, D.W., Gehrchen, M., Dahl, B., Harris, J.A., Gudipally, M., Jenkins, S., Wu, A.M., Bucklen, B.S., 2016. Use of supplemental short pre-contoured accessory rods and cobalt chrome alloy posterior rods reduces primary rod strain and range of motion across the pedicle subtraction osteotomy level: an in vitro biomechanical study. Spine 41 (7), E388–95.

Hato, T., Kawahara, N., Tomita, K., Murakami, H., Akamaru, T., Tawara, D., Sakamoto, J., Oda, J., Tanaka, S., 2007. Finite-element analysis on closing-opening correction osteotomy for angular kyphosis of osteoporotic vertebral fractures. J. Orthop. Sci. 12 (4), 354–360.

Huang, J.H., Yang, W.Z., Shen, C., Chang, M.S., Li, H., Luo, Z.J., Tao, H.R., 2015. Surgical treatment of congenital scoliosis associated with tethered cord by thoracic spine-shortening osteotomy without cord detethering. Spine 40 (20), E1103–9.

Ilharreborde, B., Dubousset, J., Le Huec, J.C., 2014. Use of EOS imaging for the assessment of scoliosis deformities: application to postoperative 3D quantitative analysis of the trunk. Eur. Spine J. 23 (Suppl 4), S397–405.

Jackson, R.P., McManus, A.C., 1994. Radiographic analysis of sagittal plane alignment and balance in standing volunteers and patients with low back pain matched for age, sex, and size. A prospective controlled clinical study. Spine 19 (14), 1611–1618.

Jackson, R.P., Peterson, M.D., McManus, A.C., Hales, C., 1998. Compensatory spinopelvic balance over the hip axis and better reliability in measuring lordosis to the pelvic radius on standing lateral radiographs of adult volunteers and patients. Spine 23 (16), 1750–1767.

Lafage, V., Diebo, B.G., Schwab, F., 2015. Sagittal Spino-Pelvic Alignment: From Theory to Clinical Application. Editorial Medica Panamericana, Madrid, Spain.

Le Huec, J.C., Leijssen, P., Duarte, M., Aunoble, S., 2011a. Thoracolumbar imbalance analysis for osteotomy planification using a new method: FBI technique. Eur. Spine J. 20 (Suppl 5), 669–680.

Le Huec, J.C., Saddiki, R., Franke, J., Rigal, J., Aunoble, S., 2011b. Equilibrium of the human body and the gravity line: the basics. Eur. Spine J. 20 (Suppl. 5), 558–563.
Legaye, J., Duval-Beaupere, G., Hecquet, J., Marty, C., 1998. Pelvic incidence: a fundamental pelvic parameter for three-dimensional regulation of spinal sagittal curves. Eur. Spine J. 7 (2), 99–103.
Lenke, L.G., Sides, B.A., Koester, L.A., Hensley, M., Blanke, K.M., 2010. Vertebral column resection for the treatment of severe spinal deformity. Clin. Orthop. Relat. Res. 468 (3), 687–699.
Luca, A., Ottardi, C., Sasso, M., Prosdocimo, L., La Barbera, L., Brayda-Bruno, M., Galbusera, F., Villa, T., 2017. Instrumentation failure following pedicle subtraction osteotomy: the role of rod material, diameter, and multi-rod constructs. Eur. Spine J. 26, 764–770.
Mac-Thiong, J.M., Berthonnaud, E., Dimar 2nd, J.R., Betz, R.R., Labelle, H., 2004. Sagittal alignment of the spine and pelvis during growth. Spine 29 (15), 1642–1647.
Mangione, P., Gomez, D., Senegas, J., 1997. Study of the course of the incidence angle during growth. Eur. Spine J. 6 (3), 163–167.
Murata, Y., Takahashi, K., Yamagata, M., Hanaoka, E., Moriya, H., 2003. The knee-spine syndrome. Association between lumbar lordosis and extension of the knee. J. Bone Joint Surg. 85 (1), 95–99.
Obeid, I., Hauger, O., Aunoble, S., Bourghli, A., Pellet, N., Vital, J.M., 2011. Global analysis of sagittal spinal alignment in major deformities: correlation between lack of lumbar lordosis and flexion of the knee. Eur. Spine J. 20 (Suppl. 5), 681–685.
Ottardi, C., Galbusera, F., Luca, A., Prosdocimo, L., Sasso, M., Brayda-Bruno, M., Villa, T., 2016. Finite element analysis of the lumbar destabilization following pedicle subtraction osteotomy. Med. Eng. Phys. 38 (5), 506–509.
Ponte, A., Vero, B., Siccardi, G.L., 1984. Surgical Treatment of Scheuermann's Hyperkyphosis. Aulo Gaggi, Bologna.
Roussouly, P., Gollogly, S., Berthonnaud, E., Dimnet, J., 2005. Classification of the normal variation in the sagittal alignment of the human lumbar spine and pelvis in the standing position. Spine 30 (3), 346–353.
Safain, M.G., Burke, S.M., Riesenburger, R.I., Zerris, V., Hwang, S.W., 2015. The effect of spinal osteotomies on spinal cord tension and dural buckling: a cadaveric study. J. Neurosurg. Spine 23 (1), 120–127.
Sangiorgio, S.N., Borkowski, S.L., Bowen, R.E., Scaduto, A.A., Frost, N.L., Ebramzadeh, E., 2013. Quantification of increase in three-dimensional spine flexibility following sequential ponte osteotomies in a cadaveric model. Spine Deform. 1 (3), 171–178.
Savage, J.W., Patel, A.A., 2014. Fixed sagittal plane imbalance. Global Spine J. 4 (4), 287–296.
Scheer, J.K., Tang, J.A., Deviren, V., Buckley, J.M., Pekmezci, M., McClellan, R.T., Ames, C.P., 2011. Biomechanical analysis of revision strategies for rod fracture in pedicle subtraction osteotomy. Neurosurgery 69 (1), 164–172. discussion 172.
Schwab, F., Lafage, V., Boyce, R., Skalli, W., Farcy, J.P., 2006. Gravity line analysis in adult volunteers: age-related correlation with spinal parameters, pelvic parameters, and foot position. Spine 31 (25), E959–67.
Schwab, F.J., Lafage, V., Farcy, J.P., Bridwell, K.H., Glassman, S., Shainline, M.R., 2008. Predicting outcome and complications in the surgical treatment of adult scoliosis. Spine 33 (20), 2243–2247.
Schwab, F., Patel, A., Ungar, B., Farcy, J.P., Lafage, V., 2010. Adult spinal deformity-postoperative standing imbalance: how much can you tolerate? An overview of key parameters in assessing alignment and planning corrective surgery. Spine 35 (25), 2224–2231.
Schwab, F.J., Blondel, B., Bess, S., Hostin, R., Shaffrey, C.I., Smith, J.S., Boachie-Adjei, O., Burton, D.C., Akbarnia, B.A., Mundis, G.M., Ames, C.P., Kebaish, K., Hart, R.A., Farcy, J.P., Lafage, V., International Spine Study Group (ISSG), 2013. Radiographical spinopelvic parameters and disability in the setting of adult spinal deformity: a prospective multicenter analysis. Spine 38 (13), E803–12.
Schwab, F.J., Diebo, B.G., Smith, J.S., Hostin, R.A., Shaffrey, C.I., Cunningham, M.E., Mundis Jr., G.M., Ames, C.P., Burton, D.C., Bess, S., Gupta, M.C., Hart, R.A., Protopsaltis, T.S., Lafage, V., 2014a. Fine-tuned surgical planning in adult spinal deformity: determining the lumbar lordosis necessary by accounting for both thoracic kyphosis and pelvic incidence. Spine J. 14 (11), S73.

Schwab, F., Blondel, B., Chay, E., Demakakos, J., Lenke, L., Tropiano, P., Ames, C., Smith, J.S., Shaffrey, C.I., Glassman, S., Farcy, J.P., Lafage, V., 2014b. The comprehensive anatomical spinal osteotomy classification. Neurosurgery 74 (1), 112–120. discussion 120.

Smith-Petersen, M.N., Larson, C.B., Aufranc, O.E., 1969. Osteotomy of the spine for correction of flexion deformity in rheumatoid arthritis. Clin. Orthop. Relat. Res. 66, 6–9.

Thomasen, E., 1985. Vertebral osteotomy for correction of kyphosis in ankylosing spondylitis. Clin. Orthop. Relat. Res. 194 (194), 142–152.

Vialle, R., Levassor, N., Rillardon, L., Templier, A., Skalli, W., Guigui, P., 2005. Radiographic analysis of the sagittal alignment and balance of the spine in asymptomatic subjects. J. Bone Joint Surg. 87 (2), 260–267.

Wang, C., Bell, K., McClincy, M., Jacobs, L., Dede, O. Roach, J., Bosch, P., 2015. Biomechanical comparison of ponte osteotomy and discectomy. Spine 40 (3), E141–5.

Xia, L., Li, N., Wang, D., Liu, M., Li, J.W., Bao, D.M., Li, P., 2017. One-stage posterior spinal osteotomy in severe spinal deformities: a total of 147 cases. Clin. Spine Surg. 30, E448–E453.

TRAUMA AND TUMORS

CHAPTER

22 BIOMECHANICS OF VERTEBRAL FRACTURES AND THEIR TREATMENT

Richard A. Lindtner, Werner Schmoelz
Medical University of Innsbruck, Innsbruck, Austria

INTRODUCTION

Traumatic lesions of the spine are relatively common and encompass a wide spectrum of injuries ranging from minor fractures of spinous or transverse processes to complex, multidirectionally unstable injuries. These lesions may involve osseous or disco-ligamentous structures or both. Spinal injuries usually are the result of high-energy trauma, and associated injuries are common. Moreover, traumatic injury to the spine frequently leads to neurological deficits with potentially devastating consequences for patients, ranging from mild, transient symptoms to complete tetraplegia. Other potential sequelae of spinal injuries include progressive deformity, persistent pain, and disability.

The most appropriate treatment strategy for spinal injuries varies depending on the degree to which spinal structures have been compromised. Treatment options thus range from conservative to combined posteroanterior surgical approaches depending on the individual injury pattern. General goals of treatment, however, include restoration of anatomic alignment, decompression of neural elements (if required), adequate stabilization to allow for early mobilization, and maximization of clinical outcome and neurologic recovery.

The majority of spinal injuries affect the thoracic and lumbar spine. This chapter will therefore focus on thoracolumbar trauma, and special reference will be made to important aspects of injury mechanism, classification, and biomechanics of fracture stabilization.

INJURY MECHANISMS

In young and middle-aged individuals, thoracolumbar fractures typically result from high-energy blunt trauma. Common injury mechanisms include falls from height, motor vehicle accidents, and sports-related injuries. In elderly people, low-energy mechanisms and even spontaneous fractures without relevant trauma are more common as a consequence of diminished bone quality caused by osteoporosis. Bone disorders other than osteoporosis may also increase the susceptibility to fracture even after a low-energy impact. Ankylosing spinal disorders (i.e., ankylosing spondylitis and diffuse idiopathic skeletal hyperostosis), for example, lead to ankylosis/fusion of normally mobile spinal segments via progressive ossification of spinal ligaments and/or intervertebral discs. As a consequence of ankylosis, the normal biomechanics of the spine is massively altered, and forces can act on longer lever arms during trauma. The ankylosed thoracolumbar spine is thus much more susceptible to injury even after low-energy trauma and typically shows an otherwise uncommon hyperextension injury pattern.

Biomechanics of the Spine. https://doi.org/10.1016/B978-0-12-812851-0.00022-7

In high-energy trauma, usually complex and multidirectional forces act on the spine at the time of injury. These forces may involve compressive and distractive forces; rotational moments in the sagittal (flexion and extension), coronal (left and right side bending), and axial (left and right axial rotation) planes; as well as shear forces. For the most part, the spine is exposed to a combination of several of these forces and moments during trauma. The forces and moments acting on the spine can result in a great variety of injury patterns, including bony injuries without ligamentous disruption, ligamentous injury without bony injury, or often a combination of both. Depending on the direction of the resultant force vector, characteristic types of injury patterns result. The extent of disco-ligamentous and/or bony disruption again varies as a function of the magnitude of the resulting force vector acting on the spine.

From a biomechanical point of view, the spinal column can be divided into an anterior and a posterior column, as suggested by Holdsworth (1970). The anterior column consists of the vertebra and intervertebral discs (including the anterior and posterior longitudinal ligaments), resists compression and axial loading, and carries about 60%–90% of the load in the normal upright position (Adams and Dolan, 2005). The posterior column consists of the bone structures posterior to the posterior longitudinal ligament (posterior vertebral arch) as well as the posterior ligamentous complex (PLC), maintains tension, and carries about 10%–40% of the load in the normal upright position. The PLC is critical for spinal stability and consists of the supraspinous and interspinous ligaments, the ligamentum flavum, and the facet capsules.

Depending on the resultant force vector and resultant moments, characteristic patterns of anterior and/or posterior column involvement result. A traumatic axial load (i.e., a load applied along the spinal long axis) results in excessive compression and subsequent failure of the anterior column in the form of vertebral body fracture without significant injury to the posterior column. A traumatic forward bending moment leads to flexion of the spinal column in the sagittal plane and causes tension of the posterior column as well as compressive loading of the anterior part of the vertebral body. In contrast, a traumatic backward bending moment leads to hyperextension of the spinal column and causes tension of the anterior column (particularly of the anterior annulus fibrosus and anterior longitudinal ligament) as well as compression of the posterior column. Excessive and complex traumatic loading may cause a complete disruption of both columns that permits fracture dislocation with translational displacement in one or more planes.

The thoracolumbar junction (T11-L2) is the most common site of spinal fractures and is particularly vulnerable because it constitutes an anatomical and biomechanical transition zone between the subjacent lumbar and superjacent thoracic spine. At the thoracolumbar junction, the relatively rigid thoracic kyphosis transitions to the more flexible lumbar lordosis. The rib cage constrains motion of the thoracic spine and increases resistance to bending moments in the sagittal and coronal planes as well as axial rotation. The "free floating" ribs of T11 and T12 provide markedly less stability at the thoracolumbar junction because they are not connected to the sternum. Furthermore, the coronal orientation of the thoracic facet joints limits flexion and extension and provides significant resistance to anterior translation, whereas the more sagittal-oriented lumbar facet joints allow for increased motion in flexion and extension.

DIAGNOSTIC IMAGING

Initial evaluation of a patient with suspected thoracolumbar injury consists of clinical assessment including a thorough neurological examination as well as diagnostic imaging. Imaging modalities include conventional anteroposterior and lateral radiographs, computed tomography (CT), and magnetic resonance imaging (MRI).

CT is considered the "gold standard" for imaging of spinal trauma. It offers higher spatial resolution than conventional plain radiographs and allows visualization of the spine in the axial, sagittal, and coronal planes.

CT is superior for quantifying the extent of vertebral body comminution as well as of retropulsion of bone fragments into the spinal canal, for distinguishing wedge compression fractures from burst fractures, and for detecting subtle fractures of posterior column elements. In general, high-quality CT imaging is the imaging method of choice to characterize the bony injury pattern and forms the basis for injury classification.

MRI is time-consuming and is not routinely used for initial evaluation of a trauma patient. Nevertheless, MRI allows for direct visualization of spinal soft tissues, including intervertebral discs, spinal ligaments, and the spinal cord, and may provide valuable information in selected cases, such as suspected but undetermined disc and/or ligament disruption after CT imaging, presence of neurological deficits without evidence of spinal injury in CT imaging, and neurological dysfunction not correlating with level of spinal injury. Moreover, MRI permits differentiation between recent and old fractures, particularly in osteoporotic patients, which can be difficult when relying on CT images alone, and allows detection of bone marrow edema caused by microtrabecular fractures without cortical breaks, which are not detectable by CT imaging.

SPINAL FRACTURE CLASSIFICATION

Numerous classification systems have been developed in the past decades in an attempt to better define and characterize thoracolumbar injury patterns as well as to aid in making decisions about treatment. However, none of these classification systems has been uniformly adopted.

The two classification systems that have gained the most widespread use in the past two decades are the AO Magerl classification and the Thoracolumbar Injury Classification and Severity Score (TLICS).

Magerl et al. (1994) proposed a comprehensive, hierarchical classification system for thoracic and lumbar spinal injuries after analysis of 1445 consecutive cases over a 10-year period. This systematic and detailed system is based primarily on the pathomorphological characteristics of the injuries and consists of three main injury types:

- *Type A*: Compression injuries (vertebral body compression injuries with intact PLC; result from axial compression; 66% of all fractures analyzed);
- *Type B injuries*: Distraction injuries (either posterior or anterior element injury with distraction; result from a flexion-distraction or a hyperextension-distraction mechanism, respectively; 15% of all fractures analyzed); and
- *Type C*: Rotational injuries (anterior and posterior element injury with rotation; result from axial torque; 19% of all fractures analyzed).

Each of these three main types is divided into three groups (resulting in nine basic injury types); each of these groups is further divided into subgroups (resulting in 25 subgroup patterns), and each subgroup is divided into subdivisions. Injury severity and incidence of neurological deficit increases from type A to B to C as well as within groups. The Magerl system has been criticized for being overly complex and for being based only on a morphology and injury mechanism and not taking into account the degree of neurological injury and other clinical factors that impact decisions about treatment.

In 2005, Vaccaro et al. (2005) introduced the Thoracolumbar Injury Classification and Severity Score (TLICS) system as a practical guide to facilitate making decisions about operative versus nonoperative care and in an effort to overcome the shortcomings of the AO Magerl classification. The TLICS is based on three injury characteristics: (1) injury morphology determined by radiographic appearance (compression fracture, burst fracture, translational/rotational injury and distraction injury), (2) integrity of the posterior ligamentous complex (PLC), and (3) neurologic status of the patient. Each characteristic is assigned points according to the severity of injury, and based on the final score (sum of points), the proposed treatment algorithm guides decisions regarding operative versus nonoperative care. The TLICS takes into account the neurological status and has been shown to provide improved intra- and interobserver reliability compared to the more detailed and complex Magerl system. The TLICS, however, has been criticized for oversimplifying fracture morphology and for not sufficiently reflecting its influence on instability and treatment decisions. Moreover, the feasibility and reproducibility of evaluating PLC integrity by MRI have been questioned (Vaccaro et al., 2009; Rihn et al., 2010).

Therefore in 2013 the AOSpine Spinal Cord Injury and Trauma Knowledge Forum, an international group of academic spine surgeons, developed the AOSpine thoracolumbar spine injury classification system (Vaccaro et al., 2013) (Fig. 1). This new classification system is based on the TLICS and Magerl systems and is intended to combine the strengths of both as well as to overcome their shortcomings. This new classification system consists of a morphological injury classification, a grading system for the neurological status, and two clinical modifiers. It was developed in an attempt to provide a simple and reproducible combined morphologic and clinical classification system of thoracolumbar injuries for universal international adoption. This new classification system has been shown to provide better inter- and intraobserver reliability than the existing TLICS and Magerl system (Vaccaro et al., 2013; Kepler et al., 2016a; Kaul et al., 2017; Wood et al., 2005). Moreover, the thoracolumbar AOSpine Injury score was developed (Kepler et al., 2016b), and a surgical algorithm was established (Vaccaro et al., 2016).

The morphological injury classification of this new system basically constitutes an adapted and simplified version of the AO Magerl classification and distinguishes three main types of injuries according to the predominant mode of failure of the spinal column: type A (compression injuries; subtypes A0–A4), type B (distraction injuries; subtype B1–B3), and type C (translation injuries; no subtypes). These three basic types and eight subtypes have been defined as follows (Fig. 1):

Type A—Compression injuries of the vertebral body: Type A injuries are injuries of the anterior elements (vertebral body and/or disc) without tension band involvement as well as without any translation/displacement. Type A injuries result from failure under axial compression. This type also includes minor, insignificant fractures of the vertebra (subtype A0). The term "burst fracture" refers to a specific subtype of compression fractures. The pathognomonic feature of burst fractures is fracture involvement of the posterior wall of the vertebral body. The resulting posterior wall fragment(s) are commonly retropulsed into the spinal canal leading to spinal canal compromise and may cause spinal cord or cauda equina compression. A burst fracture may involve only one endplate (incomplete burst fracture) or both endplates (complete burst fracture), and the degree of vertebral body comminution varies. Type A injuries are divided into five subtypes:

- *Subtype A0:* Minor, insignificant fractures of the vertebra that do not compromise the structural integrity of the spinal column (such as transverse process or spinous process fractures).
- *Subtype A1: Wedge-compression or impaction fractures:* Vertebral fracture involving a single endplate but without involvement of the posterior wall of the vertebral body.
- *Subtype A2: Split fractures or pincer-type fractures:* Vertebral fracture involving both endplates but without involvement of the posterior wall of the vertebral body.
- *Subtype A3: Incomplete burst fracture:* Vertebral fracture involving the posterior wall as well as only a single endplate.
- *Subtype A4: Complete burst fracture:* Vertebral fracture involving the posterior wall as well as both endplates.

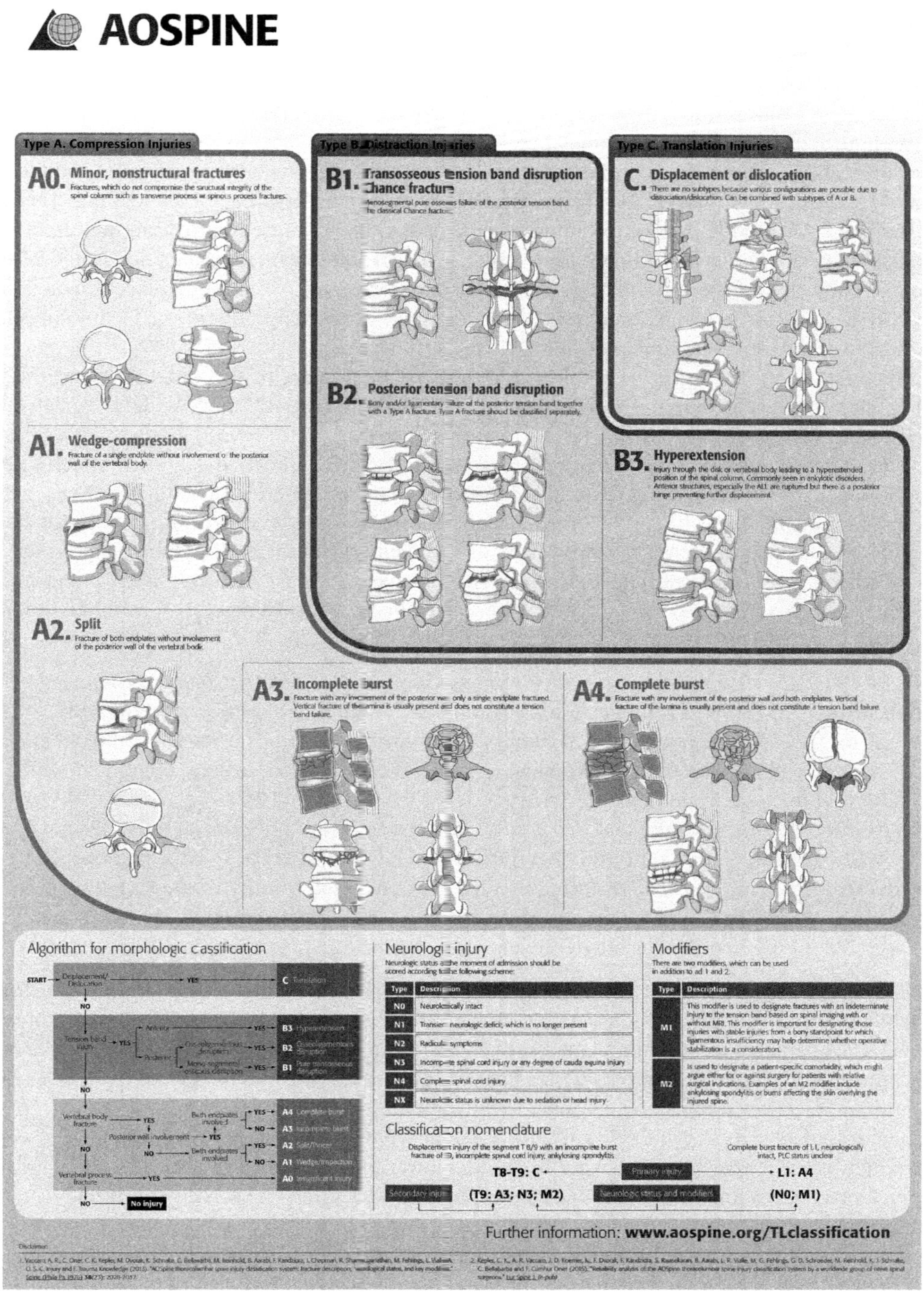

FIG. 1

AOSpine thoracolumbar spine injury classification system.

Image reproduced with permission from AO Foundation, Switzerland.

In subtypes A3 and A4, a vertical fracture of the lamina is often present. In contrast to type B injuries (with horizontal osseo-ligamentous disruption of the PLC or facet joints and resulting posterior tension band failure), the vertical lamina fracture does not constitute a posterior tension band failure; it is rather a consequence of splaying of the posterior arch of the vertebra under axial loading. The splaying of the posterior arch and resulting interpedicular widening is caused by a wedging effect of the inferior articulate processes of the upper next vertebra. As a result of excessive axial loads, the vertebral body bursts, and the inferior articulate processes of the upper next vertebra displace downward and wedge into the superior articulate processes of the fractured vertebra because of the facet joint orientation at the lower thoracic and lumbar region.

If a type B or a type C injury is combined with a type A fracture of the vertebral body, the latter should be mentioned separately (e.g., "T12/L1 (=level of injury) *type B2* injury with L1 *type A3* vertebral body fracture").

Type B—Tension band injuries (distraction injuries): Type B injuries are injuries affecting either the posterior tension band (i.e., PLC, and posterior arch) or the anterior tension band (i.e., anterior longitudinal ligament, or ALL, and anterior part of the disc) without evidence of (gross) translational displacement. The posterior tension band resists forward bending moments (flexion), whereas the anterior tension band resists backward bending moments (extension). Consequently, posterior tension band injuries typically result from a flexion-distraction mechanism, whereas anterior tension band injuries result from a hyperextension-distraction mechanism. Subtypes of type B injuries include

- *Subtype B1: Transosseous posterior tension band disruption (Chance fracture):* Monosegmental pure transosseous failure of the posterior tension band extending into the vertebral body.
- *Subtype B2: Posterior tension band disruption:* Disruption of the posterior tension band (i.e., PLC) with or without osseous involvement. Subtype B2 injuries are usually combined with a type A fracture of the vertebral body.
- *Subtype B3: Anterior tension band disruption:* Hyperextension injury through the disc or vertebral body with disruption of the anterior tension band (particularly of the anterior longitudinal ligament, or ALL) resulting in a hyperextension deformity of the spinal column. Typically seen in ankylosing spinal disorders. Anterior structures, especially the ALL, are ruptured but an intact posterior hinge prevents translational displacement.

Type C—Translation injuries: Type C injuries are injuries involving dislocation or translational displacement beyond the physiological range of the cranial and caudal parts of the spinal column in any plane. These injuries usually implicate disruption of the anterior as well as posterior tension bands. Type C injuries are not divided into subtypes because of various possible injury configurations due to dissociation/dislocation and may be combined with subtypes of type A or B.

The neurological status is graded as follows: N0: neurologically intact; N1: transient neurological deficit, which is no longer present; N2: symptoms or signs of radiculopathy; N3: incomplete spinal cord injury or cauda equina injury; N4: complete spinal cord injury; and Nx: neurological status is unobtainable.

Moreover, two patient-specific clinical modifiers relevant to treatment decision-making were included: M1: indeterminate injury to the tension band based on spinal imaging (with or without MRI), and M2: patient-specific comorbidity which might argue either for or against surgery for patients with relative surgical indications (e.g., ankylosing spondylitis). These two modifiers may not be relevant to every patient and therefore were stated to be optional.

TREATMENT PRINCIPLES AND BIOMECHANICS OF SPINAL FRACTURE STABILIZATION

GENERAL REMARKS

The optimal treatment for thoracolumbar injuries is still controversial. Treatment options range from conservative to combined posteroanterior surgical stabilization depending on the individual fracture pattern. Despite substantial regional and institutional variations in treatment, the general aims of spinal fracture treatment include fracture reduction and restoration of sagittal alignment, decompression of neural elements (in patients with neurological deficits caused by neural impingement), adequate stabilization to allow for early mobilization and to prevent short- and long-term deformity, and maximization of clinical outcome and neurologic recovery.

There is a general consensus that unstable injuries must be stabilized. However, spinal instability is difficult to assess based on imaging that captures only an isolated moment in time and typically with the patient in a supine position. Furthermore, "spinal instability" continues to be one of the most debated terms among spine surgeons, although numerous definitions have been proposed. White and Panjabi perhaps provided the most comprehensive yet practical definition: "*Clinical instability is defined as a loss in the ability of the spine under physiologic loads to maintain relationships between vertebrae in such a way that there is neither damage nor subsequent irritation to the spinal cord or nerve roots. In addition there is no development of incapacitating deformity or pain due to structural changes*" (White and Panjabi, 1978).

As a general rule of thumb, conservative treatment is recommended for subtypes A0, A1, and A2, whereas operative treatment is indicated in type B and type C injuries (Vaccaro et al., 2016). However, considerable controversy still exists regarding the optimal treatment for A3 and A4 fractures; depending on the neurological status, presence or absence of clinical modifiers, and the individual fracture configuration, either conservative or operative treatment may be more appropriate.

If surgical treatment is envisioned, several issues must be considered prior to surgery, including, among others: Is neural decompression required due to neurological deficits owing to spinal canal compromise? How will adequate reduction be achieved and maintained? Which column needs to be stabilized (posterior and/or anterior column)? Which surgical approach will be most appropriate? How many spinal segments will be included to ensure sufficient stability of the construct and to prevent loss of reduction (monosegmental versus bisegmental; short-segment versus long-segment instrumentation)? Is fusion via bone grafting necessary as a supplement to instrumentation?

Therefore, in the following sections, the different options for surgical stabilization of an injured thoracolumbar spine will be outlined and discussed.

POSTERIOR-ONLY STABILIZATION

Posterior pedicle screw-based instrumentation with and without fusion represents the most commonly used technique to stabilize thoracolumbar fractures. Modern pedicle screw/rod systems allow for reliable fixation and also facilitate fracture reduction and indirect decompression of the spinal canal via distraction and ligamentotaxis (i.e., reduce retropulsed bone fragments encroaching on the spinal canal by an intact posterior longitudinal ligament and posterior anulus fibrosus via distraction and/or traction).

One of the most fundamental considerations in posterior spinal instrumentation is the number of levels to be included in the instrumentation. On the one hand, saving motion segments by limiting the number of fixed and/or fused segments to preserve the normal biomechanics of the spine constitutes a fundamental principle of spinal surgery. On the other hand, the posterior fixation construct has to provide sufficient stability to prevent construct failure and to maintain reduction. Motion segment preservation of shorter constructs and improved resistance to deforming forces of longer constructs therefore have to be weighed against each other. Preserving motion segments is critical in the more mobile lumbar and thoracolumbar junction region, but of less importance in the less mobile thoracic spine because little motion is lost. Thus posterior instrumentation of thoracic fractures typically involves two levels above and below the injured vertebra (Fig. 2D). In the lumbar and thoracolumbar junction region, however, short segment fixation (i.e., bisegmental stabilization with pedicle screws placed in the vertebra one level cephalad and one level caudal to the fractured vertebra) (Fig. 2B) has become the standard of care. Adding additional transpedicular screws in the fractured vertebra (six-screw construct) (Fig. 2C) has been reported to provide additional construct stiffness, to enable improved reduction and to reduce postoperative loss of correction (Norton et al., 2014; Li et al., 2016). In selected cases, such as monosegmental pure ligamentous type B2 injuries or type A3 injuries with minor vertebral body comminution, monosegmental stabilization may be feasible and sufficient (Wei et al., 2010) (Fig. 2A).

Fixed-angle screw/rod systems using monoaxial screws are standard for spinal fracture stabilization in order to prevent loss of correction caused by polyaxial screw slippage (Marino, 2010; Kubosch et al., 2016). Cross links connecting both rods add torsional stability to the screw/rod construct (Wahba et al., 2010) and therefore are deemed advisable in rotationally unstable type C injuries, where there is severe anterior column instability, after vertebral body replacement and if facet joint resection is performed as part of spinal canal decompression or anterior column reconstruction from the posterior column.

Posterior or posterolateral fusion of the injured spinal segment using bone grafting is indicated in type B injuries with PLC disruption due to the poor healing ability of the PLC, but may not provide an advantage with regard to postoperative loss of reduction and functional outcome in burst fractures (type A injuries) with an intact PLC (Chou et al., 2014; Dai et al., 2009b). However, if satisfactory neural decompression and spinal canal clearance necessitate facet joint resection (at the expense of spinal stability), fusion is generally considered necessary even in type A injuries.

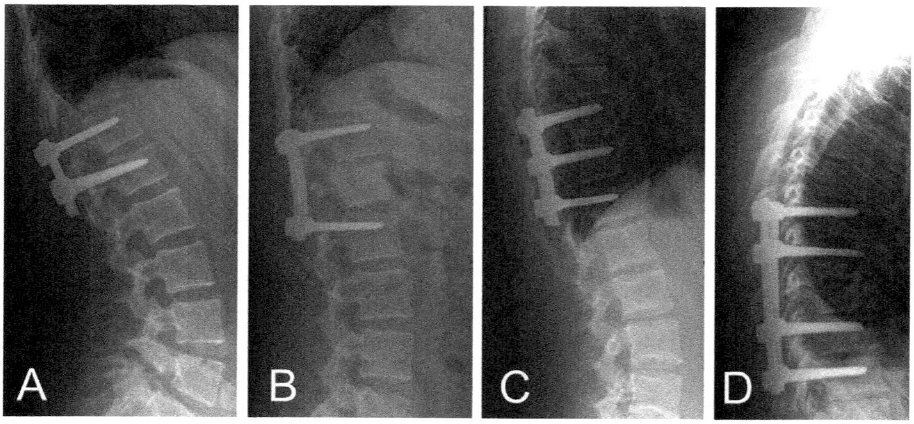

FIG. 2

Posterior fixation constructs: monosegmental (A), bisegmental (B), bisegmental with additional transpedicular screws in the fractured vertebra (six-screw construct) (C), and long-segment (two levels above and two levels below) (D) posterior fixation constructs.

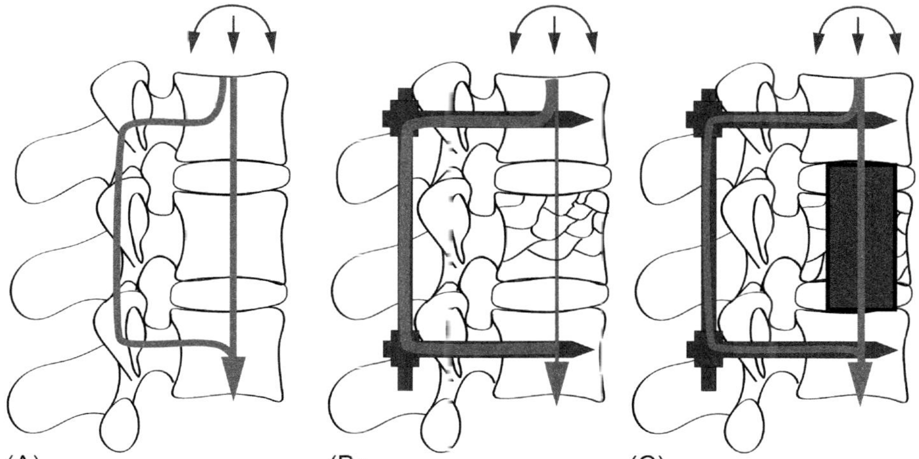

FIG. 3

Anterior und posterior load-sharing characteristics under physiological conditions (A), after posterior only fixation (B), as well as after combined posteroanterior stabilization using a vertebral body replacement device (C).

Minimally invasive percutaneous posterior instrumentation techniques may be considered to reduce blood loss and approach-related morbidity (McAnany et al., 2016) and may be appropriate particularly if fusion and decompression are not required.

From a biomechanical point of view, pedicle screw/rod constructs act as a tension band (or compression) fixation in type B1 and type B2 injuries without significant vertebral body comminution: The construct compresses the posterior column and thereby resists flexion moments and prevents displacement of the posterior column. In burst fractures with severe vertebral body comminution, however, pedicle screw/rod constructs act as a cantilever beam fixation. The construct alters the physiological anterior and posterior column load sharing (Fig. 3A) by unloading the anterior column (Fig. 3B) and shifting the loads posteriorly until healing of the vertebral body fracture has occurred. The axial load applied to the screws creates a bending moment in the sagittal plane that is resisted at the bone-screw interface. Excessive axial loading and/or too little anterior column load sharing caused by severe vertebral body comminution thus results in bone-screw interface failure and subsequent kyphotic deformity. Using longer and larger diameter screws improves the resistance to the sagittally oriented bending moment by improving the bending resistance and by increasing the screw-bone contact area.

COMBINED POSTEROANTERIOR STABILIZATION

In cases of severe vertebral body comminution (Fig. 4A), such as subtype A4 vertebral body injury, posterior-only stabilization is at risk of failure and is usually not able to maintain sagittal alignment postoperatively. Therefore, in these cases, anterior column reconstruction (ACR) has been advocated in order to prevent posterior fixation failure due to insufficient anterior column support and to ensure adequate stability. ACR enables restoration of the load-bearing capacity of the anterior column. Consequently, a more physiological load sharing between the posterior and anterior spinal column is reestablished, and stresses at the bone-pedicle screw interface are reduced (Fig. 3C).

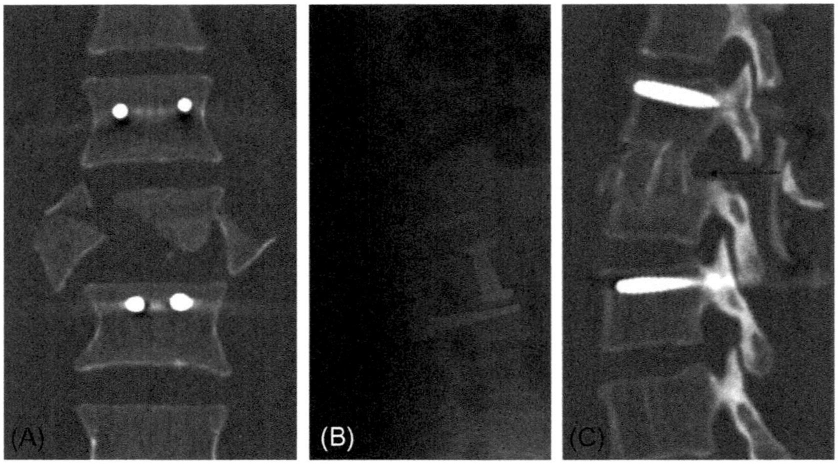

FIG. 4

Combined posteroanterior stabilization. Exemplary severe anterior column compromise (type A4 vertebral body injury) (A), combined posteroanterior stabilization using a vertebral body replacement device (B), and alternative anterior column reconstruction from posterior using monocortical strut grafts from the posterior iliac crest (C).

ACR may be achieved by different strategies, with combined posteroanterior stabilization being one of the most commonly used strategy. First, posterior fixation with or without neural decompression is performed to reduce and stabilize the unstable spinal injury. Second, the comminuted vertebral body as well as the adjacent intervertebral discs are partially resected and replaced by a vertebral body replacement device via an anterior thoracoscopic approach (Knop et al., 2005). The vertebral body replacement device is usually placed between the inferior endplate of the cephalad intact vertebra and the superior endplate of the caudal intact vertebra (Fig. 4B). This combined posteroanterior stabilization offers high primary biomechanical stability and is associated with only minimal postoperative loss of kyphosis correction (Knop et al., 2009; Schnake et al., 2014).

Alternatively, the anterior column can be reconstructed from a posterior approach via a PLIF (posterior lumbar interbody fusion) or TLIF (transforaminal interbody fusion) procedure (Schmid et al., 2010a,b). This strategy allows obviating the need for an additional anterior approach and achieves mechanical anterior column support by placing monocortical strut grafts harvested from the posterior iliac crest in the disc space adjacent to the fractured vertebra (Fig. 4C). Similarly to combined posteroanterior stabilization, this procedure aims at intervertebral fusion of the fractured with the adjacent vertebra. The clinical and radiological outcome has been reported to be comparable to that of combined posteroanterior stabilization (Schmid et al., 2012).

Another viable strategy may be to augment the compromised anterior column using calcium phosphate cement in addition to posterior fixation (Verlaan et al., 2015; Marco, 2009).

Anterior-only stabilization using a vertebral body replacement device, titanium mesh cage, or bone autograft (without posterior instrumentation) (Dai et al., 2009a), however, has not gained widespread acceptance, and biomechanical studies have indicated that anterior-only stabilization is less favorable compared to a combined posteroanterior approach (Wilke et al., 2001; Pflugmacher et al., 2004).

SPECIFIC ASPECTS OF OSTEOPOROTIC FRACTURE STABILIZATION

For patients suffering from osteoporosis and/or having reduced bone quality, the problem of fracture stabilization is not satisfactorily solved and is still a controversial topic. Standard established implants for the treatment of fractures in younger patients (e.g., pedicle screws, vertebral body replacements, and plates) cannot always be anchored sufficiently in the compromised bone stock of the elderly. Therefore various approaches to improving the treatment of fractures in geriatric patients have been developed. New developments to improve pedicle screw anchorage range from expandable screw designs to pedicle screw augmentations with various types of bone cement to changes in the screw trajectory. Techniques have been developed to help with fracture stabilization, ranging from minimally invasive procedures such as vertebroplasty and kyphoplasty (Fig. 5A) with various types of bone cements to extensions of short to long instrumentations of multiple segments with augmentation of pedicle screws (Fig. 5B). Despite all these improvements in techniques and implants, an optimal solution for the stabilization of fractures in geriatric patients is still a strong focus in research studies.

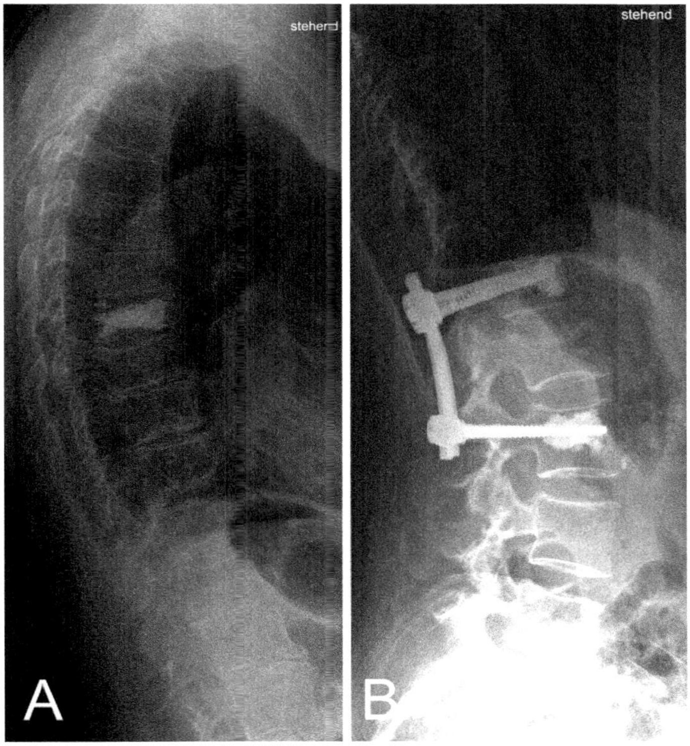

FIG. 5

Treatment options for osteoporotic fractures: vertebroplasty (A), short posterior instrumentation with augmented pedicle screws (B).

Acknowledgments

We would like to acknowledge the AO Foundation, Switzerland, and AO Spine for permission to reproduce the AOSpine thoracolumbar spine injury classification system (Fig. 1). AOSpine is a clinical division of the AO Foundation, an independent medically guided nonprofit organization. The AOSpine Knowledge Forums are pathology-focused working groups acting on behalf of AOSpine in its domain of scientific expertise. Each forum consists of a steering committee of up to 10 international spine experts who meet on a regular basis to discuss research, assess the best evidence for current practices, and formulate clinical trials to advance spine care worldwide. Study support is provided directly through AOSpine's Research department and AO's Clinical Investigation and Documentation unit.

REFERENCES

Adams, M.A., Dolan, P., 2005. Spine biomechanics. J. Biomech. 38 (10), 1972–1983.

Chou, P.-H., et al., 2014. Fusion may not be a necessary procedure for surgically treated burst fractures of the thoracolumbar and lumbar spines. J. Bone Joint Surg. Am. 96 (20), 1724–1731.

Dai, L.Y., Jiang, L.S., Jiang, S.D., 2009a. Anterior-only stabilization using plating with bone structural autograft versus titanium mesh cages for two-or three-column thoracolumbar burst fractures: a prospective randomized study. Spine 34 (14), 1429–1435.

Dai, L.Y., Jiang, L.S., Jiang, S.D., 2009b. Posterior short-segment fixation with or without fusion for thoracolumbar burst fractures. A five to seven-year prospective randomized study. J. Bone Joint Surg. 91 (5), 1033–1041.

Holdsworth, F., 1970. Fractures, dislocations, and fracture-dislocations of the spine. J. Bone Joint Surg. Am. 52 (8), 1534–1551.

Kaul, R., et al., 2017. Reliability assessment of AOSpine thoracolumbar spine injury classification system and thoracolumbar injury classification and severity score (TLICS) for thoracolumbar spine injuries: results of a multicentre study. Eur. Spine J. 26 (5), 1470–1476.

Kepler, C.K., Vaccaro, A.R., Koerner, J.D., et al., 2016a. Reliability analysis of the AOSpine thoracolumbar spine injury classification system by a worldwide group of naïve spinal surgeons. Eur. Spine J. 25 (4), 1082–1086.

Kepler, C., Vaccaro, A., Schroeder, G., et al., 2016b. The thoracolumbar AOSpine injury score. Global Spine J. 6 (4), 329–334.

Knop, C., et al., 2005. Vertebral body replacement with Synex in combined posteroanterior surgery for treatment of thoracolumbar injuries. Oper. Orthop. Traumatol. 17 (3), 249–280.

Knop, C., et al., 2009. Combined posterior–anterior stabilisation of thoracolumbar injuries utilising a vertebral body replacing implant. Eur. Spine J. 18 (7), 949–963.

Kubosch, D., et al., 2016. Biomechanical investigation of a minimally invasive posterior spine stabilization system in comparison to the universal spinal system (USS). BMC Musculoskelet. Disord. 17 (1), 134.

Li, K., et al., 2016. Pedicle screw fixation combined with intermediate screw at the fracture level for treatment of thoracolumbar fractures. Medicine 95 (33), e4574–8.

Magerl, F., et al., 1994. A comprehensive classification of thoracic and lumbar injuries. Eur. Spine J. 3 (4), 184–201.

Marco, R.A.W., 2009. Thoracolumbar burst fractures treated with posterior decompression and pedicle screw instrumentation supplemented with balloon-assisted vertebroplasty and calcium phosphate reconstruction. J. Bone Joint Surg. Am. 91 (1), 20.

Marino, J.F., 2010. Nonfusion short fixation of A3 burst fractures, loss of fixation attributable to polyaxial screw slippage? Spine J. 10 (5), 459–460. Author reply 460.

McAnany, S., et al., 2016. Open versus minimally invasive fixation techniques for thoracolumbar trauma: a meta-analysis. Global Spine J. 6 (2), 186–194.

Norton, R.P., et al., 2014. Biomechanical analysis of four- versus six-screw constructs for short-segment pedicle screw and rod instrumentation of unstable thoracolumbar fractures. Spine J. 14 (8), 1734–1739.

Pflugmacher, R., et al., 2004. Biomechanical comparison of expandable cages for vertebral body replacement in the thoracolumbar spine. Spine 29 (13), 1413–1419.

Rihn, J.A., et al., 2010. Using magnetic resonance imaging to accurately assess injury to the posterior ligamentous complex of the spine: a prospective comparison of the surgeon and radiologist. J. Neurosurg. Spine 12 (4), 391–396.

Schmid, R., et al., 2010a. Mid-term results of PLIF/TLIF in trauma. Eur. Spine J. 20 (3), 395–402.

Schmid, R., et al., 2010b. PLIF in thoracolumbar trauma: technique and radiological results. Eur. Spine J. 19 (7), 1079–1086.

Schmid, R., et al., 2012. Combined posteroanterior fusion versus transforaminal lumbar interbody fusion (TLIF) in thoracolumbar burst fractures. Injury 43 (4), 475–479.

Schnake, K.J., Stavridis, S.I., Kandziora, F., 2014. Five-year clinical and radiological results of combined anteroposterior stabilization of thoracolumbar fractures. J. Neurosurg. Spine 20 (5), 497–504.

Vaccaro, A.R., et al., 2005. A new classification of thoracolumbar injuries: the importance of injury morphology, the integrity of the posterior ligamentous complex, and neurologic status. Spine 30 (20), 2325–2333.

Vaccaro, A.R., et al., 2009. Injury of the posterior ligamentous complex of the thoracolumbar spine a prospective evaluation of the diagnostic accuracy of magnetic resonance imaging. Spine 34 (23), E841–E847.

Vaccaro, A.R., et al., 2013. AOSpine thoracolumbar spine injury classification system. Spine 38 (23), 2028–2037.

Vaccaro, A.R., et al., 2016. The surgical algorithm for the AOSpine thoracolumbar spine injury classification system. Eur. Spine J. 25 (4), 1087–1094.

Verlaan, J.-J., et al., 2015. Clinical and radiological results 6 years after treatment of traumatic thoracolumbar burst fractures with pedicle screw instrumentation and balloon assisted endplate reduction. Spine J. 15 (6), 1172–1178.

Wahba, G.M., et al., 2010. Biomechanical evaluation of short-segment posterior instrumentation with and without crosslinks in a human cadaveric unstable thoracolumbar burst fracture model. Spine 35 (3), 278–285.

Wei, F.-X., et al., 2010. Transpedicular fixation in management of thoracolumbar burst fractures: monosegmental fixation versus short-segment instrumentation. Spine 35 (15), E714–20.

White, A.A., Panjabi, M.M., 1978. Clinical Biomechanics of the Spine. Lippincott, Philadelphia.

Wilke, H.J., et al., 2001. Combined anteroposterior spinal fixation provides superior stabilisation to a single anterior or posterior procedure. J. Bone Joint Surg. Br. 83 (4), 609–617.

Wood, K.B., et al., 2005. Assessment of two thoracolumbar fracture classification systems as used by multiple surgeons. J. Bone Joint Surg. Am. 87 (7), 1423–1429.

CHAPTER

SPINE TUMORS

23

Francesco Costa*, Fabio Galbusera†

Humanitas Clinical and Research Center, Rozzeno, Italy IRCCS Galeazzi Orthopedic Institute, Milan, Italy†*

TUMORS INVOLVING THE SPINE

Spinal axis tumors are relatively frequent, as they represent about 16.4% of all bone neoplasms. These lesions can be primary, or they can be secondary; that is, they can be due to the spread of cells from the original (primary) cancer located in another part of the body. Whereas primary tumors involving the spine are relatively rare, constituting around 5% of bone tumor cases (Wilartratsami et al., 2014), metastatic lesions are rather common.

Primary tumors can be divided into benign or malignant tumors. In children, benign spinal tumors are more frequent than malignant ones, whereas malignant tumors such as multiple myelomas, osteosarcomas, and chordomas are more prevalent in adults (Boos and Aebi, 2008) (Fig. 1). Osseous tumors involving the spine are more frequently located in the vertebral body, but tumors such as osteoblastomas and osteoid osteomas in the posterior regions are not infrequent.

Metastatic lesions can have a slow- or fast-growing pattern, depending on a tumor's original source, such as breast, prostate, lung, and kidney cancers (Boos and Aebi, 2008; Brihaye et al., 1988). Common metastatic pathways include the venous system (Boos and Aebi, 2008; Batson, 1940); the arterial system, common for lung cancer (Coman et al., 1951); as well as direct extension to the spine from tumors in the pelvis or the mediastinum (Boos and Aebi, 2008). Similarly to primary bone tumors, metastatic lesions of the spine are more frequent in the vertebral bodies than in the posterior elements of the spine. Involvement of the intervertebral space is rare.

From a biomechanical point of view, the main consequence of a spinal tumor, independent of its origin, is an increased risk of vertebral fractures caused by deterioration of the quality of the involved bone, possibly with subsequent spinal cord and nerve root compression and thus neurological deficits. This aspect is the main focus of the following discussion and has a marked clinical relevance, because the prevention of vertebral fractures and the efficacy of the treatment strategies play a critical role in determining the quality of the lives of patients with spinal tumors. However, it should be noted that the clinical picture of such patients is generally complex, involving fever, weight loss, and a general deterioration of health, and other medical aspects that are distinct from those of patients without spinal tumors (Boos and Aebi, 2008). In addition to considering the risk of fracture, treatment for neoplastic diseases involving the spine is selected based on the patient's general health, the prognosis for survival, the type of tumor, pain levels, possible spinal cord and/or nerve root compression, as well as the possible presence of spinal instability (Whyne, 2014).

Biomechanics of the Spine. https://doi.org/10.1016/B978-0-12-812851-0.00023-9

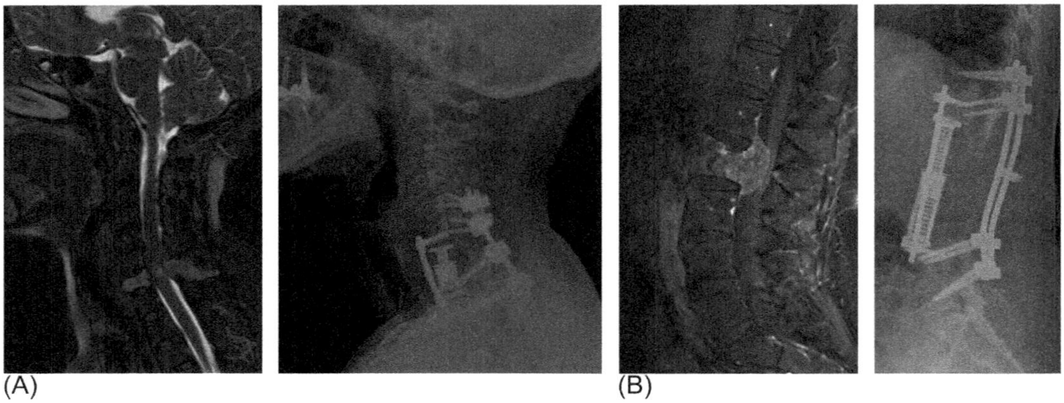

FIG. 1

Preoperative and postoperative images of patients suffering from primary tumors of the spine: (A) chondrosarcoma at C7; (B) chordoma at L2.

VERTEBRAL LESIONS

In healthy subjects, living bone is a highly optimized structure continually undergoing processes of growth, reinforcement, and resorption. The remodeling processes in living bone are the mechanisms by which the bone adapts its morphological structure to changes in long-term loading (Cowin and Hegedus, 1976). Bone formation and resorption are continuous throughout life and are necessary for the maintenance of a normal functional skeleton. The formation of bone by the osteoblast lineage cells results from the synthesis of the bone matrix followed by mineralization, whereas resorption by osteoclasts involves bone matrix degradation and dissolution of the mineral phase. The two processes are tightly coupled and allow bone to adapt itself to a variety of mechanical loads.

Spinal tumors, including vertebral metastases, alter the dynamic equilibrium between the bone formation and resorption processes, thus leading to changes in the density, the architecture, and in turn the mechanical strength of the bone tissue, which may be critical in many patients (Whyne, 2014). Lesions are generally distinguished according to two main classes: osteolytic, in which there is a general loss of bone tissue and reduced density due to predominance of the resorption process (Fig. 2); and osteoblastic, in which bone formation prevails, resulting in dense but disorganized bone tissue. Studies have shown that the risk of vertebral fracture is increased in all cases of vertebral lesions (Vassiliou et al., 2007). In addition, mixed lesions showing both osteolytic and osteoblastic features have frequently been observed (Fig. 3). The irregular distribution of tissue density has been attributed to the existence of local regions (pockets) of bone resorption and subsequent formation (Clarke, 2008; Burke et al., 2017); the pathological alteration of such mechanism might influence bone mineralization, leading to an inhomogeneous bone density distribution. In addition, recent studies conducted on animal models showed that metastatic involvement affects bone crystal dimensions as well (Burke et al., 2017).

The mechanical properties of spine neoplastic tissue have been investigated in a relatively small number of studies. In general, it was found that the pathological changes caused a decrease in bone stiffness and hardness, even in comparison with osteoporotic bone (Nazarian et al., 2008; Maimoun

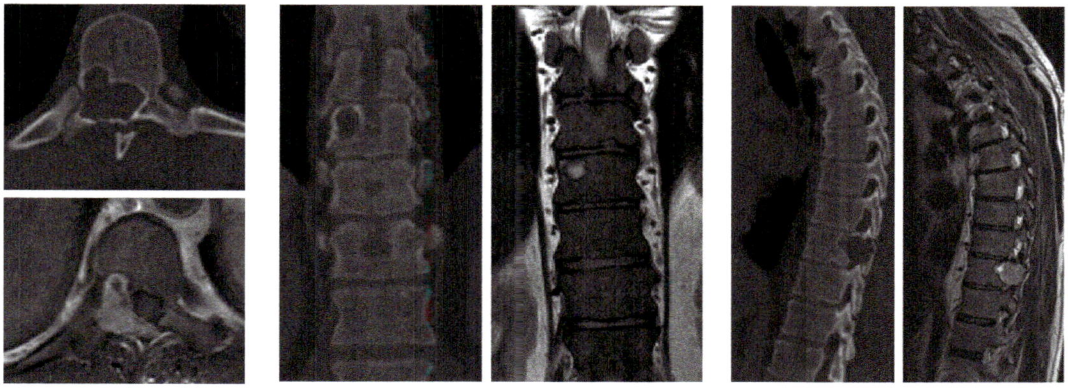

FIG. 2

CT and MRI scans showing a metastatic osteolytic lesion at T9, in axial (left), coronal (middle), and sagittal (right) views.

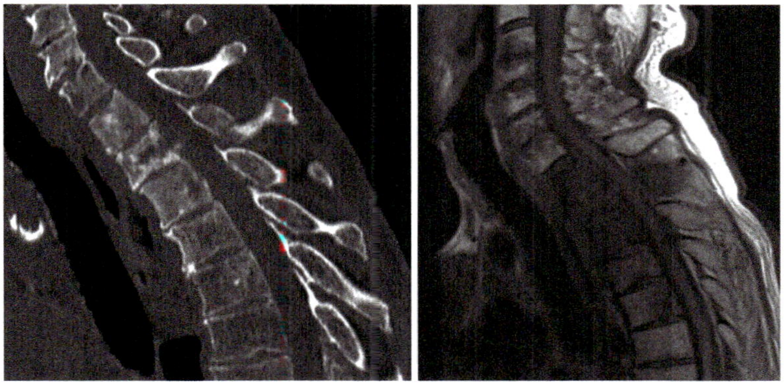

FIG. 3

CT (left) and MRI (right) scans showing several metastatic lesions from gastric cancer, with the major ones in T1, T2, and T3, with a mixed nature, including both osteolytic and osteoblastic areas. The posterior elements of T2 are also involved.

et al., 2012). Nazarian and colleagues tested metastatic material retrieved from the spines and femurs of human patients, analyzed the specimens by means of microCT and backscatter emission imaging, and finally tested them in uniaxial compression and nanoindentation. The authors found that the bone volume density (BV/TV) was the major determinant of the mechanical strength of the tissue, which agreed with previous investigations conducted on femoral specimens only (Kaneko et al., 2004). However, the study did not make a distinction between osteolytic and osteoblastic lesions. Studies conducted in animal models showed a reduction in tissue stiffness and strength both for osteolytic (Richert et al., 2015) and mixed osteolytic/osteoblastic lesions (Hojjat et al., 2012). Burke and colleagues (Burke et al., 2017) analyzed in detail the relevance of several features of osteoblastic lesions by means of backscatter emission imaging and nanoindentation in an animal model. The findings revealed a decrease in the average mineral content, density distribution, and crystal size and confirmed the negative impact of the osteoblastic process on the mechanical strength of the bone tissue.

IMAGING OF SPINAL TUMORS

Clinical CT and MR imaging is commonly used for the diagnosis and characterization of spinal tumors. In patients, investigations are aimed at identifying the location and size of the lesion, the involvement of the posterior elements, the degree of spinal canal invasion, the presence of fluid in the lesions, as well as possible sclerosis of the adjacent tissue (Boos and Aebi, 2008). Specifically, MRI scans are the modality of choice for the detection of alterations to bone marrow and the infiltration of the surrounding soft tissue as well as of the spinal cord and nerve roots. CT scans provide better information about the tissue calcification and sclerotic zones and can be used as guidance during biopsies. Nuclear imaging, which is typically a technetium-99m bone scan, is also frequently used in the diagnostic process because it highlights alterations in bone metabolism caused by the pathology.

From a biomechanical point of view, CT and microCT are the imaging modalities most commonly employed to investigate the neoplastic spine (Whyne, 2014). Both techniques provide information about the local tissue density and the extent of the lesions, but only limited information about the tissue composition. The high resolution of microCT allows analyzing the trabecular microarchitecture in order to extract parameters such as BV/TV, trabecular number, and thickness, and is commonly done in studies about osteoporosis (Chappard et al., 2008; Lespessailles et al., 2006). In contrast with the latter pathology, which has been studied extensively, only a few studies have investigated bone quality at the microscale level in the case of spinal tumors. Tamada and colleagues analyzed trabecular samples retrieved from the lumbar spine of six deceased patients and quantitatively compared normal, degenerative, sclerotic, and metastatic samples by means of microCT (Tamada et al., 2005). The analysis revealed an increased trabecular number and BV/TV in the osteoblastic regions of the metastatic specimens with respect to normal ones, whereas the trabecular thickness was not altered. The trabeculae showed a low degree of anisotropy in the osteoblastic lesions, indicating that the physiological remodeling process inducing a trabecular orientation following the main loading directions was compromised (Fig. 4). Three-dimensional renderings confirmed the disordered appearance with multiple concave resorption areas of the osteoblastic metastases, which are compatible with a decrease in the mechanical strength (Sone et al., 2004).

FRACTURE MECHANISMS

As previously mentioned, in the case of either osteolytic, osteoblastic, or mixed lesions, the risk of vertebral fracture was shown to be increased (Whyne, 2014; Vassiliou et al., 2007). Frequently, the fracture involves the anterior wall of the vertebra and not the posterior one, resulting in a relatively stable configuration named compression fracture, which is often treated conservatively in oncologic patients. On the contrary, in a burst fracture, the posterior wall is compromised, and there is a risk of spinal instability and displacement of bone fragments and/or pathological tissue into the spinal canal, with possible neurological consequences. In such cases, surgical stabilization is

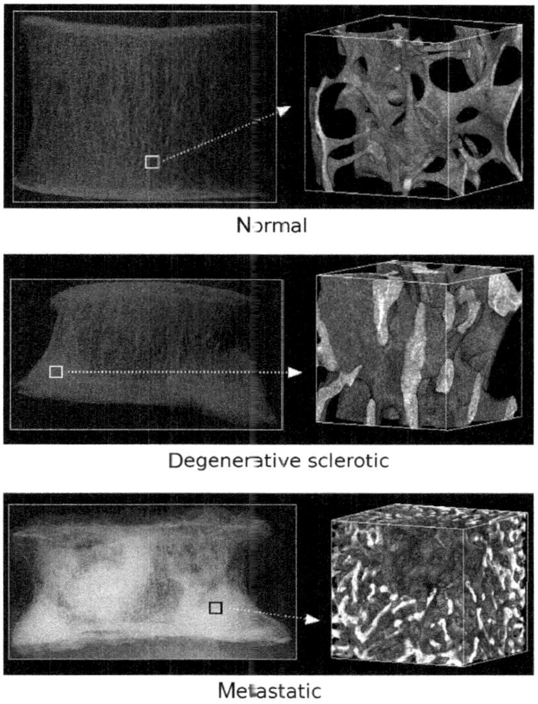

FIG. 4

X-rays images (left) and 3D reconstructions from microCT (right) of normal vertebral bone (top), degenerative sclerotic tissue (middle), and an osteoblastic metastasis from prostate cancer (bottom).

Reproduced with permission from Sone, T., Tamada, T., Jo, Y., Miyoshi, H., Fukunaga, M., 2004. Analysis of three-dimensional microarchitecture and degree of mineralization in bone metastases from prostate cancer using synchrotron microcomputed tomography. Bone 35(2), 432–438.

usually necessary. In healthy subjects, the mechanism of burst fractures has been related to the sudden pressurization of the vertebral fluid content under impact loading, which creates high tensile hoop stresses in the vertebral bodies (Holdsworth, 1970). It has been hypothesized that in vertebrae with cancer-related lesions, the loss in the tissue's mechanical strength described in the previous paragraph causes a similar fracture pattern even under physiologic or low-loading scenarios (Whyne, 2014; Whyne et al., 2003b). This concept was tested by means of parametric finite element analyses in which metastatically involved vertebrae were simulated by poroelastic models under impact loading in compression (Whyne et al., 2003b). In fact, osteolytic lesions have a high water content and a clear time-dependent mechanical response (Dimar et al., 1998; Whyne et al., 2000), which justifies the use of a poroelastic and biphasic modeling approach (Whyne et al., 2001). Various sizes of the lesion were simulated, as well as other variables such as bone density, pedicle involvement, loading rate, and condition of the intervertebral disc. Estimates of the vertebral bulge and the consequent narrowing of the spinal canal highlighted the fundamental role of the tumor's size and of the bone density in determining the mechanical response, and the possible protective role of disc degeneration,

Table 1 Summary of the Fracture Mechanisms in Metastatically Involved Vertebrae Depending on the Size of the Lesions and the Loading Rate (Whyne et al., 2003b)

	Low Loading Rate	High Loading Rate
Small lesion	• Lowest risk of fracture	• Pressurization of the intervertebral disc • High risk of endplate fracture
Large lesion	• High vertebral bulge • High stress in bone tissue	• Pressurization of the fluid inside the lesion • High risk of burst fracture

which may slightly reduce the risk of fracture. In addition, higher loading rates caused a reduction in vertebral strains and pressurization of the inner vertebral fluid, likely due to the fluid flow through the endplate, which is limited by its low permeability. In such cases, the pressurization of the intervertebral disc might involve a fracture of the endplate rather than a vertebral burst fracture. Similar findings were described via in vitro tests conducted on intact vertebrae (Tran et al., 1995). However, in the presence of a large osteolytic lesion, its pressurization might cause a burst fracture before the endplate reaches a critical stress level (Table 1). Other finite element studies using poroelastic models focused on the location and shape of the lesion in the thoracic spine (Tschirhart et al., 2004, 2007) (Fig. 5). The tumor's location was found to be a critical factor, but still of lower relevance than the bone density and tumor size. Tumors located close to the posterior wall had a higher negative impact on the risk of fracture initiation, whereas the involvement of the cortical shell resulted in a decreased risk of burst fracture with respect to centrally located lesions but in a largely increased vertebral bulge because of the compromised anterior wall (Alkalay, 2015). The authors also determined that the upper vertebrae are more at risk of burst fractures than lower ones and that a high kyphotic angle as well as the presence of the rib cage act as protecting factors.

FRACTURE RISK PREDICTION

In clinical practice, the Spinal Instability Neoplastic Score (SINS) is used to characterize the stability of a spine with tumor lesions (Table 2) (Fisher et al., 2010). This classification system is aimed at supporting management of the patient and selection of treatment, and it incorporates biomechanical aspects as well as neurological, oncological, and systemic factors. The score is calculated for each lesion; that is, values obtained for distinct lesions should not be summed. Mechanical factors taken into account include the location of the lesion, either in a stiff or flexible region of the spine; the spinal alignment; the possible presence of compression fractures; the nature of the lesion (osteolytic, osteoblastic, or mixed); as well as the size of the lesion. If a high score indicating an unstable spine is assessed, surgical consultation should be performed prior to planning radiation treatment. The classification system has gained good interobserver agreement among both radiologists and radiation oncologists (Fisher et al., 2014a,b).

Findings from biomechanical investigations have been used to create guideline equations able to estimate the risk of a burst fracture in metastatically involved spines based on parameters that can be collected by means of imaging techniques, such as the size of the lesion, the bone density, the degenerative conditions of the intervertebral disc, and the involvement of the pedicles (Whyne et al., 2003a; Roth et al.,

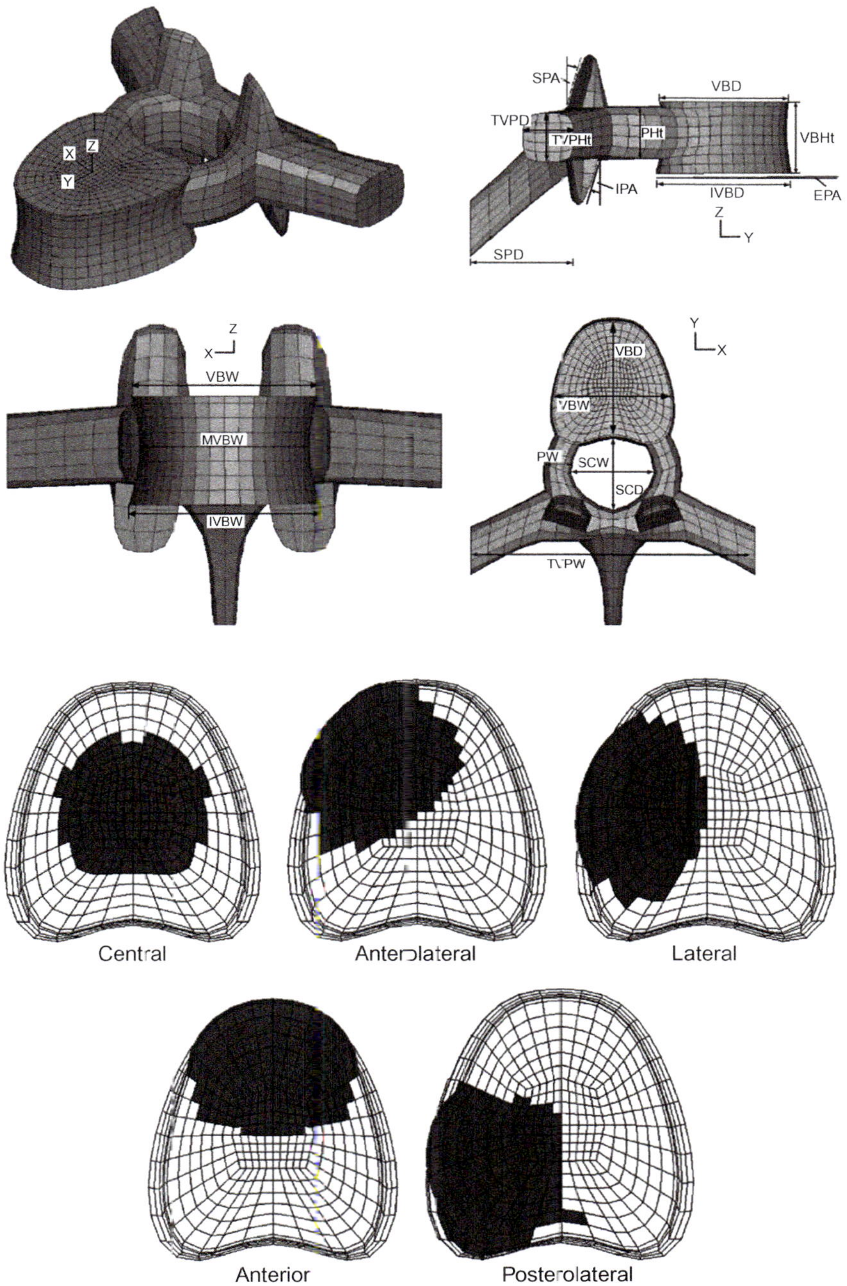

FIG. 5

Parametric finite element model of a thoracic vertebra (top) used to investigate the relevance of the location of the lesion and the involvement of the cortical shell (bottom). *IVBD*, inferior vertebral body depth; *IVBW*, inferior vertebral body width; *IPA*, inferior process angle; *MVBW*, mid-vertebral body width; *PHt*, pedicle height; *PW*, pedicle width; *SCD*, spinal canal depth; *SCW*, spinal canal width; *SPA*, superior process angle; *SPD*, spinous process depth; *TVPD*, transverse process depth; *TVPHt*, transverse process height; *TVPW*, transverse process width; *VBD*, vertebral body depth; *VBHt*, vertebral body height; *VBW*, vertebral body width.

Reproduced with permission from Tschirhart, C.E., Finkelstein, J.A., Whyne, C.M., 2007. Biomechanics of vertebral level, geometry, and transcortical tumors in the metastatic spine. J. Biomech. 40(1), 46–54.

Table 2 The Spinal Instability Neoplastic Score (SINS) Used to Determine the Risk of Instability of the Neoplastic Spine (Fisher et al., 2010)

Location	
Junctional (C0-C2, C7-T2, T11-L1, L5-S1)	3 points
Mobile spine (C3-C6, L2-L4)	2 points
Semirigid spine (T3-T10)	1 point
Rigid spine (S2-S5)	0 points
Pain	
Mechanical pain (with movement or in standing, improves with recumbency)	3 points
Occasional pain but not mechanical	1 point
No pain	0 points
Bone lesion	
Osteolytic	2 points
Mixed	1 point
Osteoblastic	0 points
Spinal alignment	
Subluxation/translation	4 points
De novo deformity	2 points
Normal alignment	0 points
Vertebral body collapse	
>50% collapse	3 points
<50% collapse	2 points
No collapse with >50% vertebral body involved	1 point
No collapse with <50% vertebral body involved	0 points
Involvement of the posterior elements	
Bilateral	3 points
Unilateral	1 point
None	0 points

Score 0–6: no instability; 7–12: potential instability; 13–18: instability.

2004). Algebraic equations that can be used directly to estimate the values of vertebral bulge and axial displacement under the application of an impact load were developed based on the results of validated finite element models. The authors estimated that predicted values corresponding to strains higher than 0.8% indicate a high risk of burst fracture. Despite the fact that the validity of such equations was confirmed by a retrospective analysis of vertebrae with osteolytic lesions, to our knowledge the guideline equations are not widely used in clinical practice. It should be noted that these equations can predict only the risk of fracture for osteolytic lesions, whereas osteoblastic and mixed lesions are not covered.

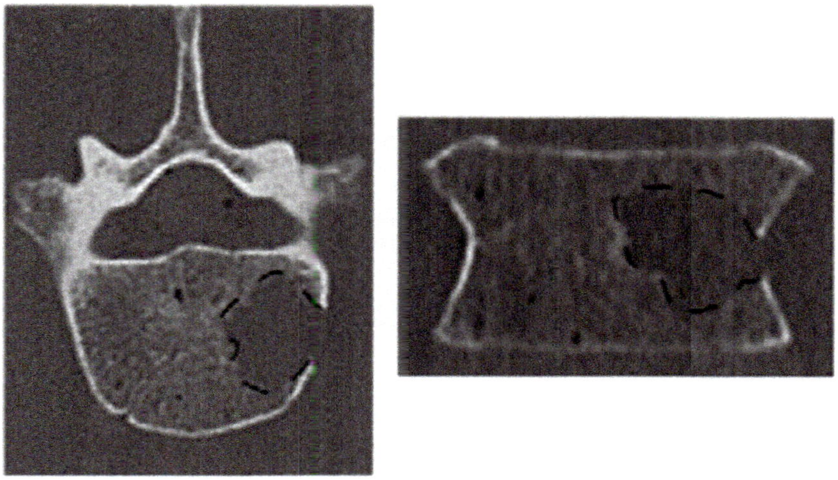

FIG. 6

Example of an artificial defect mimicking an osteolytic lesion in a human cadaver specimen, aimed at investigating the initiation and progression of vertebral failure.

Reproduced with permission from Alkalay, R.N., 2015. Effect of the metastatic defect on the structural response and failure process of human vertebrae: an experimental study. Clin. Biomech. 30(2), 121–128.

EXPERIMENTAL TESTING OF NEOPLASTIC SPINES

Cadaveric specimens have been used extensively to study the biomechanics of spines with neoplasms and metastatic involvement. Although such specimens with tumors are not difficult to retrieve, the variability of size, location, and biomechanical characteristics of the lesions are a barrier to the specimens' use in in vitro testing, in which reproducibility of the results should be assessed.

Methods to artificially create lesions with biomechanical features similar to those of the tumoral tissue have been established. Osteolytic lesions have been simulated by removing trabecular bone from the center of the vertebral bone through a hole drilled from the lateral side (Alkalay, 2015; Matsuura et al., 2014) (Fig. 6). Other authors have simulated lytic defects in other locations (Whealan et al., 2000; Windhagen et al., 1997, 2000). This approach assumed that the lesions were not able to sustain any load—which is in contrast with studies that highlighted the relevance of the pressurization of the fluid contained in the lesion under impact loading (Whyne et al., 2003b)—but could be deemed to have a minor impact in quasistatic testing. As a matter of fact, burst fractures could not be replicated by using such cadaver models, whereas compression fractures not involving the posterior vertebral wall were demonstrated (Whealan et al., 2000; Windhagen et al., 1997). An improvement to this approach was achieved by Whyne and colleagues (Whyne et al., 2003b), who filled the artificial lesions with an agarose gel and sealed the hole used to create the lesion by means of a cap of bone tissue and polymethylmethacrylate. Indeed, this setup allowed obtaining fracture patterns more closely resembling the typical burst fractures observed clinically.

Animal models have also been used extensively to investigate spinal tumors (Fig. 7). Neoplasms have been induced in rodents by xenotransplantation of human cancer cells in primary tumor sites

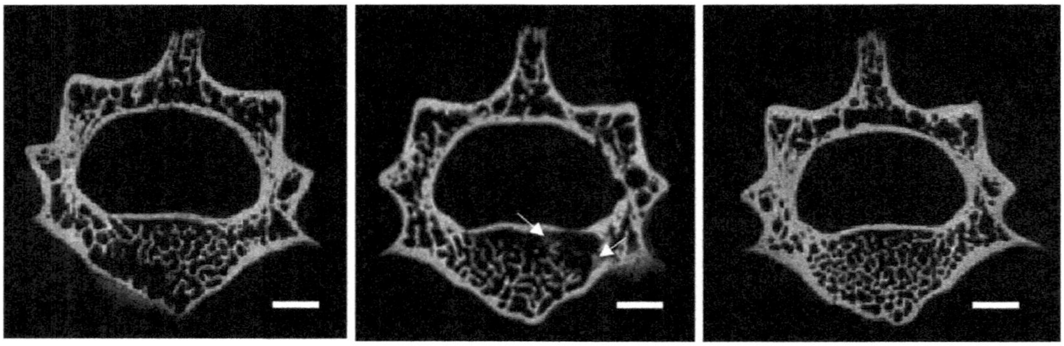

FIG. 7

microCT images of rat vertebrae inoculated with human cervical cancer cells (left), with canine prostate cancer cells (middle), and from the healthy control group (right), clearly showing osteolytic lesions in the inoculated specimens.

Reproduced with permission from Burke, M., Atkins, A., Kiss, A., Akens, M., Yee, A., Whyne, C., 2017. The impact of metastasis on the mineral phase of vertebral bone tissue. J. Mech. Behav. Biomed. Mater. 69, 75–84.

such as the breast and prostate glands in order to create bone metastases or into the circulatory system. Direct injection into bones has been used to investigate the lytic lesions of long bones (Kurth and Muller, 2001; Arrington et al., 2006), but is deemed to have little utility for spinal tumors because of the model's small size and therefore the bone damage caused by the direct injection (Whyne, 2014). Intracardiac injection of cancer cells is now the most commonly employed technique and has been used for the morphological and biomechanical investigation of the tumor tissue (Hojjat et al., 2012; Hardisty et al., 2012) as well as to evaluate the efficacy of various treatment options (Lo et al., 2012, 2013; Won et al., 2010; Akens et al., 2010).

REFERENCES

Akens, M.K., Hardisty, M.R., Wilson, B.C., Schwock, J., Whyne, C.M., Burch, S., Yee, A.J., 2010. Defining the therapeutic window of vertebral photodynamic therapy in a murine pre-clinical model of breast cancer metastasis using the photosensitizer BPD-MA (Verteporfin). Breast Cancer Res. Treat. 119 (2), 325–333.

Alkalay, R.N., 2015. Effect of the metastatic defect on the structural response and failure process of human vertebrae: an experimental study. Clin. Biomech. 30 (2), 121–128.

Arrington, S.A., Schoonmaker, J.E., Damron, T.A., Mann, K.A., Allen, M.J., 2006. Temporal changes in bone mass and mechanical properties in a murine model of tumor osteolysis. Bone 38 (3), 359–367.

Batson, O.V., 1940. The function of the vertebral veins and their role in the spread of metastases. Ann. Surg. 112 (1), 138–149.

Boos, N., Aebi, M., 2008. Spinal Disorders—Fundamentals of Diagnosis and Treatment. Springer, Heidelberg.

Brihaye, J., Ectors, P., Lemort, M., Van Houtte, P., 1988. The management of spinal epidural metastases. Adv. Tech. Stand. Neurosurg. 16, 121–176.

Burke, M., Atkins, A., Kiss, A., Akens, M., Yee, A., Whyne, C., 2017. The impact of metastasis on the mineral phase of vertebral bone tissue. J. Mech. Behav. Biomed. Mater. 69, 75–84.

Chappard, D., Baslé, M., Legrand, E., Audran, M., 2008. Trabecular bone microarchitecture: a review. Morphologie 92 (299), 162–170.
Clarke, B., 2008. Normal bone anatomy and physiology. Clin. J. Am. Soc. Nephrol. 3 (Suppl. 3), S131–9.
Coman, D.R., de Long, R.P., MccUtcheon, M., 1951. Studies on the mechanisms of metastasis; the distribution of tumors in various organs in relation to the distribution of arterial emboli. Cancer Res. 11 (8), 648–651.
Cowin, S.C., Hegedus, D.H., 1976. Bone remodeling I: theory of adaptive elasticity. J. Elast. 6 (3), 313–326.
Dimar, J.R., Voor, M.J., Zhang, Y.M., Glassman, S.D., 1998. A human cadaver model for determination of pathologic fracture threshold resulting from tumorous destruction of the vertebral body. Spine 23 (11), 1209–1214.
Fisher, C.G., DiPaola, C.P., Ryken, T.C., Bilsky, M.H., Shaffrey, C.I., Berven, S.H., Harrop, J.S., Fehlings, M.G., Boriani, S., Chou, D., Schmidt, M.H., Polly, D.W., Biagini, R., Burch, S., Dekutoski, M.B., Ganju, A., Gerszten, P.C., Gokaslan, Z.L., Groff, M.W., Liebsch, N.J., Mendel, E., Okuno, S.H., Patel, S., Rhines, L.D., Rose, P.S., Sciubba, D.M., Sundaresan, N., Tomita, K., Varga, P.P., Vialle, L.R., Vrionis, F.D., Yamada, Y., Fourney, D.R., 2010. A novel classification system for spinal instability in neoplastic disease: an evidence-based approach and expert consensus from the Spine Oncology Study Group. Spine 35 (22), E1221–9.
Fisher, C.G., Schouten, R., Versteeg, A.L., Boriani, S., Varga, P.P., Rhines, L.D., Kawahara, N., Fourney, D., Weir, L., Reynolds, J.J., 2014a. Reliability of the spinal instability neoplastic score (SINS) among radiation oncologists: an assessment of instability secondary to spinal metastases. Radiat. Oncol. 9 (1), 69.
Fisher, C.G., Versteeg, A.L., Schouten, R., Boriani, S., Varga, P.P., Rhines, L.D., Heran, M.K., Kawahara, N., Fourney, D., Reynolds, J.J., 2014b. Reliability of the spinal instability neoplastic scale among radiologists: an assessment of instability secondary to spinal metastases. Am. J. Roentgenol. 203 (4), 869–874.
Hardisty, M.R., Akens, M.K., Hojjat, S.P., Yee, A., Whyne, C.M., 2012. Quantification of the effect of osteolytic metastases on bone strain within whole vertebrae using image registration. J. Orthop. Res.: Off. Publ. Orthop. Res. Soc. 30 (7), 1032–1039.
Hojjat, S., Foltz, W., Wise-Milestone, L., Whyne, C.M., 2012. Multimodal μCT/μMR based semiautomated segmentation of rat vertebrae affected by mixed osteolytic/osteoblastic metastases. Med. Phys. 39 (5), 2848–2853.
Holdsworth, F., 1970. Fractures, dislocations, and fracture-dislocations of the spine. J. Bone Joint Surg. Am. 52 (8), 1534–1551.
Kaneko, T.S., Bell, J.S., Pejcic, M.R., Tehranzadeh, J., Keyak, J.H., 2004. Mechanical properties, density and quantitative CT scan data of trabecular bone with and without metastases. J. Biomech. 37 (4), 523–530.
Kurth, A.A., Muller, R., 2001. The effect of an osteolytic tumor on the three-dimensional trabecular bone morphology in an animal model. Skelet. Radiol. 30 (2), 94–98.
Lespessailles, E., Chappard, C., Bonnet, N., Benhamou, C.L., 2006. Imaging techniques for evaluating bone microarchitecture. Joint Bone Spine 73 (3), 254–261.
Lo, V.C., Akens, M.K., Moore, S., Yee, A.J., Wilson, B.C., Whyne, C.M., 2012. Beyond radiation therapy: photodynamic therapy maintains structural integrity of irradiated healthy and metastatically involved vertebrae in a pre-clinical in vivo model. Breast Cancer Res. Treat. 135 (2), 391–401.
Lo, V.C., Akens, M.K., Wise-Milestone, L., Yee, A.J., Wilson, B.C., Whyne, C.M., 2013. The benefits of photodynamic therapy on vertebral bone are maintained and enhanced by combination treatment with bisphosphonates and radiation therapy. J. Orthop. Res. 31 (9), 1398–1405.
Maimoun, L., Brennan-Speranza, T.C., Rizzoli, R., Ammann, P., 2012. Effects of ovariectomy on the changes in microarchitecture and material level properties in response to hind leg disuse in female rats. Bone 51 (3), 586–591.
Matsuura, Y., Giambini, H., Ogawa, Y., Fang, Z., Thoreson, A.R., Yaszemski, M.J., Lu, L., An, K.N., 2014. Specimen-specific nonlinear finite element modeling to predict vertebrae fracture loads after vertebroplasty. Spine 39 (22), E1291–6.
Nazarian, A., von Stechow, D., Zurakowski, D., Muller, R., Snyder, B.D., 2008. Bone volume fraction explains the variation in strength and stiffness of cancellous bone affected by metastatic cancer and osteoporosis. Calcif. Tissue Int. 83 (6), 368–379.

Richert, L., Keller, L., Wagner, Q., Bornert, F., Gros, C., Bahi, S., Clauss, F., Bacon, W., Clézardin, P., Benkirane-Jessel, N., 2015. Nanoscale stiffness distribution in bone metastasis. World J. Nano Sci. Eng. 5 (04), 219.

Roth, S.E., Mousavi, P., Finkelstein, J., Chow, E., Kreder, H., Whyne, C.M., 2004. Metastatic burst fracture risk prediction using biomechanically based equations. Clin. Orthop. Relat. Res. 419, 83–90.

Sone, T., Tamada, T., Jo, Y., Miyoshi, H., Fukunaga, M., 2004. Analysis of three-dimensional microarchitecture and degree of mineralization in bone metastases from prostate cancer using synchrotron microcomputed tomography. Bone 35 (2), 432–438.

Tamada, T., Sone, T., Jo, Y., Imai, S., Kajihara, Y., Fukunaga, M., 2005. Three-dimensional trabecular bone architecture of the lumbar spine in bone metastasis from prostate cancer: comparison with degenerative sclerosis. Skelet. Radiol. 34 (3), 149–155.

Tran, N.T., Watson, N.A., Tencer, A.F., Ching, R.P., Anderson, P.A., 1995. Mechanism of the burst fracture in the thoracolumbar spine: the effect of loading rate. Spine 20 (18), 1984–1988.

Tschirhart, C.E., Nagpurkar, A., Whyne, C.M., 2004. Effects of tumor location, shape and surface serration on burst fracture risk in the metastatic spine. J. Biomech. 37 (5), 653–660.

Tschirhart, C.E., Finkelstein, J.A., Whyne, C.M., 2007. Biomechanics of vertebral level, geometry, and transcortical tumors in the metastatic spine. J. Biomech. 40 (1), 46–54.

Vassiliou, V., Kalogeropoulou, C., Petsas, T., Leotsinidis, M., Kardamakis, D., 2007. Clinical and radiological evaluation of patients with lytic, mixed and sclerotic bone metastases from solid tumors: is there a correlation between clinical status of patients and type of bone metastases? Clin. Exp. Metastas. 24 (1), 49–56.

Whealan, K.M., Kwak, S.D., Tedrow, J.R., Inoue, K., Snyder, B.D., 2000. Noninvasive imaging predicts failure load of the spine with simulated osteolytic defects. J. Bone Joint Surg. Am. 82 (9), 1240–1251.

Whyne, C.M., 2014. Biomechanics of metastatic disease in the vertebral column. Neurol. Res. 36 (6), 493–501.

Whyne, C.M., Hu, S.S., Workman, K.L., Lotz, J.C., 2000. Biphasic material properties of lytic bone metastases. Ann. Biomed. Eng. 28 (9), 1154–1158.

Whyne, C.M., Hu, S.S., Lotz, J.C., 2001. Parametric finite element analysis of verterbral bodies affected by tumors. J. Biomech. 34 (10), 1317–1324.

Whyne, C.M., Hu, S.S., Lotz, J.C., 2003a. Biomechanically derived guideline equations for burst fracture risk prediction in the metastatically involved spine. J. Spinal Disord. Tech. 16 (2), 180–185.

Whyne, C.M., Hu, S.S., Lotz, J.C., 2003b. Burst fracture in the metastatically involved spine: Development, validation, and parametric analysis of a three-dimensional poroelastic finite-element model. Spine 28 (7), 652–660.

Wilartratsami, S., Muangsomboon, S., Benjarassameroj, S., Phimolsarnti, R., Chavasiri, C., Luksanapruksa, P., 2014. Prevalence of primary spinal tumors: 15-year data from Siriraj Hospital. J. Med. Assoc. Thailand 97 (Suppl. 9), S83–7.

Windhagen, H.J., Hipp, J.A., Silva, M.J., Lipson, S.J., Hayes, W.C., 1997. Predicting failure of thoracic vertebrae with simulated and actual metastatic defects. Clin. Orthop. Relat. Res. 344 (344), 313–319.

Windhagen, H., Hipp, J.A., Hayes, W.C., 2000. Postfracture instability of vertebrae with simulated defects can be predicted from computed tomography data. Spine 25 (14), 1775–1781.

Won, E., Wise-Milestone, L., Akens, M.K., Burch, S., Yee, A.J., Wilson, B.C., Whyne, C.M., 2010. Beyond bisphosphonates: photodynamic therapy structurally augments metastatically involved vertebrae and destroys tumor tissue. Breast Cancer Res. Treat. 124 (1), 111–119.

Glossary of Spine Biomechanics

Axis of Rotation In 3D, the *instantaneous axis of rotation* (IAR) is a straight line fixed to a moving body around which all points of the body rotate at a specific instant of time. In practical applications, 3D finite motions (over a non-infinitesimal time period) that are not pure translations are often decomposed in a rotation around an axis and a translation around the same axis, which is in this case referred to as *helical axis* or *screw axis*. Several computational methods based on optimization procedures can be used to estimate the screw axis of a finite motion. For the 2D equivalent, see "Center of Rotation."

Center of Rotation In 2D, the *instantaneous center of rotation* (ICR) is a point fixed to a moving body that has zero velocity at a specific instant of time (Fig. 1); that is, all the points of the body are moving in a circular trajectory centered on the ICR. In practical applications such as in vitro testing, the center of rotation is usually estimated over a finite time period rather for a specific time instant, by means of least squares procedures or other algorithms based on the coordinates of landmarks/markers at the beginning and at the end of the time period under consideration. In such cases, the center of rotation should be more rigorously named *pole of the displacement*, but the expression *center of rotation* is also common. The robustness of the estimation of the pole of the displacement increases with the magnitude of the rotation. For the 3D equivalent, see "Axis of Rotation."

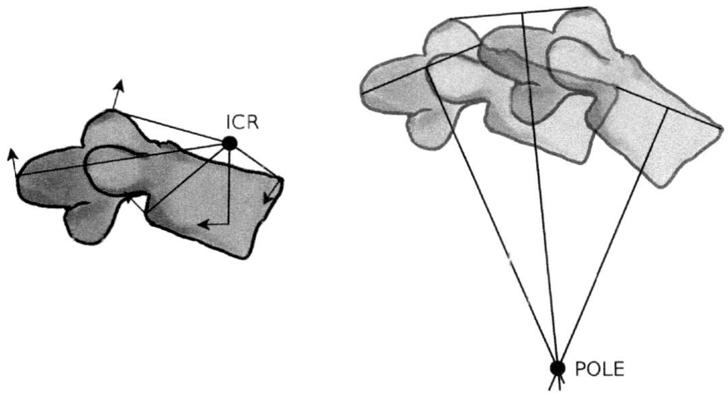

FIG. 1

Instantaneous center of rotation (ICR) of a body moving in a 2D space *(left)*; all points of the body are rotating around the ICR in an infinitesimal motion at a specific time instant. Pole of the displacement (POLE) of a finite planar motion *(right)*, which is commonly used as an estimate of the ICR in practical applications.

Clinical Instability Clinical instability refers to loss of the ability of the spine, or of a specific motion segment, to maintain its patterns of displacement under physiological loads. Clinical instability is typically associated with degenerative conditions of the spine and is believed to be related to pain and neurologic dysfunction. Although it is commonly associated with hypermobility, its precise biomechanical definition and its diagnosis in patients are frequently matters of debate. An increase of the *neutral zone* (NZ) has been hypothesized as a good indicator of clinical instability.

Cobb Angle With reference to a specific spine region, the Cobb angle is defined as the angle between a line parallel to the superior endplate to the most cranial vertebra of the region and a line parallel to the lower endplate of the most caudal vertebra in a radiographic projection. The Cobb angle was introduced to assess spinal deformities in the coronal plane and is now also widely used for sagittal views. It can be measured manually or with the support of computer software on digital images.

Coordinate System A conventional 3D coordinate system, compliant with the standard ISO/DIS 2631, is generally accepted to describe loading directions and motions of the spine (Fig. 2). In in vitro tests, the origin of the coordinate system is in the middle of the lower side of the most cranial vertebra.

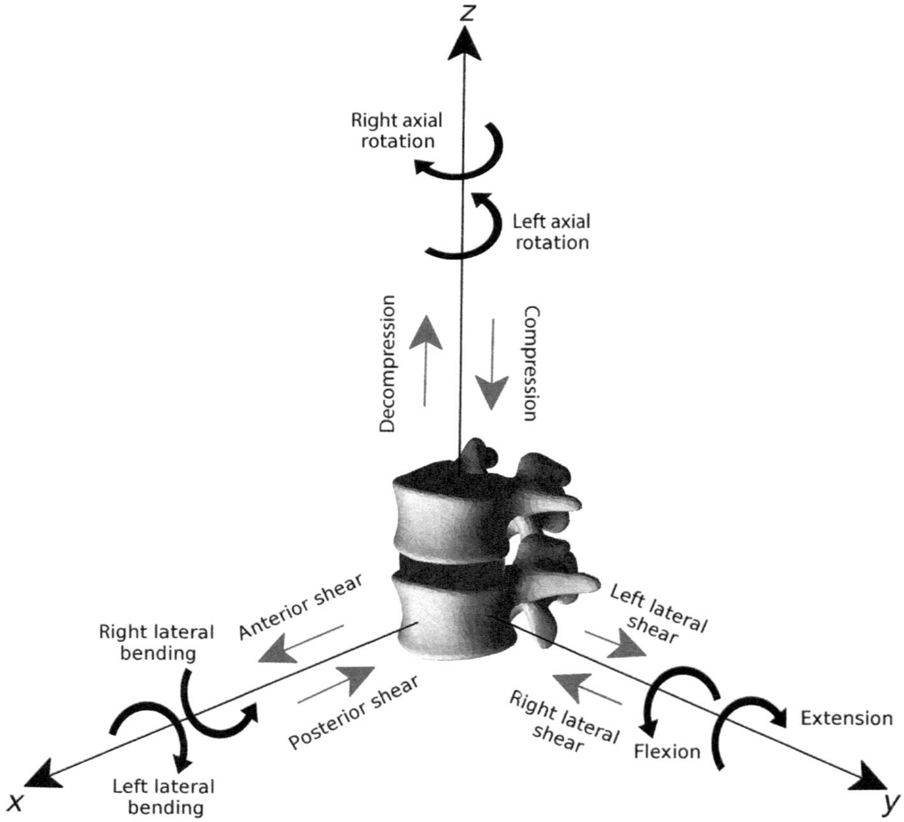

FIG. 2

Conventional 3D coordinate system and naming of the loading directions and motions of the spine.

Coupled Motion Under pure moment unconstrained loading, the *coupled motion* is a rotation occurring around a physiological axis different from the axis of the applied load. In the cervical and lumbar spine regions, lateral bending and axial rotation are coupled; that is, if a lateral bending moment is applied, a coupled axial rotation also occurs. The extent of the coupled motions is related to the posterior elements, in particular with the orientation of the facet joints.

Creep Creep refers to the viscoelastic phenomenon in which a material subjected to a constant stress exhibits a time-dependent strain (Fig. 3). Creep tests in compression, which frequently involve load cycles constituted by multiple loading and resting phases, are commonly performed to investigate the time-dependent response of the intervertebral disc as well as the fluid exchange through the endplates, see "Relaxation."

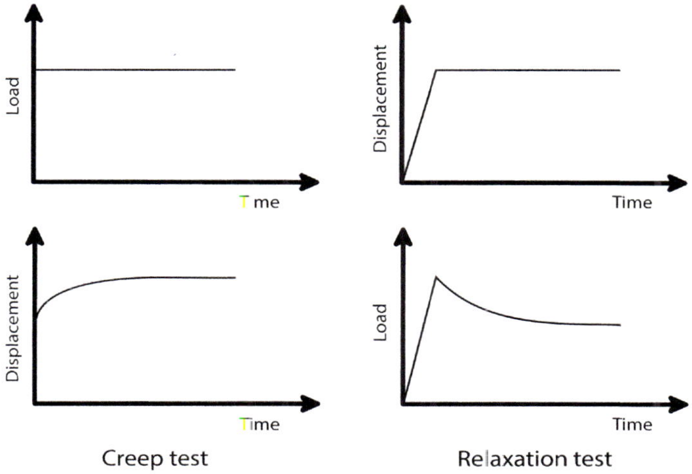

FIG. 3

Creep *(left)* and relaxation tests *(right)* that can be used to investigate the time-dependent response of a viscoelastic or of a porous fluid-filled (poroelastic) material. In a creep test, the load is constant and the specimen exhibits a time-dependent strain; in a relaxation test, the strain is kept fixed and the load and stress decrease over time.

Dynamics Dynamics refers to the branch of mechanics describing the effects of forces and moments on the motion of bodies. It differs from kinematics, in which the motion of bodies is studied without reference to its mechanical causes. In spine research, dynamic testing refers to experiments conducted in conditions in which inertia effects cannot be disregarded, such as the simulation of injury and whiplash. Cyclic dynamic testing (*Dynamic Mechanical Analysis*, DMA) is also frequently used to evaluate the viscoelastic and poroelastic properties of biological tissues.

Elasticity Elasticity refers to the capability of a body subjected to mechanical loading to resist deformation and to return to its original shape after the load is removed. An ideal elastic material does not exhibit a time-dependent response (such as *viscoelastic* materials) and permanent deformations, that is, deformations that do not disappear after unloading (see "Plasticity"). Perfectly elastic materials do not dissipate energy when a load is applied or removed.

Elastic Zone Elastic zone (EZ) refers to the amplitude of the motion in a specific direction between the end of the *neutral zone* and the application of the maximal load. EZ is usually determined from the load-displacement curve (Fig. 4). The slope of the latter curve in the EZ corresponds to the *stiffness* of the specimen in that specific loading direction.

Flexibility Flexibility refers to the capability of the spine to deform in response to an applied load in a specific loading direction (see "Stiffness"), typically rotational and measured under the application of a *pure moment* in quasi-static conditions (see "Statics").

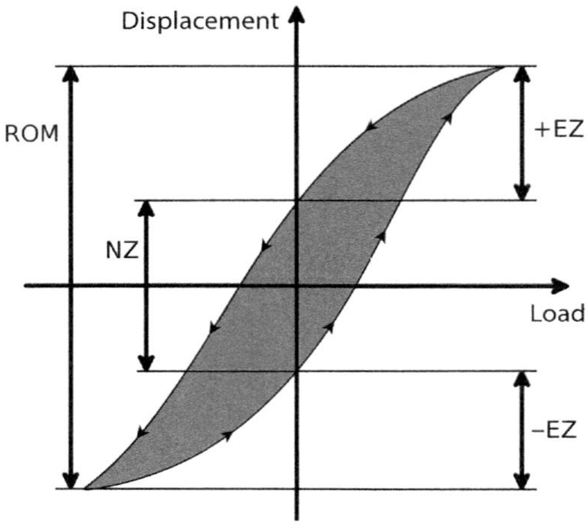

FIG. 4

Schematic load-displacement curve obtained for a loading cycle (e.g., in flexion-extension), showing the range of motion (ROM), the neutral zone (NZ), and the elastic zone (sum of +EZ and −EZ). The area of the hysteresis curve (in gray) represents the dissipated energy in form of heat.

Follower Load A follower load is a compressive load, which can be applied in both in vitro tests and numerical simulations, mimicking the effect of the body weight and of the back muscles acting to keep the standing posture. The follower load should induce a pure compression of the motion segments, with no or negligible intervertebral rotations.

Hybrid Protocol Hybrid protocol refers to a loading protocol used in in vitro tests and numerical simulations to assess the effect of implants on the adjacent segments. It consists of two steps: (1) application of a pure moment to the intact spine in order to determine its global range of motion (ROM), and (2) application of a pure moment to the implanted spine until the global rotation equals the ROM previously measured. The accuracy of the hybrid protocol is debated because its results are dependent on the number of motion segments in the specimen and because there is no clear recommendation for selection of the correct moment to be applied in the intact conditions.

Intradiscal Pressure Intradiscal pressure is the hydrostatic pressure in the center of the nucleus pulposus and is usually measured by means of pressure transducers during in vitro testing. The intradiscal pressure is affected by external loads, but is not null even in unloaded conditions caused by disc *swelling*.

Kinematics Kinematics is the branch of mechanics describing the motion of bodies. In spinal cord science, the definition of kinematics is not rigorously interpreted, and the expression "spinal kinematics" usually refers to flexibility-related qualities such as ROM, NZ, and so on.

Loading Directions Based on the conventional 3D coordinate system, three primary loading directions aligned with the three axes of the coordinate system are defined (Fig. 2). Translational load and motion components are named: anterior and posterior shear, left and right lateral shear, and decompression and compression. Rotational load and motion components are named: flexion and extension, left and right lateral bending, and left and right axial rotation.

Motion Segment/Functional Spine Unit This unit refers to a set of two adjacent vertebrae, the in-between intervertebral disc and the adjoining ligaments. It is the smallest motion unit of the spine, the mechanical behavior of which represents that of a longer spine region or of the entire spine.

Neutral Zone NZ refers to the amplitude of the motion that can be obtained by applying a minimal load in an in vitro test, which describes the laxity of the specimen. NZ is usually determined from the load-displacement curve as the amplitude of the hysteresis area in correspondence with the zero-load axis (Fig. 4), and is considered to be an indicator of clinical instability.

Plasticity Plasticity refers to the characteristics of a material to show nonreversible deformation after being subjected to a mechanical load. In biological tissues, plastic deformations are associated with changes in the microstructure of the material such as microcracks, and thus to damage.

Permeability Permeability is the capability of a porous material to allow fluid to flow through it. Permeability is related to a material's porosity, that is, to its void fraction, as well as to the shape and connectivity of the pores. In Darcy's law, which is widely used in finite element modeling, permeability acts as a proportionality coefficient relating the fluid discharge and the pressure gradient.

Poroelasticity Poroelasticity refers to the branch of physics describing the mechanical behavior of fluid-filled porous materials, that is, the interactions between the elastic behavior of the solid porous matrix and the flow of the fluid filling it. *Poromechanics* is the generalization of poroelasticity that covers the possible inelastic mechanical response of the solid phase. In poroelasticity, the elastic deformation of the solid matrix influences the fluid flow and vice versa. In the most common theoretical framework of poroelasticity, fluid flow is governed by Darcy's law (see "Permeability"). Poroelastic models are widely used to simulate the mechanics of the intervertebral disc, as the disc is indeed a fluid-filled porous structure and such models are capable of capturing its creep and relaxation response.

Preconditioning Preconditioning refers to loads applied to a specimen prior to an in vitro test with the aim of minimizing viscoelastic effects. In flexibility testing, a few loading cycles are usually applied, and the first ones (at least two) are considered as preconditioning and are discarded from data analysis. Prior to *creep* or *relaxation* testing, a compressive sustained load may be applied in order to reduce the hydration of the specimen to physiological levels. Preconditioning can also be simulated in finite element analysis when a time-dependent response is predicted, such as with viscoelastic or poroelastic models.

Pure Moment Pure moment refers to a mechanical load, typically applied in in vitro tests by means of electric motors or pneumatic actuators, consisting of a moment with a zero resultant force. Pure moment loading is the de facto standard for flexibility testing of the spine. A pure moment application can be combined with a *follower load* mimicking the actions of the body weight and back muscles. Pure moments are not suitable for the investigation of the effects of implants on the adjacent motion segments (see "Hybrid Protocol"), which should not be affected by the implantation if the loading protocol is correctly implemented.

Radiological Instability Radiological instability refers to the hypermobility of the motion segment, either translational or rotational, observed by means of radiological methods. It is typically diagnosed by means of functional X-rays in flexion-extension (dynamic X-rays) that can be used to calculate the ROM, or by comparing medical images in the orthostatic and recumbent postures. Threshold values to define radiological instability have been published for both the rotational ROM and the anteroposterior translation in flexion-extension. Radiological instability is not necessarily associated with symptoms.

Range of Motion ROM refers to the potential mobility of a motion segment or of a spine region in physiologic conditions, usually rotational and measured in degrees. The ROM is associated with a loading direction; that is, ROMs in flexion-extension, lateral bending, and axial rotation are typically assessed. In in vitro tests as well as in finite element analysis, the ROM of a motion segment is defined as the difference between the relative rotations of the two vertebrae, that is, under the application of a physiological load with respect to the unloaded condition. The ROM is the sum of the *neutral zone* and the *elastic zone* (Fig. 4); for a correct measurement during in vitro tests, the viscoelastic behavior of the specimen and energy dissipation effects should be taken into account (see "Neutral Zone and Elastic Zone").

Relaxation Relaxation refers to the viscoelastic phenomenon in which a material subjected to a constant deformation exhibits a time-dependent stress (Fig. 3). Relaxation tests in compression, usually involving load cycles with multiple loading and resting phases, can be performed to investigate the time-dependent response of the intervertebral disc as well as the fluid exchange through the endplates. However, creep tests are usually preferred for this purpose, see "Creep."

Statics Statics refers to the branch of mechanics describing the effect of loads acting on bodies that are in equilibrium, that is, do not experience an acceleration but are at rest or move with constant velocity. In practical application, *quasi-static* tests are performed by applying loads to a structure with very low loading rates so that inertia effects are negligible. Typical quasi-static experiments are tensile tests of biological tissues such as ligaments and tendons. Prior to such tests, *preconditioning* should be considered in order to minimize viscoelastic effects.

Stiffness Stiffness refers to the capability of the spine to resist deformation in response to an applied load in a specific loading direction (see "Flexibility"), typically rotational and assessed under the application of a moment load. Stiffness can be determined from the load-deformation curve because its slope is in the *elastic zone*.

Swelling In general, swelling refers to the accumulation of fluid in biological tissues. In spine research and especially finite element modeling, swelling is associated with hydration of the intervertebral disc, which, because of its proteoglycan content and the consequent osmotic gradient, absorbs fluid from the external environment. Disc swelling is responsible for the existence of a non-null *intradiscal pressure* in absence of mechanical loading.

Viscoelasticity A *viscoelastic* material exhibits both the elastic (see "Elasticity") and the viscous behaviors in which the strain increases linearly with time under constant loading. Viscoelastic materials have therefore a time-dependent mechanical response. If a loading cycle is applied to a viscoelastic material, that is, a load is applied and then removed, energy is lost in the form of heat (Fig. 4). The value of the dissipated energy can be estimated from the stress-strain curve of the material as the area of the hysteresis loop. Most biological tissues such as bone and ligaments exhibit a time-dependent response similar to that of a viscoelastic material under physiological loading, that is, in absence of damage or failure (see "Plasticity"). The viscoelastic properties of a material can be assessed by means of *creep* and *relaxation* testing.

Index

Note: Page numbers followed by *f* indicate figures, and *t* indicate tables.

J

K

L

M

T

9780128128510